VOLUME ONE HUNDRED AND TWELVE

ADVANCES IN COMPUTERS

VOLUME ONE HUNDRED AND TWELVE

ADVANCES IN COMPUTERS

Edited by

ATIF M. MEMON
College Park, MD,
United States

Academic Press is an imprint of Elsevier
50 Hampshire Street, 5th Floor, Cambridge, MA 02139, United States
525 B Street, Suite 1650, San Diego, CA 92101, United States
The Boulevard, Langford Lane, Kidlington, Oxford OX5 1GB, United Kingdom
125 London Wall, London, EC2Y 5AS, United Kingdom

First edition 2019

Notices
Knowledge and best practice in this field are constantly changing. As new research and experience broaden our understanding, changes in research methods, professional practices, or medical treatment may become necessary.

Practitioners and researchers must always rely on their own experience and knowledge in evaluating and using any information, methods, compounds, or experiments described herein. In using such information or methods they should be mindful of their own safety and the safety of others, including parties for whom they have a professional responsibility.

To the fullest extent of the law, neither the Publisher nor the authors, contributors, or editors, assume any liability for any injury and/or damage to persons or property as a matter of products liability, negligence or otherwise, or from any use or operation of any methods, products, instructions, or ideas contained in the material herein.

ISBN: 978-0-12-815121-1
ISSN: 0065-2458

For information on all Academic Press publications
visit our website at https://www.elsevier.com/books-and-journals

Publisher: Zoe Kruze
Acquisition Editor: Zoe Kruze
Editorial Project Manager: Shellie Bryant
Production Project Manager: James Selvam
Cover Designer: Greg Harris

Typeset by SPi Global, India

CONTENTS

PREFACE

This volume of *Advances in Computers* is the 112th in this series. This series, which has been continuously published since 1960, presents in each volume four to eight chapters describing new developments in software, hardware, or uses of computers. For each volume, I invite leaders in their respective fields of computing to contribute a chapter about recent advances.

Volume 112 focuses on eight topics. In Chapter 1, entitled "Mobile Application Quality Assurance," the authors, Konstantin Holl and Frank Elberzhager, rightly observe that mobile applications have become highly pervasive in recent years. Their quality is essential since application failures can lead to serious consequences, such as damage of corporate reputation or financial loss. The authors identify and expose approaches that address the issue of quality assurance for mobile applications. In order to drive their systematic mapping study, they derived eight research questions. They systematically identified 311 articles based on 4607 captured records, created clustered views to answer the research questions, and used existing surveys to complement our overview of current challenges. Their results show an overall upward trend of publications since 2003. Hot topics include automation of GUI tests and assurance of nonfunctional qualities. Aspects of future research could be the integration of review techniques into existing approaches and focusing more strongly on defects addressing the specific characteristics of mobile applications.

In Chapter 2, "Advances in Combinatorial Testing," Rachel Tzoref-Brill observes that combinatorial methods for test design gathered popularity as a testing best practice and as a prominent software testing research area. The chapter reviews recent advances in combinatorial testing, with special focus on the research since 2011. It provides a brief background on the theory behind combinatorial testing and on its use in practice. Requirements from industry usage have led to advances in various areas examined in this chapter, including constraints handling in combinatorial algorithms, support for the combinatorial modeling process, and studies on metrics to support the effectiveness of combinatorial testing. The author also highlights recent case studies describing novel use cases for test and field quality improvement in the context of system test, and for optimization of test data. Finally, the author examines recent developments in advanced topics such as utilization of existing tests, test case prioritization, fault localization, and evolution of combinatorial models.

Chapter 3 is entitled "Advances in Applications of Object Constraint Language for Software Engineering." The authors, Atif Aftab Jilani, Muhammad Zohaib Iqbal, Muhammad Uzair Khan, and Muhammad Usman, observe that Object Constraint Language (OCL) has become a standard defined by Object Management Group for specifying constraints on models. Since its introduction as part of Unified Modeling Language (UML), OCL has received significant attention by researchers with works in the literature ranging from temporal extensions of OCL to automated test generation by solving OCL constraints. They provide a survey of various works discussed in the literature related to OCL with the aim of highlighting the advances made in the field. They have classified the literature into five broad categories and provide summaries for various works in the literature. They also provide insights and highlight the potentials areas of further research in the field.

In Chapter 4, "Advances in Techniques for Test Prioritization," Hadi Hemmati posits that with the increasing size of software systems and the continuous changes that are committed to the software's codebase, regression testing has become very expensive for real-world software applications. Test case prioritization is a classic solution in this context. Test case prioritization is the process of ranking existing test cases for execution with the goal of finding defects sooner. It is useful when the testing budget is limited and one needs to limit their test execution cost, by only running top *N* test cases, according to the testing budget. There are many heuristics and algorithms to rank test cases. This chapter sees some of the most common test case prioritization techniques from software testing literature as well as trends and advances in this domain.

In Chapter 5, "Data Warehouse Testing," the authors Hajar Homayouni, Sudipto Ghosh, and Indrakshi Ray discuss the use of data warehouses to accumulate data from multiple sources for data analysis and research. Since organizational decisions are often made based on the data stored in a data warehouse, all its components must be rigorously tested. Researchers have proposed a number of approaches and tools to test and evaluate different components of data warehouse systems. The authors present a comprehensive survey of data warehouse testing techniques. They start with a classification framework that can categorize the existing testing approaches, followed by a discussion of open problems in the field, finally proposing new research directions.

In Chapter 6, "Mutation Testing Advances: An Analysis and Survey," the authors Mike Papadakis, Marinos Kintis, Jie Zhang, Yue Jia, Yves Le Traon, and Mark Harman discuss mutation testing, which realizes the idea of using artificial defects to support testing activities. Mutation is typically used as a way to evaluate the adequacy of test suites, to guide the generation of test cases, and to support experimentation. Mutation has reached a maturity phase and gradually gains popularity both in academia and in industry. In this chapter, the authors present a survey of recent advances, over the past decade, related to the fundamental problems of mutation testing and set out the challenges and open problems for the future development of the method. They also collect advice on best practices related to the use of mutation in empirical studies of software testing, thereby giving the reader a "mini-handbook"-style roadmap for the application of mutation testing as an experimentation methodology.

In Chapter 7, "Event-Based Concurrency: Applications, Abstractions, and Analyses," Aditya Kanade observes that due to the increased emphasis on responsiveness, an event-based design has become mainstream in software development. Software applications are required to maintain responsiveness even while performing multiple tasks simultaneously. This has resulted in the adoption of a combination of thread and event-based concurrency in modern software such as smartphone applications. The chapter presents the fundamental programming and semantic concepts in the combined concurrency model of threads and events. The paradigm of event-based concurrency cuts across programming languages and application frameworks. The chapter begins with a flavor of event-driven programming in a few languages and application frameworks; the mix of threads and events complicates reasoning about correctness of applications under all possible interleavings; followed by a discussion of advances in the core concurrency analysis techniques for event-driven applications with focus on happens-before analysis, race detection, and model checking; it also surveys other analysis techniques and related programming abstractions.

In Chapter 8, "A Taxonomy of Software Integrity Protection Techniques," the authors Mohsen Ahmadvand, Alexander Pretschner, and Florian Kelbert revisit the idea that tampering with software by Man-At-The-End (MATE) attackers is an attack that can lead to security circumvention, privacy violation, reputation damage, and revenue loss. In this model, adversaries are end users who have full control over software as well as its execution environment. This full control enables them to tamper with programs to their benefit and to the detriment of software vendors or other end users. Software integrity protection research seeks for means to mitigate those attacks. Since the seminal work of Aucsmith, a great deal of research effort has been devoted to fight MATE attacks, and many protection schemes were designed by both academia and industry. Advances in trusted hardware, such as TPM and Intel SGX, have also enabled researchers to utilize such technologies for additional protection. Despite the introduction of various protection schemes, there is no comprehensive comparison study that points out advantages and disadvantages of different schemes. Constraints of different schemes and their applicability in various industrial settings have not been studied. There is no taxonomy of integrity protection techniques. These limitations have left practitioners in doubt about effectiveness and applicability of such schemes to their infrastructure. In this chapter, the authors propose a taxonomy that captures protection processes by encompassing system, defense, and attack perspectives. Later, they carry out a survey and map reviewed papers on the taxonomy. Finally, they correlate different dimensions of the taxonomy and discuss observations along with research gaps in the field.

I hope that you find these articles of interest. If you have any suggestions of topics for future chapters, or if you wish to be considered as an author for a chapter, I can be reached at atif@cs.umd.edu.

PROF. ATIF M. MEMON, PHD
College Park, MD, USA

Mobile Application Quality Assurance

Konstantin Holl, Frank Elberzhager
Fraunhofer Institute for Experimental Software Engineering IESE, Kaiserslautern, Germany

Contents

Abstract

Mobile applications have become highly pervasive in recent years. Their quality is essential since application failures can lead to serious consequences, such as damage of corporate reputation or financial loss. The goal of this work is to identify and expose approaches that address the issue of quality assurance for mobile applications. In order to drive our systematic mapping study, we derived eight research questions based on the stated goal. Ultimately, we systematically identified 311 articles based on 4607 captured records. We created clustered views to answer the research questions and used existing surveys to complement our overview of current challenges. The results show an overall upward trend of publications since 2003. Hot topics include automation of GUI tests and assurance of nonfunctional qualities. Aspects of future research could be the integration of review techniques into existing approaches and focusing more strongly on defects addressing the specific characteristics of mobile applications.

Advances in Computers, Volume 112
ISSN 0065-2458
https://doi.org/10.1016/bs.adcom.2017.12.001

1. INTRODUCTION

The market for mobile devices is growing rapidly and new mobile applications are being developed and shipped continuously. The number of mobile phone users in the world is expected to pass the 5 billion mark by 2019 [1]. Furthermore, a projection of mobile app revenues until 2020 done by Statista [2] shows that global mobile app revenues, which amounted to 69.7 billion US dollars in 2015, will generate 188.9 billion US dollars in revenues by 2020, via app stores and in-app advertising. The high pervasiveness of mobile devices and their constant presence have led to mobile applications being developed in almost all domains. This is true for mobile applications for private use as well as for mobile applications for business use, which benefit from the mobility potential of business processes to provide greater availability or faster distribution of information—which again leads to optimized response times when it comes to the performance of the corresponding business tasks. A mobile application is often more than just a transferred desktop application, a game application, or a content-viewing application. Mobile applications are usually tailored to a mobile device such as a smartphone or a tablet (not a laptop), are integrated into an existing IT infrastructure, are task-oriented and focused on a clear and limited scope of functionality, and are based on the mobility potential of a company's business processes [3]. Hence, depending on the scope of the business process, failures of such mobile applications can lead to serious consequences, such as concrete monetary costs. High quality of mobile applications is the key factor for preventing such costs.

Mobile application experts of Capgemini [4] indicate that the quality of IT solutions has a much stronger and more immediate impact on business results nowadays. Application failures often translate into business process failures. In the context of connected and mobile businesses and consumers, application faults are not likely to go unnoticed and can damage corporate reputation. The authors concluded that organizations have to invest in mobile application quality to remain competitive. Considering a broader scope than business applications, a study with about 3500 subjects has shown that users will not accept a problem-ridden mobile application and will abandon it after only one or two failed attempts; only 16% would give it more than two attempts [5]. In another study with about 500 subjects, 96% said that they would write a negative review and 44% said that they would delete the application immediately [6]. According to various statements, this is even true for minor problems, which means that small matters may affect the penetration of a mobile application as well as its failure in the market [7]. Whether making an online purchase, streaming videos and music, or checking a bank balance, consumers demand flawless execution of the applications and technology they use. Even small digital moments have a big impact according to a study based on 2000 subjects [8]. Also, practitioners' portals proclaim that quality, not quantity, is the real mobile application problem [9] and that users have low tolerance for defect-prone applications [10].

Consequently, the quality of mobile applications is essential for the success of companies that provide such applications. Even minor failures can lead to the application being discarded by the users. Typical failures of mobile applications that are caused by a lack of mobile-specific quality assurance may include:

- Changing the orientation of the device leads to a missing button (compare issue "Sign-in button not visible in landscape mode" [11])
- Low energy of the device leads to transaction issues (compare issue "If battery is low, outgoing transactions get discarded without warning" [12])
- Local events disturb current actions (compare issue "Voice announcements during a call" [13])
- Switches between applications destroy context (compare issue "Loses context of the current message being viewed or composed after switching away and back again" [14])

One of the main reasons identified as causes for insufficient quality in mobile applications is the lack of specialized methods for mobile testing [15], as well as the lack of methods that are most effective during mobile testing [16]. Due to missing or insufficient quality assurance approaches for mobile development, tailored quality assurance approaches for mobile applications may be necessary.

The goal of this contribution is to provide an overview of the state of the art regarding quality assurance for mobile applications. A systematic map is presented that shows different aspects of quality assurance in the mobile area, such as the qualities addressed with quality assurance techniques. This information can be used by researchers to determine future research directions, as we derived both lacks and challenges. The results may also help practitioners get an overview of existing approaches, which may lead to optimization of their own quality assurance.

Section 2 describes the methodology for retrieving the state of the art. Section 3 presents the results, including the identified challenges and trends answering the research questions. Section 4 discusses the results and explains the identified implications for practitioners and researchers. Finally, Section 5 concludes the contribution.

2. METHODOLOGY FOR RETRIEVING THE STATE OF THE ART

We performed this systematic mapping study according to Petersen et al. [17] and enhanced our procedure with concepts used in systematic literature reviews according to Kitchenham [18], such as the use of review protocols and measurement of the reviewer's reliability of inclusion decisions based on Cohen's kappa. The methodology for retrieving the state of the art is based on the definition of eight research questions as part of the objective, the definition of selection criteria, the specification of selection phases, and, as part of the selection performance, the creation of a search string and the determination of reference databases.

2.1 Objective

In order to reduce failures of mobile applications, it is important to know the existing quality assurance approaches for mobile applications. This also includes detailed knowledge about these approaches, such as how and where in the development process such approaches are applied, and what kind of qualities and defects they are able to address. Identifying the challenges and lacks of these approaches makes it possible to improve them. Hence, we defined the following main research question (RQ):

How is quality assurance performed for mobile applications?

This overall and coarse-grained research question led us to derive further, more fine-grained research questions regarding quality assurance for mobile applications.

Besides answering the research questions, we also provide some metainformation about the identified result set, such as a timeline of when the papers were published and data about the publication channels. The outcome of answering the research questions of Table 1 is intended to be a comprehensive systematic mapping study that provides information about existing approaches and a basis for future development and research.

Table 1 Research Questions

#	Research Question	Description
RQ1	Which types of quality assurance approaches exist?	This provides a first general classification of the identified approaches, e.g., with respect to testing or reviews. We present an initial overview of the kinds of quality assurance activities conducted in mobile application development
RQ2	Which test levels are addressed?	This assigns the approaches found to the test levels in the development process and provides an overview of where current quality assurance activities are concentrated during the development of mobile applications
RQ3	Which test phases are addressed?	This assigns the approaches found to the phases of quality assurance processes, i.e., we can check whether planning, preparation, execution, or other steps are considered more strongly during quality assurance approaches
RQ4	Which qualities of the mobile application are addressed?	This relates the approaches to the qualities they address and provides—to the extent possible—background information on which mobile-specific failures are addressed
RQ5	Which kinds of automation are implemented?	This first distinguishes approaches into manual or automatic ones and further provides information about the type of automation if a particular automation solution is given. As the trend for automation is relevant for almost every step in the software development process to make it more efficient, this might also be true for our specific mobile quality assurance context, which we are investigating with this research question
RQ6	How are the approaches evaluated?	This maps the evaluation types of each approach to their performance in the academic or industrial field. We are interested in whether several established and well-evaluated approaches exist or whether only initial adaptations of quality assurance techniques will be found that are not well evaluated and, consequently, cannot be assessed sufficiently regarding their power for addressing mobile-specific failures
RQ7	Which trends exist?	This describes trends that were identified in quality assurance for mobile applications to check whether the focus has changed during the last decade regarding research initiatives and to identify the main research topics during the last decade. Furthermore, the answer will describe current and new trends we found, especially from a practitioner's point of view
RQ8	Which challenges exist?	This explains current challenges, which provide another source for future research directions

```
test* OR verif* OR valid* OR assurance OR inspect*
                    OR review

                       AND

       mobile AND (app* OR software OR device*)

                        OR

 (smartphone* OR ios OR android OR "windows phone"
           OR blackberry OR symbian)

                        OR

    "mobile test*" OR "mobile quality assurance"

                        OR

         "app test*" OR "app quality assurance"
```

Fig. 1 Search string for capturing the relevant articles.

2.2 Selection Scope

The search string in the context of the research questions was created on the basis of two fundamental aspects: *quality assurance* and *mobile*.

The search string consists of four basic parts (see Fig. 1). These parts were derived from the general research question and present our scope. Of course, several synonymous terms were considered within each of the four parts. The first part is related to terms that are related to quality assurance (like *testing*). We explicitly considered static and dynamic quality assurance terms. Testing is the main term considered here, including its variations. Furthermore, inspections and reviews are classical terms for static quality assurance. Finally, verification and validation were selected as well as the general term "assurance" in order to cover quality assurance. This first part was combined logically with the second part, which describes the mobile area with several terms, such as mobile applications or mobile software-specific devices that are used. A presearch we performed while defining the final search term revealed that some papers that were relevant for us did not show up due to missing reasonable search terms. That is why we included two more specific parts that explicitly add those terms, for example, the terms "mobile quality assurance" or "app testing."

The search string was applied to capture articles from the following four reference databases:

- Elsevier's Scopus[a]
- ScienceDirect[b]

[a] http://www.scopus.com.

[b] http://www.sciencedirect.com.

- IEEE Xplore Digital Library[c]
- ACM Digital Library[d]

Scopus, ScienceDirect, and ACM required the submission of our search string as *one* string with minor changes in its structure according to the requirements of each search engine. In contrast, the IEEE Xplore Digital Library required the division of our search string in order to cover all defined keywords. This is due to the limitation regarding the number of keywords and is especially crucial when using wildcards. In our case, it was not possible to apply more than five wildcards in one search string. Furthermore, the command search we used did not offer the definition of general search parameters regarding which metadata of an article has to be checked. Hence, we needed to expand our search string by targeting certain data fields like "Document Title" and "Author Keywords." Considering the limitation of the number of keywords, we finally separated our search string into five separate search strings for this reference database. Overall, the equivalence of the applied search strings with the search string of Fig. 1 was verified by a second researcher.

2.3 Selection Criteria

Since the set of potentially relevant articles might be very comprehensive, concrete inclusion and exclusion criteria were specified before the selection was performed. Based on these criteria, two researchers were able to select those articles that fit our defined scope.

The inclusion criteria were specified as follows:

- The addressed applications are intended for usage on mobile devices (such as smartphones or tablets).
- Quality assurance for software is addressed.
- The article is related to analytical quality assurance (such as testing or reviews).
- A concrete approach is described or a survey about mobile-specific quality assurance is provided.
- The paper is written in English.

The exclusion criteria were specified as follows:

- The approach focuses on applications for desktop devices such as laptops.
- The paper is sketchy or underreported, such as a one-pager or a vision paper.
- The paper provides only a summary of a conference workshop.
- Only a slide version or an abstract is documented.

---

[c] http://ieeexplore.ieee.org.
[d] http://dl.acm.org.

- The paper has only the character of a (advertisement) brochure without details.
- The paper describes constructive quality assurance, such as coding guidelines.

These criteria formed the rationale for the homogeneous selection of articles by both researchers during several selection phases. We did not restrict the selection regarding the publication date in order to identify all available articles with respect to our topic, and to explicitly find the oldest relevant publications based on the defined selection criteria.

2.4 Selection Phases

The selection criteria formed an essential part for the selection of relevant articles. They were used in the defined selection phases.

The following phases describe the procedures used to find, identify, and select relevant articles:

- Phase 0: Preparation of systematic search

  During the preparation, the search string was created carefully as described above with the intention of getting a high ratio of relevant articles in the search results set as well as minimizing the set of irrelevant articles. The approach used to do this was a so-called presearch, where search string variations were applied, considering specified evaluation criteria for the search results. The evaluation criteria were (i) the ratio of relevant articles regarding the first 100 search results and (ii) the inclusion of 20 articles already known to be relevant within the results set.
- Phase 1, first search: Application of search string

  The application of the created search string comprised the capture of the metadata (e.g., title, publication year, authors, keywords, and abstract) of all articles found by the selected reference databases. Due to the different options and character limitations of these reference databases, we needed to assemble the search string in different ways to make it usable. This was done without changing its logic expression.

  The results were stored locally in an Excel sheet to harmonize the heterogeneous data schemes provided by the four reference databases. We stored the metadata of each reference in this Excel sheet but did not persist the full paper yet. One researcher performed this step, which resulted in an initial set of 3629 papers.
- Phase 2, first search: Removal of duplicates

  Duplicates in the results set were removed. On the one hand, real duplicates were removed and, on the other hand, articles that describe the same contents, e.g., a journal publication containing all the contents of its corresponding conference publication led to the removal of the conference publication from the results list. A single researcher also performed this step. This reduced the number of papers to 3491.

- Phase 3, first search: Selection based on titles

  Two researchers selected relevant articles based on their titles, taking into account the selection criteria. This means that each researcher investigated the full results list from phase 2 and determined an identified set of relevant articles by himself, which were afterward merged in a joint discussion. Each researcher tagged each article with either "include" or "not include." This assessment was done simultaneously in separate work places. The results of the assessment were merged in a joint meeting. Deviations were handled according to the rule that the approval of one researcher was enough to accept an article. That is, if one researcher approved an article that had not been approved by the other researcher, we captured it for the next phase—the selection based on abstracts—where it could still be sorted out. The Cohen's kappa coefficient, a measure of interrater agreement, was about 0.69 considering all articles selected during this phase. This means according to Landis and Koch [19] that there was substantial agreement between the two researchers regarding the identification of relevant articles. This shows, on the one hand, their common understanding of the relevance of an article and, on the other hand, the benefits of parallel assessment by two researchers, as the probability of missing relevant articles was reduced. Typical topics of excluded articles were mobile robots or other types of autonomous systems with mobility aspects. After this step, the number of remaining papers was 315.

- Phase 4, first search: Selection based on abstracts

  Using the results set from phase 3, the two researchers selected relevant articles based on their abstracts, still considering the selection criteria. Again, each researcher investigated the full search results list from the previous phase and determined an identified set of relevant articles by himself. These sets were then merged in a joint discussion. 247 papers remained after this.

- Phase 5, first search: Selection based on full text

  The articles referenced by the results list from phase 4 were stored as full documents in addition to the metadata. Selecting articles based on their full text reduces the probability of missing an article compared to selection based on titles or abstracts. This is reasoned by the higher informational input for the researcher and the lower number of articles to check. Hence, the final selection of relevant articles was done by one researcher reading the full text of the articles, again taking into account the selection criteria. Afterward, the second researcher reviewed the validity of the results. The final results set of this first search was 221 papers.

- Phases 1–5, Updating search

  Since analyzing and documenting the results regarding the selected articles from phase 5 took a certain amount of time, phases 1–5 were repeated 1 year after the completion of the first analysis and documentation. This was done with regard to new articles added since the last application of the search string as part of phase 1. After checking and processing the identified articles, we performed the further search as an update. This search led to an additional 978 records and finally to 80 selected articles. Overall, we selected 301 articles out of 4607 captured records. The updating search was performed by one researcher and cross-checked by the other researcher for validity purposes. Fig. 2 shows the results of the selection approach based on phases 0–5.

- Phase 6: Consideration of SMS results

  265 of the selected 301 articles describe concrete mobile-specific quality assurance approaches and 36 describe surveys where specific topics regarding quality assurance for mobile applications are addressed. However, three of the surveys turned out to be SMS (systematic mapping studies) related to our search scope: the contributions of Corral et al. [20], Sahinoglu et al. [21], and Zein et al. [22]. We therefore matched the results of the two identified SMS to our results. This matching led to a delta of 63 articles we had not found or selected during our systematic mapping study, i.e., these articles were not part of our results set. Performing phases 3–5 with this delta led to 10 articles that are relevant according to our search criteria, all describing approaches. This resulted in a final set of 311 relevant articles. We also analyzed whether these SMS cover our results set or already answer our research questions, which turned out not to be the case. Consequently, we proceeded with our analysis.

3. RESULTS OF SELECTION

The results of the application of the six phases established the information basis for answering our eight research questions. This section provides an overview of the findings, introduces eight views referencing each of the research questions, and presents a description of the threats to validity.

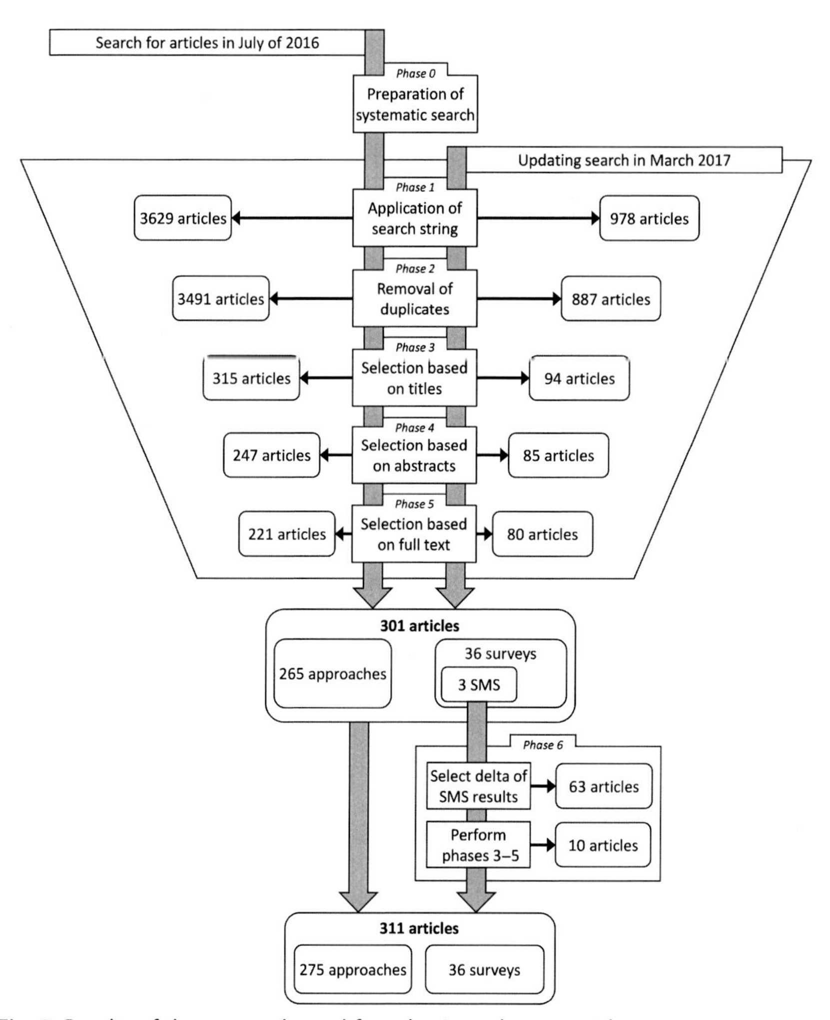

Fig. 2 Results of the approach used for selecting relevant articles.

3.1 Overview

This overview shows details about the findings regarding publication timeline, frequently used keywords, and types of sources.

3.1.1 *Timeline*

A temporal restriction as a search criterion was not necessary due to the topicality of the search strings. This means that more than 99% of the records found by the search string referred to articles published within the last decade. The identified relevant articles were published between 2003 and 2016 (see Fig. 3). The peak of relevant publications was in 2016. Due to the delayed registration of records in the reference databases for 2016, there is a high likelihood that the number of relevant publications from 2016 will be even higher when the search results are updated at the end of 2017. With regard to the years 2004–2016, there was a continuous increase in the number of relevant publications, with two exceptions. In particular, the number of relevant publications in the year 2010 was significantly lower than in 2009. Assuming the average of the years 2009 and 2011 as the expected number of articles for 2010, there are 44% fewer articles than expected. No special reason was found for this low number.

3.1.2 Frequent Keywords

The distribution of keywords regarding the relevant articles represents the most frequently mentioned fields of research according to these articles. The keywords were counted using a formula that considers different

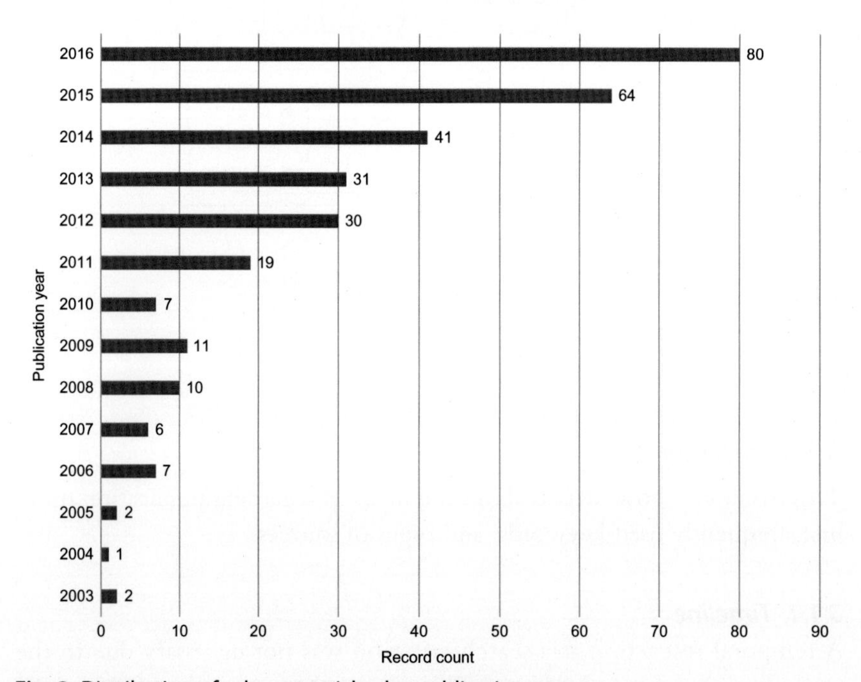

Fig. 3 Distribution of relevant articles by publication year.

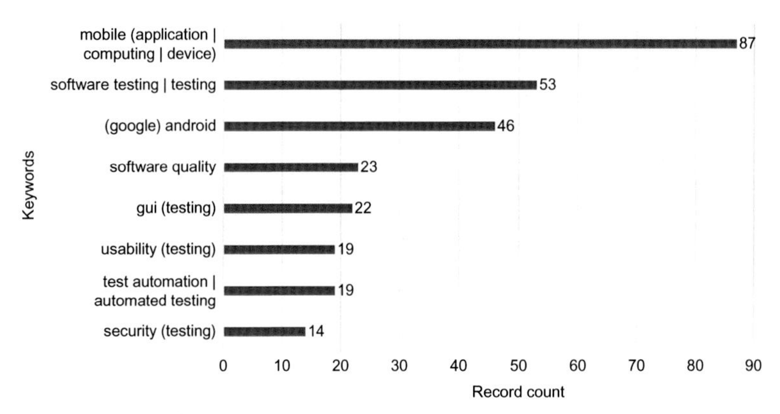

Fig. 4 Distribution of the clustered articles' keywords with more than 10 counts.

spellings, such as the usage of singular and plural terms or the usage of hyphens and spaces between words, as the same keyword. Furthermore, we clustered the keywords manually according to their meaning. *Mobile application* and *mobile device*, for instance, are part of the same cluster. Fig. 4 shows the clustered keywords, ranked on the basis of their count value and limited to count values greater than 10.

Most of these high-ranked keywords are equivalent to the search terms defined in phase 0. The keywords "testing" and "mobile," for example, did not give any new implications about the found articles because these words were already part of the search term. High-ranked keywords that are not obviously related to the defined search terms are:

- GUI (testing)
- usability (testing)
- test automation/automated testing, respectively, test case generation
- security (testing)

These keyword clusters offer some first indications regarding the findings. Quality assurance for mobile applications is often related to the qualities usability and security. Furthermore, testing the GUI level and aspects of automation such as test case generation may play a major role.

3.1.3 Types of Sources and Contents of the Articles

The identified results set comprises conference publications and journal publications. Figs. 5 and 6 show that mainly conference articles, including workshops (88%), were identified as relevant. Furthermore, 8% journal articles and 4% other articles, such as book chapters, were identified.

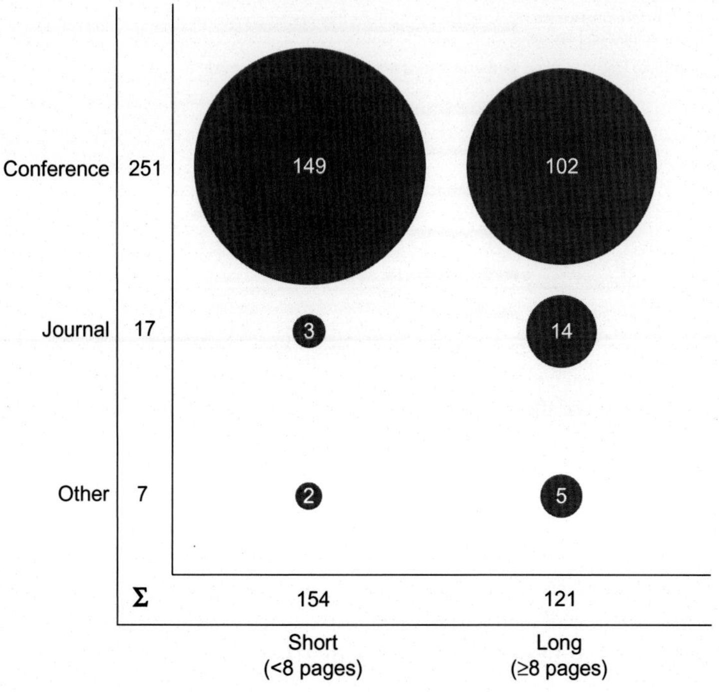

Fig. 5 Mapping of approaches to the length of their publication.

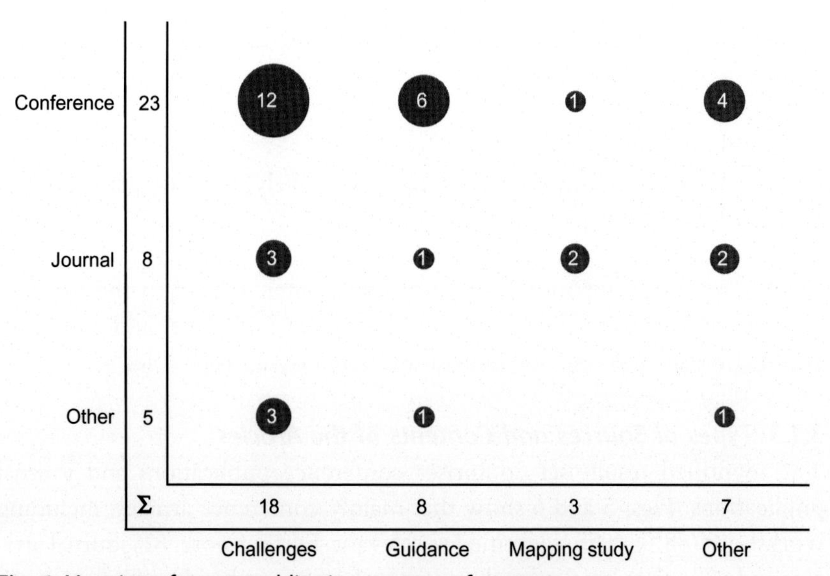

Fig. 6 Mapping of survey publications to type of content.

Two types of content are described in the selected articles. Either an article contains some kind of approach for quality assurance for mobile applications, such as a method, a technique, or a tool, or it contains a survey, an overview of challenges, or approaches considered as part of a literature survey or based on various experiences of the authors.

We clustered the identified surveys into overviews of challenges (i.e., sets of common issues faced during quality assurance for mobile applications), guidance proposals (i.e., proposals of important aspects to be considered during quality assurance for mobile applications), mapping studies (i.e., systematic reviews of quality assurance for mobile applications), and other types of surveys (i.e., presentations of unsystematic reviews or experiences). Furthermore, we clustered approaches published as short papers (fewer than eight pages) and those published as long papers (eight or more pages).

Fig. 5 shows the overview that distinguishes approaches mapped to their length and Fig. 6 the surveys mapped to their content type. Furthermore, both are related to their publication type (such as conference or journal). The majority of the approaches (54%) were published as short conference papers, whereas short journal papers are rare (1%), as expected. The large number of short papers and the number of long papers will be broken down into different types of evaluation later on in the evaluation view in Section 3.2.6.

Surveys published at conferences mostly present challenges (33%) regarding quality assurance for mobile applications or guidance on how to perform quality assurance (17%). Table 2 shows the 36 survey articles, with a brief summary of the addressed topics.

3.2 Approaches to Performing Quality Assurance for Mobile Applications

As the survey articles did not focus on any one special approach, we did not include these articles to answer our research questions. Hence, the following sections refer only to those articles that describe approaches and not surveys: this is true for 275 papers (88%) of the identified articles. The surveys we found will be used later on to complement the identified challenges of the state of the art.

3.2.1 RQ1: Type View

Research question RQ1 (Which types of quality assurance approaches exist?) was answered by dividing the identified approaches into

Table 2 Identified Survey Articles Listed by Type, Topic, and Year

Type	Topic	Year	Reference
Challenge	Cloud testing	2016	Kaur and Kaur [23]
Challenge	Mobile application testing	2016	Rajasekaran [24]
Challenge	Mobile application testing	2016	Samuel and Pfahl [25]
Challenge	Mobile security testing	2015	Wang et al. [26]
Challenge	Testing-as-a-service for mobile applications	2015	Starov et al. [27]
Challenge	Mobile testing within agile development	2015	Santos and Correia [28]
Challenge	Mobile testing in industrial contexts	2015	Zein et al. [29]
Challenge	Testing-as-a-service for mobile applications	2014	Gao et al. [30]
Challenge	Cloud computing	2013	Al-Ahmad et al. [31]
Challenge	Testing in mobile development	2013	Dubinsky and Abadi [32]
Challenge	Testing android applications	2013	Amalfitano et al. [33]
Challenge	Test automation for mobile applications	2013	Kirubakaran and Karthikeyani [34]
Challenge	Mobile application testing	2012	Voas and Miller [35]
Challenge	Model-based GUI testing of mobile applications	2012	Janicki et al. [36]
Challenge	Open issues in mobile application testing	2012	Muccini et al. [15]
Challenge	Android unit testing	2011	Sadeh et al. [37]
Challenge	Wireless application testing	2008	Ding and Chang [38]
Challenge	Usability testing of mobile applications	2005	Betiol and de Abreu Cybis [39]

Guidance	Analyzing user issues for testing consideration	2016	McIlroy et al. [40]
Guidance	Android application development considering testability	2014	Knych and Baliga [41]
Guidance	Mobile application testing tutorial	2014	Gao et al. [42]
Guidance	Using test automation tools	2014	Saad and Awang Abu Bakar [43]
Guidance	Android network security testing	2013	Hunt [44]
Guidance	Overview of testing practices	2011	Sekanina et al. [45]
Guidance	Strategy for testing cell phone software	2009	Cao et al. [46]
Guidance	Definition of testing requirements	2009	Dantas et al. [47]
Mapping	Mobile application testing techniques	2016	Zein et al. [22]
Mapping	Mobile application verification	2015	Sahinoglu et al. [21]
Mapping	Mobile applications assurance practices	2014	Corral et al. [20]
Other	Security assessment taxonomy	2016	Sadeghi et al. [48]
Other	Survey about security for android	2015	Xu et al. [49]
Other	Test input generation for android	2015	Choudhary et al. [50]
Other	Insights of source code and platform dependencies	2014	Syer et al. [51]
Other	Complaints about mobile applications	2014	Khalid et al. [52]
Other	Survey of user perceptions and expectations	2013	Kritikos et al. [53]
Other	Overview of usability testing approaches	2005	Zhang and Adipat [54]

dynamic approaches and static approaches. This covers all identified approaches regarding quality assurance for mobile applications. Dynamic testing requires the execution of a mobile application while static testing does not. Static testing relies on manual examination or automated analysis of the code or other project artifacts without executing the code. Unlike dynamic testing, static testing finds defects in the sense of "causes of failures," while dynamic testing aims at finding failures themselves [55].

We classified the results set into dynamic and static approaches on the one hand (first category) and approaches that address functional and non-functional requirements on the other hand. Some approaches could not be classified into either one of the aforementioned sets, which resulted in a third set for the second category. Fig. 7 depicts the number of classified papers per category in an overview map.

The majority of the identified approaches (93%) represent dynamic approaches. More specifically, the majority of dynamic approaches can be classified as functional testing (53%). Typical examples of this category are model-based approaches, such as those described by Griebe and Gruhn [56] for test case generation and automated execution. This contribution is

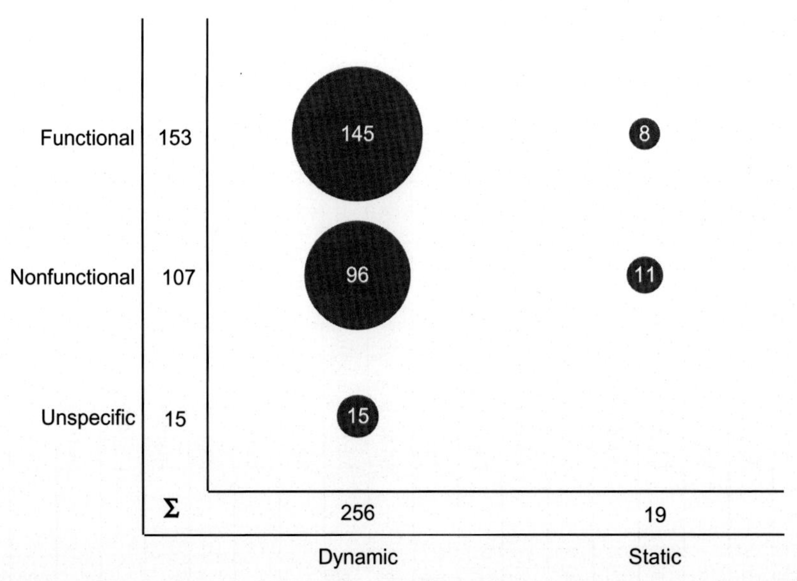

Fig. 7 Distribution of types of approaches used to perform quality assurance for mobile applications.

suitable for context-aware mobile applications because their model-based approach addresses the improvement of testing context-aware mobile applications by deducing test cases from design-time system models.

Dynamic approaches that address nonfunctionalities represent 35% of all approaches in this view. For instance, Qian and Zhou [57] describe an efficiency testing approach focusing on test case prioritization. They build a prediction model to determine whether a test can possibly lead to memory leaks. This model is based on machine learning on selected code features. The contribution of Wan [58] proposes an automated test generation tool that detects energy bugs and hotspots in Android applications as part of efficiency testing. The approach leverages display power modeling and automated display transformation techniques to detect hotspots and prioritize them for developers.

5% of the approaches can be used to address any type of quality. An example of these is iTest by Yan et al. [59], which represents a testing software with a mobile crowdsourcing focus. The framework enables software developers to submit their mobile application and conveniently get the test results, which can address any quality, from the crowd testers.

In addition, there are static approaches that focus on functionality, such as the methodology for retrieving review data based on the code review tool Gerrit by Mukadam et al. [60]. These include, for instance, the static analysis of Android programs by Payet and Spoto [61] resulting in an extension of Julia, which used to be the first sound static analyzer for Android programs, on a formal basis.

Static approaches represent about 7% of the identified approaches. Static approaches in the area of nonfunctional approaches were found especially in the context of security, such as the approach of Krishnan et al. [62], which describes, among other things, a static analysis followed by code inspection to reveal security issues. This kind of approach for quality assurance of mobile applications is one without a functional focus as part of the static approaches (like 58% of all static approaches).

We identified no static approaches that are suitable for addressing all types of quality during quality assurance for mobile applications.

Other possible types of approaches for quality assurance, such as the application of coding guidelines, are not part of this mapping study according to the selection criteria (see Section 2.2).

3.2.2 RQ2: Level View

Research question RQ2 (Which levels are addressed?) was answered by classifying the identified approaches according to their test levels as system, unit, and integration level. Furthermore, there are approaches that can be applied independent of the test level. This led to the distribution shown in Fig. 8.

The most frequently addressed test level is the system level. This is the case in 74% of all approaches, meaning they are intended for testing a complete system in order to verify that it meets its specified requirements. For example, most GUI-test-related approaches (which are very frequent according to the keyword distribution in Section 3.1.2) are based on system-level testing. Such an approach is described by Amalfitano et al. [63], who presented the AndroidRipper. This automated technique is based on a user-interface-driven decomposer (called "ripper") that explores the graphical user interface of the mobile application in order to derive test cases. These rippers are often based on the dynamic crawling and reverse engineering of a model of the application to create test cases based on events. An approach related to such rippers was presented by Tao and Gao [64]. Its mobile app test coverage analysis is performed on the basis of GUI ripping models. Other popular approaches on the system level use the capture and replay principle, which consists of capturing test scripts, usually via user interactions, and replaying these scripts to represent test case execution (see Liu et al. [65]).

The unit level was addressed in 20% of all approaches. These approaches consider testing of individual software components, e.g., to test their efficiency or to perform static analysis with regard to leaks. Wilke et al. [66] presented JouleUnit, which represents a generic framework for energy

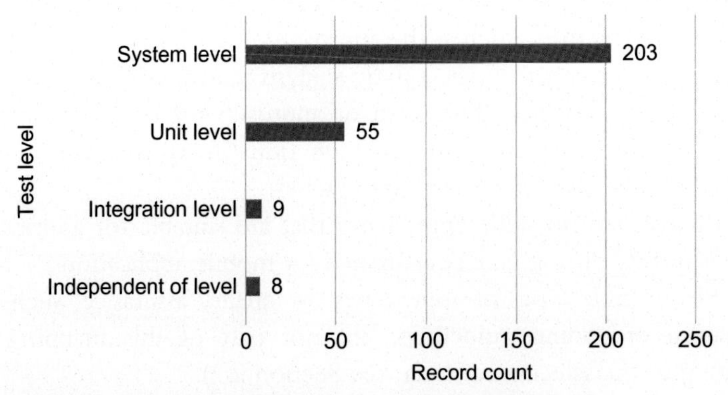

Fig. 8 Distribution of relevant approaches by addressed test level.

profiling and testing. This allows unit testing with regard to energy consumption for several devices and applications.

The integration level was addressed in 3% of all approaches. These approaches referred either to the testing of the interfaces, to the interactions between integrated components, or to the integration into the mobile environment. For instance, Weiss and Zduniak [67] developed automated integration tests for mobile applications in Java 2 micro edition.

Furthermore, 3% of all approaches did not address any one test level in particular, but can be applied in different phases. These approaches were classified as independent. One example of this kind of approach is the Mobile Applications Quality Assurance Tool by Kim [68]. This approach comprises several tools and components that address, among others, program-analysis-based techniques, including automated software inspection for mobile applications, software visualization, testing coverage analysis, performance evaluation, and concurrent program debugging.

3.2.3 RQ3: Phases View

Research question RQ3 (Which phases are addressed?) was answered by assigning the 275 papers to test phases or to static quality assurance phases, depending on their content. The 256 dynamic approaches found are presented according to the fundamental test process (see Fig. 9). According to Spillner et al. [55] and the International Software Testing Qualifications Board (ISTQB), there are five test phases: test planning and control, test analysis and design, test implementation and execution, evaluating exit criteria and reporting, and test closure activities.

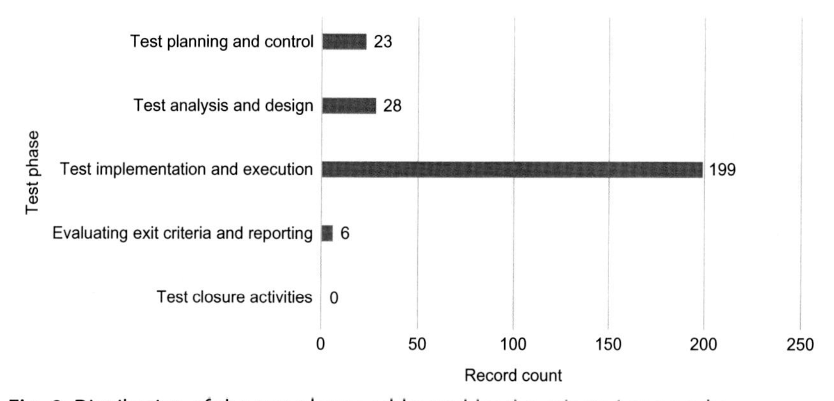

Fig. 9 Distribution of the test phases addressed by the selected approaches.

The majority of the approaches, 78%, address test implementation and execution. In addition, frequent capture and replay tools (e.g., Liu et al. [65] and Lin et al. [69]) approaches using mobile-based testing techniques were identified. For instance, de Cleva Farto and Endo [70] adopted an event sequence graph technique for generating test cases. Furthermore, this approach supports automated test case execution using the Robotium framework.

11% of the approaches address test analysis and design. One example is the approach of Jabbarvand et al. [71]. Here, energy-aware test suites are minimized by relying on energy-aware coverage criteria that indicate the degree to which energy-greedy segments of a mobile application are tested. Another approach regarding test analysis and design was developed by Morgado et al. [72] and is aimed at automated pattern-based testing of mobile applications through the usage of previously defined behavior patterns.

Test planning and control is considered by 9% of all approaches. One example is from Vilkomir and Amstutz [73], which proposes considering combinatorial approaches as part of the test strategy in order to cover various device characteristics. This is intended to prevent device-specific failures in the field, which are currently very common in mobile software applications.

Evaluating exit criteria and reporting is the main aspect of 2% of the approaches. These are two publications. The one by Gawkowski et al. [74] developed a fault injection tool for mobile software, which simulates fault effects and observes their propagation using a fault injection technique. The other one was published by Ma et al. [75] and is a toolkit for usability analysis.

No approaches were identified that address test closure activities. An approach that addresses this phase needs to consider mainly the collection of data from completed test activities in order to consolidate testware, facts and numbers, or experience.

The 19 static approaches were excluded from the representation based on the fundamental test process because they cannot be assigned to the test phases. Test case design, for example, is not part of static approaches. Hence, we grouped the static approaches into activity clusters that are adequate for static approaches. As different static approaches were identified (reviews, static analysis), which are, to a certain extent, also heterogeneous with respect to the phases in which they are conducted, we extracted the main phases directly from the articles instead of using a predefined process. The identified phases are still similar to the fundamental test process: execution, analysis, and closure (see Fig. 10).

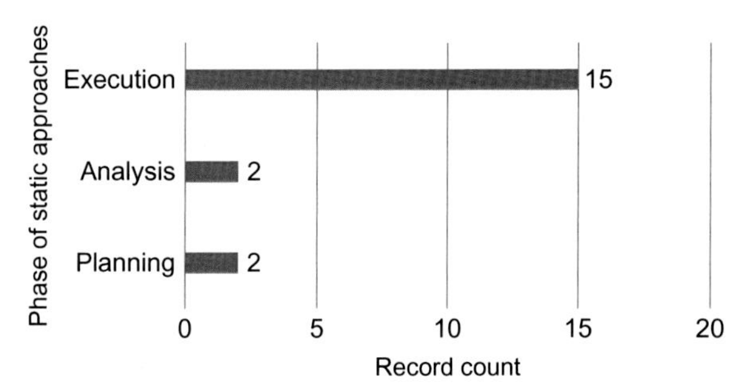

Fig. 10 Distribution of static approaches clustered by addressed phases.

The majority of the approaches (79%) address the execution phase. An example is the mobile application compatibility test system design for Android fragmentation by Ham and Park [76], which compares code analysis results with API pretesting results to reveal Android fragmentation in order to reduce time and cost for subsequent dynamic testing. Another approach is presented by Delamaro et al. [77]. It provides static analysis and offers the generation of control flow graphs, which can be used afterward to perform coverage testing of mobile applications.

Furthermore, 11% of the approaches focus on the analysis phase in order to prepare the execution phase, such as the tool Gerrit by Mukadam et al. [60] already described above, which retrieves review data in order to apply the gathered information within a static approach. The other approaches of this cluster consider interapplication security in Android (Guo et al. [78]) and the enhancement of perspective-based reading for mobile quality assurance (Holl et al. [79]).

11% of the static approaches address the planning phase, such as the already mentioned initial approach by Krishnan et al. [62], which shows how to plan security assurance by considering existing vulnerability databases and which offers application development guidelines to reveal security issues. Furthermore, Ham and Park [76] presented a compatibility test system design for Android fragmentation.

To summarize: with respect to the phases addressed by quality assurance activities in the mobile domain, mainly the execution phase is addressed (overall about 78%). Certain activities before the execution are also supported, but almost no activities are conducted after the execution phase, for example, to systematically evaluate the results or store test cases for easier reuse.

3.2.4 RQ4: Quality View

Research question RQ4 (Which qualities are addressed?) was answered by segregating the identified approaches into approaches whose main objectives regard functionality, those that regard nonfunctionalities, and those that regard unspecific objectives with respect to quality (i.e., approaches that can address any quality). Furthermore, each nonfunctional quality mentioned by an article as a main objective was added in order to cluster the approaches addressing nonfunctional qualities, which were found to be security, usability, efficiency, compatibility, performance, and reliability. This led to the distribution shown in Fig. 11.

Overall, 56% of all approaches focus on the functionality of the mobile application, e.g., on test case generation based on the functional specification or on a test model. One such approach was published by Griebe and Gruhn [56] and provides a framework for model-based test case generation and execution focusing on context-aware mobile applications. Functional failures that the approach can reveal are related to location services in order to verify correct handling by an application if there is no GPS connection. Other functional failures to be revealed by these kinds of approaches were described by Hu et al. [80] in their publication about failure detection with a tool named appDoctor. Failures that were found included, for instance, crashes due to the incorrect assumption of the presence of Google services, and information about installed software that could not be gathered due to an orientation change of the device.

Considering only those approaches that do not address functionality, 33% of them focus on security, such as the verification of privacy leakage. MobiLeak by Stirparo et al. [81] focuses on the management of users' data when these are loaded in the volatile memory of the device, e.g., processes

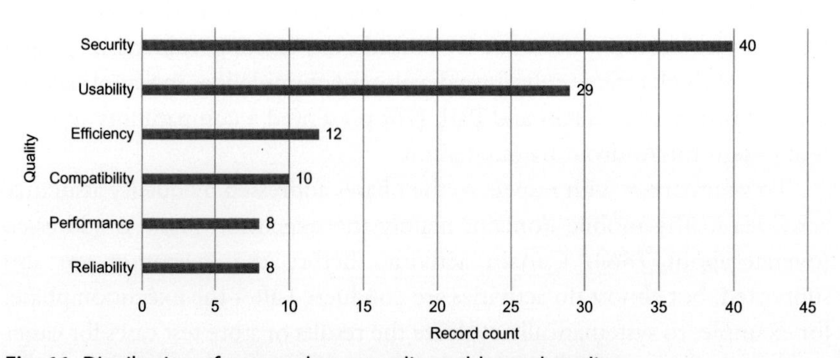

Fig. 11 Distribution of approaches regarding addressed quality.

that were quit via the home button are still running in the background and still keep data allocated in the memory, which can lead to privacy issues.

Usability is the focus of 24% of all approaches, which consider automating usability tests in order to be able to do with fewer experts and subjects. An approach for automated usability testing for mobile applications was proposed by Enriquez and Casas [82]. It is based on a framework called FUsAM (Framework of Usability for Mobile Applications). This framework is extensible and can generate and integrate usability testing in mobile applications. Its design and implementation is based on software product lines combined with feature-oriented programming and aspect-oriented programming.

The efficiency of mobile applications is addressed by 10% of all approaches, which mention resource-related aspects such as energy performance testing. Naik et al. [83] categorized the configuration parameters of smartphones to obtain an approach for energy performance testing. Issues such as battery drains during application usage are typical issues revealed by such approaches.

8% of the approaches focus on compatibility, such as the consideration of heterogeneous devices during testing. AppACTS is an approach by Huang [84] that offers developers the possibility of uploading their mobile applications, selecting the platforms they want for testing, and receiving a compatibility test report in the end. Typical issues revealed by such approaches are functional failures due to the missing compatibility of an application with a device or the impossibility to even install an application.

A focus on performance is found in 7% of all approaches, mentioning ways to test an application in terms of responsiveness and stability under a particular workload, e.g., addressing performance metrics for use-case verification based on benchmark testing. An approach for performance testing of mobile applications was proposed by Kim et al. [85]. It utilizes a database that is based on the results of benchmark testing and is used for emulator-based test environments at the unit test level. Detected issues include latencies that are often too high for the user or even timeouts between the application and the backend system.

Furthermore, 7% of the approaches focus on reliability. For instance, the approach VanarSena described by Ravindranath et al. [86] tests a mobile application within a cloud based on an automated fault finder running several "monkeys" to emulate user, sensor data, and network behavior. This is intended to evaluate failure occurrence during usage of the application for a specified period of time. Consequently, the reliability aspect is considered as quality over time.

The remaining 12% of all approaches do not focus on any specific quality and are thus unspecific, such as the derivation of a test model based on a quality model for mobile applications. Franke and Weise [87] described this kind of approach as part of a framework that provides patterns for mobile application development and metrics for testing mobile applications.

3.2.5 RQ5: Automation View

Research question RQ5 (Which kinds of automation are implemented?) was answered by clustering the identified approaches into the categories tool, framework, testbed, and manual approaches. Furthermore, we considered the manual approaches in comparison to the automated approaches. Manual approaches comprise no automation aspects at all. Fig. 12 shows the distribution of the clusters regarding automation.

The majority of the approaches, i.e., the approaches described by 48% of the relevant publications, were realized as tools. Here, tool stands for an executable program that supports or realizes the quality assurance approach, for instance, a tool for automated GUI model generation of mobile applications. One example is the MZoltar tool presented by Machado et al. [88], which represents a toolset for automated debugging of Android applications. This is done by means of a dynamic analysis via spectrum-based fault location, which results in the generation of a diagnostic report to identify defects.

To get a better idea of the number of published tools, it should be noted that the majority of them have a prototype-like character. This means that a basic tool was developed for internal use and is usually not available to others. Furthermore, seven of the identified tools (i.e., prototypes and non-prototypes) are open-source tools such as AppACTS, a mobile application automated compatibility testing service published by Huang [84], and

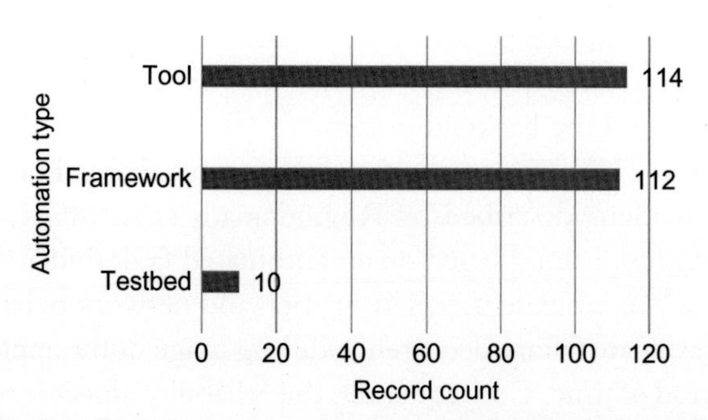

Fig. 12 Distribution of relevant approaches by type of automation addressed.

Dynodroid, an input generation system for Android applications presented by MacHiry et al. [89]. 5% of all identified tools are commercial or only usable with a commercial software, such as MATE (Mobile Analysis Tool for Usability Experts) by Porat et al. [90], which supports the employees of a well-known telecommunications company. Another approach that was classified as commercial was published by Payet and Spoto [61] and extends the commercial tool Julia for static analysis of Android applications.

47% of the approaches were realized as a framework. This means that several programs, libraries, or other components were implemented to realize the approach. One example of such a framework is the GUICC framework by Baek and Bae [91]. This is a model-based Android GUI testing framework using multilevel GUI comparison criteria. It provides selection of multiple abstraction levels for GUI model generation in order to reduce the inherent state explosion problems of existing single-level approaches.

A testbed, in terms of a test environment consisting of hardware and software to test mobile applications on real devices, was part of 4% of all approaches. Such a testbed typically contains instructions for simulating the communication between human–machine interfaces and the system under test. For instance, Hargassner et al. [92] present a testbed for mobile software frameworks that adapts the APOXI (Application Programming Object-oriented eXtendable Interface) framework, which is an application framework for rapid development of human–machine interfaces and for the integration of diverse applications for mobile communication products.

Manual approaches form the remaining set of the selected articles. 14% of the approaches do not represent any kind of automated solutions. Typically, these kinds of approaches address test strategies for mobile software such as the inclusion of testability factors during the development of mobile applications proposed by Knych and Baliga [41]. They claim that the fact that the universe of possible failures is just too large is not only true for classical desktop applications but also for mobile applications. Hence, a mobile-specific test strategy is required to reduce the number of test cases.

3.2.6 RQ6: Evaluation View

Research question RQ6 (How are the approaches evaluated?) was answered by (i) separating the identified approaches into academic and industrial works, (ii) separating them into short and long papers (as done in Section 3.1.3), and (iii) classifying their evaluation into experiences, experiments, case studies, and empirical studies (see Fig. 13). Experience means that there was no systematic evaluation and that the approach is documented

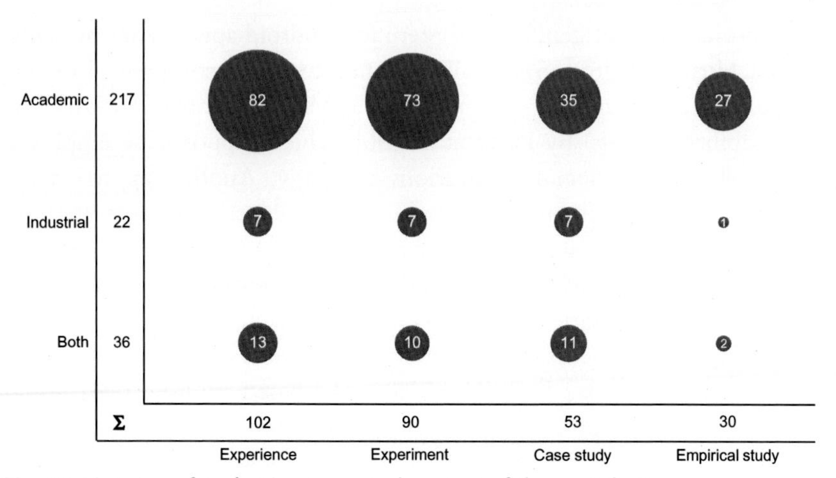

Fig. 13 Mapping of evaluation types and context of the contribution.

based on tryouts or impressions. Experiments are usually run in a laboratory setting and are highly controlled. In contrast, a case study describes an evaluation in a real project and consequently in a typical situation. Empirical studies are based on existing facts and are used for analysis and interpretation.

The distinction between academic and industrial contexts was made based on the affiliations of the authors. Universities and research centers were considered as academic, whereas software companies and telecommunications organizations, for example, were considered as industrial.

The majority of the articles contain academic results (79%), while purely industrial results are represented by only 8% of the articles. Furthermore, 13% of the articles were produced by a combination of academic work and industrial work. Overall, 37% of the articles were not evaluated but written based on experience. In 33% of the articles, at least one experiment was performed, in 19% at least one case study, and 11% were based on an empirical study.

Considering the mapping of evaluation type and context of the contribution, the most frequent type of evaluation were academic experiments. In contrast, empirical studies in the industrial context were not identified at all among the relevant articles.

As presented in Section 3.1.3, the majority of the 275 articles that describe approaches are classified as short papers, which represents a proportion of 56%. Fig. 14 shows the mapping of the classified long and short papers to their evaluation types. 47% of the short papers are based on experiences of the authors, while 53% are based on evaluations such as experiments, case

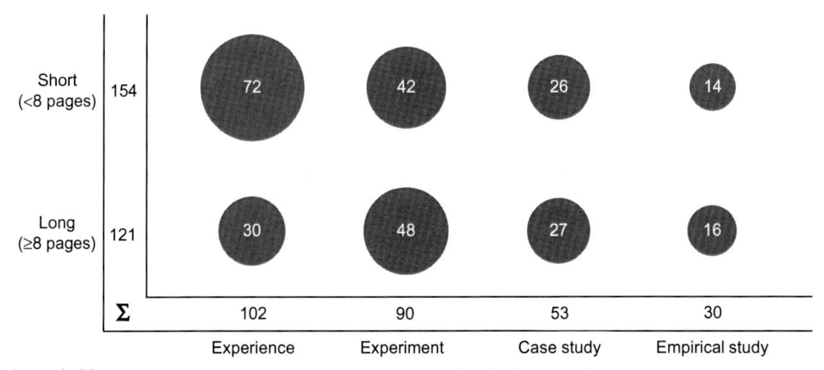

Fig. 14 Mapping of evaluation types and length of the publication.

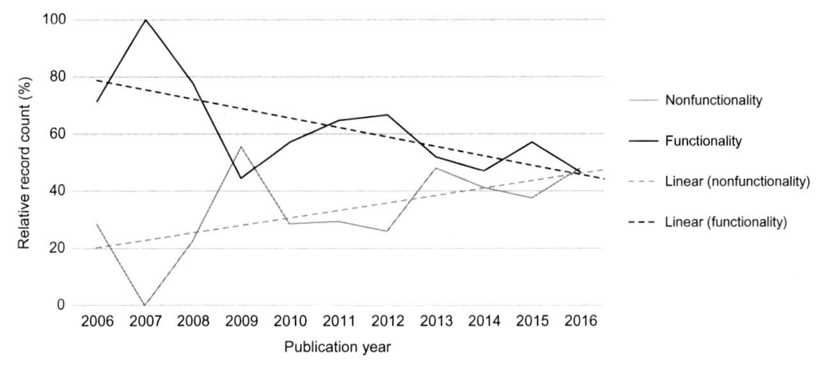

Fig. 15 Trends of approaches that address functionality and those that address nonfunctionality.

studies, or empirical studies. In contrast to the short papers, just 25% of the long papers are based on experiences, while 75% are based on particular evaluations.

3.2.7 RQ7: Trends

Research question RQ7 (Which trends exist?) was answered by performing a trend analysis regarding the views of RQ1 to RQ6. As a result, we identified trends regarding the quality view (RQ4), the automation view (RQ5), and the evaluation view (RQ6). The graphs in Fig. 15 (as well as the other trend graphs in Figs. 16 and 17) show the record count of two clusters in percentage points in proportion to the total number of articles identified for each year. Hence, there is a relative distribution of each view based on the publication year of the identified approaches separated into two clusters. Furthermore, we inserted a linear trend line (dashed line) for each

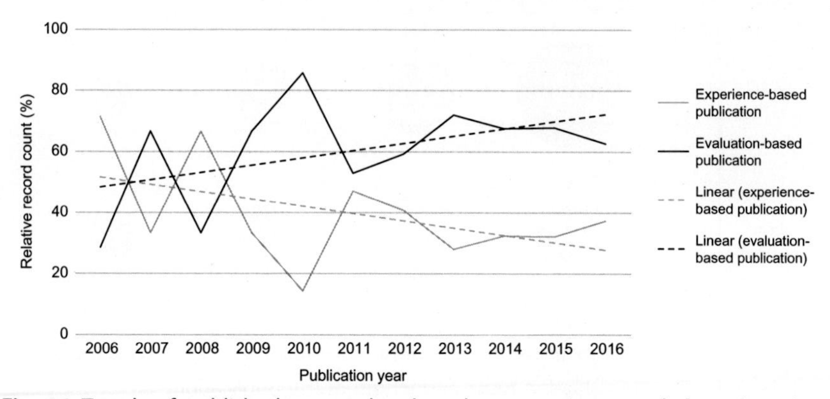

Fig. 16 Trends of published approaches based on experience and those based on evaluation.

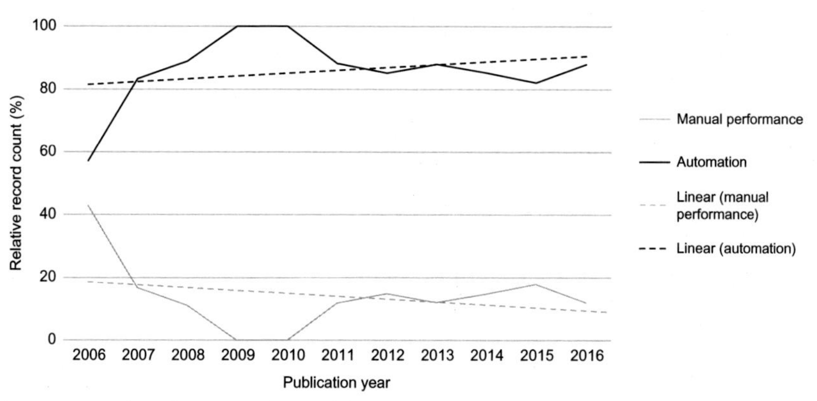

Fig. 17 Trends of approaches addressing manual performance and approaches addressing automation.

cluster, which is based on the relative distribution and shows linear regression to support the identification of a trend. We excluded the articles that were published before 2006 because the number of identified articles from these years was too low for the trend analyses.

We considered the yearly number of approaches that address functionality and those that address nonfunctionality in relation to the total number of approaches from each year. The total number of approaches also comprises the approaches that were classified as unspecific (see Section 3.2.4). Overall, we identified 153 approaches that address functionality and 107 approaches that address nonfunctionality (15 approaches were unspecific).

Except for 2007, when none of the identified approaches addressed nonfunctionality, the ratio of approaches addressing nonfunctionality was always between 20% and 60% and lower than the ratio regarding approaches addressing functionality. We identified a significant trend that the rate of approaches addressing nonfunctionality is increasing.

Furthermore, we considered the approaches that had been published based on experiences and the approaches that had been evaluated prior to publication. Since 2009, the relative number of experience-based publications has always been lower than the relative number of evaluation-based publications.

Consequently, a significant trend can be recognized, namely, that most approaches are evaluated (e.g., by means of experiments or case studies) before they are published.

Another trend was identified regarding the approaches that address manual performance compared to those that are automated. The relative number of manual approaches was lower than that of automated approaches throughout all years.

There is just a slight but not significant trend that the number of automated approaches (e.g., using a tool or a framework) being published is slightly higher than the number of approaches that have to be performed manually.

The presented trends reveal insights regarding the past and may imply future progress. To determine the approaches used in the world of today, we considered the latest topics presented by practitioners. This was done by (i) separating the identified approaches into pure academic works and works with industrial participation based on the distinction in Section 3.2.6 and (ii) considering approaches found in articles published from 2011 to today, since approaches described by articles published before 2011 (represented by 25% of the identified articles) often discuss aspects that are of minor interest today, mainly due to technological evolution.

The restriction imposed on all 275 approaches according to our definition of the latest industry-related approaches led to a set of 65 approaches. Table 3 shows the topics of these articles mapped to eight topical clusters and references representative articles. *NF Automation* comprises approaches that address automated testing of nonfunctional qualities (e.g., usability or security). *Test Base Improvement* clusters approaches that enhance the given test bases (e.g., the requirements specification) with further information in order to derive mobile-specific test cases. *Monkey Testing* is about approaches that integrate a component that performs automated tests with any kind of random testing. Approaches as part of the cluster *UI Ripping* use traversing and recognition techniques to derive and execute test cases automatically.

Table 3 Latest Industry-Related Articles Identified

Cluster	Topic	Year	Reference
NF Automation	Usability testing	2016	Filho et al. [93]
NF Automation	Compatibility testing	2015	Zhang et al. [94]
NF Automation	Performance testing	2014	Raghavan et al. [95]
NF Automation	Usability testing	2013	Canfora et al. [96]
NF Automation	Usability testing	2013	Porat et al. [90]
NF Automation	Security testing	2013	Matsumoto and Sakurai [97]
NF Automation	Security testing	2011	Gilbert et al. [98]
Test Base Improvement	Test case prioritization	2016	Qian and Zhou [57]
Test Base Improvement	Location-based testing	2015	Aktouf et al. [99]
Test Base Improvement	Test oracle generation	2014	Zaeem et al. [100]
Test Base Improvement	Contextual fuzzing	2014	Liang et al. [101]
Test Base Improvement	Code review data usage	2013	Mukadam et al. [60]
Test Base Improvement	Client–server testing	2013	Tulpule [102]
Monkey Testing	Scalable fault detection	2014	Ravindranath et al. [86]
Monkey Testing	Model-based GUI testing	2014	Yeh et al. [103]
Monkey Testing	Selective symbolic execution	2011	Takala et al. [104]
UI Ripping	Model-based crawling	2016	Tao and Gao [105]
UI Ripping	Crawling- and OCR-based testing	2012	Yumei and Zhifang [106]
UI Ripping	Crawling- and activity-page-based testing	2012	Lu et al. [107]
Remote Testing	Remote UI testing	2012	Dhanapal et al. [108]
Remote Testing	Remote UI testing	2012	Kaasila et al. [109]

Table 3 Latest Industry-Related Articles Identified—cont'd

Cluster	Topic	Year	Reference
Fault Simulation	Network impact testing	2012	Dhanapal et al. [110]
Fault Simulation	Fault injection	2011	Gawkowski et al. [74]
Conformance Testing	Testing of lifecycle-dependent properties	2012	Franke et al. [111]
Conformance Testing	Model-based conformance testing	2012	Jing et al. [112]
Cross-Platform Testing	Hybrid application testing	2016	Brucker and Herzberg [113]
Cross-Platform Testing	Testing on heterogeneous mobile platforms	2011	Song et al. [114]

Remote Testing comprises approaches that offer a remote access interface for mobile devices, which allows the tester access to various devices without the need for a physical presence. The *Fault Simulation* cluster is about approaches that provide the possibility to simulate certain conditions for the system under test, which may lead to failures of the application. Approaches as part of *Conformance Testing* systematically consider the compliance of a mobile application with its requirements specification. *Cross-Platform Testing* is about the ability to test mobile applications on different operating systems.

The eight identified approaches that consider test automation regarding nonfunctionalities confirm the overall trends presented in Figs. 15 and 17. These figures show that there is a significant trend toward more articles being published that focus on nonfunctionalities rather than functionalities. In addition, there is a slight trend that more published articles describe automated approaches than manual approaches. Especially approaches for automated usability testing show up in current articles by practitioners.

The six approaches that consider test base improvement imply that testing of mobile applications calls for a mobile-specific test design due to the peculiarities of mobile applications. This comprises especially contextual aspects that can be supplemented by simulation approaches, e.g., for changing the network conditions during test execution.

The seven approaches altogether for *Monkey Testing, UI Ripping,* and *Remote Testing* are based on the graphical user interface of mobile devices running the mobile applications. Considering the usability appearance of the *NF Automation* cluster as well may imply a user-centered testing focus.

While the articles of the cluster *Conformance Testing* represent more classical testing approaches, the *Cross-Platform Testing* article presents a typical mobile-specific issue: testing of basically the same applications on fundamentally different operating systems such as iOS and Android. This approach may benefit from *Remote Testing* approaches in order to access multiple platforms efficiently.

3.2.8 RQ8: Challenges

Research question RQ8 (Which challenges exist?) was answered by referring to the results of other comprehensive articles offering surveys of mobile quality assurance and correlating them with the challenges mentioned in these articles that describe concrete approaches. Just as we did for determining approaches used in the world of today (see previous section), we focused on challenges mentioned in articles published in 2011 or later due to recent crucial technological changes, which make challenges reported earlier less interesting. For instance, ZhiFang and Xiao-Peng [115] describe future work issues referring to operating system constraints for applying a test automation framework. Here, the considered operating system is Symbian OS on Nokia devices, which has no relevant pervasiveness nowadays and will no longer be deployed on mobile phones. Also, challenging coding issues for stress testing as described by Mazlan [116], who considered MIDlets as part of J2ME development, are no longer of great interest regarding a solution because the applied technical approach and the devices it focused on (such as the Motorola i88s) have no relevant pervasiveness anymore.

Dubinsky and Abadi [32] identified a list of 45 challenges for testing in mobile development. They grouped and prioritized these challenges, which resulted in the definition of the following unanswered research questions: "How to automatically simulate factors affecting the environment?" "How to leverage crowdsourcing to increase quality?" and "How can we simulate the network behavior, latency, connectivity, protocols?"

Gao et al. [42] claim that the construction of mobile test environments is one of the major challenges due to the high costs and levels of complexity. Setting up a mobile test environment is very expensive and time consuming. This is associated with another major challenge: test automation. This is challenging due to a lack of standardization regarding mobile test infrastructure, scripting languages, and connectivity protocols between mobile test tools and platforms. Furthermore, there are issues regarding unified test automation infrastructures and solutions that cross platforms and browsers on most mobile devices.

Muccini et al. [15] identified several fundamental challenges, such as test selection. The state-of-the-art test process does not provide testing criteria according to which mobile test cases are selected in the scope of mobile application development. Mobile-specific issues are, among others, context awareness and the multitude of mobile devices. Other identified challenges are GUI testing with regard to automated test execution and quality-related testing, e.g., for security and performance. The challenges described by Muccini et al. [15] are in accordance with the recently performed survey by Rajasekaran [24].

Based on a survey, Janicki et al. [36] revealed challenges in deploying model-based GUI testing of mobile software. It is challenging to compare different testing approaches due to missing metrics regarding the test results. Furthermore, it is a challenge to realize quick bug localization.

Further challenges regarding special topics, such as mobile cloud computing applications, include dynamic assignment of cloud resources to applications that run on mobile devices (mentioned by Al-Ahmad et al. [31] and Kaur and Kaur [23]). This is related to challenges of mobile testing-as-a-service, which lacks well-defined and well-developed mobile-device-based test clouds and infrastructures (mentioned by Gao et al. [30]).

The results of the mapping study of Sahinoglu et al. [21] indicate that there are current research challenges regarding the testing of mobile applications in the cloud with respect to test execution automation or test environment management for system-level functional testing. Furthermore, the authors observed the general importance of system-level issues.

The challenges mentioned in the survey articles and the challenges mentioned in the articles describing concrete approaches are in accordance with each other. Further challenges that are not described in the surveys are:

- Context simulation in order to execute mobile application tests under real-time conditions, e.g., regarding location services (mentioned in Rege et al. [117] and implied in Aktouf et al. [99])
- Security-by-design aspects, especially due to the uniqueness of the mobile devices (mentioned in Stirparo et al. [81] and Wang et al. [26])
- Device-specific test coverage and fault analyses for compatibility testing, especially with regard to the relations between mobile device features and compatibility faults (mentioned in Zhang and Adipat [54] and in Cheng et al. [118])
- Usage of user reviews as input for mobile quality assurance, e.g., for device selection as part of test planning (mentioned in McIlroy et al. [40] and in Khalid et al. [119])

- Generation of test cases that focus on energy issues (mentioned in Wan et al. [58] and implied in Yan et al. [120])

3.3 Threats to Validity

Several factors may influence the concrete results set of a systematic mapping study, from the researchers doing the selection and analysis of articles via the databases used to the definition of the search term. In the following, we will describe the main threats to the validity of our research and how we mitigated them.

First of all, the definition of the search term and the selected databases have a huge impact on the results set. For the definition of our search term, we first carefully defined different parts and derived several synonyms to cover a large set of relevant papers. Furthermore, the search term was reviewed by two additional researchers and a presearch was performed prior to the real search in order to consolidate our search term until a reasonable level of quality was achieved. With respect to the databases, we considered four well-known databases that cover several conferences and workshop papers, journal papers, and magazine articles. The results of the metasearch offered by Elsevier's Scopus, which currently has more than 53 million records, were supplemented with findings from ScienceDirect (which has about 11 million records), IEEE Xplore, and ACM. We did not consider open-source databases as the heterogeneity in such databases is often very high, which would have resulted in many papers that would have been excluded afterward. Of course, some papers might not have been found during our search, but the general conclusions drawn from our final results set should not be significantly influenced by that, as the basis for our general conclusions was always a large number of papers (e.g., with respect to static vs dynamic approaches, focus on functionality, or automation conclusions). In addition, two identified mapping studies that are related to our research were used to include some further relevant papers and improve the validity of our results set.

Furthermore, the researchers who performed the selection might have had an influence on the final set of articles. In order to ensure that both had the same understanding of the topic and performed their selection similarly, explicit inclusion and exclusion criteria were defined and both researchers made their selection based on all papers (i.e., the papers were not divided into two sets, one for each researcher). Cohen's kappa value showed high agreement regarding common understanding. If only one reviewer selected a paper, this paper was included in the next step. All these means helped to ensure that all relevant papers were selected. However, if other researchers had made the selection, a slightly different final results set might have been possible, which, however, would not have had much impact on the general conclusions.

We only considered articles published in English. This may also have led to some articles not being included in the results set, but the value of articles not written in English, which are thus not available for the whole research community, is rather low.

4. DISCUSSION AND IMPLICATIONS

First of all, the topic of quality assurance for mobile applications has gained increasing attention over the last decade. Though we did not limit the timeframe for finding relevant papers in our search, almost every relevant paper was published in the last 10 years. Of course, the mobile market has increased heavily during this time, which provides a simple explanation. However, only for the last 5 years has the number of papers increased significantly, and it can be assumed that this trend will continue.

Most of the published papers focus on concrete approaches to how quality assurance can be conducted with respect to mobile applications. No completely new quality assurance approaches are usually discussed, but the issue is mainly how to adapt existing ones to the mobile domain. Testing was the main technique we identified as being used to assure the quality of mobile applications, and only about 7% of the approaches we found describe static approaches such as reviews or static analysis. Besides such approach papers, 12% of the complete results set describe a certain kind of survey, e.g., summarizing current challenges, providing some general guidelines, or presenting other mapping studies.

When we consider the approaches themselves, most of them are applied on the system level, followed by about 20% on the unit level and a minor number on the integration level. GUI testing is very popular on the system level, though it might be time consuming (due to, e.g., finding and implementing a capture/replay tool, selecting test cases for automation, creating automated test cases, updating test cases) or incomplete (because not every test case can be automated feasibly). The fact that integration testing is rarely considered might result in additional risks, as there is either no major focus on interface and communication testing, or this is done at the unit or system level where the focus is usually on other test aspects. Especially when applications communicate with a backend, a database, or with other devices and are thus part of a certain ecosystem, neglecting dedicated integration testing may lead to serious problems (e.g., functional problems, data problems, performance problems).

Regarding the considered phases, a strong focus on the execution phase can be observed. This makes sense as the main goal of quality assurance is to find failures and faults, and this is done during execution. However, the systematic preparation of quality assurance also got some attention (planning and design), whereas phases after execution are neglected. This might also be a risk in the sense that no systematic analysis of results or closing activities (e.g., how to store test cases in such a way that they can be reused more easily) could be observed. Especially analyzing test results and generalizing them in order to be able to control quality assurance better and identify improvement potential is important for further optimization and prevents stagnation of test activities.

Overall, more papers focus on functional problems than on nonfunctional problems. Considering the trend of the last years, there was a focus on functionality especially in the years 2006–2008, whereas afterward, the difference was not that great. Indeed, in the last few years, an increasing number of papers have focused on nonfunctional properties. Here, we could identify security and usability as the most important nonfunctional properties, followed by efficiency, compatibility, performance, and reliability aspects. Many papers discuss specific defect patterns that can be addressed using the described quality assurance technique. However, we could not find any comprehensive list of typical mobile-specific defects nor any general classification of mobile-specific defect types.

About 86% of the approach papers consider automation in various ways, either by a tool, a set of tools (i.e., a framework), or a testbed that also takes hardware components into account. The majority of such tools have a prototype character and are thus less mature, meaning that based on our results set, only a limited number of useful automation solutions exist. However, as we focused more on research databases and, in addition, also excluded commercial brochures we found, we are aware that a lot more open-source and commercial test tools exist that offer support for quality assurance engineers. The potential for new tools is high, and many conceptual ideas found in approaches have been evaluated at least partially with tool prototypes, which allows drawing more realistic conclusions than simply providing adapted, but theoretical models, approaches, or techniques. However, considering explicitly how many approaches have been evaluated, we observed that some kind of evaluation had been done in about 63% of the cases, but most of that evaluation had been performed in an academic environment (and thus it can be questioned how realistic this is).

So, to conclude our main findings, we can state that quality assurance in mobile application development plays a crucial role and a lot of research has been done in this field in recent years, but several open issues remain. In addition to those mentioned in Section 3.2.8 based on our results set, we derived the following challenges. For each, we will briefly discuss how they affect researchers and practitioners.

- A comprehensive list of mobile-specific failures, faults, and triggers that can be used for quality assurance could not be found. Such a list, which of course would have to be updated over time, could offer great support for practitioners when doing their quality assurance, allowing them to know what to focus on and reducing the risk that typical defects are missed.
- Considering stronger static quality assurance techniques. Many publications in the field of static quality assurance substantiate their benefits and, depending on the study, appear to be more effective and efficient than testing techniques. Therefore, when a practitioner has to define an overall quality assurance strategy, it is worth considering a mix of static and dynamic quality assurance. Our research showed that (1) static quality assurance techniques are of minor interest and (2) no stronger combination of static and dynamic quality assurance is being considered, which might result in further improvement potential. Researchers have to find new ways of improving quality assurance for the mobile domain. However, for practitioners it is good to know that they can still use the review and testing techniques they know.
- Results from quality assurance activities have to be analyzed carefully to identify further improvement potential. This means that not only the planning, design, and execution phases are relevant, but also the analysis and closing activities. Although such analyses take time, this time should be invested by practitioners (e.g., in a Sprint review meeting). Researchers should show ways for analyzing such data and drawing conclusions based on data such as number of defects (per defect type) and time needed for quality assurance.
- Tool prototypes should be developed into useful tools. We are aware that several test tools exist that can also be applied for conducting quality assurance for mobile applications. However, only a small number of such tools are really being developed to address the specifics of the mobile

domain. Therefore, more dedicated tool prototypes should be developed into sophisticated tools.

- GUI testing is often a time-consuming task. Usually, today's capture–replay tools use the GUI to do system tests. Functional problems can be found in this way, whereas nonfunctional properties such as usability are difficult to assess. However, the application of these tools is usually very time consuming. One solution could be to test "below" the GUI, directly using interfaces or communication channels. Further solutions have to be found in order to improve the system test.
- There is only little consideration of integration testing, but integration testing is relevant in terms of (1) how to integrate and test the application under development regarding the communication between internal classes and components, and (2) how to integrate and test the communication of the application with external devices and components. Practitioners should think about including integration testing in an overall quality assurance strategy, while researchers should focus on how to do integration tests in current development environments, considering recent trends such as continuous integration.
- A good test basis supports comprehensive and complete tests. However, such a test basis often does not exist, which results in incomplete tests and overlooked faults. Ways have to be found for handling missing or insufficient test bases, or for creating a minimal test basis with little effort that serves the needs of a tester.
- The number of test experts, especially those with thorough knowledge of the mobile domain, is limited and needs to be increased to ensure that modern applications continue to have high quality. Practitioners have to invest into building up this knowledge, while researchers have to provide information about what to consider during quality assurance for mobile applications.

5. CONCLUSION AND FUTURE WORK

We performed a systematic mapping study addressing analytic quality assurance in mobile application development. From an initial total of 4607 papers, we selected 311 articles for detailed analyses conducted by two researchers. We focused on the kinds of quality assurance that exist, on the levels they address and the phases in which they are conducted, on the qualities on which they focus and on the extent to which automation is performed, and on how well they have been evaluated. In addition, we derived trends and challenges.

We were able to confirm that quality assurance for mobile applications is a relevant topic with an increasing number of publications over the last years. We expect that this topic will gain even more attention as the mobile market continues to grow rapidly and quality will play an increasing role, especially when applications become more complex and business relevant. The main techniques we identified are reviews, static analyses, and especially testing. These techniques have been in existence for several decades now, but were mainly adapted to the specifics of mobile quality assurance, e.g., by considering specific defect types or test environments. With respect to our research questions, we found that mainly system testing is addressed, and that the execution phase is considered most often. Both functional and non-functional properties are addressed during quality assurance, with a slightly stronger focus on the former. Moreover, automation also plays an important role for mobile-specific quality assurance; however, the maturity of such tools is low.

We furthermore derived and identified several challenges in this field, for example, consolidating the types of failures and faults, improving automation maturity, or analyzing results with more rigor to be able to identify further improvement potential. Researchers can use our results to identify further research directions and as motivation for performing a more rigorous evaluation of their approaches. Practitioners get a comprehensive overview of existing quality assurance techniques, of the qualities that these techniques address, and about where these techniques can be applied. In the future, we will focus particularly on a more mature and complete defect type classification and on greater integration of static and dynamic quality assurance techniques.

A reasonable next step is the derivation of requirements considering the mobile application development. These include requirements due to the typically applied agile process, like minimal overhead and high degree of automation. The mobile application itself yields requirements such as consideration of context awareness and autonomy. The definition of testing activities as part of the fundamental test process according to ISTQB (Spillner et al. [55]) provides requirements such as review of the test bases during test analysis and design, or prioritization of test cases as part of test implementation and execution. These requirements will be used to reveal weaknesses and strengths of the state-of-the-art approaches in order to enable their systematic improvement.

ACKNOWLEDGMENTS

The research described in this chapter was performed as part of the research project Opti4Apps (BMBF project number 02K14A182) and in the context of the Fraunhofer Project Center for Software and Systems Engineering at UFBA, a joint initiative of the *Fraunhofer-Gesellschaft* and the Federal University of Bahia in Brazil, with support from the Bahia State Government. Furthermore, we would like to thank Sonnhild Namingha for proofreading.

APPENDIX: SELECTED ARTICLES

Abogharaf, A., Palit, R., Naik, K., and Singh, A., A methodology for energy performance testing of smartphone applications. 7th International Workshop on Automation of Software Test (AST), pp. 110–116, 2012.

Adamsen, C.Q., Mezzetti, G., and Møller, A., Systematic execution of android test suites in adverse conditions. ISSTA 2015 Proceedings of the 2015 International Symposium on Software Testing and Analysis, pp. 83–93, 2015.

Ahmed, M., Ibrahim, R., and Ibrahim, N., Adaptation model for testing android application. Second International Conference on Computing Technology and Information Management (ICCTIM), 2015, pp. 130–133, 2015.

Aktouf, O.-E.-K., Zhang, T., Gao, J., and Uehara, T., Testing Location-Based Function Services for Mobile Applications. 2015 IEEE Symposium on, pp. 308–314, 2015.

Al-Ahmad, A.S., Aljunid, S.A., and Sani, A.S.A., Mobile cloud computing testing review. International Conference on Advanced Computer Science Applications and Technologies (ACSAT), pp. 176–180, 2013.

Amalfitano, D., Amatucci, N., Fasolino, A. R., and Tramontana, P., A Conceptual Framework for the Comparison of Fully Automated GUI Testing Techniques. 30th IEEE/ACM International Conference on Automated Software Engineering Workshop (ASEW), pp. 50–57, 2015.

Amalfitano, D., Amatucci, N., Fasolino, A. R., Tramontana, P., Kowalczyk, E., and Memon, A. M., Exploiting the Saturation Effect in Automatic Random Testing of Android Applications. 2nd ACM International Conference on Mobile Software Engineering and Systems (MOBILESoft), pp. 33–43, 2015.

Amalfitano, D., Amatucci, N., Fasolino, A.R., Gentile, U., Mele, G., Nardone, R., and Marrone, S., Improving code coverage in android apps testing by exploiting patterns and automatic Exploiting Patterns and Automatic Test Case Generation. WISE 14 Workshop, pp. 29–34, 2014.

Amalfitano, D., Fasolino, A., Tramontana, P., Ta, B., and Memon, A., MobiGUITAR - A Tool for Automated Model-Based Testing of Mobile Apps. IEEE Computer Society Press, pp. 1–6, 2014.

Amalfitano, D., Fasolino, A.R., and Tramontana, P., A GUI Crawling-Based Technique for Android Mobile Application Testing. Verification and Validation Workshops (ICSTW), 2011 IEEE Fourth International Conference on Software Testing, pp. 252–261, 2011.

Amalfitano, D., Fasolino, A.R., Tramontana, P., and Amatucci, N., Considering Context Events in Event-Based Testing of Mobile Applications. IEEE Sixth International Conference on Software Testing, Verification and Validation Workshops (ICSTW), pp. 126–133, 2013.

Amalfitano, D., Fasolino, A.R., Tramontana, P., and Robbins, B., Testing Android Mobile Applications: Challenges, Strategies, and Approaches. Advances in Computers, pp. 1–52, 2013.

Amalfitano, D., Fasolino, A.R., Tramontana, P., De Carmine, S., and Imparato, G., A toolset for GUI testing of Android applications. 28th IEEE International Conference on Software Maintenance (ICSM), pp. 650–653, 2012.

Amalfitano, D., Fasolino, A.R., Tramontana, P., De Carmine, S., and Memon, A.M., Using GUI ripping for automated testing of Android applications. Proceedings of the 27th IEEE/ACM International Conference on Automated Software Engineering (ASE '12), pp. 258–261, 2012.

Anand, S., Naik, M., Harrold, M.J., and Yang, H., Automated concolic testing of smartphone apps. Proceedings of the ACM SIGSOFT 20th International Symposium on the Foundations of Software Engineering (FSE '12), pp. 1–11, 2012.

Anbunathan, R. and Basu, A., A recursive crawler algorithm to detect crash in Android application. Computational Intelligence and Computing Research (ICCIC), 2014 IEEE International Conference on, pp. 1–4, 2015.

Anbunathan, R. and Basu, A., An event based test automation framework for Android mobiles. IEEE Conference Publications, pp. 76–79, 2014.

Armando, A., Bocci, G., Costa, G., Mammoliti, R., Merlo, A., Ranise, S., Traverso, R., and Valenza, A., Mobile App Security Assessment with the MAVeriC Dynamic Analysis Module. MIST 2015 Proceedings of the 7th ACM CCS International Workshop on Managing Insider Security Threats, pp. 41–49, 2015.

Arora, H. and Jaliminche, L.N., Design and implementation of test harness for device drivers in SOC on mobile platforms. Conference, pp. 1–6, 2015.

Avancini, A. and Ceccato, M., Security testing of the communication among Android applications. 8th International Workshop on Automation of Software Test (AST), pp. 57–63, 2013.

Ayyal Awwad, A.M. and Slany, W., Automated bidirectional languages localization testing for android apps with rich GUI. Mobile Information Systems, pp. 1–13, 2016.

Azim, T. and Neamtiu, I., Targeted and depth-first exploration for systematic testing of android apps. ACM SIGPLAN Notices - OOPSLA '13, pp. 641–660, 2013.

Badura, T. and Becher, M., Testing the Symbian OS Platform Security Architecture. International Conference on Advanced Information Networking and Applications (AINA '09), pp. 838–844, 2009.

Baek, Y. M. and Bae, D. H., Automated model-based Android GUI testing using multi-level GUI comparison criteria. International Conference on Automated Software Engineering, pp. 238–249, 2016.

Bagheri, H., Sadeghi, A., Garcia, J. and Malek, S., COVERT: Compositional Analysis of Android Inter-App Permission Leakage. IEEE Transactions on Software Engineering (Volume:41, Issue: 9), pp. 866–886, 2015.

Baluda, M., Pistoia, M., Castro, P., and Tripp, O., A framework for automatic anomaly detection in mobile applications. Proceedings - International Conference on Mobile Software Engineering and Systems, pp. 297–298, 2016.

Bao, L., Lo, D., Xia, X., and Li, S., What permissions should this android app request? Proceedings - 2016 International Conference on Software Analysis, Testing and Evolution, pp. 36–41, 2016.

Baride, S., and Dutta, K., A cloud based software testing paradigm for mobile applications. SIGSOFT Software Engineering Notes, pp. 1–4, 2011.

Bastani, O., Anand, S., and Aiken A., An interactive approach to mobile app verification. MobileDeLi 2015, Proceedings of the 3rd International Workshop on Mobile Development Lifecycle, pp. 45–46, 2015.

Benli, S., Habash, A., Herrmann, A., Loftis, T., and Simmonds, D., A Comparative Evaluation of Unit Testing Techniques on a Mobile Platform. New Generations (ITNG), 2012 Ninth International Conference on Information Technology, pp. 263–268, 2012.

Bentes, L., Rocha, H., Valentin, E., and Barreto, R., JFORTES: Java Formal Unit TESt Generation. VI Brazilian Symposium on Computing Systems Engineering, pp. 16–23, 2016.

Bernacki, J., Błażejczyk, I., Indyka-Piasecka, A., Kopel, M., Kukla, E., and Trawiński, B., Responsive web design: Testing usability of mobile web applications. Lecture Notes in Computer Science (including subseries Lecture Notes in Artificial Intelligence and Lecture Notes in Bioinformatics), pp. 257–269, 2016.

Betiol, A.H. and de Abreu Cybis, W., Usability testing of mobile devices: A comparison of three approaches. Human-Computer Interaction - INTERACT 2005, IFIP TC13 International Conference, pp. 470–481, 2005.

Billi, M., Burzagli, L., Catarci, T., Santucci, G., Bertini, E., Gabbanini, F., and Palchetti, E., A unified methodology for the evaluation of accessibility and usability of mobile applications. Springer, Universal Access in the Information Society, pp. 337–356, 2010.

Birgitta, B. and Larsson, S., A case study of real-world testing. MUM'08 - Proceedings of the 7th International Conference on Mobile and Ubiquitous Multimedia, pp. 113–116, 2008.

Bo, J., Xiang, L., and Xiaopeng, G., MobileTest: A Tool Supporting Automatic Black Box Test for Software on Smart Mobile Devices. Second International Workshop on Automation of Software Test (AST '07), pp. 1–7, 2007.

Bojjagani, S. and Sastry, V.N., STAMBA: Security testing for android mobile banking apps. Advances in Intelligent Systems and Computing, pp. 671–683, 2016.

Borys, M. and Milosz, M., Mobile application usability testing in quasi-real conditions. 8th International Conference on Human System Interaction (HSI), pp. 381–387, 2015.

Braghin, C., Sharygina, N., and Barone-Adesi, K., A model checking-based approach for security policy verification of mobile systems. Formal Aspects of Computing, pp. 627–648, 2011.

Brucker, A.D. and Herzberg, M., On the static analysis of hybrid mobile apps: A report on the state of Apache Cordova nation. Lecture Notes in Computer Science (including subseries Lecture Notes in Artificial Intelligence and Lecture Notes in Bioinformatics), pp. 1–17, 2016.

Calpur, M.C. and Yilmaz, C., Towards having a cloud of mobile devices specialized for software testing. Proceedings - International Conference on Mobile Software Engineering and Systems, pp. 9–10, 2016.

Candra, A., Kurniawan, Y., and Rhee, K. H., Security analysis testing for secure instant messaging in android with study case: Telegram. 6th International Conference on System Engineering and Technology, pp. 92–96, 2016.

Canfora, G., Di Sorbo, A., Mercaldo, F., and Visaggio, C.A., Exploring mobile user experience through code quality metrics. Lecture Notes in Computer Science (including subseries Lecture Notes in Artificial Intelligence and Lecture Notes in Bioinformatics), pp. 705–712, 2016.

Canfora, G., Mercaldo, F., Visaggio, C.A., D'Angelo, M., Furno, A., and Manganelli, C., A Case Study of Automating User Experience-Oriented Performance Testing on Smartphones. Verification and Validation (ICST), 2013 IEEE Sixth International Conference on Software Testing, pp. 66–69, 2013.

Cao, G., Yang, J., Tu, Z.-H., and Tang, Y., Research on cell phone software testing strategy. International Conference on Computational Intelligence and Software Engineering (CiSE '09), pp. 1–4, 2009.

Carbone, R., Compagna, L., Panichella, A., and Ponta, S.E., Security threat identification and testing. IEEE 8th International Conference on Software Testing, Verification and Validation (ICST), pp. 1–8, 2015.

Carvalho, S.A.L., Lima, R.N., Cunha, D.C., and Silva-Filho, A.G., A hardware and software Web-based environment for Energy Consumption analysis in mobile devices. Proceedings - 2016 IEEE 15th International Symposium on Network Computing and Applications, pp. 242–245, 2016.

Chandra, R., Karlsson, B., Lane, N., Liang, C., Nath, S., Padhye, J., Ravindranath, L., and Zhao, F., How to the Smash Next Billion Mobile App Bugs? GetMobile: Mobile Computing and Communications Volume 19 Issue 1, pp. 34–38, 2015.

Chen, J., Xue, Y., and Chen, Z., An empirical study on test driven development process for Android applications. Proceedings of the IASTED International Conference on Software Engineering and Applications, SEA 2011, pp. 137–144, 2011.

Cheng, J., Zhu, Y., Zhang, T., Zhu, C., and Zhou, W., Mobile Compatibility Testing Using Multi-objective Genetic Algorithm. IEEE Conference Publications, pp. 302–307, 2015.

Cheng, L.C., The mobile app usability inspection (MAUi) framework as a guide for minimal viable product (MVP) testing in lean development cycle. The 2nd International Human Computer Interaction and User Experience Conference in Indonesia: Bridging the Gaps in the HCI and UX World, pp. 1–11, 2016.

Choi, W., Necula, G., and Sen, K., Guided GUI testing of android apps with minimal restart and approximate learning. ACM SIGPLAN Notices - OOPSLA '13, pp. 623–640, 2013.

Choudhary, S. R., Gorla, A., and Orso, A., Automated Test Input Generation for Android: Are We There Yet? 30th IEEE/ACM International Conference on Automated Software Engineering (ASE), pp. 429–440, 2015.

Choudhary, S.R., Cross-platform testing and maintenance of web and mobile applications. Companion Proceedings of the 36th International Conference on Software Engineering, pp. 642–645, 2014

Chynał, P., Szymański, J.M., and Sobecki, J., Using eyetracking in a mobile applications usability testing. 4th Asian Conference on Intelligent Information and Database Systems (ACIIDS '12), pp. 178–186, 2012.

Coelho, T., Lima, B., and Faria, J. P. MT4A: a no-programming test automation framework for Android applications. Proceedings of the 7th International Workshop on Automating Test Case Design, Selection, and Evaluation, pp. 59–65, 2016.

Coiana, M., Conconi, A., Nigay, L., and Ortega, M., Test-bed for multimodal games on mobile devices. Lecture Notes in Computer Science (including subseries Lecture Notes in Artificial Intelligence and Lecture Notes in Bioinformatics), pp. 75–87, 2008.

Coimbra Morgado, I., Paiva, A.C.R., and Faria, J.P., Automated Pattern-Based Testing of Mobile Applications. IEEE Conference Publications, pp. 294–299, 2014.

Colley, A., Tikka, P., Huhtala, J., and Häkkilä, J., Investigating text legibility in mobile UI - A case study comparing automated vs. user study based evaluation. Proceedings of the 17th International Academic MindTrek Conference: "Making Sense of Converging Media", MindTrek 2013, pp. 304–306, 2013.

Corral, L., Sillitti, A., and Succi, G., Software assurance practices for mobile applications. Springer Vienna, Computing Journal, pp. 1–24, 2014.

Costa, G., Merlo, A., Verderame, L., and Armando, A., Automatic security verification of mobile app configurations. Future Generation Computer Systems, pp. 1–18, 2016.

Dantas, V.L.L., Marinho, F.G., da Costa, A.L., and Andrade, R.M.C., Testing requirements for mobile applications. 24th International Symposium on Computer and Information Sciences (ISCIS '09), pp. 555–560, 2009.

de Cleva Farto, G. and Endo, A.T., Evaluating the model-based testing approach in the context of mobile applications. Electronic Notes in Theoretical Computer Science, pp. 3–21, 2015.

de Sá, M. and Carrico, L., An Evaluation Framework for Mobile User Interfaces. Human-Computer Interaction – INTERACT 2009, 12th IFIP TC 13 International Conference, Uppsala, Sweden, August 24–28, 2009, Proceedings, Part I, pp. 708–721, 2009.

Delamaro, M.E., Vincenzi, A.M.R., and Maldonado, J.C., A strategy to perform coverage testing of mobile applications. Proceedings - International Conference on Software Engineering, pp. 118–124, 2006.

Deng, L., Offutt, J., Ammann, P., and Mirzaei, N., Mutation operators for testing Android apps. IEEE Eighth International Conference on Software Testing, Verification and Validation Workshops (ICSTW), pp. 1–10, 2015.

Dhanapal, K.B., Deepak, K.S., Sharma, S., Joglekar, S.P., Narang, A., Vashistha, A., Salunkhe, P., Rai, H.G.N., Somasundara, A.A., and Paul, S., An Innovative System for Remote and Automated Testing of Mobile Phone Applications. Annual SRII Global Conference (SRII), pp. 44–54, 2012.

Dhanapal, K.B., Sankaran, S., Somasundara, A.A., and Paul, S., WindTunnel: A Tool for Network Impact Testing of Mobile Applications. Annual SRII Global Conference (SRII), pp. 34–43, 2012.

DiFilippo, K., Andrade, J., Huang, W. H., and Chapman-Novakofski, K., Reliability Testing of a Mobile App Quality Assessment Tool. Journal of Nutrition Education and Behavior, pp. 42–43, 2016.

Ding, Z. and Chang, K.H., Issues related to wireless application testing. Proceedings of the 46th Annual Southeast Regional Conference on XX, pp. 513–514, 2008.

Do Nascimento, L.H.O. and MacHado, P.D.L., An experimental evaluation of approaches to feature testing in the mobile phone applications domain. DoSTA 2007: Workshop on Domain-Specific Approaches to Software Test Automation - In conjunction with the 6th ESEC/FSE Joint Meeting, pp. 27–33, 2007.

Do, Q., Yang, G., Che, M., Hui, D., and Ridgeway, J., Redroid: A regression test selection approach for android applications. Proceedings of the International Conference on Software Engineering and Knowledge Engineering, pp. 486–491, 2016.

Do, Q., Yang, G., Che, M., Hui, D., and Ridgeway, J., Regression test selection for android applications. Proceedings - International Conference on Mobile Software Engineering and Systems, pp. 27–28, 2016.

Dubinsky, Y. and Abadi, A., Challenges and research questions for testing in mobile development: Report on a mobile testing activity. MobileDeLi 2013 - Proceedings of the 2013 ACM Workshop on Mobile Development Lifecycle, pp. 37–38, 2013.

Enriquez, J. and Casas, S., Development and evaluation of a framework for generation usability testing for mobile application. Latin American Computing Conference, pp. 1–10, 2016.

Esipchuk, I. and Vavilov, D., PTF-based Test Automation for JAVA Applications on Mobile Phones. Consumer Electronics, 2006. ISCE '06. 2006 IEEE Tenth International Symposium on, pp. 1–3, 2006.

Espada, A.R., del Mar Gallardo, M., Salmerón, A., and Merino, P., Runtime verification of expected energy consumption in smartphones. 22nd International Symposium, SPIN 2015, Springer, pp. 132–149, 2015.

Eugster, P., Pervaho: A Development & Test Platform for Mobile Ad hoc Applications. Mobile and Ubiquitous Systems: Networking & Services, 2006 Third Annual International Conference on, pp. 1–5, 2006.

Fang, Z., Liu, Q., Zhang, Y., Wang, K., and Wang, Z., IVDroid: Static detection for input validation vulnerability in Android inter-component communication. ISPEC 2015, 11th International Conference, ISPEC 2015, Springer, pp. 378–392, 2015.

Feijó Filho, J., Prata, W., and Oliveira, J., Where-how-what am I feeling: User context logging in automated usability tests for mobile software. Lecture Notes in Computer Science (including subseries Lecture Notes in Artificial Intelligence and Lecture Notes in Bioinformatics), pp. 14–23, 2016.

Feng, H. and Shin, K.G., Understanding and defending the binder attack surface in android. ACM International Conference Proceeding Series, pp. 398–409, 2016.

Feng, Y., Chen, Z., Jones, J.A., Fang, C., and Xu, B., Test report prioritization to assist crowdsourced testing. ESEC/FSE 2015 Proceedings of the 2015 10th Joint Meeting on Foundations of Software Engineering, pp. 225–236, 2015.

Feng, Y., Jones, J. A., Chen, Z., and Fang, C., Multi-objective test report prioritization using image understanding. Proceedings of the 31st IEEE/ACM International Conference on Automated Software Engineering, pp. 202–213, 2016.

Fetaji, M., Usability testing and evaluation of a mobile software solution: A case study. Information Technology Interfaces, 2008. ITI 2008. 30th International Conference on, pp. 501–506, 2008.

Figueiredo, A., Andrade, W., and Machado, P., Generating interaction test cases for mobile phone systems from use case specifications. ACM SIGSOFT Software Engineering Notes, pp. 1–10, 2006.

Filho, J.F., Prata, W., and Oliveira, J., Affective-ready, contextual and automated usability test for mobile software. Proceedings of the 18th International Conference on Human-Computer Interaction with Mobile Devices and Services Adjunct, pp. 638–644, 2016.

Filho, J.F., Valle, T., and Prata, W., Automated usability tests for mobile devices through live emotions logging. MobileHCI 2015 Proceedings of the 17th International Conference on Human-Computer Interaction with Mobile Devices and Services Adjunct, pp. 636–643, 2015.

Franke, D. and Weise, C., Providing a Software Quality Framework for Testing of Mobile Applications. IEEE Fourth International Conference on Software Testing, Verification and Validation (ICST), pp. 431–434, 2011.

Franke, D., Kowalewski, S., Weise, C., and Prakobkosol, N., Testing Conformance of Life Cycle Dependent Properties of Mobile Applications. IEEE Fifth International Conference on Software Testing, Verification and Validation (ICST), pp. 241–250, 2012.

Gandini, S., Ravotto, D., Ruzzarin, W., Sanchez, E., Squillero, G., and Tonda, A., Automatic detection of software defects: An industrial experience. Proceedings of the 11th Annual Genetic and Evolutionary Computation Conference, GECCO-2009, pp. 1921–1922, 2009.

Gao, J., Bai, X., Tsai, W., and Uehara, T., Mobile Application Testing: A Tutorial. IEEE Computer Society, pp. 46–55, 2014.

Gao, J., Tsai, W.-T., Paul, R., Bai, X., and Uehara, T., Mobile testing-as-a-service (MTaaS) - Infrastructures, issues, solutions and needs. Proceedings of the 2014 IEEE 15th International Symposium on High-Assurance Systems Engineering (HASE '14), pp. 158–167, 2014.

Gatsou, C., Politis, A., and Zevgolis, D., Exploring inexperienced user performance of a mobile tablet application through usability testing. Federated Conference on Computer Science and Information Systems (FedCSIS), pp. 557–564, 2013.

Gawkowski, P., Pawełczyk, P., Sosnowski, J., Cabaj, K., and Gajda, M., LRFI - Fault injection tool for testing mobile software. Studies in Computational Intelligence, pp. 269–282, 2011.

Gilbert, P., Chun, B.-G., Cox, L.P., and Jung, J., Vision: Automated security validation of mobile apps at app markets. Proceedings of the Second International Workshop on Mobile Cloud Computing and Services (MCS '11), pp. 21–26, 2011.

Gómez, M., Rouvoy, R., Adams, B., and Seinturier, L., Mining Test Repositories for Automatic Detection of UI Performance Regressions in Android Apps. 13th Working Conference on Mining Software Repositories. pp. 13–24, 2016.

Griebe, T. and Gruhn, V., A model-based approach to test automation for context-aware mobile applications. Proceedings of the 29th Annual ACM Symposium on Applied Computing (SAC '14), pp. 420–427, 2014.

Griebe, T., Hesenius, M., and Gruhn, V., Towards automated UI-tests for sensor-based mobile applications. 14th International Conference, SoMet 2015, Naples, Italy, Springer, pp. 3–17, 2015.

Grønli, T. M. and Ghinea, G., Meeting Quality Standards for Mobile Application Development in Businesses: A Framework for Cross-Platform Testing. 49th Hawaii International Conference on System Sciences, pp. 5711–5720, 2016.

Gudmundsson, V., Lindvall, M., Aceto, L., Bergthorsson, J., and Ganesan, D., Model-based testing of mobile systems - An empirical study on QuizUp Android app. Electronic Proceedings in Theoretical Computer Science, pp. 16–30, 2016.

Guo, C., Xu, J., Yang, H., Zeng, Y., and Xing, S., An automated testing approach for inter-application security in Android. Proceedings of the 9th International Workshop on Automation of Software Test (AST 2014), pp. 8–14, 2014.

Gustarini, M., Scipioni, M. P., Fanourakis, M., and Wac, K., Differences in smartphone usage: Validating, evaluating, and predicting mobile user intimacy. Original Research Article. Pervasive and Mobile Computing, pp. 50–72, 2016.

Hale, M.L. and Hanson, S., A Testbed and Process for Analyzing Attack Vectors and Vulnerabilities in Hybrid Mobile Apps Connected to Restful Web Services. 2015 IEEE World Congress on Services, pp. 181–188, 2015.

Ham, H.K. and Park, Y.B., Designing knowledge base mobile application compatibility test system for android fragmentation. International Journal of Software Engineering and its Applications, pp. 303–314, 2014.

Ham, H.K. and Park, Y.B., Mobile application compatibility test system design for Android fragmentation. Communications in Computer and Information Science, pp. 314–320, 2011.

Hargassner, W., Hofer, T., Klammer, C., Pichler, J., and Reisinger, G., A Script-Based Testbed for Mobile Software Frameworks. 1st International Conference on Software Testing, Verification, and Validation, pp. 448–457, 2008.

Hay, R., Tripp, O., and Pistoia, M., Dynamic detection of inter-application communication vulnerabilities in android. ISSTA 2015 Proceedings of the 2015 International Symposium on Software Testing and Analysis, pp. 118–128, 2015.

Hesenius, M., Griebe, T., and Gruhn, V., Towards a behavior-oriented specification and testing language for multimodal applications. Proceedings of the 2014 ACM SIGCHI Symposium on Engineering Interactive Computing Systems (EICS '14), pp. 117–122, 2014.

Holl, K. and Vieira, V., Focused Quality Assurance of Mobile Applications: Evaluation of a Failure Pattern Classification. 41st Euromicro Conference on Software Engineering and Advanced Applications, pp. 349–356, 2015.

Holl, K. and Elberzhager, F., Mobile Application Quality Assurance: Reading Scenarios as Inspection and Testing Support. 42th Euromicro Conference on Software Engineering and Advanced Applications, pp. 245–249, 2016.

Holl, K., Elberzhager, F., and Vieira, V., Towards a Perspective-Based Usage of Mobile Failure Patterns to Focus Quality Assurance. Springer International Publishing, pp. 20–31, 2015.

Holl, K., Vieira, V., and Faria, I., An Approach for Evaluating and Improving the Test Processes of Mobile Application Developments. Procedia Computer Science, pp. 33–40, 2016.

Holland, B., Deering, T., Kothari, S., Mathews, J., and Ranade, N., Security Toolbox for Detecting Novel and Sophisticated Android Malware. IEEE/ACM 37th IEEE International Conference on Software Engineering (Volume: 2), pp. 733–736, 2015.

Holzmann, C. and Hutflesz, P., Multivariate Testing of Native Mobile Application. Conference, pp. 85–94, 2014.

Hu, C. and Neamtiu, I., A GUI bug finding framework for Android applications. Proceedings of the ACM Symposium on Applied Computing, pp. 1490–1491, 2011.

Hu, C. and Neamtiu, I., Automating GUI testing for android applications. Proceedings - International Conference on Software Engineering, pp. 77–83, 2011.

Hu, G., Yuan, X., Tang, Y., and Yang, J., Efficiently, effectively detecting mobile app bugs with appDoctor. Proceedings of the Ninth European Conference on Computer Systems, pp. 1–15, 2014.

Hu, Y. and Neamtiu, I., Fuzzy and cross-app replay for smartphone apps. Proceedings of the 11th International Workshop on Automation of Software Test, pp. 50–56, 2016.

Hu, Y. and Neamtiu, I., VALERA: An effective and efficient record-and-replay tool for android. Proceedings - International Conference on Mobile Software Engineering and Systems, pp. 285–286, 2016.

Hu, Y., Azim, T., and Neamtiu, I., Versatile yet lightweight record-and-replay for android. OOPSLA 2015, Proceedings of the 2015 ACM SIGPLAN International Conference on Object-Oriented Programming, Systems, Languages, and Applications, pp. 349–366, 2015.

Huang, J.-F. and Gong, Y.-Z., Remote mobile test system: a mobile phone cloud for application testing. IEEE 4th International Conference on Cloud Computing Technology and Science (CloudCom), pp. 1–4, 2012.

Huang, J.-F., AppACTS: Mobile App Automated Compatibility Testing Service. 2nd IEEE International Conference on Mobile Cloud Computing, Services, and Engineering (MobileCloud), pp. 85–90, 2014.

Humayoun, S. and Dubinsky, Y., MobiGolog: formal task modelling for testing user gestures interaction in mobile applications. MOBILESoft 2014 Proceedings of the 1st International Conference on Mobile Software Engineering and Systems, pp. 46–49, 2014.

Hunt, R., Security testing in Android networks - A practical case study. 19th IEEE International Conference on Networks (ICON), pp. 1–6, 2013.

Hwang, S.-M. and Chae, H.-C., Design and implementation of mobile GUI testing tool. Proceedings - 2008 International Conference on Convergence and Hybrid Information Technology, ICHIT 2008, pp. 704–707, 2008.

Imparato, G., A Combined Technique of GUI Ripping and Input Perturbation Testing for Android Apps. 2015 IEEE/ACM 37th IEEE International Conference on Software Engineering (Volume: 2), pp. 760–762, 2015.

Jaaskelainen, A., Katara, M., Kervinen, A., Maunumaa, M., Paakkonen, T., Takala, T., and Virtanen, H., Automatic GUI test generation for smartphone applications - an evaluation. 31st International Conference on Software Engineering - Companion Volume, pp. 112–122, 2009.

Jaaskelainen, A., Kervinen, A., and Katara, M., Creating a Test Model Library for GUI Testing of Smartphone Applications. The Eighth International Conference on Quality Software, pp. 276–282, 2008.

Jabbarvand, R., Sadeghi, A., Bagheri, H., and Malek, S., Energy-aware test-suite minimization for Android apps. Proceedings of the 25th International Symposium on Software Testing and Analysis, pp. 425–436, 2016.

Jabbarvand, R., Sadeghi, A., Garcia, J., Malek, S., and Ammann, P., EcoDroid: An Approach for Energy-Based Ranking of Android Apps.

IEEE/ACM 4th International Workshop on Green and Sustainable Software (GREENS), pp. 8–14, 2015.

Jaber, M., Falcone, Y., Dak-Al-Bab, K., Abou-Jaoudeh, J., and El-Katerji, M., A high-level modeling language for the efficient design, implementation, and testing of Android applications. International Journal on Software Tools for Technology Transfer, pp. 1–18, 2016.

Jamrozik, K. and Zeller, A., DroidMate: A Robust and Extensible Test Generator for Android. International Conference on Mobile Software Engineering and Systems, pp. 293–294, 2016.

Janicki, M., Katara, M., and Pääkkönen, T., Obstacles and opportunities in deploying model-based GUI testing of mobile software: A survey. Software Testing Verification and Reliability, pp. 313–341, 2012.

Jha, A. K., Lee, S., and Lee, W. J., Modeling and Test Case Generation of Inter-component Communication in Android. 2nd ACM International Conference on Mobile Software Engineering and Systems (MOBILESoft), pp. 113–116, 2015.

Jing, Y., Ahn, G.-J., and Hu, H., Model-based conformance testing for android. Lecture Notes in Computer Science (including subseries Lecture Notes in Artificial Intelligence and Lecture Notes in Bioinformatics), pp. 1–18, 2012.

Kaasila, J., Ferreira, D., Kostakos, V., and Ojala, T., Testdroid: Automated remote UI testing on android. Proceedings of the 11th International Conference on Mobile and Ubiquitous Multimedia (MUM '12), pp. 1–4, 2012.

Kallio, T. and Kekäläinen, A., Improving the effectiveness of mobile application design: User-pairs testing by non-professionals. Lecture Notes in Computer Science (including subseries Lecture Notes in Artificial Intelligence and Lecture Notes in Bioinformatics), pp. 315–319, 2004.

Kaur, K. and Kaur, A., Cloud era in mobile application testing. Proceedings of the 10th INDIACom, pp. 1057–1060, 2016.

Kawakami, L., Knabben, A., Rechia, D., Bastos, D., Pereira, O., Pereira e Silva, R., and Dos Santos, L.C.V., An object-oriented framework for improving software reuse on automated testing of mobile phones. Lecture Notes in Computer Science (including subseries Lecture Notes in Artificial Intelligence and Lecture Notes in Bioinformatics), pp. 199–201, 2007.

Keng, J. C. J., Jiang, L., Wee, T. K., and Balan, R. K., Graph-aided directed testing of Android applications for checking runtime privacy behaviours. Proceedings of the 11th International Workshop on Automation of Software Test, pp. 57–63, 2016.

Keng, J.C.J., Automated testing and notification of mobile app privacy leak-cause behaviours. Proceedings of the 31st IEEE/ACM International Conference on Automated Software Engineering, pp. 880–883, 2016.

Keng, J.C.J., Jiang, L., Wee, T.K., and Balan, R.K., Graph-aided directed testing of android applications for checking runtime privacy behaviours. Proceedings - 11th International Workshop on Automation of Software Test, pp. 57–63, 2016.

Khalid, H., Nagappan, M., Shihab, E., and Hassan, A.E., Prioritizing the Devices to Test Your App on A Case Study of Android Game Apps. ACM Symposium, pp. 610–620, 2014.

Khalid, H., Shihab, E., Nagappan, M., and Hassan, A.E., What do mobile app users complain about? IEEE Conference Publications, pp. 70–77, 2014.

Khan, A.I., Al-Khanjari, Z., and Sarrab, M., Crowd sourced testing through end users for Mobile Learning application in the context of Bring Your Own Device. 7th IEEE Annual Information Technology, Electronics and Mobile Communication Conference, pp. 1–6, 2016.

Kim, H., Choi, B., and Wong, W.E., Performance Testing of Mobile Applications at the Unit Test Level. Third IEEE International Conference on Secure Software Integration and Reliability Improvement (SSIRI '09), pp. 171–180, 2009.

Kim, H., Choi, B., and Yoon, S., Performance testing based on test-driven development for mobile applications. Proceedings of the 3rd International Conference on Ubiquitous Information Management and Communication (ICUIMC '09), pp. 612–617, 2009.

Kim, H.-K., Hybrid model based testing for mobile applications. International Journal of Software Engineering and its Applications, pp. 223–238, 2013.

Kim, H.-K., Mobile applications software testing methodology. Communications in Computer and Information Science, pp. 158–166, 2012.

Kim, H.K., Test driven mobile applications development. Proceedings of the World Congress on Engineering and Computer Science (WCECS '13), pp. 785–789, 2013.

Kirubakaran, B. and Karthikeyani, V., Mobile application testing-Challenges and solution approach through automation. Proceedings of the 2013 International Conference on Pattern Recognition, Informatics and Mobile Engineering, PRIME 2013, pp. 79–84, 2013.

Kluth, W., Krempels-H, K., and Samsel, C., Automated usability testing for mobile applications. WEBIST 2014 - Proceedings of the 10th

International Conference on Web Information Systems and Technologies, pp. 149–156, 2014.

Knych, T.W. and Baliga, A., Android application development and testability. Proceedings of the 1st International Conference on Mobile Software Engineering and Systems, pp. 37–40, 2014.

Ko, J.-W., Sim, S.-H., and Song, Y.-J., Test Based Model Transformation Framework for Mobile Application. International Conference on Information Science and Applications (ICISA), pp. 1–7, 2011.

Kobashi, T., Yoshizawa, M., Washizaki, H., Fukazawa, Y., Yoshioka, N., Okubo, T., and Kaiya, H., TESEM: A tool for verifying security design pattern applications by model testing. IEEE 8th International Conference on Software Testing, Verification and Validation (ICST), pp. 1–8, 2015.

Krishnan, P., Hafner, S., and Zeiser, A., Applying Security Assurance Techniques to a Mobile Phone Application: An Initial Approach. IEEE Fourth International Conference on Software Testing, Verification and Validation Workshops (ICSTW), pp. 545–552, 2011.

Kritikos, K., Pernici, B., Plebani, P., Cappiello, C., Comuzzi, M., Benrernou, S., Brandic, I., Kertész, A., Parkin, M., and Carro, M., A survey on mobile users' software quality perceptions and expectations. Journal Article, 2013.

Kuo, J.Y., Liu, C.-H., and Yu, W.T., The Study of Cloud-Based Testing Platform for Android. 2015 IEEE International Conference on Mobile Services, pp. 197–201, 2015.

Kwon, O.-H. and Hwang, S.-M., Mobile GUI testing tool based on image flow. Seventh IEEE/ACIS International Conference on Computer and Information Science (ICIS 08), pp. 508–512, 2008.

Laurencot, P., Testing mobile and distributed systems: method and experimentation. 8th International Conference, OPODIS 2004, pp. 37–51, 2014.

Le, H.V., Modeling human behavior during touchscreen interaction in mobile situations. Proceedings of the 18th International Conference on Human-Computer Interaction with Mobile Devices and Services Adjunct, pp. 901–902, 2016.

Lee, J. and Kim, H., Poster: Framework for automated power estimation of android applications. Proceeding of the 11th Annual International Conference on Mobile Systems, Applications, and Services (MobiSys '13), pp. 541–542, 2013.

Lee, J., Kim, Y., and Kim, S., Design and Implementation of a Linux Phone Emulator Supporting Automated Application Testing. Convergence and Hybrid Information Technology, 2008. ICCIT '08. Third International Conference on, pp. 256–259, 2008.

Lee, M. and Kim, G.J., On applying experience sampling method to A/B testing of mobile applications: A case study. 15th IFIP TC 13 International Conference on Human-Computer Interaction, INTERACT 2015, pp. 203–210, 2015.

Leem, D.-W., Jung, H.-J., Hwang, M.-S., Shim, J.-A., and Kwon, H. J., Design of a work process and implementation of a prototype for the development of an automation tool for android application vulnerability inspection. International Journal of Software Engineering and its Applications, pp. 29–38, 2016.

Li, A., Qin, Z., Chen, M., and Liu, J., ADAutomation: An Activity Diagram Based Automated GUI Testing Framework for Smartphone Applications. IEEE Conference Publications, pp. 68–77, 2014.

Li, D., Lyu, Y., Wan, M., and Halfond, W.G.J., String analysis for Java and android applications. ESEC/FSE 2015 Proceedings of the 2015 10th Joint Meeting on Foundations of Software Engineering, pp. 661–672, 2015.

Liang, C.-J.M., Lane, N.D., Brouwers, N., Zhang, L., Karlsson, B.F., Liu, H., Liu, Y., Tang, J., Shan, X., Chandra, R., and Zhao, F., Caiipa Automated Large-scale Mobile App Testing through Contextual Fuzzing. Conference, pp. 519–530, 2014.

Liang, H., Song, H., Fu, Y., Cai, X., and Zhang, Z., A remote usability testing platform for mobile phones. Proceedings - 2011 IEEE International Conference on Computer Science and Automation Engineering, CSAE 2011, pp. 312–316, 2011.

Ligman, J., Pistoia, M., Tripp, O., and Thomas, G., Improving design validation of mobile application user interface implementation. Proceedings - International Conference on Mobile Software Engineering and Systems, pp. 277–278, 2016.

Lin, B.T. and Green, P.A., A simple method to record keystrokes on mobile phones and other devices for usability evaluations. Lecture Notes in Computer Science (including subseries Lecture Notes in Artificial Intelligence and Lecture Notes in Bioinformatics), pp. 434–444, 2016.

Lin, Y., Rojas, J., Chu, E., and Lai, Y., On the Accuracy, Efficiency, and Reusability of Automated Test Oracles for Android Devices. IEEE Computer Society, pp. 957–970, 2014.

Linares-Vásquez, M., Enabling Testing of Android Apps. IEEE/ACM 37th IEEE International Conference on Software Engineering (Volume: 2), pp. 763–765, 2015.

Liu, C. H., Chen, W. K., and Chen, S. L., A Concurrent Approach for Improving the Efficiency of Android CTS Testing. International Computer Symposium, pp. 611–615, 2016.

Liu, C.-H., Lu, C.-Y., Cheng, S.-J., Chang, K.-Y., Hsiao, Y.-C., and Chu, W.-M., Capture-Replay Testing for Android Applications. International Symposium on Computer, Consumer and Control (IS3C), pp. 1129–1132, 2014.

Liu, Y. and Xu, C., VeriDroid: automating Android application verification. Proceedings of the 2013 Middleware Doctoral Symposium (MDS '13), pp. 1–6, 2011.

Liu, Y., Lu, Y., and Li, Y., An Android-based approach for automatic unit test. IET Conference Publications, pp. 1–4, 2014.

Liu, Z., Gao, X., and Long, X., Adaptive random testing of mobile application. 2nd International Conference on Computer Engineering and Technology (ICCET), pp. 297–301, 2010.

Liu, Z., Liu, B., and Gao, X., Test automation on mobile device. Proceedings of the 5th Workshop on Automation of Software Test (AST '10), pp. 1–7, 2010.

Lu, L., Hong, Y., Huang, Y., Su, K., and Yan, Y., Activity Page Based Functional Test Automation for Android Application. Third World Congress on Software Engineering (WCSE), pp. 37–40, 2012.

Ma, X., Wang, N., Xie, P., Zhou, J., Zhang, X., and Fang, C., An Automated Testing Platform for Mobile Applications. Proceedings - 2016 IEEE International Conference on Software Quality, Reliability and Security-Companion, pp. 159–162, 2016.

Ma, X., Yan, B., Chen, G., Zhang, C., Huang, K., and Drury, J., A toolkit for usability testing of mobile applications. Lecture Notes of the Institute for Computer Sciences, Social-Informatics and Telecommunications Engineering, pp. 226–245, 2012.

Ma, X., Yan, B., Chen, G., Zhang, C., Huang, K., Drury, J., and Wang, L., Design and implementation of a toolkit for usability testing of mobile apps. Mobile Networks and Applications, pp. 81–97, 2013.

Ma, Y. and Choi, E.M., Hook-based mobile software testing by using aspect-oriented programming. International Conference on Systems and Informatics (ICSAI), pp. 2528–2532, 2012.

Machado, P., Campos, J., and Abreu, R., MZoltar: Automatic debugging of android applications. Proceedings of the 2013 International Workshop on Software Development Lifecycle for Mobile (DeMobile '13), pp. 9–16, 2013.

MacHiry, A., Tahiliani, R., and Naik, M., Dynodroid: an input generation system for Android apps. Proceedings of the 2013 9th Joint Meeting on Foundations of Software Engineering (ESEC/FSE '13), pp. 224–234, 2013.

Mahmood, R., Esfahani, N., Kacem, T., Mirzaei, N., Malek, S., and Stavrou, A., A whitebox approach for automated security testing of Android applications on the cloud. 7th International Workshop on Automation of Software Test (AST), pp. 22–28, 2012.

Mahmood, R., Mirzaei, N., and Malek, S., EvoDroid segmented evolutionary testing of Android apps. FSE 2014 Proceedings of the 22nd ACM SIGSOFT International Symposium on Foundations of Software Engineering, pp. 599–609, 2014.

Majchrzak, T.A. and Schulte, M., Context-Dependent app testing. Proceedings of the 27th Conference on Advanced Information Systems Engineering (CAiSE) Forum, pp. 27–39, 2015.

Malek, S., Esfahani, N., Kacem, T., Mahmood, R., Mirzaei, N., and Stavrou, A., A Framework for Automated Security Testing of Android Applications on the Cloud. SERE-C '12 Proceedings of the 2012 IEEE Sixth International Conference on Software Security and Reliability Companion, pp. 35–36, 2012.

Matsumoto, S. and Sakurai, K., A proposal for the privacy leakage verification tool for Android application developers. Proceedings of the 7th International Conference on Ubiquitous Information Management and Communication (ICUIMC '13), pp. 1–8, 2013.

Mazlan, M.A., Stress Test on J2ME Compatible Mobile Device. Innovations in Information Technology, 2006, pp. 1–5, 2006.

McIlroy, S., Ali, N., Khalid, H., and Hassan, A.E., Analyzing and automatically labelling the types of user issues that are raised in mobile app reviews. Journal Article, pp. 1–40, 2015.

Méndez-Porras, A., Nieto Hidalgo, M., García-Chamizo, J.M., Jenkins, M., and Porras, A.M., A top-down design approach for an automated testing framework. UCAmI 2015, 9th International Conference, pp. 37–49, 2015.

Mirzaei, N., Bagheri, H., Mahmood, R., and Malek, S., SIG-Droid: Automated system input generation for Android applications. IEEE 26th

International Symposium on Software Reliability Engineering (ISSRE), pp. 461–471, 2015.

Mirzaei, N., Malek, S., Păsăreanu, C.S., Esfahani, N., and Mahmood, R., Testing android apps through symbolic execution. SIGSOFT Software Engineering Notes, pp. 1–5, 2013.

Moran, K., Linares-Vásquez, M., Bernal-Cárdenas, C., and Poshyvanyk, D., FUSION: a tool for facilitating and augmenting android bug reporting. In Proceedings of the 38th International Conference on Software Engineering Companion, pp. 1–4, 2016.

Moran, K., Linares-Vásquez, M., Bernal-Cárdenas, C., Vendome, C., and Poshyvanyk, D., Automatically Discovering, Reporting and Reproducing Android Application Crashes. IEEE International Conference on Software Testing, Verification and Validation, pp. 33–44, 2016.

Morgado, I. C. and Paiva, A. C. R., The iMPAcT Tool: Testing UI Patterns on Mobile Applications. 30th IEEE/ACM International Conference on Automated Software Engineering (ASE), pp. 876–881, 2015.

Muccini, H., Di Francesco, A., and Esposito, P., Software testing of mobile applications: Challenges and future research directions. 7th International Workshop on Automation of Software Test (AST), pp. 29–35, 2012.

Mukadam, M., Bird, C., and Rigby, P.C., Gerrit software code review data from Android. Proceedings of the 10th Working Conference on Mining Software Repositories, pp. 45–48, 2013.

Munro, D., Calitz, A.P., and Vogts, D., AARemu: An outdoor mobile augmented reality emulator for android. ACM International Conference Proceeding Series, pp. 1–9, 2016.

Murugesan, L. and Balasubramanian, P., Cloud based mobile application testing. ICIS 2014, pp. 287–289, 2014.

Nagowah, L. and Sowamber, G., A novel approach of automation testing on mobile devices. International Conference on Computer & Information Science (ICCIS), pp. 924–930, 2012.

Naik, K., Ali, Y., Mahinthan, V., Singh, A., and Abogharaf, A., Categorizing configuration parameters of smartphones for energy performance testing. Proceedings of the 9th International Workshop on Automation of Software Test (AST '14), pp. 15–21, 2014.

Nascimento, I., Silva, W., Lopes, A., Rivero, L., Gadelha, B., Oliveira, E., and Conte, T., An empirical study to evaluate the feasibility of a UX and usability inspection technique for mobile applications. Proceedings of the International Conference on Software Engineering and Knowledge Engineering, pp. 372–383, 2016.

Payet, É. and Spoto, F., Static analysis of Android programs. Information and Software Technology, pp. 1192–1201, 2012.

Pesonen, J., Extending Software Integration Testing Using Aspects in Symbian OS. Proceedings of Testing: Academic and Industrial Conference - Practice And Research Techniques, pp. 147–151, 2006.

Porat, T., Schclar, A., and Shapira, B., MATE: a mobile analysis tool for usability experts. CHI '13 Extended Abstracts on Human Factors in Computing Systems, pp. 265–270, 2013.

Prathibhan, C.M., Malini, A., Venkatesh, N., and Sundarakantham, K., An automated testing framework for testing Android mobile applications in the cloud. IEEE Conference Publications, pp. 1216–1219, 2014.

Prongsang, C. and Suwannasart, T., A tool for test case impact analysis from user interface changes in android mobile application. Lecture Notes in Engineering and Computer Science, pp. 483–486, 2016.

Püschel, G., Seiger, R., and Schlegel, T., Test modeling for context-aware ubiquitous applications with feature petri nets. Proceedings of the 2nd International Workshop on Model-based Interactive Ubiquitous Systems, pp. 37–40, 2012.

Puspika, B.N., Hendradjaya, B., and Danar Sunindyo, W., Towards an automated test sequence generation for mobile application using colored Petri Net. International Conference on Electrical Engineering and Informatics (ICEEI), pp. 445–449, 2015.

Qian, J. and Zhou, D., Prioritizing Test Cases for Memory Leaks in Android Applications. Journal of Computer Science and Technology, pp. 869–882, 2016.

Qin, Z., Tang, Y., Novak, E., and Li, Q., MobiPlay: A remote execution based record-and-replay tool for mobile applications. Proceedings - International Conference on Software Engineering, pp. 571–582, 2016.

Raghavan, G., Salomaki, A., and Lencevicius, R., Model based estimation and verification of mobile device performance. Proceedings of the 4th ACM International Conference on Embedded Software (EMSOFT '04), pp. 34–43, 2014.

Rajan, V.S., Malini, A., and Sundarakantham, K., Performance evaluation of online mobile application using Test My App. IEEE Conference Publications, pp. 1148–1152, 2014.

Rajasekaran, M. J., Challenges in mobile application testing: A survey. International Journal of Control Theory and Applications, pp. 159–163, 2016.

Ravindranath, L., Nath, S., Padhye, J., and Balakrishnan, H., Automatic and scalable fault detection for mobile applications. MobiSys 2014 - Proceedings of the 12th Annual International Conference on Mobile Systems, Applications, and Services, pp. 190–203, 2014.

Rege, M.R., Handziski, V., and Wolisz, A., A Context Simulation Harness for Realistic Mobile App Testing. Conference, pp. 489, 2015.

Reichstaller, A., Eberhardinger, B., Knapp, A., Reif, W., and Gehlen, M., Risk-based interoperability testing using reinforcement learning. Lecture Notes in Computer Science (including subseries Lecture Notes in Artificial Intelligence and Lecture Notes in Bioinformatics), pp. 52–69, 2016.

Ricky, M. Y., Purnomo, F., and Yulianto, B., Mobile Application Software Defect Prediction. IEEE Symposium on Service-Oriented System Engineering, pp. 307–313, 2016.

Ridene, Y. and Barbier, F., A model-driven approach for automating mobile applications testing. ACM International Conference Proceeding Series, pp. 1–7, 2011.

Ridene, Y., Barbier, F., Belloir, N., and Couture, N., A DSML for Mobile Phone Applications Testing. Proceeding DSM '10 Proceedings of the 10th Workshop on Domain-Specific Modeling, pp. 25–30, 2010.

Rodriguez, R.A., Vera, P.M., Valles, G.Y., and Martinez, M.R., Creating a usability lab for testing on mobile devices using free tools, CACIDI 2016 - Congreso Aergentino de Ciencias de la Informatica y Desarrollos de Investigacion, pp. 1–6, 2016.

Rodríguez-Mota, A., Escamilla-Ambrosio, P. J., Morales-Ortega, S., Salinas-Rosales, M., and Aguirre-Anaya, E., Towards a 2-hybrid Android malware detection test framework. International Conference on Electronics, Communications and Computers, pp. 54–61, 2016.

Rojas, I. K. V., Meireles, S., and Dias-Neto, A. C., Cloud-based mobile app testing framework: Architecture, implementation and execution. ACM International Conference Proceeding Series, pp. 1–10, 2016.

Saad, N.H. and Awang Abu Bakar, N.S., Automated testing tools for mobile applications. IEEE Conference Publications, pp. 1–5, 2014.

Sadeghi, A., Bagheri, H., Garcia, J., and Malek, S., A Taxonomy and Qualitative Comparison of Program Analysis Techniques for Security Assessment of Android Software. IEEE Transactions on Software Engineering, pp. 1–48, 2016.

Sadeh, B., Ørbekk, K., Eide, M.M., Gjerde, N.C.A., Tønnesland, T.A., and Gopalakrishnan, S., Towards unit testing of user interface code for android mobile applications. Second International Conference on Software Engineering and Computer Systems (ICSECS '11), pp. 163–175, 2011.

Sahinoglu, M., Incki, K., and Aktas, M., Mobile Application Verification: A Systematic Mapping Study. ICCSA 2015, pp. 147–163, 2015.

Salva, S. and Zafimiharisoa, S.R., Data vulnerability detection by security testing for Android applications. Information Security for South Africa, pp. 1–8, 2013.

Sama, M., Elbaum, S., Raimondi, F., Rosenblum, D.S., and Wang, Z., Context-aware adaptive applications: Fault patterns and their automated identification. IEEE Transactions on Software Engineering, pp. 644–661, 2010.

Samuel, T. and Pfahl, D., Problems and solutions in mobile application testing. Lecture Notes in Computer Science (including subseries Lecture Notes in Artificial Intelligence and Lecture Notes in Bioinformatics), pp. 249–267, 2016.

Santos, A. and Correia, I., Mobile testing in software industry using agile: Challenges and opportunities. IEEE 8th International Conference on Software Testing, Verification and Validation (ICST), pp. 1–2, 2015.

Satoh, I., A testing framework for mobile computing software. Journal of IEEE Transactions on Software Engineering, pp. 1112–1121, 2003.

Satoh, I., Software testing for mobile and ubiquitous computing. Proceedings of the 6th International Symposium on Autonomous Decentralized Systems (ISADS'03), pp. 185–192, 2003.

Schulz, S., Honkola, J., and Huima, A., Towards model-based testing with architecture models. 14th Annual IEEE International Conference and Workshops on the Engineering of Computer-Based Systems (ECBS '07), pp. 495–502, 2007.

Sekanina, A., Prokopova, Z., and Silhavy, R., Mobile applications testing. Annals of DAAAM and Proceedings of the International DAAAM Symposium, pp. 1605–1606, 2011.

Shabtai, A., Fledel, Y., and Elovici, Y., Automated static code analysis for classifying android applications using machine learning. Proceedings - 2010 International Conference on Computational Intelligence and Security, CIS 2010, pp. 329–333, 2010.

Shahriar, H., North, S., and Mawangi, E., Testing of Memory Leak in Android Applications. IEEE 15th International Symposium on High-Assurance Systems Engineering (HASE), pp. 176–183, 2014.

Shan, Z., Azim, T., and Neamtiu, I., Finding resume and restart errors in android applications. Proceedings of the Conference on Object-Oriented Programming Systems, Languages, and Applications, pp. 864–880, 2016.

She, S., Sivapalan, S., and Warren, I., Hermes: A Tool for Testing Mobile Device Applications. Software Engineering Conference, ASWEC '09, pp. 121–130, 2009.

Shi, Y., You, W., Qian, K., Bhattacharya, P., and Qian, Y., A hybrid analysis for mobile security threat detection. 2016 IEEE 7th Annual Ubiquitous Computing, Electronics and Mobile Communication Conference, pp. 1–7, 2016.

Silva, D. B., Endo, A. T., Eler, M. M., and Durelli, V. H., An analysis of automated tests for mobile Android applications. Latin American Computing Conference (CLEI), pp. 1–9, 2016.

Soad, G.W., Duarte Filho, N.F., and Barbosa, E.F., Quality evaluation of mobile learning applications. Proceedings - Frontiers in Education Conference, pp. 280–283, 2016.

Song, H., Ryoo, S., and Kim, J.H., An Integrated Test Automation Framework for Testing on Heterogeneous Mobile Platforms. First ACIS International Symposium on Software and Network Engineering (SSNE), pp. 141–145, 2011.

Song, K., Han A.-R., Jeong, S., and Cha, S., Generating various contexts from permissions for testing Android applications. 27th International Conference on Software Engineering and Knowledge Engineering, SEKE 2015, pp. 87–92, 2015.

Srirama, S., Kakumani, R., Aggarwal, A., and Pawar, P., Effective Testing Principles for the Mobile Data Services Applications. First International Conference on Communication System Software and Middleware, Comsware 2006, pp. 1–5, 2006.

Starov, O., Vilkomir, S., Gorbenko, A., and Kharchenko, V., Testing-as-a-service for mobile applications: State-of-the-art survey. Journal Article, pp. 55–71, 2015.

Stirparo, P., Fovino, I.N., and Kounelis, I., Data-in-use leakages from Android memory - Test and analysis. IEEE 9th International Conference on Wireless and Mobile Computing, Networking and Communications (WiMob), pp. 701–708, 2013.

Subramanian, S., Singleton, T., and El Ariss, O., Class coverage GUI testing for Android applications. International Conference on System Reliability and Science, pp. 84–89, 2016.

Sun, C., Zhang, Z., Jiang, B., and Chan, W.K., Facilitating Monkey Test by Detecting Operable Regions in Rendered GUI of Mobile Game Apps. Proceedings - 2016 IEEE International Conference on Software Quality, Reliability and Security, pp. 298–306, 2016.

Sun, Q., Xu, L., Chen, L., and Zhang, W., Replaying Harmful Data Races in Android Apps. IEEE International Symposium on Software Reliability Engineering Workshops, pp. 160–166, 2016.

Syer, M.D., Nagappan, M., Adams, B., and Hassan, A.E., Studying the relationship between source code quality and mobile platform dependence. Software Quality Journal, pp. 485–508, 2014.

Szabo, C., Samuelis, L., Ivanovic, M., and Fesic, T., Database refactoring and regression testing of Android mobile applications. IEEE 10th Jubilee International Symposium on Intelligent Systems and Informatics (SISY), pp. 135–139, 2012.

Takala, T., Katara, M., and Harty, J., Experiences of System-Level Model-Based GUI Testing of an Android Application. IEEE Fourth International Conference on Software Testing, Verification and Validation (ICST), pp. 377–386, 2011.

Tang, H., Wu, G., Wei, J., and Zhong, H., Generating test cases to expose concurrency bugs in android applications. Proceedings of the 31st IEEE/ACM International Conference on Automated Software Engineering, pp. 648–653, 2016.

Tao, C. and Gao, J., Building a Model-Based GUI Test Automation System for Mobile Applications. International Journal of Software Engineering and Knowledge Engineering, pp. 238–249, 2016.

Tao, C. and Gao, J., Cloud-based mobile testing as a service. International Journal of Software Engineering and Knowledge Engineering, pp. 147–152, 2016.

Tao, C. and Gao, J., On building test automation system for mobile applications using GUI ripping. Proceedings of the International Conference on Software Engineering and Knowledge Engineering, pp. 480–485, 2016.

Tsakiltsidis, S., Miranskyy, A., and Mazzawi, E., On Automatic Detection of Performance Bugs. Proceedings - 2016 IEEE 27th International Symposium on Software Reliability Engineering Workshops, pp. 132–139, 2016.

Tulpule, N., Strategies for testing client-server interactions in mobile applications: How to move fast and not break things. Proceedings of the

2013 ACM Workshop on Mobile Development Lifecycle (MobileDeLi '13), pp. 19–20, 2013.

van der Merwe, H., Tkachuk, O., van der Merwe, B., and Visser, W., Generation of Library Models for Verification of Android Applications. Journal Article, pp. 1–5, 2015.

van der Merwe, H., van der Merwe, B., and Visser, W., Verifying android applications using Java PathFinder. SIGSOFT Software Engineering Notes, pp. 1–5, 2012.

Vemuri, R., Testing Predictive Software in Mobile Devices. 1st International Conference on Software Testing, Verification, and Validation, pp. 440–447, 2008.

Vieira, V., Holl, K., and Hassel, M., A context simulator as testing support for mobile apps. SAC 2015 Proceedings of the 30th Annual ACM Symposium on Applied Computing, pp. 535–541, 2015.

Vilkomir, S. and Amstutz, B., Using Combinatorial Approaches for Testing Mobile Applications. IEEE Seventh International Conference on Software Testing, Verification and Validation Workshops (ICSTW), pp. 78–83, 2014.

Vilkomir, S., Marszalkowski, K., Perry, C., and Mahendrakar, S., Effectiveness of Multi-device Testing Mobile Applications. 2nd ACM International Conference on Mobile Software Engineering and Systems (MOBILESoft), pp. 44–47, 2015.

Villanes, I.K., Costa, E.A.B., and Dias-Neto, A.C., Automated Mobile Testing as a Service (AM-TaaS). 2015 IEEE World Congress on Services, pp. 79–86, 2015.

Villarroel, L., Bavota, G., Russo, B., Oliveto, R., and Di Penta, M., Release planning of mobile apps based on user reviews. Proceedings - International Conference on Software Engineering, pp. 14–24, 2016.

Voas, J. and Miller, K.W., Software Testing: What Goes Around Comes Around. IEEE Computer Society, IT Professional, pp. 4–5, 2012.

Wan, M., Jin, Y., Li, D., and Halfond, W.G.J., Detecting Display Energy Hotspots in Android Apps. IEEE Conference Publications, pp. 1–10, 2015.

Wang, L., and Li, R., The research of orthogonal experiment applied in mobile phone's software test case generation. Proceedings of the 2010 International Forum on Information Technology and Applications (IFITA '10), pp. 345–348, 2010.

Wang, P., Liang, B., You, W., Li, J., and Shi, W., Automatic Android GUI Traversal with High Coverage. Fourth International Conference on

Communication Systems and Network Technologies (CSNT), pp. 1161–1166, 2014.

Wang, Y., An automated virtual security testing platform for android mobile apps. First Conference on Mobile and Secure Services (MOBISECSERV), pp. 1–2, 2015.

Wang, Y. and Alshboul, Y., Mobile security testing approaches and challenges. Conference, pp. 1–5, 2015.

Wei, O.K. and Ying, T.M., Knowledge management approach in mobile software system testing. IEEE International Conference on Industrial Engineering and Engineering Management, pp. 2120–2123, 2007.

Weiss, D. and Zduniak, M., Automated integration tests for mobile applications in Java 2 micro edition. Lecture Notes in Computer Science (including subseries Lecture Notes in Artificial Intelligence and Lecture Notes in Bioinformatics), pp. 478–487, 2007.

Wen, H.-L., Lin, C.-H., Hsieh, T.-H., and Yang C.-Z., PATS: A Parallel GUI Testing Framework for Android Applications. Computer Software and Applications Conference (COMPSAC), IEEE 39th Annual (Volume:2), pp. 210–215, 2015.

Wilke, C., Götz, S., and Richly, S., JouleUnit: a generic framework for software energy profiling and testing. Proceedings of the 2013 Workshop on Green in/by Software Engineering (GIBSE '13), pp. 9–14, 2013.

Wu, J., Wu, Y., Wu, Z., Yang, M., Luo, T., and Wang, Y., AndroidFuzzer: Detecting android vulnerabilities in fuzzing cloud. IEEE Sixth International Conference on Cloud Computing, pp. 954–955, 2015.

Xu, Y. and Chen, N., Evaluating mobile apps with A/B and quasi A/B tests. Proceedings of the ACM SIGKDD International Conference on Knowledge Discovery and Data Mining, pp. 313–322, 2016.

Xu, Y.-P., Ma, Z.-F., Wang, Z.-H., Niu, X.-X., and Yang, Y.-X., Survey of security for Android smart terminal. Tongxin Xuebao/Journal on Communications, pp. 1–4, 2016.

Yan, D., Yang, S., and Rountev, A., Systematic testing for resource leaks in Android applications. IEEE 24th International Symposium on Software Reliability Engineering (ISSRE), pp. 411–420, 2013.

Yan, M., Sun, H., and Liu, X., iTest Testing Software with Mobile Crowdsourcing. Conference, pp. 19–24, 2014.

Yang, S., Yan, D., and Rountev, A., Testing for poor responsiveness in android applications. 1st International Workshop on the Engineering of Mobile-Enabled Systems (MOBS), pp. 1–6, 2013.

Yang, W., Prasad, M., and Xie, T., A grey-box approach for automated GUI-model generation of mobile applications. Lecture Notes in Computer Science (including subseries Lecture Notes in Artificial Intelligence and Lecture Notes in Bioinformatics), pp. 250–265, 2013.

Yeh, C.-C. and Huang, S.-K., CovDroid: A black-box testing coverage system for android. Computer Software and Applications Conference (COMPSAC), IEEE 39th Annual (Volume:3), pp. 447–452, 2015.

Yeh, C.C., Lu, H.L., Chen, C.Y., Khor, K.K., and Huang, S.K., CRAXDroid: Automatic Android System Testing by Selective Symbolic Execution. IEEE Conference Publications, pp. 140–148, 2014.

Yeh, C.-C., Huang, S.-K., and Chang, S.-Y., A black-box based android GUI testing system. Proceeding of the 11th Annual International Conference on Mobile Systems, Applications, and Services (MobiSys '13), pp. 529–530, 2013.

Yu, S. and Takada, S., External event-based test cases for mobile application. C3S2E 2015 Proceedings of the Eighth International C* Conference on Computer Science & Software Engineering, pp. 148–149, 2015.

Yu, S. and Takada, S., Mobile application test Case generation focusing on external events. Proceedings of the 1st International Workshop on Mobile Development, pp. 41–42, 2016.

Yumei, W. and Zhifang, L., A Model Based Testing Approach for Mobile Device. International Conference on Industrial Control and Electronics Engineering (ICICEE), pp. 1885–1888, 2012.

Yusop, N., Kamalrudin, M., Sidek, S., and Grundy, J., Automated support to capture and validate security requirements for mobile apps. Communications in Computer and Information Science, pp. 97–112, 2016.

Zaeem, R.N., Prasad, M.R., and Khurshid, S., Automated Generation of Oracles for Testing User-Interaction Features of Mobile Apps. IEEE Seventh International Conference on Software Testing, Verification and Validation (ICST), pp. 183–192, 2014.

Zein, S., Salleh, N., and Grundy, J., A systematic mapping study of mobile application testing techniques. Journal of Systems and Software, pp. 334–356, 2016.

Zein, S., Salleh, N., and Grundy, J., Mobile application testing in industrial contexts: An exploratory multiple case-study. 14th International Conference, SoMet 2015, Naples, Italy, Springer, pp. 30–41, 2015.

Zhang, D. and Adipat, B., Challenges, methodologies, and issues in the usability testing of mobile applications. International Journal of Human-Computer Interaction, pp. 293–308, 2005.

Zhang, H., Wu, H., and Rountev, A., Automated Test Generation for Detection of Leaks in Android Applications. 11th International Workshop in Automation of Software Test, pp. 64–70, 2016.

Zhang, P. and Elbaum, S., Amplifying Tests to Validate Exception Handling Code: An Extended Study in the Mobile Application Domain. Journal Article - ACM Transactions on Software Engineering and Methodology (TOSEM) - Special Issue International Conference on Software Engineering (ICSE 2012) and Regular Papers, pp. 1–28, 2012.

Zhang, T., Gao, J., Aktouf, O.-E.-K., and Uehara T., Test model and coverage analysis for location-based mobile services. SEKE2015, pp. 1–7, 2015.

Zhang, T., Gao, J., Cheng, J., and Uehara, T., Compatibility Testing Service for Mobile Applications. IEEE Symposium, pp. 179–186, 2015.

Zhang, X., Chen, Z., Fang, C., and Liu, Z., Guiding the crowds for Android testing. Proceedings of the 38th International Conference on Software Engineering Companion, pp. 752–753, 2016.

Zhao, H., Sun, J., and Hu, G., Study of Methodology of Testing Mobile Games Based on TTCN-3. 10th ACIS International Conference on Software Engineering, Artificial Intelligences, Networking and Parallel/Distributed Computing (SNPD '09), pp. 579–584, 2009.

Zhauniarovich, Y., Philippov, A., Gadyatskaya, O., Crispo, B., and Massacci F., Towards black box testing of android apps. 2015 10th International Conference on Availability, Reliability and Security (ARES), pp. 501–510, 2015.

Zhi-Fang, L. and Xiao-Peng, G., SOA Based Mobile Device Test. Second International Conference on Intelligent Computation Technology and Automation (ICICTA '09), pp. 641–644, 2009.

Zhong, J., Huang, J., and Liang, B., Android Permission *Re*-delegation Detection and Test Case Generation. International Conference on Computer Science & Service System (CSSS), pp. 871–874, 2012.

Zhu, H., Ye, X., Zhang, X., and Shen, K., A Context-Aware Approach for Dynamic GUI Testing of Android Applications. Computer Software and Applications Conference (COMPSAC), IEEE 39th Annual (Volume: 2), pp. 248–253, 2015.

Zivkov, D., Kastelan, I., Neborovski, E., Miljkovic, G., and Katona, M., Touch screen mobile application as part of testing and verification system. Proceedings of the 35th International Convention (MIPRO), pp. 892–895, 2012.

Zun, D., Qi, T., and Chen, L., Research on automated testing framework for multi-platform mobile applications. Proceedings of 2016 4th IEEE International Conference on Cloud Computing and Intelligence System, pp. 82–87, 2016.

REFERENCES

[1] Statista, Number of smartphone users worldwide from 2014 to 2020. The Statistics Portal, https://www.statista.com/statistics/330695/number-of-smartphone-users-worldwide/, 2015. (accessed 03/15/2017).
[2] Statista. Worldwide mobile app revenues in 2015, 2016 and 2020 (in billion U.S. dollars), The Statistics Portal. https://www.statista.com/statistics/269025/worldwide-mobile-app-revenue-forecast/, 2017. (accessed 03/15/2017).
[3] S. Hess, F. Kiefer, R. Carbon, In: Quality by construction through mConcAppt: towards using UI-construction as driver for high quality mobile app engineering, Eighth International Conference on the Quality of Information and Communications Technology (QUATIC), 2012, pp. 313–318.
[4] Capgemini, Testing and SMAC technologies: ensuring a seamless and secure customer experience, in: World Quality Report 2014–15, sixth ed., Capgemini, HP, Sogeti, 2014.
[5] Compuware, Mobile apps: what consumers really need and want—a global study of consumers' expectations and experiences of mobile applications, in: Compuware Survey, 2013.
[6] Apigee, Apigee survey: users reveal top frustrations that lead to bad mobile app reviews, Survey Conducted by uSamp. http://apigee.com/about/press-release/apigee-survey-users-reveal-top-frustrations-lead-bad-mobile-app-reviews, 2012. (accessed 03/15/2017).
[7] B. Arizanov, 30 Pitiful reasons why users hate and uninstall your mobile app, Mobiloitte Technologies, 2013. http://blog.mobiloitte.com/30-reasons-for-uninstalling-a-mobile-app. Accessed 1 May 2016.
[8] AppDynamics, The app attention span, research report, in partnership with the Institute of Management Studies (IMS) at Goldsmiths, University of London, 2014, p. 4.
[9] G. Gruman, Quality, not quantity, is the real mobile app problem, InfoWorld, 2014. http://www.infoworld.com/article/2607985/mobile-apps/quality--not-quantity–is-the-real-mobile-app-problem. Accessed 15 March 2017.
[10] S. Perez, Users have low tolerance for buggy apps, Crunch Network, 2013. http://techcrunch.com/2013/03/12/users-have-low-tolerance-for-buggy-apps-only-16-will-try-a-failing-app-more-than-twice. Accessed 15 March 2017.
[11] Persona, GitHub, Persona Identification System, https://github.com/mozilla/persona/issues/1325, 2015. Accessed 9 January 2015.
[12] Bitcoin Wallet, Google Project Hosting, Bitcoin Wallet, 2015. http://code.google.com/p/bitcoin-wallet/issues/detail?id=63. Accessed 9 January 2015.
[13] MyTracks, Google Project Hosting, MyTracks for Android, https://code.google.com/p/mytracks/issues/detail?id=34, 2015. Accessed 9 January 2015.
[14] K-9 Mail, Google Project Hosting, K-9 Mail Email Client for Android, http://code.google.com/p/k9mail/issues/detail?id=4528, 2015. Accessed 9 January 2015.
[15] H. Muccini, A. Di Francesco, P. Esposito, In: Software testing of mobile applications: challenges and future research directions, 7th International Workshop on Automation of Software Test (AST), 2012, pp. 29–35.

[16] Capgemini, Digital transformation: disrupting business models for a better customer experience, in: World Quality Report 2016–17, eighth ed., Capgemini, HP, Sogeti, 2016.
[17] K. Petersen, R. Feldt, S. Mujtaba, M. Mattsson, In: Systematic mapping studies in software engineering, Proceedings of the 12th International Conference on Evaluation and Assessment in Software Engineering (EASE'08), 2008, pp. 68–77.
[18] B. Kitchenham, Guidelines for performing systematic literature reviews in software engineering, in: EBSE Technical Report, Version 2.3, Software Engineering Group, 2007.
[19] J.R. Landis, G.G. Koch, The measurement of observer agreement for categorical data, Biometrics 33 (1) (1977) 159–174.
[20] L. Corral, A. Sillitti, G. Succi, Software assurance practices for mobile applications, Computing (2014) 1–24. Springer Vienna.
[21] M. Sahinoglu, K. Incki, M. Aktas, In: Mobile application verification: a systematic mapping study, International Conference on Computational Science and Its Applications 2015, 2015, pp. 147–163.
[22] S. Zein, N. Salleh, J. Grundy, A systematic mapping study of mobile application testing techniques, J. Syst. Softw. 117 (2016) 334–356.
[23] K. Kaur, A. Kaur, In: Cloud era in mobile application testing, Proceedings of the 10th INDIACom, 2016, pp. 1057–1060.
[24] M.J. Rajasekaran, Challenges in mobile application testing: a survey, Int. J. Control Theory Appl. 9 (2016) 159–163.
[25] T. Samuel, D. Pfahl, Problems and solutions in mobile application testing, in: Lecture Notes in Computer Science (Including Subseries Lecture Notes in Artificial Intelligence and Lecture Notes in Bioinformatics), 2016, pp. 249–267.
[26] Y. Wang, In: An automated virtual security testing platform for android mobile apps, First Conference on Mobile and Secure Services (MOBISECSERV), 2015, pp. 1–2.
[27] O. Starov, S. Vilkomir, A. Gorbenko, V. Kharchenko, Testing-as-a-service for mobile applications: state-of-the-art survey, in: Dependability Problems of Complex Information Systems, Springer, 2015, pp. 55–71.
[28] A. Santos, I. Correia, In: Mobile testing in software industry using agile: challenges and opportunities, IEEE 8th International Conference on Software Testing, Verification and Validation (ICST), 2015, pp. 1–2.
[29] S. Zein, N. Salleh, J. Grundy, In: Mobile application testing in industrial contexts: an exploratory multiple case-study, 14th International Conference, SoMet 2015, Springer, Naples, Italy, 2015, pp. 30–41.
[30] J. Gao, W.-T. Tsai, R. Paul, X. Bai, T. Uehara, In: Mobile testing-as-a-service (MTaaS)—infrastructures, issues, solutions and needs, Proceedings of the 2014 IEEE 15th International Symposium on High-Assurance Systems Engineering (HASE '14), 2014, pp. 158–167.
[31] A.S. Al-Ahmad, S.A. Aljunid, A.S.A. Sani, In: Mobile cloud computing testing review, International Conference on Advanced Computer Science Applications and Technologies (ACSAT), 2013, pp. 176–180.
[32] Y. Dubinsky, A. Abadi, In: Challenges and research questions for testing in mobile development: report on a mobile testing activity, MobileDeLi 2013—Proceedings of the 2013 ACM Workshop on Mobile Development Lifecycle, 2013, pp. 37–38.
[33] D. Amalfitano, A.R. Fasolino, P. Tramontana, B. Robbins, Testing android mobile applications: challenges, strategies, and approaches, Adv. Comput. 89 (2013) 1–52.
[34] B. Kirubakaran, V. Karthikeyani, In: Mobile application testing—challenges and solution approach through automation, Proceedings of the 2013 International Conference on Pattern Recognition, Informatics and Mobile Engineering, PRIME 2013, 2013, pp. 79–84.

[35] J. Voas, K.W. Miller, Software Testing: What Goes Around Comes Around, IEEE Computer Society, IT Professional, 2012, pp. 4–5.
[36] M. Janicki, M. Katara, T. Pääkkönen, Obstacles and opportunities in deploying model-based GUI testing of mobile software: a survey, Softw. Test. Verification Reliab. 22 (2012) 313–341.
[37] B. Sadeh, K. Ørbekk, M.M. Eide, N.C.A. Gjerde, T.A. Tønnesland, S. Gopalakrishnan, In: Towards unit testing of user interface code for android mobile applications, Second International Conference on Software Engineering and Computer Systems (ICSECS '11), 2011, pp. 163–175.
[38] Z. Ding, K.H. Chang, In: Issues related to wireless application testing, Proceedings of the 46th Annual Southeast Regional Conference on XX, 2008, pp. 513–514.
[39] A.H. Betiol, W. de Abreu Cybis, In: Usability testing of mobile devices: a comparison of three approaches, Human-Computer Interaction—INTERACT 2005, IFIP TC13 International Conference, 2005, pp. 470–481.
[40] S. McIlroy, N. Ali, H. Khalid, A.E. Hassan, Analyzing and automatically labelling the types of user issues that are raised in mobile app reviews, Empir. Softw. Eng. 21 (2016) 1067–1106.
[41] T.W. Knych, A. Baliga, In: Android application development and testability, Proceedings of the 1st International Conference on Mobile Software Engineering and Systems, 2014, pp. 37–40.
[42] J. Gao, X. Bai, W. Tsai, T. Uehara, In: Mobile application testing: a tutorial, IEEE Computer Society, 2014, pp. 46–55.
[43] N.H. Saad, N.S. Awang Abu Bakar, In: Automated testing tools for mobile applications, IEEE Conference Publications, 2014, pp. 1–5.
[44] R. Hunt, In: Security testing in android networks—a practical case study, 19th IEEE International Conference on Networks (ICON), 2013, pp. 1–6.
[45] A. Sekanina, Z. Prokopova, R. Silhavy, In: Mobile applications testing, Annals of DAAAM and Proceedings of the International DAAAM Symposium, 2011, pp. 1605–1606.
[46] G. Cao, J. Yang, Z.-H. Tu, Y. Tang, In: Research on cell phone software testing strategy, International Conference on Computational Intelligence and Software Engineering (CiSE '09), 2009, pp. 1–4.
[47] V.L.L. Dantas, F.G. Marinho, A.L. da Costa, R.M.C. Andrade, In: Testing requirements for mobile applications, 24th International Symposium on Computer and Information Sciences (ISCIS '09), 2009, pp. 555–560.
[48] A. Sadeghi, H. Bagheri, J. Garcia, S. Malek, A taxonomy and qualitative comparison of program analysis techniques for security assessment of android software, IEEE Trans. Softw. Eng. (2016) 1–48.
[49] Y.-P. Xu, Z.-F. Ma, Z.-H. Wang, X.-X. Niu, Y.-X. Yang, Survey of security for android smart terminal, J. Commun. 7 (2016) 1–4.
[50] S.R. Choudhary, A. Gorla, A. Orso, In: Automated test input generation for android: are we there yet? 30th IEEE/ACM International Conference on Automated Software Engineering (ASE), 2015, pp. 429–440.
[51] M.D. Syer, M. Nagappan, B. Adams, A.E. Hassan, Studying the relationship between source code quality and mobile platform dependence, Softw. Qual. J. 23 (2014) 485–508.
[52] H. Khalid, E. Shihab, M. Nagappan, A.E. Hassan, In: What do mobile app users complain about? IEEE Conference Publications, 2014, pp. 70–77.
[53] K. Kritikos, B. Pernici, P. Plebani, C. Cappiello, M. Comuzzi, S. Benrernou, I. Brandic, A. Kertész, M. Parkin, M. Carro, A survey on mobile users' software quality perceptions and expectations, 2013.

[54] D. Zhang, B. Adipat, Challenges, methodologies, and issues in the usability testing of mobile applications, Int. J. Hum. Comput. Interact. 18 (2005) 293–308.
[55] A. Spillner, T. Linz, H. Schaefer, Software Testing Foundations: A Study Guide for the Certified Tester Exam (Rocky Nook Computing), fourth ed., Rocky Nook Computing, 2014.
[56] T. Griebe, V. Gruhn, In: A model-based approach to test automation for context-aware mobile applications, Proceedings of the 29th Annual ACM Symposium on Applied Computing (SAC '14), 2014, pp. 420–427.
[57] J. Qian, D. Zhou, Prioritizing test cases for memory leaks in android applications, J. Comput. Sci. Technol. 31 (2016) 869–882.
[58] M. Wan, Y. Jin, D. Li, W.G.J. Halfond, In: Detecting display energy hotspots in android apps, IEEE Conference Publications, 2015, pp. 1–10.
[59] M. Yan, H. Sun, X. Liu, In: iTest testing software with mobile crowdsourcing, Proceedings of the 1st International Workshop on Crowd-based Software Development Methods and Technologies, 2014, pp. 19–24.
[60] M. Mukadam, C. Bird, P.C. Rigby, In: Gerrit software code review data from android, Proceedings of the 10th Working Conference on Mining Software Repositories, 2013, pp. 45–48.
[61] É. Payet, F. Spoto, Static analysis of android programs, Inf. Softw. Technol. 54 (2012) 1192–1201.
[62] P. Krishnan, S. Hafner, A. Zeiser, In: Applying security assurance techniques to a mobile phone application: an initial approach, IEEE Fourth International Conference on Software Testing, Verification and Validation Workshops (ICSTW), 2011, pp. 545–552.
[63] D. Amalfitano, A.R. Fasolino, P. Tramontana, S. De Carmine, A.M. Memon, In: Using GUI ripping for automated testing of android applications, Proceedings of the 27th IEEE/ACM International Conference on Automated Software Engineering (ASE '12), 2012, pp. 258–261.
[64] C. Tao, J. Gao, In: On building test automation system for mobile applications using GUI ripping, Proceedings of the International Conference on Software Engineering and Knowledge Engineering, 2016, pp. 480–485.
[65] C.-H. Liu, C.-Y. Lu, S.-J. Cheng, K.-Y. Chang, Y.-C. Hsiao, W.-M. Chu, In: Capture-replay testing for android applications, International Symposium on Computer, Consumer and Control (IS3C), 2014, pp. 1129–1132.
[66] C. Wilke, S. Götz, S. Richly, In: JouleUnit: a generic framework for software energy profiling and testing, Proceedings of the 2013 Workshop on Green in/by Software Engineering (GIBSE '13), 2013, pp. 9–14.
[67] D. Weiss, M. Zduniak, Automated integration tests for mobile applications in Java 2 micro edition, in: Lecture Notes in Computer Science (Including Subseries Lecture Notes in Artificial Intelligence and Lecture Notes in Bioinformatics), 2007, pp. 478–487.
[68] H.-K. Kim, Mobile applications software testing methodology, in: Communications in Computer and Information Science, 2012, pp. 158–166.
[69] Y. Lin, J. Rojas, E. Chu, Y. Lai, In: On the accuracy, efficiency, and reusability of automated test oracles for android devices, IEEE Computer Society, 2014, pp. 957–970.
[70] G. de Cleva Farto, A.T. Endo, Evaluating the model-based testing approach in the context of mobile applications, Electron. Notes Theor. Comput. Sci. 314 (2015) 3–21.
[71] R. Jabbarvand, A. Sadeghi, H. Bagheri, S. Malek, In: Energy-aware test-suite minimization for android apps, Proceedings of the 25th International Symposium on Software Testing and Analysis, 2016, pp. 425–436.

[72] I.C. Morgado, A.C.R. Paiva, J.P. Faria, In: Automated pattern-based testing of mobile applications, IEEE Conference Publications, 2014, pp. 294–299.
[73] S. Vilkomir, B. Amstutz, In: Using combinatorial approaches for testing mobile applications, IEEE Seventh International Conference on Software Testing, Verification and Validation Workshops (ICSTW), 2014, pp. 78–83.
[74] P. Gawkowski, P. Pawełczyk, J. Sosnowski, K. Cabaj, M. Gajda, LRFI—fault injection tool for testing mobile software, in: Studies in Computational Intelligence, vol. 369, Springer, 2011, pp. 269–282.
[75] X. Ma, B. Yan, G. Chen, C. Zhang, K. Huang, J. Drury, A toolkit for usability testing of mobile applications, in: Lecture Notes of the Institute for Computer Sciences, Social-Informatics and Telecommunications Engineering, 2012, pp. 226–245.
[76] H.K. Ham, Y.B. Park, Mobile application compatibility test system design for android fragmentation, in: Communications in Computer and Information Science, 2011, pp. 314–320.
[77] M.E. Delamaro, A.M.R. Vincenzi, J.C. Maldonado, In: A strategy to perform coverage testing of mobile applications, Proceedings—International Conference on Software Engineering, 2006, pp. 118–124.
[78] C. Guo, J. Xu, H. Yang, Y. Zeng, S. Xing, In: An automated testing approach for inter-application security in android, Proceedings of the 9th International Workshop on Automation of Software Test (AST 2014), 2014, pp. 8–14.
[79] K. Holl, F. Elberzhager, V. Vieira, Towards a Perspective-Based Usage of Mobile Failure Patterns to Focus Quality Assurance, Springer International Publishing, 2015, pp. 20–31.
[80] G. Hu, X. Yuan, Y. Tang, J. Yang, In: Efficiently, effectively detecting mobile app bugs with appDoctor, Proceedings of the Ninth European Conference on Computer Systems, 2014, pp. 1–15.
[81] P. Stirparo, I.N. Fovino, I. Kounelis, In: Data-in-use leakages from android memory—test and analysis, IEEE 9th International Conference on Wireless and Mobile Computing, Networking and Communications (WiMob), 2013, pp. 701–708.
[82] J. Enriquez, S. Casas, In: Development and evaluation of a framework for generation usability testing for mobile application, Latin American Computing Conference, 2016, pp. 1–10.
[83] K. Naik, Y. Ali, V. Mahinthan, A. Singh, A. Abogharaf, In: Categorizing configuration parameters of smartphones for energy performance testing, Proceedings of the 9th International Workshop on Automation of Software Test (AST '14), 2014, pp. 15–21.
[84] J.-F. Huang, In: AppACTS: mobile app automated compatibility testing service, 2nd IEEE International Conference on Mobile Cloud Computing, Services, and Engineering (MobileCloud), 2014, pp. 85–90.
[85] H. Kim, B. Choi, S. Yoon, In: Performance testing based on test-driven development for mobile applications, Proceedings of the 3rd International Conference on Ubiquitous Information Management and Communication (ICUIMC '09), 2009, pp. 612–617.
[86] L. Ravindranath, S. Nath, J. Padhye, H. Balakrishnan, In: Automatic and scalable fault detection for mobile applications, MobiSys 2014—Proceedings of the 12th Annual International Conference on Mobile Systems, Applications, and Services, 2014, pp. 190–203.
[87] D. Franke, C. Weise, In: Providing a software quality framework for testing of mobile applications, IEEE Fourth International Conference on Software Testing, Verification and Validation (ICST), 2011, pp. 431–434.
[88] P. Machado, J. Campos, R. Abreu, In: MZoltar: automatic debugging of android applications, Proceedings of the 2013 International Workshop on Software Development Lifecycle for Mobile (DeMobile '13), 2013, pp. 9–16.

[89] A. MacHiry, R. Tahiliani, M. Naik, In: Dynodroid: an input generation system for android apps, Proceedings of the 2013 9th Joint Meeting on Foundations of Software Engineering (ESEC/FSE '13), 2013, pp. 224–234.
[90] T. Porat, A. Schclar, B. Shapira, In: MATE: a mobile analysis tool for usability experts, CHI '13 Extended Abstracts on Human Factors in Computing Systems, 2013, pp. 265–270.
[91] Y.M. Baek, D.H. Bae, In: Automated model-based android GUI testing using multi-level GUI comparison criteria, International Conference on Automated Software Engineering, 2016, pp. 238–249.
[92] W. Hargassner, T. Hofer, C. Klammer, J. Pichler, G. Reisinger, In: A script-based testbed for mobile software frameworks, 1st International Conference on Software Testing, Verification, and Validation, 2008, pp. 448–457.
[93] J.F. Filho, W. Prata, J. Oliveira, In: Affective-ready, contextual and automated usability test for mobile software, Proceedings of the 18th International Conference on Human-Computer Interaction With Mobile Devices and Services Adjunct, 2016, pp. 638–644.
[94] T. Zhang, J. Gao, J. Cheng, T. Uehara, In: Compatibility testing service for mobile applications, IEEE Symposium, 2015, pp. 179–186.
[95] G. Raghavan, A. Salomaki, R. Lencevicius, In: Model based estimation and verification of mobile device performance, Proceedings of the 4th ACM International Conference on Embedded Software (EMSOFT '04), 2014, pp. 34–43.
[96] G. Canfora, F. Mercaldo, C.A. Visaggio, M. D'Angelo, A. Furno, C. Manganelli, In: A case study of automating user experience-oriented performance testing on smartphones, Verification and Validation (ICST), 2013 IEEE Sixth International Conference on Software Testing, 2013, pp. 66–69.
[97] S. Matsumoto, K. Sakurai, In: A proposal for the privacy leakage verification tool for android application developers, Proceedings of the 7th International Conference on Ubiquitous Information Management and Communication (ICUIMC '13), 2013, pp. 1–8.
[98] P. Gilbert, B.-G. Chun, L.P. Cox, J. Jung, In: Vision: automated security validation of mobile apps at app markets, Proceedings of the Second International Workshop on Mobile Cloud Computing and Services (MCS '11), 2011, pp. 21–26.
[99] O.-E.-K. Aktouf, T. Zhang, J. Gao, T. Uehara, In: Testing location-based function services for mobile applications, IEEE Symposium on Service-Oriented System Engineering (SOSE), 2015, pp. 308–314.
[100] R.N. Zaeem, M.R. Prasad, S. Khurshid, In: Automated generation of oracles for testing user-interaction features of mobile apps, IEEE Seventh International Conference on Software Testing, Verification and Validation (ICST), 2014, pp. 183–192.
[101] C.-J.M. Liang, N.D. Lane, N. Brouwers, L. Zhang, B.F. Karlsson, H. Liu, Y. Liu, J. Tang, X. Shan, R. Chandra, F. Zhao, In: Caiipa: automated large-scale mobile app testing through contextual fuzzing, MobiCom'14, 2014, pp. 519–530.
[102] N. Tulpule, In: Strategies for testing client-server interactions in mobile applications: how to move fast and not break things, Proceedings of the 2013 ACM Workshop on Mobile Development Lifecycle (MobileDeLi '13), 2013, pp. 19–20.
[103] C.C. Yeh, H.L. Lu, C.Y. Chen, K.K. Khor, S.K. Huang, In: CRAXDroid: automatic android system testing by selective symbolic execution, IEEE Conference Publications, 2014, pp. 140–148.
[104] T. Takala, M. Katara, J. Harty, In: Experiences of system-level model-based GUI testing of an android application, IEEE Fourth International Conference on Software Testing, Verification and Validation (ICST), 2011, pp. 377–386.
[105] C. Tao, J. Gao, Building a model-based GUI test automation system for mobile applications, Int. J. Softw. Eng. Knowl. Eng. 26 (2016) 238–249.

[106] W. Yumei, L. Zhifang, In: A model based testing approach for mobile device, International Conference on Industrial Control and Electronics Engineering (ICICEE), 2012, pp. 1885–1888.
[107] L. Lu, Y. Hong, Y. Huang, K. Su, Y. Yan, In: Activity page based functional test automation for android application, Third World Congress on Software Engineering (WCSE), 2012, pp. 37–40.
[108] K.B. Dhanapal, K.S. Deepak, S. Sharma, S.P. Joglekar, A. Narang, A. Vashistha, P. Salunkhe, H.G.N. Rai, A.A. Somasundara, S. Paul, In: An innovative system for remote and automated testing of mobile phone applications, Annual SRII Global Conference (SRII), 2012, pp. 44–54.
[109] J. Kaasila, D. Ferreira, V. Kostakos, T. Ojala, In: Testdroid: automated remote UI testing on android, Proceedings of the 11th International Conference on Mobile and Ubiquitous Multimedia (MUM '12), 2012, pp. 1–4.
[110] K.B. Dhanapal, S. Sankaran, A.A. Somasundara, S. Paul, In: WindTunnel: a tool for network impact testing of mobile applications, Annual SRII Global Conference (SRII), 2012, pp. 34–43.
[111] D. Franke, S. Kowalewski, C. Weise, N. Prakobkosol, In: Testing conformance of life cycle dependent properties of mobile applications, IEEE Fifth International Conference on Software Testing, Verification and Validation (ICST), 2012, pp. 241–250.
[112] Y. Jing, G.-J. Ahn, H. Hu, Model-based conformance testing for android, in: Lecture Notes in Computer Science (Including Subseries Lecture Notes in Artificial Intelligence and Lecture Notes in Bioinformatics), 2012, pp. 1–18.
[113] A.D. Brucker, M. Herzberg, On the static analysis of hybrid mobile apps: a report on the state of Apache Cordova nation, in: Lecture Notes in Computer Science (Including Subseries Lecture Notes in Artificial Intelligence and Lecture Notes in Bioinformatics), 2016, pp. 1–17.
[114] H. Song, S. Ryoo, J.H. Kim, In: An integrated test automation framework for testing on heterogeneous mobile platforms, First ACIS International Symposium on Software and Network Engineering (SSNE), 2011, pp. 141–145.
[115] L. ZhiFang, G. Xiao-Peng, In: SOA based mobile device test, Second International Conference on Intelligent Computation Technology and Automation (ICICTA '09), 2009, pp. 641–644.
[116] M.A. Mazlan, In: Stress test on J2ME compatible mobile device, Innovations in Information Technology, 2006, 2006, pp. 1–5.
[117] M.R. Rege, V. Handziski, A. Wolisz, In: Poster: a context simulation harness for realistic mobile app testing, MobiSys'15, 2015, p. 489.
[118] J. Cheng, Y. Zhu, T. Zhang, C. Zhu, W. Zhou, In: Mobile compatibility testing using multi-objective genetic algorithm, IEEE Conference Publications, 2015, pp. 302–307.
[119] H. Khalid, M. Nagappan, E. Shihab, A.E. Hassan, In: Prioritizing the devices to test your app on a case study of android game apps, ACM Symposium, 2014, pp. 610–620.
[120] D. Yan, S. Yang, A. Rountev, In: Systematic testing for resource leaks in android applications, IEEE 24th International Symposium on Software Reliability Engineering (ISSRE), 2013, pp. 411–420.

ABOUT THE AUTHORS

Konstantin Holl is a project manager at the Fraunhofer Institute for Experimental Software Engineering. His current research interests and his PhD are about the systematical quality assurance of information systems, especially testing and inspections within the development of mobile application software.

Frank Elberzhager is a senior engineer at the Fraunhofer Institute for Experimental Software Engineering. His research interests include software quality assurance, inspection and testing, software engineering processes, and software architecture. He received a PhD in computer science from the University of Kaiserslautern.

CHAPTER TWO

Advances in Combinatorial Testing

Rachel Tzoref-Brill
IBM Research, Haifa, Israel

Contents

Advances in Computers, Volume 112
ISSN 0065-2458
https://doi.org/10.1016/bs.adcom.2017.12.002

Abstract

Since their introduction into software testing in the mid-1980s, combinatorial methods for test design gathered popularity as a testing best practice and as a prominent software testing research area. This chapter reviews recent advances in combinatorial testing, with special focus on the research since 2011. It provides a brief background on the theory behind combinatorial testing and on its use in practice. Requirements from industry usage have led to advances in various areas examined in this chapter, including constraints handling in combinatorial algorithms, support for the combinatorial modeling process, and studies on metrics to support the effectiveness of combinatorial testing. We also highlight recent case studies describing novel use cases for test and field quality improvement in the context of system test, and for optimization of test data. Finally, we examine recent developments in advanced topics such as utilization of existing tests, test case prioritization, fault localization, and evolution of combinatorial models.

1. INTRODUCTION

Software bugs experienced in the field cost a fortune to the global economy. In 2002, the US National Institute of Standards and Technology (NIST) estimated that the US economy loses around $60 billion each year in costs associated with software bugs [1]. A more recent study from 2013 from the University of Cambridge suggests that the situation has not improved in the last decade, as it estimates the annual cost of software bugs to the global economy by $312 billion [2]. Expectedly, great efforts and costs are devoted to software testing and debugging. Brooks estimated back in 1995 the cost of testing to be between 40% and 80% of the development process compared to less than 20% for the coding itself [3]. In 2002, a study from IBM Research estimated that typically software vendors spend 50%–75% of their total development cost on testing, debugging, and verification activities [4]. This trend continues with the study in [2] reporting that programmers spend around 50% of their time on finding and fixing bugs.

To demonstrate the huge testing challenge that software engineers face worldwide, we consider a highly simplified example of an online shopping system. A simplified business requirements document of the shopping system is given in Fig. 1.

From the requirements document, we learn about the different choices that can be made and conditions that can occur when exercising the system under test (SUT). For example, according to the first requirement, a user can order an item which may or may not be in stock. The second requirement

1. The system shall take orders for any valid item, whether it is in stock or not.
2. The system shall support multiple pricing schemes for an order.
   a. The first scheme... [description omitted]
   b. The second scheme... [description omitted]
   c. The third scheme... [description omitted]
3. The system shall validate the current credit status of the purchaser, when known.
4. The purchaser can select one of the following timeframes for order delivery: immediate, within 1 working week, and within 1 month. Ground shipping is default, while sea shipping is allowed for orders being delivered in a week or a month, and air shipping is allowed for immediate or 1 week orders.
5. When an item is classified as export controlled, the system shall generate the appropriate work items to comply with governmental requirements.

Fig. 1 A simplified requirements document for an online shopping system.

states that there are three different pricing schemes. From the fourth requirement we learn about three different timeframes for order delivery and about three different order shipping methods. Based on the fifth requirement, export control may or may not apply. There are also choices which are implicitly mentioned in the document. For example, the first requirement mentions a *valid item*, which implies that a user can also try and order an invalid item. These different choices and conditions which influence the system's exercised logic and flow can be, for example, inputs to the system, configuration variables, and internal system states and conditions. From here on, we will collectively refer to them as *parameters*. Each parameter can take different values, which represent the different ways in which it can vary. For example, the order delivery parameter takes three different values: immediate, 1 week, and 1 month. Item validity takes two values: valid and invalid.

Given the various ways in which an SUT can be executed, developers and testers are continuously faced with a challenge of deciding what are the "right" test cases to run to assure sufficient quality outcome.

1.1 Systematic Test Space Definition

Test design, also known as test planning, refers to the process of selecting which test cases to run out of an enormous space of potential test cases, termed as *test space*. A common naive test design approach is the ad hoc approach, in which a single test is designed at a time in isolation, i.e., without considering its relation to the test cases gathered thus far and its unique contribution to them. For example, a tester can decide to order a valid item from the online shopping system, request for an immediate delivery with ground shipping. He/she can then decide to order a valid item with immediate delivery shipped via sea, and so on and so forth. The ad hoc approach for test design provides no indication of test coverage nor any notion of test progress, i.e., it is hard to tell when the collected tests are sufficient for the testing task at hand, and what portion of the SUT they cover. As a result, the test plan might contain many redundant test cases on the one hand, and many coverage gaps on the other hand.

As opposed to the ad hoc approach for test design, in which tests are locally constructed from an undefined universe, combinatorial testing relies on a systematic test space definition. Specifically, it relies on a definition that uses the notion of parameters and their values. By bringing them to the forefront, the test space can be defined. Thus, a first step is to specify the parameters and respective values that influence the function to be tested. A test case is defined as an assignment of a single value to each parameter. The test space is the Cartesian product of the parameter values. In practice, the test space is not defined in one shot, but rather there is an iterative process in which the test space is corrected and refined, e.g., by locating missing parameters or values and removing redundant ones. Sources for locating parameters and values can be requirements and other specification documents, code inspection, analysis of past failures, legacy test cases, interviews with domain experts, etc.

Table 1 lists the parameters and respective values for our online shopping system. It is easy to see the exponential growth in the number of test cases depending on the number of parameters and values. Even for such a small toy example, there are $2^4 \times 3^4 = 1296$ potential test cases, too many to execute in practice. In real-world systems, test space definitions can contain many billions of potential test cases even for the testing of a single feature of the SUT. In industrial settings, typically it takes hours for a single test case to run to completion, sometimes reaching days or even weeks of execution, depending on the domain and level of testing. The ability to quantify the size

Table 1 Example Online Shopping Model

Parameter	**Values**
Item validity	Valid, invalid
In-stock status	In stock, out of stock
Export control	Yes, no
Shipping destination	Domestic, foreign
Pricing scheme	Scheme1, scheme2, scheme3
Delivery time frame	Immediate, 1 week, 1 month
Order shipping	Ground, sea, air
Customer credit status	Approved, denied, unknown

and complexity of the testing task being faced is already a first advantage of the systematic test space definition process.

1.2 Interaction Coverage Metric for Test Design

Given a test space with a practically infinite number of test cases to execute, how can we come up with a reasonable approach for test design? We answer this question by asking another question: where do the bugs hide? Or put otherwise, what triggers software failures?

This question has been investigated in empirical studies that analyzed software failures in various domains [5–10]. The findings reveal that most failures are triggered by the interaction of a small number of parameters of the SUT. An example of an interaction failure in our online shopping system would be a failure that is always triggered when sea shipping is requested together with export control, regardless of how the other parts of the SUT are executed. Such failures are considered two-way interaction failures, since they depend on the combination of two parameter values. Similarly, if a failure is always triggered when sea shipping is requested with export control and a foreign destination, then it is a three-way interaction failure, since it depends on the combination of three parameter values.

Fig. 2 presents the distribution of interaction failures in various domains as a function of the interaction level. It is evident that in all the investigated domains, most failures occur due to a single parameter value or the interaction of two parameters, with progressively fewer failures in higher interaction levels. Another finding is that all investigated failures were triggered by

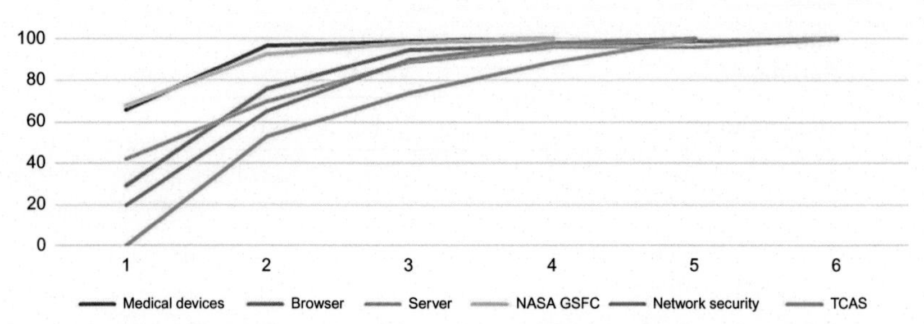

Fig. 2 Number of variables involved in triggering software faults. *Based on data available from National Institute of Standards and Technology, http://csrc.nist.gov/groups/SNS/acts/ftfi.html.*

the interaction of up to four to six parameters. One possible explanation for these findings is that branching points in software usually contain few complex interactions, and the distribution of the number of variables appearing in branches correlates well with the distribution of the number of failure-triggering parameters [11].

Going back to our original questions, given that according to empirical findings most bugs "hide" in low interaction levels, then this result suggests an approach for test design, by providing a test coverage metric called *interaction coverage*. The aim of interaction coverage is to cover all the ways in which a small number of parameters can interact, i.e., the test plan should include all value combinations of every t parameters, where t is a small number given by the user as input. The approach for test design which relies on interaction coverage is called *combinatorial testing* (CT), a.k.a. *combinatorial interaction testing* (CIT), or *combinatorial test design* (CTD). For example, if we construct a test plan that covers all parameter value pairs (two-way interactions), then empirically it will detect most of the bugs in the SUT. Two-way interaction, a.k.a., pairwise testing, is a well-known and widely accepted testing technique [12–15]. However, as the empirical data suggests, higher levels of interaction should be of interest as well. Naturally, the higher the level of interaction requested, the more tests are needed to cover all interactions of the requested level, hence in practice in addition to empirical data, test budget and resources limitations come into play as well when deciding on the requested interaction level.

One significant advantage of the interaction coverage metric over other known and commonly used coverage metrics such as code coverage is that interaction coverage is determined a priori by construction of the test plan, whereas coverage according to other metrics can usually be measured only after the tests have been implemented and executed. In this sense, CT encourages a *shift left* approach to testing. By considering test coverage during early stages of test design rather than during later stages of test execution, failures are found earlier in the development life cycle, and the potential savings is huge [16].

1.3 Combinatorial Test Design Optimization

Fig. 3 presents a pairwise test plan for our online shopping system, as generated by the combinatorial testing tool IBM Functional Coverage Unified Solution (IBM FOCUS) [17, 18]. This is only one out of many possible pairwise test plans.

The test plan achieves pairwise coverage because every pair of values for every pair of parameters appears at least once in some test case in the test plan. Pairwise coverage is achieved with only 12 out of the 1296 possible test cases. When three-way coverage is required, only 38 test cases are produced. The huge reduction in the number of required test cases is the result of applying a *combinatorial testing algorithm* to produce the test plan. Various algorithms exist, but they all achieve significant reduction in the number of test cases compared to the full test space by constructing a test plan in which each test case covers as many unique value combinations as possible, maximizing its added value to the test plan. In contrast, in the manual ad hoc approach discussed earlier, it is common to use copy and paste to produce similar test cases with very little variation between them. We further discuss combinatorial algorithms in depth in Section 3.

One topic which is challenging both for combinatorial algorithms and for test space definitions is that of constraints. Constraints are rules that define invalid parameter value combinations, i.e., combinations that should not be tested for different reasons, for example, because they cannot physically occur. Therefore, these combinations should not appear in any test

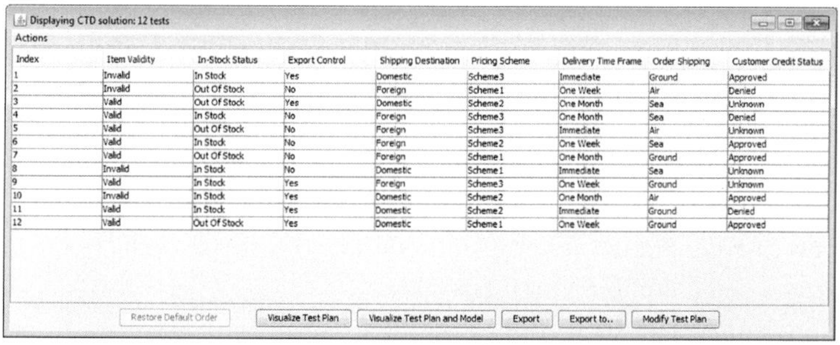

Displaying CTD solution: 12 tests

Actions

Index	Item Validity	In-Stock Status	Export Control	Shipping Destination	Pricing Scheme	Delivery Time Frame	Order Shipping	Customer Credit Status
1	Invalid	In Stock	Yes	Domestic	Scheme3	Immediate	Ground	Approved
2	Invalid	Out Of Stock	No	Foreign	Scheme1	One Week	Air	Denied
3	Valid	Out Of Stock	Yes	Domestic	Scheme2	One Month	Sea	Unknown
4	Valid	In Stock	No	Foreign	Scheme3	One Month	Sea	Denied
5	Valid	Out Of Stock	No	Foreign	Scheme3	Immediate	Air	Unknown
6	Valid	In Stock	No	Foreign	Scheme2	One Week	Sea	Approved
7	Valid	Out Of Stock	No	Foreign	Scheme1	One Month	Ground	Approved
8	Invalid	In Stock	No	Domestic	Scheme1	Immediate	Sea	Unknown
9	Valid	In Stock	Yes	Foreign	Scheme3	One Week	Ground	Unknown
10	Invalid	In Stock	Yes	Domestic	Scheme2	One Month	Air	Approved
11	Valid	In Stock	Yes	Domestic	Scheme2	Immediate	Ground	Denied
12	Valid	Out Of Stock	Yes	Domestic	Scheme1	One Week	Ground	Approved

Restore Default Order | Visualize Test Plan | Visualize Test Plan and Model | Export | Export to.. | Modify Test Plan

Fig. 3 A pairwise test plan for the online shopping system in the IBM FOCUS tool.

case in the output of the combinatorial algorithm. For example, we may want to exclude the combination (Shopping Destination = Domestic, Export Control = Yes) from the test plan, since it never occurs in real life. In practice, constraints appear in the vast majority of real-world test spaces, and hence supported by most CT tools. Constraints should be specified a priori as part of the test space definition. One may suggest that instead, we could simply skip test cases containing invalid combinations. However, if we skip such test cases, we lose not only the invalid combinations, but also valid combinations uniquely covered by those test cases. Thus, 100% coverage of the requested interaction level would no longer be guaranteed. From here on, we refer to the aggregate of parameters, their respective values, and constraints on values combinations, as a *combinatorial model.* We discuss approaches for handling constraints in combinatorial algorithms in Section 3, and approaches for specifying constraints in combinatorial models in Section 4.

As Fig. 3 shows, the test cases in a combinatorial test plan are not executable ones, but are rather assignments of values to parameters (value combinations), suggesting how to implement the corresponding tests. While in some cases, it is easy to plug these combinations as input into an automation framework; in many other cases, it is challenging to implement the resulting tests. Moreover, due to the huge variety and heterogeneous nature of software environments, test plan implementation and automation are custom in nature and existing solutions are difficult to reuse. This is yet another challenge for wide deployment of combinatorial testing in real-world settings.

2. TRENDS IN RESEARCH TOPICS

Traditionally, most research on CT concentrated on algorithms for generating a combinatorial test plan. In a survey published in early 2011, Nie and Leung examined 93 key research papers on CT from the years 1985 to 2008 and classified them according to the following categories (definitions taken from [19]):

1. Modeling (Model): studies on identifying the parameters, values, and the interrelations of parameters of SUT.
2. Test case generation (Gen.): studies on algorithms for effectively generating a small combinatorial test plan.
3. Constraints (Constr.): studies on avoiding invalid test cases in the test plan generation.

4. Failure characterization and diagnosis (Fault): studies on detecting the failure-inducing combinations and fixing the detected faults.
5. Improvement of testing procedures and the application of CT (App.): studies on practical testing procedures for CT and reporting the results of the CT application.
6. Prioritization of test cases (Prior.): studies on the order of test execution to detect faults as early as possible in the most economical way.
7. Metric (Metric): studies on measuring the combination coverage of CT and the effectiveness of fault detection.
8. Evaluation (Eval.): studies on the degree to which CT contributes to the improvement of software quality.

The results of the classification performed in [19] reflect the dominance of the test plan generation algorithms as a research topic for CT. As expected, 60% of the publications (56 out of 93) are on algorithms for test generation and on a related topic of handling model constraints in the generation problem. About 18% are on applications of CT (17 out of 93). In contrast, only about 5% of the publications are on the modeling process (5 out of 93), and only about 2% on metrics for CT (2 out of 93).

It is interesting to observe the trend of research on CT in the years following the 2011 survey. In 2012, an annual workshop on CT was established, named the International Workshop on Combinatorial Testing (IWCT) [20]. While many papers on CT were published also in other venues during the time period from 2012 to date, the topics published in IWCT reflect to a certain degree some of the current directions which the CT research community is taking, considering that many of the active research groups contributing to CT in general are also active in IWCT. To that end, we classify the 77 papers published in IWCT in the years 2012–2017 according to the same categories defined in [19]. The results of both classifications (using percentage out of total publications in each case) are presented in Fig. 4.

A comparison between the two classifications reveals a huge decline in the percentage of papers on test plan generation (from 60 to slightly over 16), while the percentage of application papers has more than doubled (from 18 to over 37), the percentage of metric papers increased by approximately a factor of 4 (from 2 to nearly 8), and the percentage of modeling papers increased by approximately a factor of 5 (from about 5 to nearly 25).

The difference in the results may be attributed to the specific venue chosen, but it may also be an indication that the increasing adoption of CT and its related procedures by industry has revealed bottlenecks that emphasize

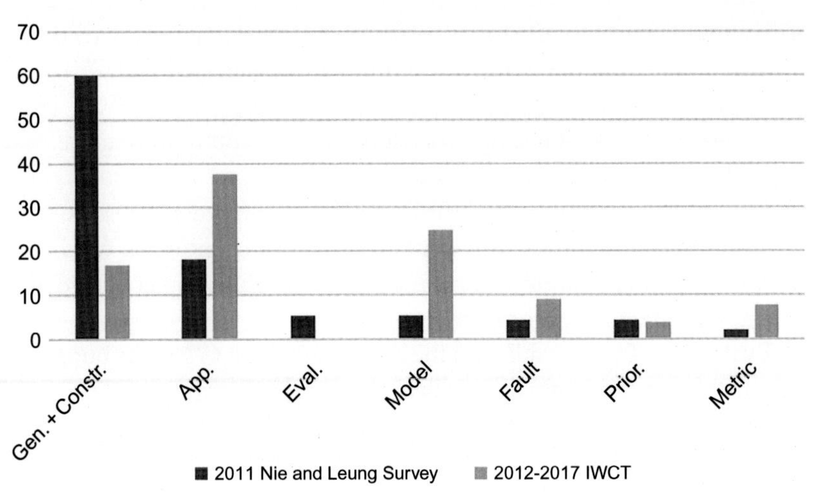

Fig. 4 Distribution of research publications during 1985–2008 according to [19] and of the IWCT workshop publications during 2012–2017.

additional research challenges beyond test plan generation, hence the rise in the number of publications about modeling for CT, test procedures and applications of CT, and metrics for CT. We note that the IWCT publications are only a portion of the CT publications in recent years. When considering all venues, the distribution of recent contributions may differ. In addition, paper count may not be the best proxy for progress in a scientific community.

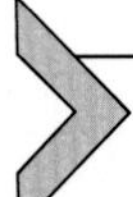

3. ALGORITHMS AND TOOLS FOR CT

3.1 Background

Historically, the CT approach to test design evolved from design of experiments (DoE) [21, 22], a methodology to learn about a cause–effect system and determine its important factors, by conducting a set of experiments. In each experiment, values of a number of test factors are changed, and the corresponding change in the output is measured. Classical DoE started in the 1920s in agricultural science and was later used also in many other industries such as medical research and the chemical industry. In the mid-1940s, orthogonal arrays (OA) were formally defined by Rao to set the experiment plan [23]. A fixed-value OA, denoted by OA(N, v^k, t), is an $N \times k$ matrix on the symbols in v with the property that every $N \times t$ submatrix contains each ordered subset of size t from the symbols in v *the same number of times*. A mixed-value OA is an extension of a fixed-value OA, where each of the k variables has a distinct set of symbols that may vary in size from the other variables.

From the mid-1980s, OAs were used for combinatorial testing of software systems [13, 14, 24–26]. It quickly became apparent that the notion of OA is impractical for software testing due to the many limitations on its structure and possible instantiations. Instead, based on the observation that the requirement for an identical number of occurrences for each interaction is redundant in the case of software testing, covering arrays (CA) were suggested for use independently by Mandl [14], Tatsumi [13, 26], and Sherwood [27], then formally defined by Sloane [28]. A fixed-value CA, denoted by CA(N, v^k, t), is an $N \times k$ matrix on the symbols in v with the property that every $N \times t$ submatrix contains each ordered subset of size t from the symbols in v *at least once*. The positive integer t is the strength of the CA. A mixed-value CA is an extension of a fixed-value CA, where each of the k variables has a distinct set of symbols that may vary in size from the other variables. According to these definitions, OAs are a special case of CAs. Finding a combinatorial test plan that covers interaction level t is mathematically equivalent to finding a CA of strength t. Table 2 presents the mixed-value covering array CA(8, $2^3 \times 4^1$, 2), that is, a CA with eight rows, achieving pairwise coverage of three parameters with two values each, and one parameter with four values. This CA is minimal, i.e., $N = 8$ is the

Table 2 Covering Array CA(8, $2^3 \times 4^1$, 2)

	1	**2**	**3**	**4**
1	1	0	1	0
2	0	0	0	1
3	1	1	0	3
4	0	1	1	2
5	1	1	1	1
6	0	0	1	3
7	0	1	0	0
8	1	0	0	2

minimal number that instantiates a CA for the parameters $2^3 \times 4^1$ and $t = 2$, since a lower bound for $t = 2$ (when no constraints are specified) is the size of the product of the two parameters with the largest number of values, hence we get that the lower bound in this case is $2 \times 4 = 8$.

3.2 Combinatorial Test Generation Strategies

The construction of a minimal CA is an NP-complete problem [29, 30]. Numerous algorithms exist to generate CAs or mixed-level CAs to represent combinatorial test plans, and none of them is superior in general to all others. In [31], Cohen et al. classified the various algorithms for constructing a CA according to three main approaches:

- *Mathematical (algebraic) constructions.* These methods are limited in use since they can be applied only when certain conditions on the combinations of t, k, v are met and are hard to accommodate for handling constraints. However, they are generally fast and produce small, often best-known, results.
- *Greedy algorithms.* This is the most commonly used approach. It is applicable to the general case, and believed to be fast yet produce suboptimal results.
- *Meta-heuristic search.* Search techniques such as simulated annealing, genetic algorithms, and tabu search start with a random matrix and convert it into a CA via a series of transformations. These methods are believed to produce near-optimal results but are often time consuming.

More recently, an empirical study was made by Petke et al. in [32], comparing between the greedy and meta-heuristic search approaches. Its results challenge the current belief that search-based methods produces more effective results, and that they cannot scale to higher strength coverage. The findings of this study indicate that constraints allow simulated annealing to achieve higher strengths; the results for the greedy algorithm were actually slightly superior; and the genetic algorithm was competitive only for pairwise testing with a small number of constraints.

In [33], Jia et al. come up with a hyperheuristic algorithm that learns search strategies across a broad range of problem instances, and selects which heuristics to apply from a set of low-level heuristics. Using this approach, it equals or outperforms the best result previously reported in the literature. In [34], Gargantini and Vavassori suggest an automated approach to select been various techniques for CT test plan generation. Using data mining techniques, it learns from a distribution of combinatorial models and their test plans, and based on cost of execution and combinatorial model characteristics predicts which algorithm performs better on a given problem.

In the following we describe in more detail advancements in several greedy approaches to construct a combinatorial test plan. Each of them uses a different formulation of the CA construction problem, which induces significant differences is constraints handling. We then describe approaches for test generation which combine higher and lower CA strengths.

3.2.1 The Expansion Approach

The traditional and most common greedy approach for constructing a combinatorial test plan is to construct a CA from scratch via expansion, i.e., starting from an empty seed and expanding in the two dimensions of the matrix, while greedily maximizing the gain in interaction coverage based on a local heuristic. Expansion at the row dimension adds new values to the current (partial) test case, and expansion at the column dimension adds new test cases (or partial test cases) to the current test plan. Two well-known examples of such algorithms are AETG [35] and IPO [29]. These algorithms represent two different classes of greedy expansion approaches. AETG follows the one-row-at-a-time approach, where a single row is added to the matrix at a time, until reaching t-way interaction coverage. A framework for greedy AETG-like methods is described in [36]. In contrast, IPO incrementally creates a CA with m columns from a CA with $m - 1$ columns, until reaching $m = k$.

Using the expansion approach, handling constraints on value combinations requires considerable effort. The reason is that regardless of the specific method for expansion, throughout the algorithm there is a need to constantly validate that the added values do not cause a violation of the constraints, in order to ultimately produce a valid CA, i.e., a CA in which all rows satisfy the constraints. One of the main challenges in handling constraints is that of *implicit constraints* [31, 37], a.k.a. *derived exclusions*. These are value combinations that are excluded from the model not explicitly by a specific user-defined constraint, but rather implicitly due to the interaction between several constraints due to the transitivity property. Detecting all implicit constraints in advance is computation-intensive. Various methods and optimizations were suggested for constraints validation in general and also specifically for handling implicit constraints [31, 38–42]. These allow for better performance and application to more complex constraints; however, constraints validation remains a challenging and computation-intensive step in algorithms that are based on the expansion approach.

3.2.2 The Subset Selection Approach

A different approach to the construction of a combinatorial test plan was introduced by Segall et al. in [18]. While it still uses the one-row-at-a-time approach similarly to AETG, its starting point is not an empty seed, but rather the entire test space as defined by the combinatorial model. Hence, the construction problem is formulated as a subset selection problem. Given a set of valid tests, denoted by $S(P, V, C)$, where P is the set of parameters, V the set of value sets, and C the constraints, the goal is to select a small as possible subset S' of S, so that all value combinations of size t that appear in S, appear also in S'. In other words, S' preserves all interactions of size t that occur in the set of valid tests.

One significant advantage of this approach compared to the expansion approach is that it considerably simplifies constraints handling. There is no longer a need to verify validity of intermediate results throughout the algorithm. Constraints are handled only at the initial stage of constructing the set $S(P, V, C)$ of valid tests, and from that point on tests are selected only from this set, hence the algorithm never steps out of the domain of valid tests. Moreover, implicit constraints are a nonissue in this approach, since the set the valid tests already captures all explicit and implicit constraints.

An obvious challenge of this approach is how to construct and represent the set of possible tests, given that real-world test spaces may contain many billions of test cases. To that end, [18] suggests the use of binary decision diagrams (BDDs) [43], a compact data structure to represent and manipulate Boolean functions. A Boolean function viewed as a characteristic function represents a set of Boolean vectors, which is the set of all assignments to the variables of the function for which it is evaluated to 1. Segall et al. [18] utilize the efficient computation of Boolean operations on BDDs, such as negation, conjunction, and disjunction, to compute the BDD representing the set of valid tests from the user-specified constraints. The set of invalid tests in the model is represented using the conjunction of the BDDs for each of the constraints. Multivalued parameters are handled using standard Boolean encoding and reduction techniques to BDDs [44]. The set of valid tests is represented by the negation of the BDD for the invalid tests, conjunct with a BDD that represents the legal multivalued to Boolean encodings of the parameter values. Obviously, tests cannot be selected from the BDD of valid tests by explicit traversal. Instead, BDDs representing uncovered valid value tuples for parameter combinations of size t are iteratively conjunct with the BDD of valid tests until reaching a subset of the test cases in which each test greedily maximizes the gain in interaction coverage. If in a certain iteration the conjunct BDD becomes too large, intermediate results are used to select the next test, and the algorithm continues to its next iteration.

Since constraints are user-specified and concisely describe the function that determines which tests are valid, this approach scales in practice for enormous test spaces. In [45], Gargantini and Vavassori suggest that the use of multivalued decision diagrams instead of BDDs to implement the same approach may improve the results in terms of test plan sizes.

3.2.3 The SAT-Based Approach

Satisfiability (SAT) solvers have risen in the last couple of decades as a powerful technique in formal verification and in hardware synthesis. Given their huge impact in these fields, using them for the CA construction problem is a natural step. In 2006, [46] suggested a constraints-based approach to CA construction. The problem is translated into a Boolean formula in conjunctive normal form (CNF) as follows. A certain size N of the CA is determined, and variables are produced to represent the matrix values, as well as to represent whether each of the t-way combinations occurs in the matrix. The requirement for t-way coverage is expressed using these latter variables. The formula is then fed into a SAT solver. A satisfying assignment means that a CA of size N exists. The algorithm searches for a minimal N using multiple SAT invocations with varying CA sizes. Several encoding improvements were suggested in recent years [47, 48].

In the presence of constraints, formulation becomes more complicated. Due to implicit constraints, some of the t-way tuples may become invalidated, hence directly encoding the constraints into the SAT formula will result in false unsatisfiability. A first approach for constraint handling in SAT-based CT generation was proposed in [49] by initially ignoring constraints, and then detecting forbidden tuples in the resulting test plan by additional local SAT invocations for each test case separately.

There are several drawbacks to the SAT-based approaches. As the size of the CA decreases, it often becomes hard to determine if the resulting formula is satisfiable. In addition, since it is hard to estimate the minimal size in advance, many invocations of the SAT solver are needed, reducing the scalability of the approach. In 2015, Yamada et al. suggested the use of incremental SAT solving and cooperation with efficient greedy algorithms to tackle these drawbacks [50]. Incremental SAT solving [51] is used to reuse the learned clauses from previous SAT invocations of closely related formulas and improve the overall performance. A greedy algorithm is used to quickly instantiate an initial test plan whose size is used as the CA size estimation in the SAT formula. In addition, all valid *t*-way tuples are collected by enumerating the tuples that appear in that initial test plan.

A different way in which researchers have utilized powerful SAT techniques for combinatorial test generation is in conjunction with greedy or search-based algorithms, forming a hybrid solution. Rather than using a SAT solver to construct a formula whose instantiation represents a potential *t*-way test plan, the SAT solver is used for performing intermediate tasks for the original algorithm, namely checking for constraints violation and pruning the search space of the algorithm. In [31], Cohen et al. suggest to translate the CT constraints into a Boolean formula representing all forbidden tuples that may occur in the test space, and to extend the test generation algorithm to call a SAT solver to check for constraints violation whenever a value is assigned in the current test set solution, and may have been part of a forbidden tuple. Representatives from the two classes of algorithms are experimented with: an AETG-like greedy algorithm and a simulated annealing algorithm. Cohen et al. further optimize their constraint handling approach by leveraging incremental SAT solving to reuse learned clauses from different SAT invocations, thus significantly reducing the run time overhead incurred due to constraint handling [52]. The coupling of SAT techniques for constraints handling with search-based algorithms for combinatorial test generation led to the creation of the CASA tool, an improved simulated annealing algorithm with efficient constraints handling [53]. Recently, Yamada et al. suggested to use the unsatisfiable cores produced by SAT solvers when the formula is unsatisfiable in order to lazily detect forbidden tuples and significantly reduce the number of SAT solving calls performed during the execution of a greedy algorithm [40].

Finally, Calvagna and Gargantini suggest a formal logic approach to test generation in the presence of constraints, in which generating the covering array and satisfying constraints are not handled separately, but rather constraints satisfaction is embedded into the generation process [54]. In this approach, logical predicates are used to describe all aspects of the combinatorial test generation problem, including the model, the tuples required to be covered, constraints on value combinations, and other advanced requirements. The formal specification is then fed into the SAL model checker, which uses an SMT solver in order to iteratively generate satisfying assignments, each representing a test case to be added to the test plan, eventually constructing a covering array.

3.2.4 Combining Lower and Higher Strengths

Naturally, the higher the requested strength t is, the more defects are detected by the resulting t-way test plan, but also the more tests are needed to construct it. Hence, the choice of t reflects a trade-off between risk and cost. To handle this trade-off, requesting for a single strength might be too rigid. The concept of *hierarchies and subrelations* was first introduced in [35] to allow requesting different strengths for different subsets of the parameters. Later formalized as *variable strength generation* by Cohen et al. [55, 56], this concept is useful in practice to express knowledge that some parameter combinations are more sensitive than others, for example, based on domain experts or analysis of historical defects. Some of the generation algorithms can easily support variable strength generation, e.g., variants of the subset selection approach [18] and the expansion approach [38]. The math theory behind variable strength covering arrays is defined in [57].

In [58], Fouché suggests a different approach to handle this trade-off. Rather than committing a priori to a certain strength, test generation starts at a low strength, then iteratively increases the strength while reusing already generated tests, i.e., complementing them to achieve coverage of the next strength. The process stops when testing resources are exhausted. In [59], Segall suggests to take advantage of the fact that testing is done in repeated iterations in order to achieve higher strength coverage. Given a requested strength t, for each iteration a test plan of strength t is constructed, however the degree of freedom in the test plan generation algorithm is directed toward achieving $(t + 1)$-way coverage over time across the different t-way test plans.

3.3 Examples of CT Tools

Various CT tools have been developed over the years. The pairwise.org website lists over 40 such tools [60], including well-known tools such as PICT [38], ACTS [61], AETG [35], and CASA [53]. There are commercial and noncommercial tools, command-line, GUI-based, and Web-based tools. They implement different algorithms, receive their input in different specification formats and languages, and support different features. CITLAB is a framework for CT tools suggested by Gargantini and Vavassori [62]. It is integrated with an Eclipse editor and proposes a common formal specification language for CT tools. Existing tools and algorithms can be added to the framework as Eclipse plug-ins. The CASA simulated annealing tool allows plug-in extensions to its state space exploration algorithm, by introducing modifications to definitions of its *energy function* (mapping of states to energy which the algorithm needs to minimize) and choice of *neighbor states* [53].

IBM Functional Unified Solution (IBM FOCUS) is a GUI-based commercial tool developed in IBM [17]. It started as a functional coverage tool which analyzed and visualized the interaction coverage of already executed test cases [63]. Since 2009 it also evolved into a CT tool, implementing a proprietary algorithm [18]. Fig. 5 displays the main screen of IBM FOCUS.

From the get go, the research and development of IBM FOCUS was directed by challenges encountered by its users in the various aspects of the application of CT in real-world settings, such as challenges in the modeling process, in test generation and prioritization, and in working with legacy test cases. For example, to assist the user define and debug constraints in models of real-world complex test spaces, IBM FOCUS visualizes projections of the test space on different parameters [18] and automatically generates explanations for excluded combinations [64]. To simplify model definition, it supports advanced modeling constructs [65]. To allow specifying complex bad path scenarios, it extends the notion of negative values suggested in [38] to negative value combinations [66]. To enable tuning the resulting test plan, it supports (1) variable strength generation, (2) definition of don't care values, (3) requirements on parameter values distribution, a.k.a. weights [18, 38], and (4) interactive modification of the test plan while viewing the resulting coverage gaps [67]. To comprehend updates made to the model following changes to the SUT, it performs semantic differencing of model versions [68]. To ease the effort of implementing executable tests, it integrates with several test automation frameworks and test management tools, and produces textual test cases from template definitions. To effectively reuse legacy test cases, it measures their interaction coverage, performs test plan minimization to preserve measured interaction coverage and test plan augmentation to achieve full interaction coverage [69], and performs hybrid test plan minimization and generation [70].

3.3.1 Visualization in CT Tools

Visualization is an important and useful instrument to assist users in comprehending and analyzing data. In the context of CT, several tools use visualization to display measured interaction coverage. For example, the CCM tool by NIST [71] presents measured coverage (according to extended definitions of the basic interaction coverage) using a graphical representation,

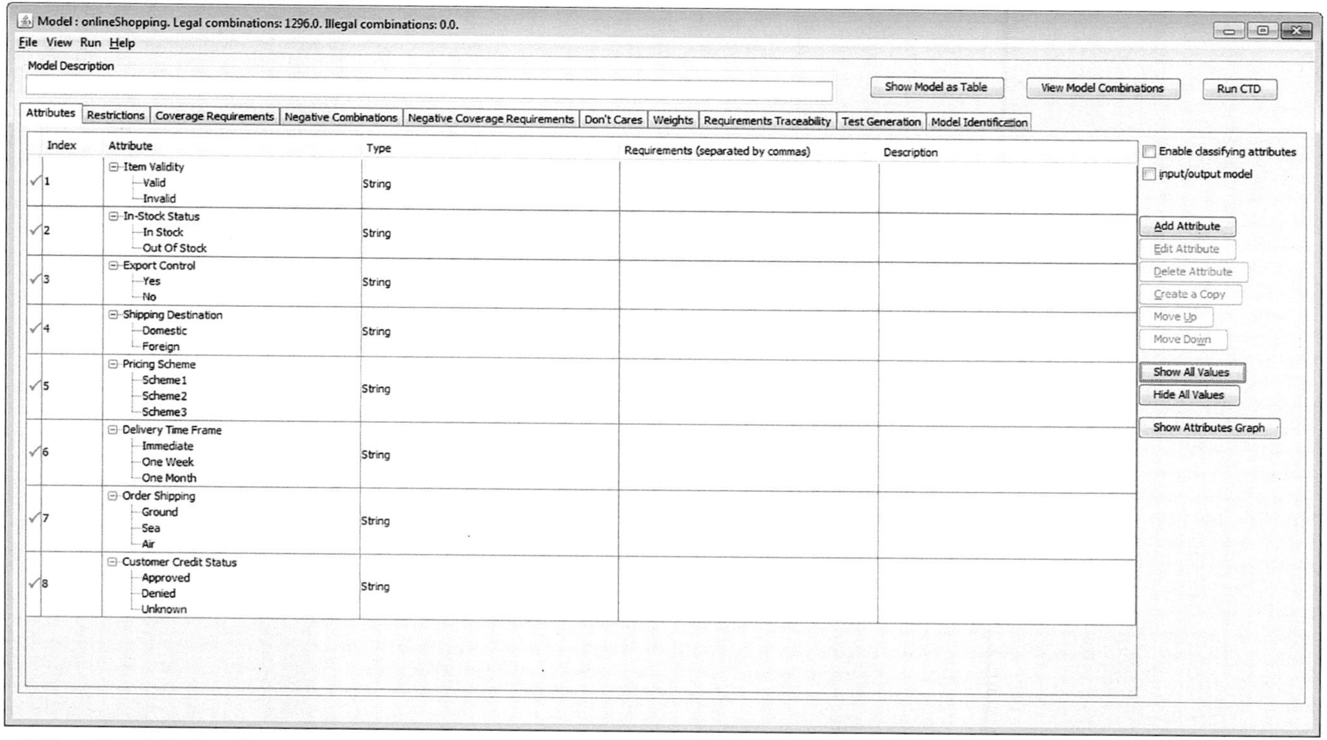

Fig. 5 The IBM FOCUS tool.

where the x-axis represents the percentage of combinations and the y-axis represents the percentage of coverage.

The IBM FOCUS tool contains various types of visualizations to assist the user in different tasks, for example: analyze test plan coverage, comprehend parameter relations as defined in the model, debug model constraints, and prioritize the test plan [72]. Fig. 6 presents some of these visualizations. Three different forms of visualization are supported: matrices, graphs, and treemaps. They are used to visualize the relationships between the different elements of the model, and to visualize the strength of each test in the test plan and the relationships between the different tests in terms of interaction coverage. For example, in the matrix view, the upper half visualizes percentage of valid value pairs for each parameter pair, and the lower half visualizes percentage of covered value pairs out of all valid value pairs for each parameter pair. There is a treemap view which visualizes this information also for higher level interactions. In the test plan treemap view, each square represents a test case, and its size represents the number of value tuples it uniquely covers in the test plan. In the test plan graph view one can also analyze the amount of overlap between different test cases (represented by the graph nodes), as the width of the edge between two nodes indicates the number of common value tuples between the corresponding test cases. Both test plan treemap and graph views have been found useful for test plan prioritization in real-world projects, as described in [72].

4. MODELING THE CT TEST SPACE

Defining the combinatorial model is the most crucial part in CT. As in other testing approaches, it is a necessary step in which human knowledge is captured and manually transferred into a structured, machine-readable form. This step is always challenging and CT is no exception. The way it is performed affects the entire effectiveness and efficiency of the CT application process. For example, if parameters or values are missing from the model, the test plan will contain coverage gaps, and defects might be missed and escaped to the field. If the model definition is too low level, i.e., without using known testing techniques such as equivalence partitioning, then no matter how powerful the CT algorithm is, the resulting test plan will be too large, containing many similar and redundant test cases. If constraints are wrongly specified, then combinations that should not be tested might appear in the test plan, or worse, combinations that should be tested might be wrongly excluded and never tested. Similar problems and challenges exist also in

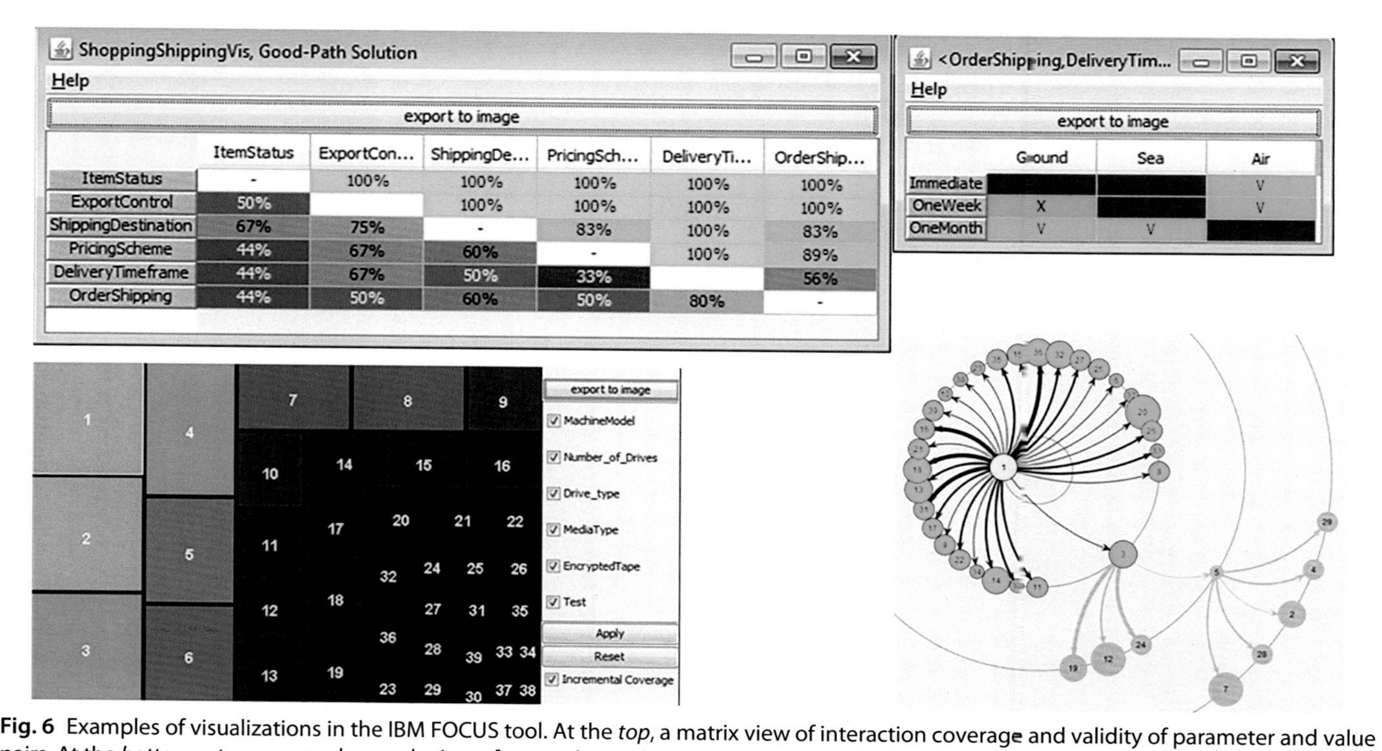

Fig. 6 Examples of visualizations in the IBM FOCUS tool. At the *top*, a matrix view of interaction coverage and validity of parameter and value pairs. At the *bottom*, a treemap and a graph view of a test plan, indicating the amount of unique interaction coverage of each test case and its overlapping coverage with other test cases.

other testing techniques (e.g., gaps in test specification, wrong level of abstraction resulting in redundant test cases, errors in test definitions), however since CT brings the notion of parameters and values to the surface in a structured manner, it helps reach a clearer definition and easier identification of these problems.

Though syntactically all combinatorial models have the same underlying structure (parameters, values and potentially constraints on value combinations) and are used in a similar manner (given as input to a CT generation algorithm), semantically they vary quite significantly. Not only in the specific domain and application logic being captured, but also in the level of testing. That is, depending on the level and scope of testing, the parameters reflect different notions and may be of different abstraction levels. For example, a model used for interface testing will contain parameters describing inputs to the SUT. A model used for configuration testing will contain parameters describing environment variables. Some models may contain both. Models for system test may contain parameters reflecting notions of higher level of abstraction such as system flow steps, usage patterns and internal states. Despite these differences, combinatorial models also share many common generic characteristics which may aid in the modeling process. We briefly touch on this topic in Section 4.1.

4.1 Modeling Patterns and Common Pitfalls

Early reports of case studies and applications of CT have already included modeling issues that were encountered and their proposed solutions, for example [37, 38, 73–75]. In 2012, a first attempt for a systematic identification and classification of recurring modeling patterns and their associated pitfalls was made [66]. Five different recurring patterns of combinatorial models (reflecting properties of the SUT) were identified and presented via simplified examples. These include, for example, capturing the fact that some parameters may be optional, conditionally excluded from the model, occur multiple times in a test case, or sensitive to the order of their presentation. These patterns are often hard to identify, resulting in common modeling mistakes which might lead to coverage gaps in the test plan. Ref. [66] presents for each pattern an erroneous model and a corrected one.

In [76], Kuhn et al. classify parameter value combinations anomalies as *missing combinations*, *infeasible combinations*, and *ineffectual combinations*. According to [76], missing combinations are combinations that may be needed to trigger a fault but are not included in the test plan. An obvious example is a combination which is of a higher strength than the one used to create the test plan. Ref. [66] demonstrates how missing combinations may occur due to erroneous constraints. Infeasible combinations are combinations which will never occur when the SUT is in use. Such combinations are extremely common and occur in almost all real-world SUTs. They are handled by constraints which exclude them from the test space. We discuss constraints specification in Section 4.2.

Finally, ineffectual combinations are combinations that are included in the test plan, but due to another combination do not actually get executed by the SUT, making them in effect missing from the test plan. The most common cause for ineffectual combinations is the presence of values or value combinations representing error cases. When an error occurs, the SUT halts, hence other values present in the test case might not get executed, and become ineffectual. One way to handle this problem is to explicitly identify the error cases using *negative values*, and create a separate test plan for them representing the bad path of the SUT [38, 66, 73, 75]. Other causes for ineffectual combinations include masking effects due to various dependencies between parts of the SUT. In [77], Yilmaz et al. present an approach to overcome masking effects without knowing them a priori, by using a feedback driven adaptive combinatorial testing approach. In each iteration, potential masking effects are detected, their likely causes are heuristically isolated, and new test cases are generate that allow the masked combinations to be tested by removing the mask causing combinations.

4.2 Advanced Modeling Constructs

The most common basic model includes parameters, their respective values, and constraints on value combinations. Constraints are usually specified as propositional formulas [38, 78]. For example, in our online shopping system from Table 1, if we want to exclude the combination `(Shopping Destination = Domestic, Export Control = Yes)` from the test space as suggested in Section 1.3, we could add the constraint `Shopping Destination = Domestic → Export Control = No`. Alternatively, we could declare `(Shopping Destination = Domestic, Export Control = Yes)` as a *forbidden tuple*. However, explicitly specifying all forbidden tuples might be quite inefficient, since there may be numerous forbidden tuples, and multiple tuples can be captured by a single propositional formula. We note that in the context of CT, constraints usually refer to *first-order constraints*, that is, they constrain the parameter values of each individual test case. Higher order constraints that constrain the behavior of a set of test cases as a whole are much harder to model and evaluate, and are rarely used in practice in CT modeling. That being said, in Section 5.2 we describe a case study in which such constraints were used in conjunction with CT.

Real-world test spaces may contain complex relationships between its parameters, resulting in the specification of dozens of propositional formulas. In such cases, the basic form of a model might be insufficient. The use of advanced constructs was already proposed in the past to simplify the modeling process and resulting model. For example, [37] suggests *auxiliary aggregates* to capture parameters and values that are common between different test spaces and can be reused, and *field groups* to support optional groups of values. If any one field from the group is present, then all fields from the group must be present. Hierarchies of parameters are described in [38]. They allow *t*-way combining lower levels first, and then *t*-way combining the resulting combinations with parameters at higher levels of the hierarchy. *Subattributes* are used in [74] to describe compound parameters that are associated with subparameters.

More recently, Segall et al. define special parameters called counters, which hold the number of occurrences in a test case of other parameter values, as specified by the practitioner [79]. They present real-world models where the use of counters significantly reduces the number of constraints that need to be specified, and an efficient BDD-based implementation to scale to real-world test space sizes. They further generalize the concept of counters to *auxiliary parameters*, i.e., parameters holding values of any predefined function on the other parameter values that may be of practical use. In [80, 81], Sherwood suggests to use embedded functions in the test case generator in order to conform to SUT constraints, and demonstrates the proposed concept through an implementation that uses the PhP scripting language. Hence, the constraints modeling task is transformed from a propositional formulation activity into a coding activity. In [82], Usaola et al. propose a framework that combines CT with regular expressions. The use of an XML file representing a model of the SUT as a starting point enables automatic generation of sequences of operations for testing, and then using CT for generating the corresponding input data.

Finally, [83] goes beyond the scope of a single model by proposing a methodology for modeling a composed SUT via a hierarchy of input models, each representing a component of the SUT. The input models are structured in a way that conforms with mathematical observations on the *t*-way coverage properties of the compound test plans that are created when merging individual test plans of different input models.

4.3 Automatic Assistance in Model Debugging

Specifying constraints may be challenging and error prone, especially when there are complex relations between the model parameters. One source of difficulty has to do with *derived exclusions*, a.k.a. *implicit constraints*, which are value combinations excluded not explicitly by a single propositional formula but rather implicitly by the interaction of several formulas due to the transitivity property. We mentioned them in Section 3.2.1 as challenging for CT generation algorithms and potentially adding to their computation effort. Both in the context of CT generation algorithms and in the context of CT modeling, the difficulty lies in the determination whether or not a given value tuple is excluded from a test space, when multiple propositional formulas are specified. In the context of CT modeling, derived exclusions might be unintuitive and/or unintended, and may potentially hint to modeling mistakes. In [64] it is suggested to allow the practitioner to review derived exclusions as part of the model review and debug process. An efficient BDD-based algorithm is used to generate all derived exclusions up to a certain interaction level.

A simplified example of a derived exclusion is shown in the following. Table 3 presents the online shopping system from Section 1.1, this time including four constraints in the form of propositional formulas. In this model, the combination `(Delivery Time Frame = Immediate, Shopping Destination = Domestic)` is a derived exclusion, due to the interaction between the first and the last constraint. Obviously, it is not an intended exclusion, bur rather stems from a mistake in one or more of the constraints. For example, a corrected last constraint which eliminates the unintended derived exclusion may be `Shopping Destination = Domestic` $\rightarrow$ `Order Shipping` $\neq$ `Sea`.

Several approaches were suggested to cope with constraint specification related challenges. In [18], a tool-assisted CT modeling methodology is described. The practitioner selects small subsets of parameters of interest, and the CT tool automatically computes the projection of the selected parameters on the test space. It then displays a view in which each of the combinations in the projection is colored according to its validity. Excluded combinations are colored with red; combinations that may be excluded depending on values of parameters missing from the projection are colored with yellow, provided that the missing parameters

Table 3 Example Online Shopping Model With Constraints

Parameter	Values
Item validity	Valid, invalid
In-stock status	In stock, out of stock
Export control	Yes, no
Shipping destination	Domestic, foreign
Pricing scheme	Scheme1, scheme2, scheme3
Delivery time frame	Immediate, 1 week, 1 month
Order shipping	Ground, sea, air
Customer credit status	Approved, denied, unknown
Constraints	
`Delivery Time Frame = Immediate → Order Shipping = Air`	
`Order Shipping = Sea → Delivery Time Frame = One Month`	
`Delivery Time Frame = One Month → Order Shipping ≠ Air`	
`Shopping Destination = Domestic → Order Shipping ≠ Air`	

have at least one common constraint with a parameter included in the projection; all other combinations are colored with green. To efficiently compute the projection and validity status, a BDD-based implementation is used.

The practitioner reviews and debugs the model by examining different value combinations, confirming their validity status, and correcting it as needed. Following an update to the model such as the addition or modification of a constraint, the practitioner can use this view to make sure that the update to the model resulted in the intended effect.

Fig. 7 presents the projection view of our example model from Table 3 on the parameters `Delivery Time Frame` and `Order Shipping`. The combinations `(Delivery Time Frame = Immediate, Order Shopping = Air)` and `(Delivery Time Frame = One Week, Order Shopping = Air)` are marked with yellow, because their validity depends on the value of `Shopping Destination`, which has a common constraint with `Order Shipping`. The practitioner can expand the projection view for a selected combination (for example, the first yellow one) to include also the parameter `Shopping Destination`. In the resulting projection, all combinations have a green or red validity, since they do

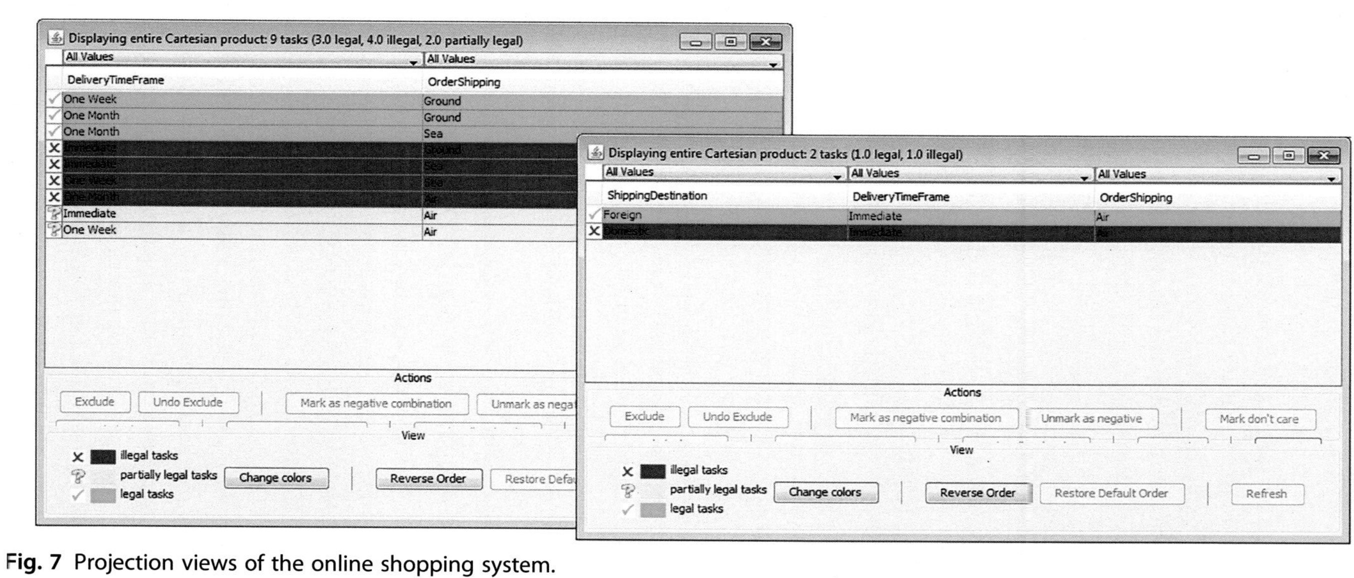

Fig. 7 Projection views of the online shopping system.

not depend on parameters outside of the projection. When the practitioner identifies an infeasible unexcluded combination in the projection view, she can directly exclude it from this view, or alternatively specify a more general constraint using a propositional formula that may capture both the infeasible combination and others that stem from the same cause, if such exist.

In [64], an "explanation" is automatically produced for unintentionally excluded combinations. When the practitioner encounters such a combination, a SAT solver is invoked on a formula consisting of the model constraints conjunct with the excluded combination. This formula is of course unsatisfiable, because the combination under review is excluded by the constraints. The unsatisfiable core produced by the SAT solver contains exactly the subset of constraints that excludes the combination under review. These constraints are displayed to the practitioner as the explanation for the faulty exclusion.

In [84], methods for validating various meta-properties of model constraints are suggested, such as validation of consistency and implication between constraints, and redundancy of constraints. Finally, in 2017, a method for automatically repairing model constraints is presented [85]. Valid and invalid test cases are generated from the model, and an oracle is used to check whether the SUT considers them valid or invalid. For each conformance fault between the model and SUT, a fault localization tool is used to isolate the failure-inducing combination, and depending on the direction of the nonconformance, the combination is either excluded or included in the model. The process is repeated until achieving conformance or near conformance.

4.4 Automatic Assistance in Model Definition

Recently, several different approaches were presented to automatically extract candidates for model elements from existing artifacts such as UML diagrams, requirement documents, and existing tests.

In [86] and [87], Satish et al. propose a rule-based approach for deriving CT models from UML activity diagrams and from UML sequence diagrams, respectively. The diagrams are automatically parsed, and different rules are applied to detect candidate parameters, values, and constraints. In 2017, Satish et al. further propose a rule-based approach for deriving CT parameters and values from UML use case diagrams and from use case specifications. To parse the latter, a natural language processing-based parser and an XMI-based parser are implemented.

Nakagawa and Tsuchiya propose a technique for semiautomated CT constraints identification from requirement documents [88]. It assumes the parameters and values are already given and tries to detect combinations of parameters that participate in the same constraint. This is achieved by measuring the distance between values of the parameters in the text of the document—the smaller the distance is, the stronger the relation is assumed to be.

Finally, in [89], Zalmanovici et al. present a cluster-based approach to analyze large textual test plans and suggest candidate parameters and values for CT models that capture them. The approach is based on the observation that in industrial settings, semi-formal specification artifacts are often unavailable while legacy test cases are both available and required to be reuse. The technique uses the text of the test case steps to define a distance metric for clustering. The reasoning is that many textual test cases are similar, and their differences may indicate points of variation that stand for parameters and their values. The approach is evaluated on real-world free text test plans from six different companies. Using test cases as the analyzed artifact enables mapping them to the CT parameter values, hence enabling their reuse in conjunction with CT, for example, using methods such as interaction-based test suite minimization [69], which we describe in Section 7.1.

Considering that modeling is a critical and challenging part of CT, there is a major potential benefit from utilizing existing artifacts for automatic assistance in CT model definition. Transforming initial steps in this direction into mature technology will be a valuable contribution to widespread adoption of CT in practice.

5. CASE STUDIES

CT is being extensively used in industry, and experience reports were published in a variety of domains. Examples include security [90], aerospace [91], web browsers [92], railway [93], database applications [94], automotive [95], and mobile applications [96], to name a few.

In this section we highlight two case studies. The first is a study reporting on the experience gathered from the continuous use of CT for system test of two large IBM products during a period of three years [97]. To the best of our knowledge, this is a first reported evidence demonstrating that the use of CT in real-world scenarios resulted not only in significant improvement in test quality but also contributed to dramatic improvement in field quality as experienced by the clients of the products.

The second case study describes the use of CT for testing of two real-world big data applications at Medidata [98]. This case study contains some unique aspects relating to the generation of the model from the database definition and to the use of CT constraints in conjunction with other types of constraints that are typically used when working with test data stored in a database. The results show dramatic reduction in the size of the test data following the CT optimization, while maintaining the same fault detection capabilities, i.e., all faults found using the original data source were also detected with the reduced one.

5.1 IBM Systems

Enterprise computing environments require constant availability. Time lost to unavailable server resources can cost financial institutions millions of dollars. When routine maintenance is performed for system hardware upgrades, code updates, or repairs, an outage of the server is scheduled. To significantly reduce the need for scheduled outages and help preserve the constant availability of the server, IBM®, POWER7®, and System z® use features collectively referred to as concurrent maintenance features. These features perform changes, additions, and repair of the system hardware and code while the system is running live business applications. Designing and testing these features is a very complex technical challenge.

Previous experience with the concurrent maintenance features on POWER6® revealed some significant deficiencies in system test, and therefore CT was used to design the system level test cases for POWER7. Following an extremely successful experience on POWER7, CT was also introduced to system test of System z concurrent upgrade [97].

System test is challenging in general, as it concentrates on the overall stress to the SUT. Multiple applications operate simultaneously exercising a number of different functions and features at the same time with interactions among them as they are all running together. As an indication of the complexity of system test, a single system test (a.k.a a trial) for the concurrent upgrade of System z can take well over 8 h. In contrast, function level testing is generally concerned with inputs and outputs of individual product features or software components. Such testing is concerned with verifying coverage of each input and output of the function, and variations are typically derived from the input and output value differences.

In the context of CT, these inherent differences between system test and function test add to the challenge of identifying parameters and values for a system level CT model, as system test lacks the formal inputs and outputs that exist in function test. Wojciak and Tzoref-Brill describe in [97] a generic methodology they came up with to model the system level test spaces for POWER7 and System z. As part of the methodology, they list a series of questions which would help derive parameters and their values for system level models. In the case of the concurrent maintenance features for POWER7, system test is concerned with performing the maintenance operations end to end including the full sequence of steps. The interactions modeled are between the firmware supporting the features and other software components executing on the system while maintenance is performed. In the case of System z, system testing for concurrent upgrade represents three distinct phases of operation with a set of preconditions occurring before the upgrade, a set of during conditions that happen while the firmware is managing the update, and a set of post conditions being done after the upgrade. The system test model treats the various functions, features, system stress, system states, and errors that can occur as part of those three distinct phases.

The IBM FOCUS tool [17] was used to define the models as well as to refine the resulting test plans. Pairwise coverage was used after experimenting also with three-way coverage and concluding that it seemed more than necessary. In addition, refinement of the pairwise requirements was done to remove insignificant interactions. The interactive refinement feature [67] in the IBM FOCUS tool was designed to support such cases. It allows educated decisions on what to exclude or modify in the test plan that results from CT by displaying the coverage gaps that are introduced for each manual modification step. It was used to reduce the appearance of lower importance values, while making sure that no coverage gaps are introduced in the process. Parameter value appearance biasing in the recommended tests was achieved by assigning weights to parameter values [18, 38] as part of the input to the CT algorithm.

Evaluation of the resulting models was done by analyzing test results, taking advantage of the fact that system test is done in waves. After each wave, the results from all test trials were analyzed and reviewed. When defects were found in trials, their root cause was analyzed and an attempt was made to correlate it to the trial parameter values. The review considered the following questions: how was the overall defect discovery quality and rate? Did all parameters seem to matter? Can some parameters be eliminated? Are new parameters necessary? The evaluation was then used to guide the next test wave.

Results from applying CT were evaluated in two dimensions: improvement in test quality, based on analysis of system test results, and improvement in the quality of the server concurrent maintenance features, based on analysis of field results. To evaluate test results, the defects per trial ratio (a.k.a. test case effectiveness metric) was measured, and its change along time was analyzed. Fig. 8A and B shows the ratio obtained in the different test cycles of each release on POWER6 and POWER7, respectively. Fig. 9A and B shows the defects per trial ratio for System z concurrent upgrade on z196 prior to the introduction of CT, and EC12 in which CT was applied, respectively. In system test of System z concurrent upgrade there was only one long test cycle, hence the analysis along time was done according to ratio per week.

In both cases, prior to CT there is no decreasing trend in the ratio as time progresses, while with CT, the decrease is evident. This trend indicates that test stability is reached, and may in turn indicate product stability. Furthermore, the first trials produced by CT contain higher numbers of unique interactions. As a result, the defects per trial ratio is at first high, then as time progresses, there are fewer interactions that remain to be tested in the CT generated tests. New interactions are still uncovering new defects but the discovery rate drops significantly. This means that with CT, more defects are found earlier during testing rather than being randomly distributed along time, and therefore stability can be reached faster.

Field results of concurrent maintenance on POWER7 were compared to those obtained on POWER6 in terms of the outcome of the concurrent maintenance operation. There are two types of possible failures of the maintenance operation, as witnessed on POWER6. The first is an abort, where the operation cannot be completed and there is a need to suspend it and retry at a later stage after consulting with the customer. The second and more severe failure type is a crash, where the entire machine comes down. As can be seen in Fig. 10, dramatic quality improvement has been experienced in the field for POWER7. While on POWER6 about 10% of the operations crashed, on POWER7 crashes were eliminated for all but a handful of cases. The percentage of aborts was reduced by half from 20% to 10%, and, respectively, the success rate increased from approximately 70% to 90%. Analyzing the aborted operations revealed that only 1.6% of the overall operations failed due to firmware errors. The other failures were due to causes on which testing (and therefore CT) has no control, e.g., insufficient

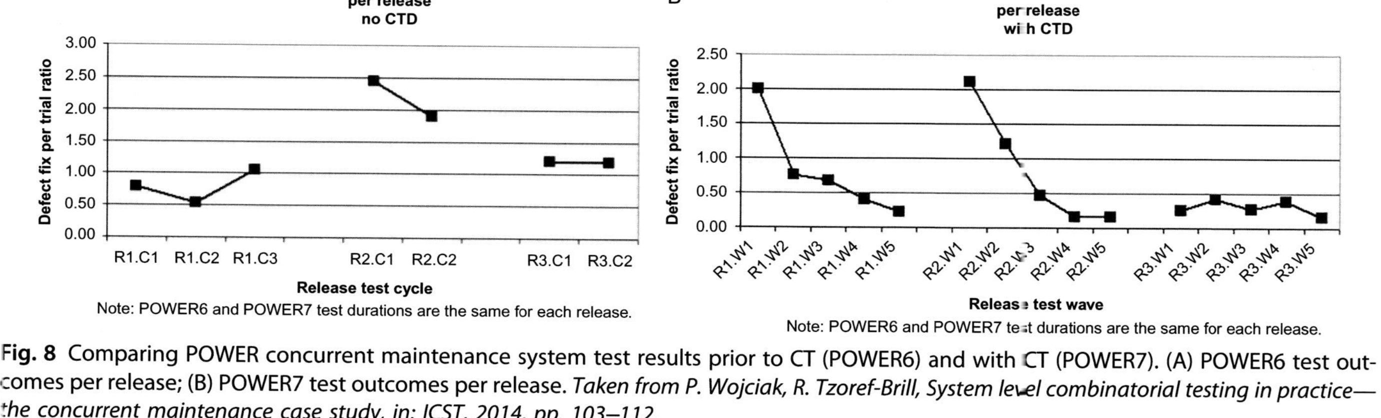

Fig. 8 Comparing POWER concurrent maintenance system test results prior to CT (POWER6) and with CT (POWER7). (A) POWER6 test outcomes per release; (B) POWER7 test outcomes per release. *Taken from P. Wojciak, R. Tzoref-Brill, System level combinatorial testing in practice—the concurrent maintenance case study, in: ICST, 2014, pp. 103–112.*

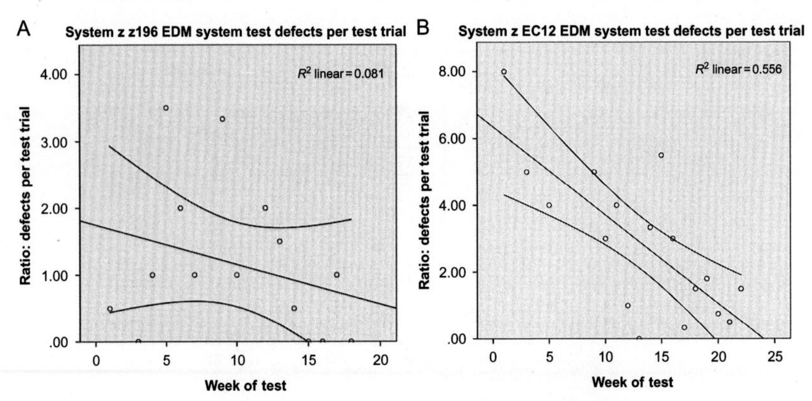

Fig. 9 Comparing System z concurrent upgrade system test results prior to CT (z196) and with CT (EC12). (A) z196 defects per test trail; (B) EC12 test defects per test trail. *Taken from P. Wojciak, R. Tzoref-Brill, System level combinatorial testing in practice—the concurrent maintenance case study, in: ICST, 2014, pp. 103–112.*

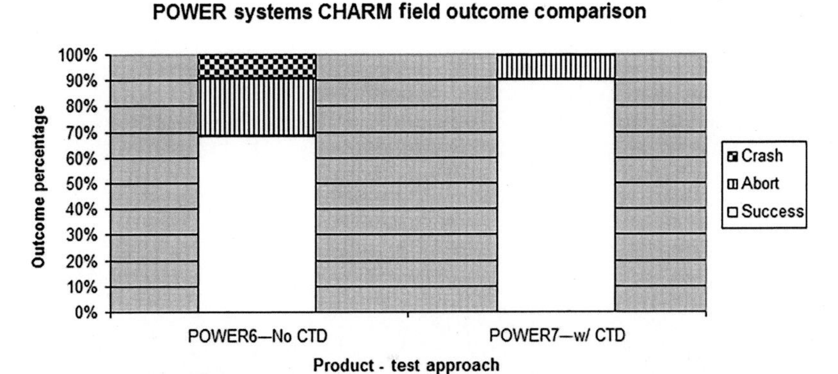

Fig. 10 Power systems concurrent maintenance field outcome comparison. *Taken from P. Wojciak, R. Tzoref-Brill, System level combinatorial testing in practice—the concurrent maintenance case study, in: ICST, 2014, pp. 103–112.*

servicer training or defective hardware. By way of comparison, on POWER6 10% of the overall operations failed due to firmware errors. Field results of concurrent upgrade on System z EC12 were absolutely remarkable. With dozens of successful operations without any failure, these were the best field results seen to date. The authors of [97] conclude from both experiences that testing with CT enabled reaching the right set of test cases to obtain the desired improvement in product quality while meeting the affordability requirements driven by fixed industrial budgets and schedules.

5.2 Big Data Applications at Medidata

Big data applications process and manage large volumes of data, often measured by terabytes or more. Handling such magnitudes of data can be extremely time-consuming, reaching up to months of processing. Naturally, for testing of these types of applications, it is desirable to use small test data sets that effectively represent the big data volumes. The current industry practice is to manually generate small test data sets for testing of big data applications. However, manual construction can be time-consuming, labor intensive, and error prone. Furthermore, there is no automated way to ensure validity and quality of the generated data sets.

The CT approach is potentially a good replacement to the manual effort of test data generation. In addition to significant size reduction, it provides a mechanism to specify and handle constraints that ensure test data validity, and it provides by construction guarantees on the coverage (and hence quality) of the generated test data. To this end, Li et al. describe the use of CT for testing of two big data application at Medidata [98]. Specifically, these two applications were of a common big data application type called Extract, Transform, and Load (ETL). The applications under test at Medidata compute, store, and analyze dozens of terabytes of clinical trial data through ETL processes using Amazon Web Services (AWS).

CT was applied in two major steps. In the first step, definitions of combinatorial models were semiautomatically generated for each database table. When the data source changed, the model was automatically adapted and updated to reflect the data changes. This is vital in industrial settings in which there are frequent data changes.

Specifically, the model definition process was conducted as follows. Each column of a database table was considered a CT model parameter. For each parameter, important test values were derived from constraints that are either automatically extracted from the database schema or manually specified by the user. Overall, 10 general types of rules for constraints and value definitions were supported. Known testing techniques such as equivalence partitioning and boundary values were used to reach representative value definitions, while also consulting the data source by statistically analyzing the stored data to derive high and low frequency values. The database structure was also used to identify important values, e.g., default values which are entered in a specific column when the SQL query does not specify a value for it.

Constraints in such settings require special consideration. In addition to regular user-defined constraints which reflect requirements and business rules, in the context of test data there are two more types of constraints. The first are structural constraints, i.e., constraints that relate to the database structure such as referential integrity constraints. These were automatically derived from the database schema. The second type of constraints does not in fact address data validity but rather imposes requirements on properties of the generated test data. In that sense it is closer to combinatorial coverage requirements than to CT model constraints. Examples are requirements on the number of rows in a generated table and on the data density, i.e., magnitude of child columns associated with a specific parent column. Referential integrity, density, and size constraints are all *higher level constraints* (which we mentioned in Section 4.2), since they impose constraints on multiple test cases and their relations rather than on each individual test case.

The second major step consisted of the actual test data generation. The complication of this step stems from the existence of special higher level constraints as described above. To handle this complication, constraints were gradually and incrementally considered, starting with foreign key constraints, then density constraints, combinatorial coverage requirements, logic constraints, and finally size constraints.

Evaluation was performed on two enterprise-level ETL projects which were remained nondisclosed. The first project contained three databases with 14 GB, 140 GB, and 1.4 TB data. The second project contained one database with 3 GB data. The CT process as described earlier was applied by practitioners who had no a priori knowledge on the faults that existed in the applications, then compared both to the use of the original data source and to the use of random data sets of the same size of the CT-based generated ones. Comparison to manual test data generation was not performed. It was initially tried out but abandoned because it was too labor intensive to keep the data sets up to date with the frequent constraints changes.

The eventual data sets generated with CT were of sizes ranging between 0.001% and 0.12% of the original data source sizes. Similarly, the time it took to generate the data sets with the CT approach was a fraction of the time it took to process the original data sources. For example, for the 1.4 TB database, the original processing took 18 h, while the CT-based data generation took 2 min. Execution time of the applications using the CT data sets was also a fraction of the execution time using the original data, with the former reaching up to 3 min, while the latter ranging between 3 h and 2 weeks.

Finally, all 9 faults in the first project and all 123 faults in the second project that were detected using the original data source were also detected using the CT-based approach. The random data sets were unable to detect any faults, since they did not satisfy the referential integrity constraints. In general, it is highly unlikely that randomly selected data sets will satisfy all specified constraints. The authors of [98] conclude from this experiment that CT can be effectively applied to big data applications, while highlighting the importance of clearly understanding the application's business requirements in order to specify logic constraints in a correct and complete manner, and the importance of refining the combinatorial coverage requirements only to those columns actually interacting with each other, in order to achieve significant size reduction without harming the fault detection capabilities. These two latter conclusions are relevant also to the application of CT in the general context.

6. METRICS FOR CT

Existing work that assesses the effectiveness of CT at testing phase commonly compares CT-created test plans with either random or manually created test plans. Two metrics are used: code coverage (usually branch, statement, or MC/DC coverage) and fault detection. Threats to validity typically include the use of a limited set of relatively small sized programs as the SUT and the use of seeded faults to create faulty mutants of the original program rather than using real faults. Both may affect the ability to generalize the results.

Comparison to random testing has conflicting results. Some reports conclude that CT is more effective than random testing, for example, [99–102] in the context of logical expressions, [103] for a Flight Guidance System, and [6] for network-centric software. Other reports found no significant differences between random testing and CT [104, 105]. One report found that random testing can perform as well as two-way testing but with significantly larger variance when run multiple times and has worse results than four-way testing [106].

It is important to note that depending on the generation method, random testing may not always be feasible. A random selection out of the Cartesian product of parameters and their values will most probably result in an invalid test case, i.e., a test case that does not satisfy the constraints. Hence, a random test generator must either incorporate constraints satisfaction capabilities, or produce the entire set of valid test cases in advance.

Recently, the question of comparison to random testing was revisited [107]. CT and random testing were applied to the Siemens suite [108], a well-known benchmark which has been used to evaluate many testing techniques. This suite contains seven programs with several faulty versions. Mutations were also used to strengthen the fault detection evaluation. For code coverage evaluation, statement and branch coverage were measured. For each program, 100 t-way test plans were generated for each strength t ranging from 2 to 5. The PICT tool [38] was used to generate multiple t-way test plans by providing its greedy algorithm with different seeds. For each t-way test plan, a random test plan of the same size was generated. The results show that in most cases, t-way testing performs better or as good as random testing; however, the differences are not as significant as expected. One explanation for this phenomena is that the random test plans had very high levels of t-way coverage, almost always over 80% and often over 95%. Two important notes with respect to the comparison between random and CT testing are as follows. First, the same CT model that was used to generate the t-way combinations was also used to generate the random tests, hence random testing was not purely random as it already leveraged the systematic test space definition provided by CT and benefited from it. Second, when using random testing in practice as opposed to this particular experimental set up, it is hard to decide when enough testing was performed, while CT is in advantage as it has a well-defined stopping criteria which is achieving full t-way coverage.

In [109], Czerwonka investigated another somewhat related question. Since many different t-way test plans exist for each given strength t, do they all provide similar code coverage and are of similar sizes? In other words, how stable are t-way test plans in terms of code coverage and size? Is stability affected by changes to t? To answer these questions, t-way and random testing were applied to four utility programs in Windows 7, and test plan sizes, statement and branch coverage were measured and analyzed. The PICT tool was used to generate multiple t-way test plans by configuring its greedy algorithm accordingly. The results show that, in general, t-way test plans provide stable coverage when $t \geq 2$. With respect to the differences experienced with different values of t, the results show that as t grows, the ranges of test plan sizes grow as well but their relative standard deviation decreases, indicating relative tightening of the range of possible test suite sizes. Code coverage also increases with t but with narrowing ranges and stable relative standard deviation, consistently with diminishing returns of testing with larger values of t.

In 2017, a new metric called *model inference* was suggested [110]. The advantage of this metric over the traditional ones is that it does not require to execute the test plan. A model is inferred by applying a decision tree learning algorithm on a test plan of unknown quality, and running it against a high quality test plan (in the case of [110], a test plan of high strength). The quality assessment is based on the probability of misclassification of the inferred model with respect to the expected outcomes of the different test cases of the high quality test plan. The results show that the model inference metric correlates well with code coverage (branch, statement, and MC/DC) and with mutation score, though there are some limitations to its applicability depending on the number and nature of SUT outcomes.

Finally, in 2017, Kuhn et al. reenforce the core reasoning for effectiveness of CT test plans [111]. Historically, it is based on empirical data showing that most failures depend on an interaction between a small number of parameters, usually 1 or 2, with progressively fewer faults depending on higher levels of interaction. Kuhn et al. strengthen the validity of the empirical data with an underlying theoretical reliability model. The model represents the distribution of t-way faults based on two assumptions: that t-way faults occur in proportion to t-way branching conditions in the code, and that t-way faults are removed when discovered without introducing new faults. Hence, the model is based on the distribution of branches in the source code and on the fault removal rate, which depends on the density of t-way tuples (the proportion of combinations of size t covered by each test). The model was run against 10 systems from various domains, and the results show that it successfully reproduces the fault distributions observed in empirical data.

7. ADVANCED TOPICS IN CT

Several advanced research topics have emerged following the use of CT in practice. In this section we review a few advanced topics which connect CT with related testing activities such as test prioritization and debugging, focusing on recent progress made in each of them.

7.1 Using CT in Conjunction With Existing Tests

In industrial settings, it is often the case that when CT is introduced, legacy test plans already exist. Considering these legacy test plans may be a requirement for the adoption of CT, especially when considerable manual effort was involved in constructing them. These requirements define a question that CT needs to handle: how to utilize existing test cases in the CT approach?

In 2004, Hartman and Riskin defined the problems of *embedding covering arrays* and *minimizing covering arrays* [112]. Embedding a covering array receives a set of test vectors and augments them to achieve full t-way coverage. Minimizing a covering array receives a covering array and finds a smallest subset that retains its t-way coverage property. Both problem definitions assume that all value tuples are valid, i.e., there are no constraints on value combinations. In 2013, Blue et al. redefine these problems to make them more widely applicable [69]. Interaction-based test-suite minimization (ITSM) receives an arbitrary set of test cases with any t-way coverage (as opposed to the covering array input as defined in [112]) and with no assumptions on the relationships between the value combinations, and finds a subset of the test cases that achieves *the same* interaction coverage as the original set, i.e., it covers the same set of tuples of size t. The set can also be augmented to reach full interaction coverage. Ref. [69] points out two major advantages of the minimization approach. First, constraints do not need to be specified (unless augmentation is performed as well), since minimization is performed on the existing test cases which are assumed to be valid. Second, there is no need to generate new executable test cases (a step which may be labor intensive, as we will describe in Section 8.3), since the existing test cases were already generated. Naturally, this approach is applicable only in cases where the quality of the existing test plan is trusted. Ref. [69] further describes the application of the ITSM approach to two large industrial case studies in which CT was too costly to apply due to the effort involved in specifying constraints and generating new executable test cases; however, the existence of a trusted legacy test plan enabled the use of ITSM.

In [70], Segall et al. suggest to combine ITSM with CT for cases in which for a part of the test space (defined by a subset of the parameters) it is easy to specify constraints and generate new test cases, while for the other part it is challenging. In such cases, they suggest to apply CT on the former part and ITSM on the latter. Accordingly, the former set of parameters is referred to as the *generated parameters*, and the latter set is referred to as the *selected parameters*. The new approach is called Minimization Generation CTD (MG-CTD). To implement MG-CTD, CT, and ITSM are not applied separately on the two subspaces and then combined, since that would lead to lose of interactions between the generated and selected parameters. Instead, a new test space is constructed in which the selected parameters are restricted to the combinations that appear in the existing tests. Then CT is applied to the new test space. The new test space is efficiently computed using BDDs. Examples of test spaces in which MG-CTD is applicable are healthcare insurance, where patient data are hard to generate while claims are relatively easily produced, and command line interface for virtual machine creation and modification, where setting up a virtual machine takes much more time than modifying parameters of a running machine. In the former case, MG-CTD minimizes the amount of generation of new patient data. In the latter case, it minimizes the number of setup operations. MG-CTD is a reasonable middle ground between CT and ITSM, as it maximizes the interaction coverage in the parts where CT can be practically applied as is, while eliminating the test generation effort and reverting to existing combinations in those parts in which it cannot.

7.2 CT-Based Test Plan Prioritization

Test plan prioritization refers to the ordering of test cases for execution according to some criteria to achieve a certain goal, usually to maximize early defect detection. A common metric to evaluate defect detection efficiency of the prioritized test suite is average percentage of faults detected (APFD) [113]. APFD measures the area under the curve of the percentage of faults found as a function of the test cases that have been executed out of the entire test plan. Interaction coverage is a natural ordering criteria when CT is used to generate the test plan. Moreover, most greedy generation algorithms already generate the test cases in a front-loaded fashion, i.e., the earlier test cases cover more unique t-way value combinations than the later ones.

Two different approaches for test plan prioritization were defined in the literature. The first is *regeneration* of prioritized test plans [114]. In this approach, every parameter value is assigned with a weight by the user, representing a utility function that should be maximized. The weight of a value pair is defined as the multiplicity of its individual value weights. A regeneration algorithm then generates a test plan from scratch while attempting to cover as much "pairs weight" as early as possible. The resulting test plan is called a *biased covering array*. In [114], Bryce et al. suggest a greedy algorithm for regeneration of biased covering arrays. In [115], Qu and Cohen generalize this algorithm to allow prioritization based on any t-way level interactions, rather than just pairs. This is achieved by extending the computation of weights from pairs to t-way value tuples.

The second prioritization approach is *reordering* of prioritized test plans, a.k.a. *pure prioritization*. Rather than generating a prioritized test plan from scratch, this approach reorders an existing test plan to maximize its early fault detection. For example, in [115], the same weights defined for regeneration are also used to reorder existing test plans by sorting the test cases in decreasing order of their interaction weights. To that aim, a test case level weight is defined based on the weights of its *t*-way value tuples.

Several case studies demonstrated that prioritization based on *t*-way interactions helps achieve early fault detection [114–116]. A prioritization tool called CPUT is available for reordering test cases according to their pairwise coverage [117].

Another goal of test plan prioritization besides early fault detection is to minimize switching cost. If a configuration setup is very costly, we would like the order of test cases to impose a minimal number of configuration changes, thus avoiding costly setups as much as possible and minimizing the overall execution time.

The problem of minimizing switching cost was originally presented by Kimito et al. in the context of generating new CT test plans [118]. They proposed two algorithms for generating pairwise test plans while attempting to minimize both their size and their switching cost. They defined a configuration change cost of a test case as the sum of its individual parameter change costs. A parameter change cost is the cost of switching from a specific value it is assigned with in the current test case to the value it will be assigned with in the next test case. Not all parameters necessarily reflect configuration changes, in which case their associated change cost is zero. In addition, not all parameters necessarily represent the same change effort. In the example given in [118], when testing a digital copier, changing the finisher may be costly, while changing the input size, paper size, and media is significantly easier. Hence, the finisher is assigned with 5 cost units while the other parameters are assigned with 1 cost unit.

In [119], Wu et al. discuss minimizing switching cost as a prioritization problem. They present greedy algorithms and graph-based algorithms that reorder a given combinatorial test plan to minimize its switching cost. Wu et al. claim that prioritization by switching cost also assists in fault localization, since in the reordered test plan, the adjacent test cases are more similar to each other than those in the original test plan. Thus, when a test case fails while its previous or subsequent test case does not fail, the smaller differences between the failing and nonfailing test cases help locate the failure causing interaction faster.

7.3 Fault Localization

Once test cases that origin from CT are executed, their unique structure consisting of parameter value combinations allows systematic fault localization. This is achieved by locating failure-triggering parameter value combinations in failing tests, known as *identifying failure-inducing combinations*. Since accurately identifying minimal failure-inducing combinations is a hard problem, most existing techniques come up with candidates for failure-inducing combinations, i.e., combinations that are likely to trigger failures but are not guaranteed to do so.

There are two major approaches for identifying failure inducing combinations. The first approach takes a single failing test case as input, and isolates a minimal value combination that triggers the observed failure. Zhang and Zhang propose a fault characterization method called *Faulty Interaction Characterization* (FIC) and its binary search alternative FIC_BS [10]. The method uses the failing test case as a seed for the generation of additional test cases that assist in locating a failure-inducing combination that occurs in the original test case. By executing the additional test cases and observing their results, one can eliminate combinations that no longer induce the failure and narrow down on the failure-inducing ones. Complexity-wise, FIC is very efficient. Its complexity is only $O(k)$ (where k is the number of parameters), while there are 2^K potential failure-inducing combinations in a failing test case. However, there are two important assumptions that reduce its applicability in practice: (1) there are no constraints in the model, and (2) no new failures are introduced when the failing test case is modified to produce new test cases. A different technique that eliminates these assumptions was proposed by Niu et al. [120]. By building a tuple relation tree (TRT) for a failing test case, they allow a clear view of its interactions and can also handle multiple, possibly overlapping, failure-inducing combinations.

The second approach takes a set of test cases as input and their execution results (pass/fail), and identifies failure-inducing combinations that may occur in them. Yilmaz et al. describe a fault localization technique based on machine learning [121]. It takes a covering array and its execution results as inputs and builds a classification tree that models the failing test cases. The model is evaluated by using it to predict the results of new test cases. Since the model encodes the information that will predict a failure, it assists developers in understanding the root cause of the failure.

Shi et al. propose a different fault localization technique, in which given a set of test cases and their execution results, candidates for failure-inducing combinations are identified [122]. These are combinations that appear only in failing test cases and not in passing ones. Next, for each failing test case, n additional test cases are generated to isolate its failure-inducing combination, where n is the number of model parameters. The initial candidate set is reduced based on the execution results of the additional test cases. More recently, another technique was suggested along with a tool called BEN [123–125]. In this technique, combinations are first ranked based on how suspicious they are in inducing failures. Next, additional test cases are created to refine the ranking, and the process repeats iteratively until either failure-inducing combinations are detected, or the set of suspicious combinations is empty.

8. FUTURE DIRECTIONS

8.1 CT Evolution

To apply the CT process in practice, a practitioner defines the CT model, refines it as necessary, produces a test plan, and implements executable test cases based on the test plan. However, this is not a one time effort. As an SUT evolves due to changes in requirements and new features being added, changed, or removed, so does its test space. Practitioners need to update and maintain the CT model that captures the evolving test space, and then update the test plan accordingly or alternatively regenerate it. This holds in any real-world environment and more so in agile software development. However, most existing CT tools do not provide special support for CT evolution, and there is only initial research and early results on this important topic.

One early work on evolution in the context of CT is that of Qu et al. [116]. It examines the fault detection effectiveness of combinatorial testing prioritization and regeneration strategies on regression testing in evolving programs with multiple versions.

More recently, Li et al. describe *Adaptive Input Domain Models* in the context of CT modeling for big data applications [98], as we discussed in Section 5.2. Models are initially generated directly from a database, and when the source data or database structure change, the models are regenerated to adapt accordingly.

The following two works explicitly handle evolution at the model level. In 2015, Tzoref-Brill and Maoz show that the Boolean semantics currently in use by CT tools for interpreting model changes is inadequate, since it provides an inconsistent interpretation of atomic changes to models which might lead to modeling errors [126]. To illustrate the problem, assume that we update the online shopping model from Table 3 by adding a new value called *Expedited Air* to the *Order Shipping* parameter. According to the model definition, the combination `(Delivery Time Frame = Immediate, Order Shopping = Expedited Air)` is excluded from the test space. However, assume that the first constraint in the original model would have been replaced by the semantically equivalent constraint: `Delivery Time Frame = Immediate` $\rightarrow$ `(Order Shipping` $\neq$ `Ground` $\wedge$ `Order Shipping` $\neq$ `Sea)`. The resulting test space would have been identical to the original one, since the model defines the same set of valid tests. However, when the new value is added, the combination `(Delivery Time Frame = Immediate, Order Shopping = Expedited Air)` is now included in the test space.

To resolve these inconsistency problems, Tzoref-Brill and Maoz suggest to replace the Boolean semantics with a lattice-based semantics for interpreting the evolution of CT models. The new semantics provides a theoretical base for a consistent interpretation of atomic changes to the model and exposes which additional parts of the model must change following an atomic change, in order to restore validity. In 2017, they further suggest a first syntactic and semantic differencing for CT models, which is based on a concise canonical representation of a model [68]. The canonical representation uses the notion of a *strongest exclusion*, which is a value combination excluded from the test space for which every strict subset is included in the test space. The syntactic differencing presents additions and removals of parameters, values, and constraints, while the semantic differencing computes and presents changes in strongest exclusions. They further evaluate their differencing technique by conducting a user study with CT practitioners. The results indicate that the proposed technique significantly improves the performance of less experienced practitioners in comprehension of CT model updates.

Despite these recent advancements, there is plenty of room left for further research on the application of CT in the context of an evolving SUT. Without the ability to reuse and evolve CT artifacts over time, it is extremely challenging to employ CT in a continuous manner in real-world settings.

8.2 CT Modeling

Modeling is the starting point and most critical step of CT. It requires expertise skills to perform correctly, and its quality determines the effectiveness of the entire CT process. As such, it remains the most prominent challenge of CT, especially when introducing CT in new settings.

As described in Section 4, work has been done to assist in the modeling process and enable applying CT to a variety of problems at different testing levels, and many case studies reporting on the application of CT have highlighted modeling issues and their resolution. However, there is a need for additional research on modeling. One direction is the continuation of research on assistance in the modeling process. Potential topics include for example model review and evaluation, and reuse of test artifacts to assist in model definitions. Another direction is the estimation of the modeling effort, for example, in terms of return on investment models that measure the benefit of CT while considering modeling costs. Such contributions will ease the introduction of CT to new teams and organizations and consequently further increase deployment of CT in practice. Finally, tools and technologies for integrating the modeling process into the general development and testing environments are in need. These should consider CT models as reusable assets to maximize their benefit and avoid wasteful efforts across an organization.

8.3 CT and Test Generation

In the context of CT, the term *test generation* is often used to describe the activation of an algorithm on a CT model to produce a combinatorial test plan (represented by a covering array). However, this form of test generation does not actually generate executable test cases, since the generated test plan is expressed in terms of parameters and their values. It may take considerable effort to translate it into an executable test plan. Many gaps exist between the CT test plan and the executable one. First, depending on the abstraction level of the model, it may not refer to concrete inputs to the SUT but rather to higher level notions, thus a single model parameter value may translate to a piece of code in the executable test case. Second, some elements which are required for test generation may be missing from the model since they do not influence the logic and flow of the SUT. The most notable examples are expected results (specified via a test oracle which determines whether a test case has passed or failed) and parts of the test data that do not drive logic. Third, for each CT test case there is a need to generate a test script for either automated or manual test execution. Closing these gaps and generating executable test cases may be labor intensive.

Some of the existing CT tools support test generation to a certain level; however, there is hardly any work published on this topic. One exception is a work by Kruse, which surveys approaches for test oracles and test script generation in the context of CT [127]. Kruse describes the current test oracle support in CT tools while classifying it according to the test oracle categories defined by Barr et al. in [128]: nonautomated, implicit, derived, and specified. Using a nonautomated oracle means that a practitioner manually assigns expected results to each test case individually. Implicit test oracles do not assign explicit test results to each test case, but rather indirectly evaluate test results, for example, via monitoring performance or detecting crashes and exceptions. In a derived test oracle, test case results are again explicit; however, only general guidance is given to determine them, e.g., via analysis of logs and results of previous executions. Finally, in a specified test oracle, the practitioner specifies rules in a formal specification language to determine the expected results for each test case. This latter category is the one used by most CT tools which support the notion of expected results.

Kruse further suggests an approach for test script generation, in which rather than generating a separate script for test case in the CT test plan, a part of a test script (a test snippet) is generated for each parameter in the model. The snippets are then assembled to create a test script for each test case. This approach considerably reduces the effort associated with test script generation by allowing reuse while considering the unique structure of CT test cases.

The huge diversity in SUT settings and environments imposes a significant challenge for automated test generation techniques. Further research on generic approaches for test generation from CT test plans can both reduce the current manual effort associated with employing CT in practice, and constitute an incentive to introduce CT in new settings.

9. CONCLUSIONS

The world of software development is rapidly changing. The emergence of Cloud, Analytics and Big Data solutions is revolutionizing the way software is being developed. This requires a fundamental change also in the way it is being tested. Methodologies and paradigms such as DevOps [129] are blurring the boundaries between testers and field users. As a result of agile development [130] and continuous delivery [131], testing needs to adjust to shorter software development cycles, with monthly and weekly deliveries. Cloud development relies on the use of open source solutions and microservices, moving the development and testing bottlenecks to integration of the various pieces of code that are being assembled together. Cloud applications are exposed to the uncontrollability and unpredictability of the Cloud—not only is their execution environment no longer controllable, it is also no longer constant, as it may change along time from one invocation to another. Analytics-based and data intensive solutions rely on the quality of their data as much as they rely on the quality of their code, but the underlying impact of data quality on the analytics engine is much more implicit and hard to comprehend than that of the code. Scale issues encountered in big data solutions have no precedence.

The impact of this revolution on the role of software testing is only just starting to become evident. Borrowing concepts from traditional testing approaches assists in handling these new challenges. However, adjustments as well as new approaches are necessary, considering that the development challenges being faced are orders of magnitude greater than those of traditional software, and old assumptions may no longer be valid. It seems that a mental leap is required for quality assurance similarly to that which the software development field in undergoing.

Specifically for CT, a valid question in light of these fundamental changes in software development is whether it is still a relevant and valid approach for software testing. The reasoning behind the CT approach, that most software bugs are triggered by a small number of parameters of the SUT, is being continuously reconfirmed by industry use. Furthermore, it is even intensified by new trends of microservices, integration efforts, and Cloud uncontrollability and unpredictability which lend themselves to feature interaction clashing. However, to realize the potential and opportunity provided by the software development revolution, CT needs to adapt. Beyond methodological challenges relating to the different mind set that practitioners need to adjust to when systematically defining the test space, there are also many technical challenges. For example, considering that test automation is becoming a prerequisite in many development environments, CT should be perceived not as an isolated one time effort during testing but rather as a continuous activity which is an integral part of the end to end development process. Similarly, given considerably shorter delivery schedules and frequently changing software, CT needs to easily and efficiently allow maintenance and reuse of its artifacts.

As the use of CT in practice progresses in a constantly evolving surroundings, additional challenges are being unveiled and new research topics are emerging.

REFERENCES

[1] M. Gallaher, B. Kropp, Economic Impacts of Inadequate Infrastructure for Software Testing, National Institute of Standards and Technology, 2002.
[2] T. Britton, L. Jeng, G. Carver, P. Cheak, T. Katzenellenbogen, Reversible Debugging Software: Quantify the Time and Cost Saved Using Reversible Debuggers, University of Cambridge, 2013.
[3] F.P. Brooks Jr., The Mythical Man-Month: Essays on Software Engineering, Addison-Wesley Longman Publishing Co., Inc., 1995.
[4] B. Hailpern, P. Santhanam, Software debugging, testing, and verification, IBM Syst. J. 41 (1) (2002) 4–12.
[5] K.Z. Bell, Optimizing Effectiveness and Efficiency of Software Testing: A Hybrid Approach (Ph.D. thesis), North Carolina State University, 2006.
[6] K.Z. Bell, M.A. Vouk, On effectiveness of pairwise methodology for testing network-centric software, in: 2005 International Conference on Information and Communication Technology, 2005, pp. 221–235.
[7] D.R. Kuhn, M.J. Reilly, An investigation of the applicability of design of experiments to software testing, in: 27th NASA/IEEE Software Engineering Workshop, NASA Goddard Space Flight Center, 2002.
[8] D.R. Kuhn, D.R. Wallace, A.M. Gallo Jr., Software fault interactions and implications for software testing, IEEE Trans. Softw. Eng. 30 (6) (2004) 418–421.
[9] D.R. Wallace, D.R. Kuhn, Failure modes in medical device software: an analysis of 15 years of recall data, in: ACS/IEEE International Conference on Computer Systems and Applications, 2001, pp. 301–311.
[10] Z. Zhang, J. Zhang, Characterizing failure-causing parameter interactions by adaptive testing, in: Proceedings of the 2011 International Symposium on Software Testing and Analysis, ISSTA '11, 2011, pp. 331–341.
[11] J.J. Chilenski, An Investigation of Three Forms of the Modified Condition Decision Coverage (MCDC) Criterion, Technical report, Office of Aviation Research, 2001.
[12] S.R. Dalal, C.L. Mallows, Factor-covering designs for testing software, Technometrics 40 (3) (1998) 234–243.
[13] Y. Takeuchi, K. Tatsumi, S. Watanabe, H. Shimokawa, Conceptual support for test case design, in: Proceedings of 11th IEEE Computer Software and Applications Conference, 1987, pp. 285–290.
[14] R. Mandl, Orthogonal Latin squares: an application of experiment design to compiler testing, Commun. ACM 28 (10) (1985) 1054–1058.
[15] K. Tatsumi, S. Watanabe, Y. Takeuchi, H. Shimokawa, Conceptual support for test case design, in: Proceedings of 11th IEEE Computer Software and Applications Conference, 1987, pp. 285–290.
[16] C. Jones, Applied Software Measurement: Global Analysis of Productivity and Quality, third ed., McGraw-Hill Osborne Media, 2008.
[17] IBM Functional Coverage Unified Solution (IBM FOCUS), http://researcher.watson.ibm.com/researcher/view_group.php?id=1871.
[18] I. Segall, R. Tzoref-Brill, E. Farchi, Using binary decision diagrams for combinatorial test design, in: Proceedings of the 2011 International Symposium on Software Testing and Analysis, 2011, pp. 254–264.
[19] C. Nie, H. Leung, A Survey of Combinatorial Testing, ACM Comput. Surv. 43 (2) (2011) 11:1–11:29.
[20] IEEE International Workshop on Combinatorial Testing, http://ieeexplore.ieee.org/xpl/conhome.jsp?reload=true&punumber=1001832.
[21] R.A. Fisher, Statistical Methods for Research Workers, Oliver and Boyd, 1925.
[22] R.A. Fisher, The Design of Experiments, Oliver and Boyd, 1935.

[23] C.R. Rao, Factorial experiments derivable from combinatorial arrangements of arrays, Stat. Soc. (Suppl.) 9 (1947) 128–139.
[24] G. Taguchi, Introduction to Quality Engineering, UNIPUB Kraus International, 1986.
[25] G. Taguchi, System of Experimental Design, vols. 1 and 2, UNIPUB Kraus International, 1987.
[26] K. Tatsumi, Test-case design support system, in: Proceedings of International Conference on Quality Control, 1985, pp. 615–620.
[27] G. Sherwood, Improving Test Case Selection With Constrained Arrays. Technical memorandum, 1990.
[28] N.J.A. Sloane, Covering arrays and intersecting codes, J. Comb. Des. 1 (1) (1993) 51–63.
[29] Y. Lei, K.-C. Tai, In-parameter-order: a test generation strategy for pairwise testing, in: The 3rd IEEE International Symposium on High-Assurance Systems Engineering, HASE '98, 1998, pp. 254–261.
[30] G. Seroussi, N.H. Bshouty, Vector sets for exhaustive testing of logic circuits, IEEE Trans. Inf. Theory 34 (3) (1988) 513–522.
[31] M.B. Cohen, M.B. Dwyer, J. Shi, Interaction testing of highly-configurable systems in the presence of constraints, in: Proceedings of the 2007 International Symposium on Software Testing and Analysis, ISSTA '07, 2007, pp. 129–139.
[32] J. Petke, M.B. Cohen, M. Harman, S. Yoo, Practical combinatorial interaction testing: empirical findings on efficiency and early fault detection, IEEE Trans. Softw. Eng. 41 (9) (2015) 901–924.
[33] Y. Jia, M.B. Cohen, M. Harman, J. Petke, Learning combinatorial interaction test generation strategies using hyperheuristic search, in: 2015 IEEE/ACM 37th IEEE International Conference on Software Engineering, 2015, pp. 540–550.
[34] A. Gargantini, P. Vavassori, Using decision trees to aid algorithm selection in combinatorial interaction tests generation, in: 2015 IEEE Eighth International Conference on Software Testing, Verification and Validation Workshops (ICSTW), 2015, pp. 1–10.
[35] D.M. Cohen, S.R. Dalal, M.L. Fredman, G.C. Patton, The AETG System: An Approach to Testing Based on Combinatorial Design, IEEE Trans. Softw. Eng. 23 (7) (1997) 437–444.
[36] R.C. Bryce, C.J. Colbourn, M.B. Cohen, A framework of greedy methods for constructing interaction test suites, in: Proceedings of the 27th International Conference on Software Engineering, ICSE '05, 2005, pp. 146–155.
[37] C. Lott, A. Jain, S. Dalal, Modeling requirements for combinatorial software testing, SIGSOFT Softw. Eng. Notes 30 (4) (2005) 1–7.
[38] J. Czerwonka, Pairwise testing in the real world: practical extensions to test-case scenarios, in: PNSQC, 2006.
[39] F. Duan, Y. Lei, L. Yu, R.N. Kacker, D.R. Kuhn, Optimizing IPOGs vertical growth with constraints based on hypergraph coloring, in: 2017 IEEE Eighth International Conference on Software Testing, Verification and Validation Workshops (ICSTW), 2017.
[40] A. Yamada, A. Biere, C. Artho, T. Kitamura, E.-H. Choi, Greedy combinatorial test case generation using unsatisfiable cores, in: Proceedings of the 31st IEEE/ACM International Conference on Automated Software Engineering, ASE 2016, 2016, pp. 614–624.
[41] L. Yu, F. Duan, Y. Lei, R.N. Kacker, D.R. Kuhn, Constraint handling in combinatorial test generation using forbidden tuples, in: 2015 IEEE Eighth International Conference on Software Testing, Verification and Validation Workshops (ICSTW), 2015, pp. 1–9.

[42] L. Yu, Y. Lei, M. Nourozborazjany, R.N. Kacker, D.R. Kuhn, An efficient algorithm for constraint handling in combinatorial test generation, in: 2013 IEEE Sixth International Conference on Software Testing, Verification and Validation, 2013, pp. 242–251.
[43] R.E. Bryant, Graph-based algorithms for Boolean function manipulation, IEEE Trans. Comp. 35 (8) (1986) 677–691.
[44] S. Minato, Graph-based representations of discrete functions, in: Representation of Discrete Functions, Springer US, 1996, pp. 1–28.
[45] A. Gargantini, P. Vavassori, Efficient combinatorial test generation based on multi-valued decision diagrams, in: 10th International Haifa Verification Conference, 2014, pp. 220–235.
[46] B. Hnich, S.D. Prestwich, E. Selensky, B.M. Smith, Constraint models for the covering test problem, Constraints 11 (2) (2006) 199–219.
[47] C. Ansótegui, I. Izquierdo, F. Manyá, J. Torres Jiménez, A max-SAT-based approach to constructing optimal covering arrays, in: CCIA 2013, vol. 256 of FAIA, 2013.
[48] M. Banbara, H. Matsunaka, N. Tamura, K. Inoue, Generating combinatorial test cases by efficient SAT encodings suitable for CDCL SAT solvers, in: Proceedings of the 17th International Conference on Logic for Programming, Artificial Intelligence, and Reasoning, LPAR'10, 2010, pp. 112–126.
[49] T. Nanba, T. Tsuchiya, T. Kikuno, Using satisfiability solving for pairwise testing in the presence of constraints, IEICE Trans. 95-A (9) (2012) 1501–1505.
[50] A. Yamada, T. Kitamura, C. Artho, E.H. Choi, Y. Oiwa, A. Biere, Optimization of combinatorial testing by incremental SAT solving, in: 2015 IEEE 8th International Conference on Software Testing, Verification and Validation (ICST), 2015, pp. 1–10.
[51] O. Shtrichman, Pruning techniques for the SAT-based bounded model checking problem, in: Proceedings of the 11th IFIP WG 10.5 Advanced Research Working Conference on Correct Hardware Design and Verification Methods, CHARME '01, 2001.
[52] M.B. Cohen, M.B. Dwyer, J. Shi, Constructing interaction test suites for highly-configurable systems in the presence of constraints: a greedy approach, IEEE Trans. Softw. Eng. 34 (5) (2008) 633–650.
[53] B.J. Garvin, M.B. Cohen, M.B. Dwyer, Evaluating improvements to a meta-heuristic search for constrained interaction testing, Empir. Softw. Eng. 16 (1) (2011) 61–102.
[54] A. Calvagna, A. Gargantini, A formal logic approach to constrained combinatorial testing, J. Autom. Reason. 45 (4) (2010) 331–358.
[55] M.B. Cohen, P.B. Gibbons, W.B. Mugridge, C.J. Colbourn, Constructing test suites for interaction testing, in: Proceedings of the 25th International Conference on Software Engineering, 2003, 2003, pp. 38–48.
[56] M.B. Cohen, P.B. Gibbons, W.B. Mugridge, C.J. Colbourn, J.S. Collofello, A variable strength interaction testing of components, in: Proceedings 27th Annual International Computer Software and Applications Conference, COMPAC 2003, 2003, pp. 413–418.
[57] L. Gargano, J. Korner, U. Vaccaro, Capacities: from information theory to extremal set theory, J. Comb. Theory Ser. A 68 (2) (1994) 296–316.
[58] S. Fouché, M.B. Cohen, A. Porter, Incremental covering array failure characterization in large configuration spaces, in: Proceedings of the Eighteenth International Symposium on Software Testing and Analysis, ISSTA '09, 2009, pp. 177–188.
[59] I. Segall, Repeated combinatorial test design—unleashing the potential in multiple testing iterations, in: 2016 IEEE International Conference on Software Testing, Verification and Validation (ICST), 2016, pp. 12–21.
[60] Pairwise testing website, http://www.pairwise.org/tools.asp.

[61] R. Kuhn, Y. Lei, R. Kacker, Practical combinatorial testing: beyond pairwise, IT Professional 10 (3) (2008) 19–23.
[62] A. Gargantini, P. Vavassori, CITLAB: a laboratory for combinatorial interaction testing, in: 2012 IEEE Fifth International Conference on Software Testing, Verification and Validation, 2012, pp. 559–568.
[63] O. Lachish, E. Marcus, S. Ur, A. Ziv, Hole analysis for functional coverage data, in: Proceedings of 2002 Design Automation Conference, 2002, pp. 807–812.
[64] E. Farchi, I. Segall, R. Tzoref-Brill, Using projections to debug large combinatorial models, in: 2013 IEEE Sixth International Conference on Software Testing, Verification and Validation Workshops, 2013, pp. 311–320.
[65] I. Segall, R. Tzoref-Brill, A. Zlotnick, Simplified modeling of combinatorial test spaces, in: 2012 IEEE Fifth International Conference on Software Testing, Verification and Validation, 2012, pp. 573–579.
[66] I. Segall, R. Tzoref-Brill, A. Zlotnick, Common patterns in combinatorial models, in: 2012 IEEE Fifth International Conference on Software Testing, Verification and Validation, 2012, pp. 624–629.
[67] I. Segall, R. Tzoref-Brill, Interactive refinement of combinatorial test plans, in: 2012 34th International Conference on Software Engineering (ICSE), 2012, pp. 1371–1374.
[68] R. Tzoref-Brill, S. Maoz, Syntactic and semantic differencing for combinatorial models of test designs, in: International Conference on Software Engineering (ICSE), 2017.
[69] D. Blue, I. Segall, R. Tzoref-Brill, A. Zlotnick, Interaction-based test-suite minimization, in: Proceedings of the 2013 International Conference on Software Engineering, ICSE '13, 2013.
[70] I. Segall, R. Tzoref-Brill, A. Zlotnick, Combining minimization and generation for combinatorial testing, in: 2015 IEEE Eighth International Conference on Software Testing, Verification and Validation Workshops (ICSTW), 2015, pp. 1–9.
[71] D.R. Kuhn, I.D. Mendoza, R.N. Kacker, Y. Lei, Combinatorial coverage measurement concepts and applications, in: 2013 IEEE Sixth International Conference on Software Testing, Verification and Validation Workshops, 2013, pp. 352–361.
[72] R. Tzoref-Brill, P. Wojciak, S. Maoz, Visualization of combinatorial models and test plans, in: 2016 31st IEEE/ACM International Conference on Automated Software Engineering (ASE), 2016, pp. 144–154.
[73] D.M. Cohen, S.R. Dalal, J. Parelius, G.C. Patton, The combinatorial design approach to automatic test generation, IEEE Softw. 13 (5) (1996) 83–88.
[74] R. Krishnan, S.M. Krishna, P.S. Nandhan, Combinatorial testing: learnings from our experience, SIGSOFT Softw. Eng. Notes 32 (3) (2007) 1–8.
[75] G.B. Sherwood, Effective testing of factor combinations, in: Proceedings of the Third International Conference on Software Testing, Analysis and Review, 1994, pp. 133–166.
[76] D.R. Kuhn, R.N. Kacker, Y. Lei, Introduction to Combinatorial Testing, first ed., Chapman & Hall/CRC, 2013.
[77] C. Yilmaz, E. Dumlu, M.B. Cohen, A. Porter, Reducing masking effects in combinatorial interaction testing: a feedback driven adaptive approach, IEEE Trans. Softw. Eng. 40 (1) (2014) 43–66.
[78] S.R. Dalal, A. Jain, N. Karunanithi, J.M. Leaton, C.M. Lott, G.C. Patton, B.M. Horowitz, Model-based testing in practice, in: Proc. 21st International Conference on Software Engineering (ICSE'99), 1999, pp. 285–294.
[79] I. Segall, R. Tzoref-Brill, A. Zlotnick, Simplified modeling of combinatorial test spaces, in: 2012 IEEE Fifth International Conference on Software Testing, Verification and Validation, 2012, pp. 573–579.

[80] G.B. Sherwood, Embedded functions in combinatorial test designs, in: 2015 IEEE Eighth International Conference on Software Testing, Verification and Validation Workshops (ICSTW), 2015, pp. 1–10.
[81] G.B. Sherwood, Embedded functions for constraints and variable strength in combinatorial testing, in: 2016 IEEE Ninth International Conference on Software Testing, Verification and Validation Workshops (ICSTW), 2016, pp. 65–74.
[82] M.P. Usaola, F.R. Romero, R.R.-B. Aranda, I. García, Test case generation with regular expressions and combinatorial techniques, in: 2017 IEEE Seventh International Conference on Software Testing, Verification and Validation Workshops, 2017.
[83] L. Kampel, B. Garn, D.E. Simos, Combinatorial methods for modelling composed software systems, in: 2017 IEEE Seventh International Conference on Software Testing, Verification and Validation Workshops, 2017.
[84] P. Arcaini, A. Gargantini, P. Vavassori, Validation of models and tests for constrained combinatorial interaction testing, in: 2014 IEEE Seventh International Conference on Software Testing, Verification and Validation Workshops, 2014, pp. 98–107.
[85] A. Gargantini, J. Petke, M. Radavelli, Combinatorial interaction testing for automated constraint repair, in: 2017 IEEE Seventh International Conference on Software Testing, Verification and Validation Workshops, 2017.
[86] P. Satish, K. Sheeba, K. Rangarajan, Deriving combinatorial test design model from UML activity diagram, in: 2013 IEEE Sixth International Conference on Software Testing, Verification and Validation Workshops, 2013, pp. 331–337.
[87] P. Satish, A. Paul, K. Rangarajan, Extracting the combinatorial test parameters and values from UML sequence diagrams, in: 2014 IEEE Seventh International Conference on Software Testing, Verification and Validation Workshops, 2014, pp. 88–97.
[88] H. Nakagawa, T. Tsuchiya, Towards automatic constraints elicitation in pair-wise testing based on a linguistic approach: elicitation support using coupling strength, in: Requirements Engineering and Testing (RET), 2015, pp. 34–36.
[89] M. Zalmanovici, O. Raz, R. Tzoref-Brill, Cluster-based test suite functional analysis, in: Proceedings of the 2016 24th ACM SIGSOFT International Symposium on Foundations of Software Engineering, FSE 2016, 2016, pp. 962–967.
[90] D.E. Simos, R. Kuhn, A.G. Voyiatzis, R. Kacker, Combinatorial methods in security testing, Computer 49 (10) (2016) 80–83.
[91] J. Hagar, R. Kuhn, R. Kacker, T. Wissink, Introducing combinatorial testing in a large organization: pilot project experience report (poster), in: 2014 IEEE Seventh International Conference on Software Testing, Verification and Validation Workshops, 2014, p. 153.
[92] X. Deng, T. Wu, J. Yan, J. Zhang, Combinatorial testing on implementations of HTML5 support, in: 2017 IEEE Seventh International Conference on Software Testing, Verification and Validation Workshops, 2017.
[93] C. Rao, J. Guo, N. Li, Y. Lei, Y. Zhang, Y. Li, Y. Cao, Applying combinatorial testing to high-speed railway track circuit receiver, in: 2017 IEEE Seventh International Conference on Software Testing, Verification and Validation Workshops, 2017.
[94] K. Tsumura, H. Washizaki, Y. Fukazawa, K. Oshima, R. Mibe, Pairwise coverage-based testing with selected elements in a query for database applications, in: 2016 IEEE Ninth International Conference on Software Testing, Verification and Validation Workshops (ICSTW), 2016, pp. 92–101.
[95] G. Dhadyalla, N. Kumari, T. Snell, Combinatorial testing for an automotive hybrid electric vehicle control system: a case study, in: Proceedings of the 2014 IEEE International Conference on Software Testing, Verification, and Validation Workshops, ICSTW '14, 2014, pp. 51–57.

[96] S. Vilkomir, B. Amstutz, Using combinatorial approaches for testing mobile applications, in: 2014 IEEE Seventh International Conference on Software Testing, Verification and Validation Workshops, 2014, pp. 78–83.
[97] P. Wojciak, R. Tzoref-Brill, System level combinatorial testing in practice—the concurrent maintenance case study, in: ICST, 2014, pp. 103–112.
[98] N. Li, Y. Lei, H.R. Khan, J. Liu, Y. Guo, Applying combinatorial test data generation to big data applications, in: Proceedings of the 31st IEEE/ACM International Conference on Automated Software Engineering, ASE 2016, 2016, pp. 637–647.
[99] W.A. Ballance, S. Vilkomir, W. Jenkins, Effectiveness of pair-wise testing for software with Boolean inputs, in: 2012 IEEE Fifth International Conference on Software Testing, Verification and Validation, 2012, pp. 580–586.
[100] N. Kobayashi, T. Tsuchiya, T. Kikuno, Applicability of non-specification-based approaches to logic testing for software, in: 2001 International Conference on Dependable Systems and Networks, 2001, pp. 337–346.
[101] S. Vilkomir, D. Anderson, Relationship between pair-wise and MC/DC testing: initial experimental results, in: 2015 IEEE Eighth International Conference on Software Testing, Verification and Validation Workshops (ICSTW), 2015, pp. 1–4.
[102] S. Vilkomir, O. Starov, R. Bhambroo, Evaluation of t-wise approach for testing logical expressions in software, in: 2013 IEEE Sixth International Conference on Software Testing, Verification and Validation Workshops, 2013, pp. 249–256.
[103] R.C. Bryce, A. Rajan, M.P.E. Heimdahl, Interaction testing in model-based development: effect on model-coverage, in: 2006 13th Asia Pacific Software Engineering Conference (APSEC'06), 2006, pp. 259–268.
[104] M. Ellims, D. Ince, M. Petre, The effectiveness of T-way test data generation, in: Proceedings of the 27th International Conference on Computer Safety, Reliability, and Security, SAFECOMP '08, 2008, pp. 16–29.
[105] P.J. Schroeder, P. Bolaki, V. Gopu, Comparing the fault detection effectiveness of N-way and random test suites, in: Proceedings of the 2004 International Symposium on Empirical Software Engineering, ISESE '04, 2004, pp. 49–59.
[106] R. Kuhn, R. Kacker, Y. Lei, Random vs. combinatorial methods for discrete event simulation of a grid computer network, in: Proceedings of ModSim World, 2009, pp. 83–88.
[107] L.S. Ghandehari, J. Czerwonka, Y. Lei, S. Shafiee, R. Kacker, R. Kuhn, An empirical comparison of combinatorial and random testing, in: 2014 IEEE Seventh International Conference on Software Testing, Verification and Validation Workshops, 2014, pp. 68–77.
[108] H. Do, S. Elbaum, G. Rothermel, Supporting controlled experimentation with testing techniques: an infrastructure and its potential impact, Empirical Softw. Eng. 10 (4) (2005) 405–435.
[109] J. Czerwonka, On use of coverage metrics in assessing effectiveness of combinatorial test designs, in: 2013 IEEE Sixth International Conference on Software Testing, Verification and Validation Workshops, 2013, pp. 257–266.
[110] H. Felbinger, F. Wotawa, M. Nica, Mutation score, coverage, model inference: quality assessment for t-way combinatorial test-suites, in: 2017 IEEE Eighth International Conference on Software Testing, Verification and Validation Workshops (ICSTW), 2017.
[111] R. Kuhn, R. Kacker, Y. Lei, Estimating t-way fault profile evolution during testing, in: 2017 IEEE Eighth International Conference on Software Testing, Verification and Validation Workshops (ICSTW), 2017.
[112] A. Hartman, L. Raskin, Problems and algorithms for covering arrays, Discrete Math. 284 (1–3) (2004) 149–156.
[113] G. Rothermel, R.H. Untch, C. Chu, M.J. Harrold, Prioritizing test cases for regression testing, IEEE Trans. Softw. Eng. 27 (10) (2001) 929–948.

[114] R.C. Bryce, S. Sampath, A.M. Memon, Developing a single model and test prioritization strategies for event-driven software, IEEE Trans. Softw. Eng. 37 (1) (2011) 48–64.
[115] X. Qu, M.B. Cohen, A study in prioritization for higher strength combinatorial testing, in: Proceedings of the 2013 IEEE Sixth International Conference on Software Testing, Verification and Validation Workshops, ICSTW '13, 2013, pp. 285–294.
[116] X. Qu, M.B. Cohen, K.M. Woolf, Combinatorial interaction regression testing: a study of test case generation and prioritization, in: 2007 IEEE International Conference on Software Maintenance, 2007, pp. 255–264.
[117] S. Sampath, R.C. Bryce, S. Jain, S. Manchester, A tool for combination-based prioritization and reduction of user-session-based test suites, in: Proceedings of the 2011 27th IEEE International Conference on Software Maintenance, ICSM '11, 2011, pp. 574–577.
[118] S. Kimoto, T. Tsuchiya, T. Kikuno, Pairwise testing in the presence of configuration change cost, in: Proceedings of the 2008 Second International Conference on Secure System Integration and Reliability Improvement, SSIRI '08, 2008, pp. 32–38.
[119] H. Wu, C. Nie, F.C. Kuo, Test suite prioritization by switching cost, in: 2014 IEEE Seventh International Conference on Software Testing, Verification and Validation Workshops, 2014, pp. 133–142.
[120] X. Niu, C. Nie, Y. Lei, A.T.S. Chan, Identifying failure-inducing combinations using tuple relationship, in: 2013 IEEE Sixth International Conference on Software Testing, Verification and Validation Workshops, 2013, pp. 271–280.
[121] C. Yilmaz, M.B. Cohen, A. Porter, Covering arrays for efficient fault characterization in complex configuration spaces, in: Proceedings of the 2004 ACM SIGSOFT International Symposium on Software Testing and Analysis, ISSTA '04, 2004, pp. 45–54.
[122] L. Shi, C. Nie, B. Xu, A software debugging method based on pairwise testing, in: Proceedings of the 5th International Conference on Computational Science—Volume Part III, ICCS'05, 2005, pp. 1088–1091.
[123] J. Chandrasekaran, L.S. Ghandehari, Y. Lei, R. Kacker, D.R. Kuhn, Evaluating the effectiveness of BEN in localizing different types of software fault, in: 2016 IEEE Ninth International Conference on Software Testing, Verification and Validation Workshops (ICSTW), 2016, pp. 26–34.
[124] L.S. Ghandehari, J. Chandrasekaran, Y. Lei, R. Kacker, D.R. Kuhn, BEN: a combinatorial testing-based fault localization tool, in: 2015 IEEE Eighth International Conference on Software Testing, Verification and Validation Workshops (ICSTW), 2015, pp. 1–4.
[125] L.S.G. Ghandehari, Y. Lei, T. Xie, R. Kuhn, R. Kacker, Identifying failure-inducing combinations in a combinatorial test set, in: 2012 IEEE Fifth International Conference on Software Testing, Verification and Validation, 2012, pp. 370–379.
[126] R. Tzoref-Brill, S. Maoz, Lattice-based semantics for combinatorial model evolution, in: ATVA, 2015, pp. 276–292.
[127] P.M. Kruse, Test oracles and test script generation in combinatorial testing, in: 2016 IEEE Ninth International Conference on Software Testing, Verification and Validation Workshops (ICSTW), 2016, pp. 75–82.
[128] E.T. Barr, M. Harman, P. McMinn, M. Shahbaz, S. Yoo, The oracle problem in software testing: a survey, IEEE Trans. Softw. Eng. 41 (5) (2015) 507–525.
[129] M. Huttermann, DevOps for Developers (Expert's voice in web development), Apress, 2012, ISBN: 9781430245704.
[130] R.C. Martin, Agile Software Development: Principles, Patterns, and Practices, Prentice Hall PTR, 2003.
[131] J. Humble, D. Farley, Continuous Delivery: Reliable Software Releases Through Build, Test, and Deployment Automation, Addison-Wesley Professional, 2010.

ABOUT THE AUTHOR

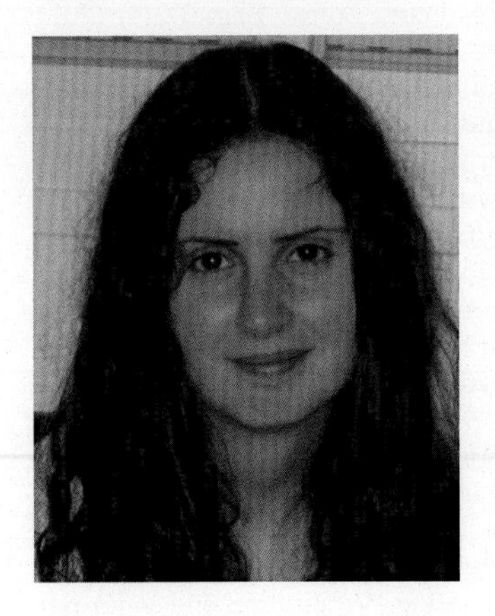

Rachel Tzoref-Brill is a lead researcher in IBM Research, Israel. She recently won an IBM Corporate Award for combinatorial test design innovations. She received her BS and MS degrees in Computer Science from the Technion, Israel Institute of Technology and is currently pursuing her PhD degree from Tel Aviv University. Her research interests include all aspects of software quality, and particularly combinatorial testing, formal methods, and empirical software engineering. She chaired the International Workshop on Combinatorial Testing (IWCT) in 2016 and 2017.

CHAPTER THREE

Advances in Applications of Object Constraint Language for Software Engineering

Atif A. Jilani, Muhammad Z. Iqbal, Muhammad U. Khan, Muhammad Usman
Quest Laboratory, FAST-National University of Computer & Emerging Sciences, Islamabad, Pakistan

Contents

Abstract

Object Constraint Language (OCL) is a standard language defined by Object Management Group for specifying constraints on models. Since its introduction as part of Unified Modeling Language, OCL has received significant attention by researchers with works in the literature ranging from temporal extensions of OCL to automated test generation by solving OCL constraints. In this chapter, we provide a survey of the various works discussed in literature related to OCL with the aim of highlighting the advances made in the field. We classify the literature into five broad categories and provide summaries for various works in the literature. The chapter also provides insights and highlights the potentials areas of further research in the field.

Advances in Computers, Volume 112
ISSN 0065-2458
https://doi.org/10.1016/bs.adcom.2017.12.003

1. INTRODUCTION

Model-driven engineering (MDE) over the last decade has emerged as a successful and a widely used approach for developing software systems where models act as the key development artifact [1]. Modeling languages ranging from the standardized Unified Modeling Language (UML) [2] to domain-specific languages (DSLs) and profiles, such as the SysML [3], are being used under the MDE umbrella. MDE has been successfully applied in a number of domains, including information systems [4], embedded systems [5,6], product-line engineering [7,8], mobile application development [9], communications [10], and games [11]. Modeling languages allow modeling of different aspects of systems, including the structural and behavioral aspects, using well-defined modeling constructs. With the use of models, the design of systems can be visualized early on in the development, errors can be minimized, early testing can be initiated, and maintenance is made easier [12,13].

A major challenge in MDE is the specification of constraints in a manner that is precise and readable from a pragmatic perspective. Object Constraint Language (OCL) [14] is a constraint specification language introduced to support writing constraints on models. OCL was initially developed by IBM [15] primarily for business modeling, however, now it is being widely used to specify constraints on models. OCL was introduced as a part of Object Management Group's (OMG) UML standard as the recommended language for specifying constraints on UML models. Currently, it is released and managed by OMG as a separate standard for specifying constraints on models and meta-models. OCL has a syntax that is relatively closer to programming languages and is therefore easier for software engineers to apply as compared to the traditional formal specification languages.

OCL has been utilized by researchers to solve a number of industrial and scientific problems, such as test data generation [16–18], model verification and validation [19–24], and constraint solving [17,25]. Extensions to OCL have also been proposed to allow writing domain-specific constraints, e.g., temporal logic extensions for time-critical systems [26,27], and extensions to specify topological constraints on spatial objects [28].

OCL has also been successfully used in industry as a part of UML standard for specifying constraints on models [29] and is supported by both open-source and proprietary tools, such as Dresden [30], Eclipse OCL [31], and IBM Rational Software Architect [32]. These tools allow a number of features for modeling OCL constraints, such as (i) syntax checking, (ii) semantic checking (i.e., type and consistency checking) [33], (iii) dynamic modeling [34,35], and (iv) test automation [36]. OCL is also integrated in most of the modeling tools to facilitate in writing specifications and validation during UML modeling [32,37,38].

In this chapter, we report the various works on OCL in the literature, its applications in industry and scientific settings, works on empirical evaluations of OCL, and various tools that support OCL. Keeping in view the broad usage of OCL in different domains, we classify the literature on OCL in various categories: *verification and validation*, *experiences and empirical evaluations*, *OCL tools*, *extensions to OCL*, and *other* works category covering diverse works that do not naturally fit in previous categories. The purpose of this chapter is to provide a comprehensive, though not exhaustive, survey of current state of the art research and advances in OCL. We highlight the areas and domains where OCL has been successfully applied and identify potential direction of future research in the field.

The chapter is divided as follows: Section 2 presents background information for researchers newly introduced to OCL and presents the language constructs with an example demonstrating constraint writing in OCL. Section 3 covers classification of works on OCL and its subsections present the survey of existing literature on OCL. Section 4 presents some insights based on authors' experiences and from the conducted survey. Finally, Section 5 concludes the chapter and provides future directions.

2. BACKGROUND

This section covers the background of the OCL and briefly describes the various OCL constructs.

OCL is a standard language defined by OMG for specifying constraints on models. OCL can be used to write constraints at various modeling levels defined in OMG's Meta-Object Facility (MOF), including the meta-model level and model level. At meta-model level, OCL is used to define constraints on elements of the language, for example, OCL is used in the OMG's document defining UML to define the constraints on various elements. At this level, OCL is important for specifying the rules of language, either for a language which is defined using a new meta-model and for a profile defined using a lightweight extension. At the model level, OCL is largely used as an integral part of modeling languages, including UML, to specify constraints on the various modeling elements.

OCL is a strongly typed and side effect-free specification language in which expressions are written in a declarative form [14]. In OCL, constraints are represented as expressions [39] that restrict the various modeling elements. Each OCL expression indicates a value or object within the model that conforms to a type, e.g., "5 + 7" is a valid OCL expression of type *Integer*, which represents the integer value "12." The primitive types supported in OCL are Boolean, Integer, Real, String, and Unlimited Natural [14]. Table 1 shows the OCL types and the various OCL operations that can manipulate them. For example, the *Boolean* OCL type supports logical operators, i.e., and, xor, and not. There are some generic type operations that are also supported by OCL for all types, such as OclInvalid and OclVoid. For example, when an OCL expression is evaluated to "undefined" then it returns OclVoid type. All types in OCL are subtypes of class OCLAny.

OCL expression comprises of four parts: (i) a *context* that defines the situation in which the statement is valid (syntactic description for a *context* is presented in Table 2), (ii) a *property* that represents characteristics of the *context* (e.g., if the *context* is a *class* then a *property* can be an *attribute*), (iii) an *operation* (e.g., arithmetic, set-oriented) that manipulates or qualifies a *property*, and (iv) *keywords* (e.g., *if, then, else, and, or, not, implies*) that are used to specify conditional expressions. For example, the OCL expression in Listing 1 shows an invariant on the *Business* class (shown in Fig. 1). As per the

Table 1 OCL Types and Their Supported Operations [14]

OCL Type	Supported Operations
Boolean	Equal (=), Not-Equal (<>), OR (or), AND (and), Exclusive-OR (xor), NOT (not), IMPLIES (implies), and If-then-else
Integer and Real	Equal (=), Not-Equal (<>), Less (<), Greater (>), Less-or-Equal (≤),Greater-or-Equal (≥), Add (+), Subtract (−), Multiply (*), and divide (/)
String	Equal (=), Not-Equal (<>), concat(), size(), toLower(), toUpper(), and substring(start, end)
Collections (Bag, Set, Sequence, and OrderedSet)	Equal (=), Not-Equal (<>), size(), count(), excludes(variable), excludesAll(collection), includes(variable), includesAll(collection), isEmpty(), notEmpty(), sum(), first(), last(), at(index), and flatten()

Table 2 Syntax for OCL Expression Parts

Item	OCL Syntax	Example	OCL Expression
Context	context <context name>	Define a context for a "Current" object	context Current
Invariant	context <classifier> inv [<constraint name>]: <boolean OCL expression>	Define an amount invariant for "Business" object where amount should be greater than zero	context Business inv AmountInvariant:self. amount > 0
Precondition	context <classifier>:: <operation> (<parameters>) pre[<constraint name>]: <boolean OCL expression>	Validate that an account should be active and amount should be greater than zero before the use of deposit operation in "Saving" object	context Saving::deposit (amount: Integer):Boolean pre: self.isActive = true and self. Amount > 0
Postcondition	context <classifier>:: <operation> (<parameters>) post[<constraint name>]: <boolean OCLexpression>	Validate that a result should be true after the completion of withdraw operation in "Saving" object	context Saving::withdraw (amount: Integer):Boolean post: result = true * here result refers to the output of the operation

Listing 1 Example of OCL expression.

```
context Account
inv self.amount > 0 implies isActive = true
```

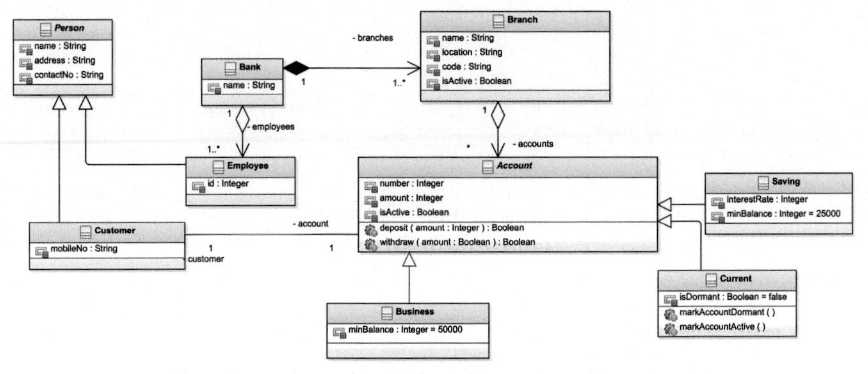

Fig. 1 Example of a simplified banking system UML class diagram.

definition of the OCL expression in Listing 1, the context is *Business*, there are two properties of *Business* class that are used as: *amount* and *isActive*, the operations are: *greater-than* (>) and *equal* (=), and the keyword is *implies*.

OCL expressions are categorized as, *invariant*, *precondition*, *postcondition*, and *guard*. In the following, we discuss each of them in detail.

Invariant. An invariant is a Boolean constraint that should hold true throughout the lifetime of all instances of the invariant's context class. Table 2 presents the syntax for the invariant definition in OCL.

Precondition. A precondition is defined on operations as a constraint that must hold true before the execution of an operation. The precondition syntax for definition is shown in Table 2.

Postcondition. A postcondition is a constraint that must hold true after the execution of an operation. Table 2 demonstrates the definition syntax for postcondition.

Guard. It is a kind of an invariant that is applicable on the transition from one state to another in a state machine.

OCL constraints are also inherited between modeling elements, however, depending on the kind of constraints it can either be further strengthened or weakened. For example, an invariant for a super class is also applicable to a subclass. A subclass may strengthen the invariant but it cannot make it weaker. A precondition may be weakened but not be strengthened in the redefinition of the operation in the subclass. A postcondition can be strengthened but cannot be weakened in the redefinition of the operation in the subclass. Relationship attributes are used to navigate from one object to another in the relationship based on its navigability. The dot (.) operator is used for this purpose. For example, consider a class diagram of a banking system shown in Fig. 1, to validate whether the associated *Account* with the *Customer* is active or not, Table 3 shows the "Navigation" example. Collections are also used for referencing more than one object in the relationships as shown in "Reference in Collection" example in Table 3. OCL also offers the iteration operations (such as any(expr), collect(expr), exists(expr), forAll(expr), isUnique(expr), and select(expr)) to facilitate iteration on a list of object, e.g., shown in "Iterator in Collection" example in Table 3. OCL allows initialization of the object value, as shown in "Initialization Value" example in Table 3. OCL also supports query operations on model that

Table 3 OCL Items and Their Examples

Item	Example	OCL Expression
Navigation	Validate whether an account associated with a customer is active or not	`context Customer inv: self.account.isActive = true`
Reference in collection	Collect all the branches of a bank where branches size is greater than one	`context Bank inv: self→collect (branches)→size() >= 1` here a Set(Branch) is returned as a result
Iterator in collection	Get the total amount for all the "Saving" accounts	`Saving. allInstances → iterate(s : Saving; sum : Real = 0 | sum + amount)`
Initialization value	Initialize an "interestRate" attribute in "Saving" object with an integer value of 12	`context Saving:: interestRate: Integer init: 12`
Query language	Create a query to get all active accounts of a branch	`context Branch:: getActiveAccounts(): Bag (Account) body: accounts→select(a: Account | a.isActive=true)`

do not change the state of the system, e.g., as shown in the example of "Query Language" in Table 3.

OCL expressions can be used in many ways as part of models (as shown in Table 3): (i) to specify the initial value of an attribute, (ii) to specify the body of an operation, (iii) to indicate an instance, (iv) to indicate a condition, and (v) to indicate actual parameter values. Fig. 1 presents the UML class diagram of a banking system as an example to demonstrate the use of OCL expressions. The diagram represents various classes (such as Bank, Branch, Account, Employee, and Customer) and their relationships. The definition of banking system is incomplete without the specification of the business rules, for example, (i) amount cannot be credited in an account if it is inactive and (ii) the amount should be greater than zero to execute the deposit and withdraw methods to ensure safe and consistent operation of the system. The OCL expression in Listing 1 shows an invariant on the *Account* class (shown in Fig. 1) for the business rule, i.e., if the amount in the account is greater than zero than it the account is active. As per the definition of the OCL expression in Listing 1, the context is *Business* class, there are two properties of *Account* class used as: *amount* and *isActive*, the operations are: *greater-than* (>) and *equal* (=), and the keyword is *implies*.

OCL can be used for writing query operations, specifying operation contracts, and as a model manipulations/query language as shown in "Query language" example in Table 3. In query operations as the name indicates, OCL expression is used to query the models and returns the results. Queries are side effect-free operations that do not change the state of the system. Operation contracts use OCL expression to specify the effects of an operation. In literature, operation effects are specified by following imperative or declarative approaches [40]. In imperative approach, the designer explicitly defines the set of structural events, for example, insert, update, or delete, that need to be applied during the execution of an operation. However, in declarative approach as is the case of OCL, a contract comprises a set of pre- and postconditions. A precondition describes a set of conditions on operation input and the system state that must hold when an operation is invoked, whereas a postcondition describes a set of conditions that must be satisfied by the system state at the end of operation. Among many other applications, OCL is also used in model transformations, where it is used to manipulate models and as a query/transformation language. In writing transformation, OCL expressions are used as a part of source and target patterns in transformation rules, such as in Atlas Transformation language (ATL) [41] and MofScript [42].

3. SURVEY ON ADVANCES IN OCL

This section discusses the works in literature on OCL. Keeping in view the diverse nature of works on OCL, we provide a high-level classification of works on OCL. Our classification places the literature in five broad categories: (i) *Verification and Validation using OCL* that includes test data generation and model verification and validation; (ii) *Experiences and Empirical Reports* on OCL that discuss the empirical evaluations aim to provide evidence on applicability and effectiveness of OCL usage in experimental and real settings; (iii) *OCL Tools* containing the works related to various OCL Tools; (iv) *OCL Extensions* that include the works that are focusing on extending OCL with new language constructs; and (v) Other works that do not fit in the above categories and apply OCL in diverse areas. Following we discuss the classification and the various papers covering the advances in OCL.

3.1 Verification and Validation Using OCL

A major area of research related to OCL has been verification and validation of systems and models using OCL. Given the nature of works, we have further classified the works on verification and validation in two broad categories: works dealing with test generation and works dealing with model verification and validation. Following we discuss the various works in these categories.

3.1.1 Test Generation From OCL Constraints

As the constraint specification language on models, OCL is widely used for test data generation during model-based testing (MBT). In MBT, the constraints on the models (for example, guards on the state machines, invariants) need to be resolved in order to generate test cases from models. As an example, consider the simple state machine shown in Fig. 2 containing two states. During testing, a test case triggers the call event operation: `markAccountDormat` with the goal of transitioning from the *Active* state to *Dormant* state. For the transition to successfully complete, the guard on the transition written as an OCL constraint: `amount < minBalance and isActive = false` has to be satisfied. The test case to be executed should contain the data that satisfies the target guard condition. Similarly other OCL

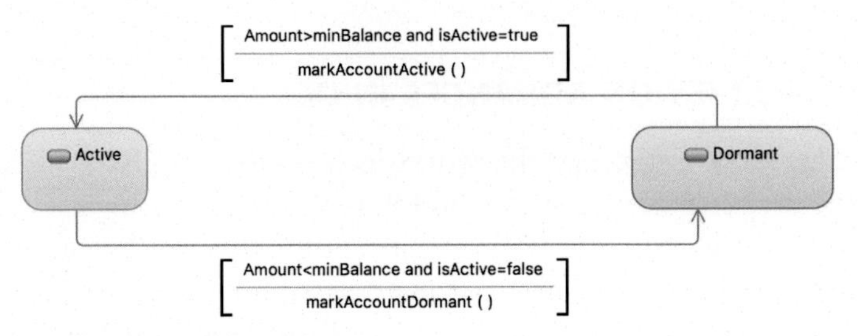

Fig. 2 Example UML state machine.

constraints written as pre- and postconditions of operations, class invariants, and guards in states also need to be satisfied during testing.

A number of works have specifically targeted test data generation from OCL constraints that can be used during MBT. Following we discuss these techniques.

One of the earlier works in this area is by Benattou et al. [18] which discusses an approach of test data generation from OCL specifications that is inspired from existing works in the area for test generation from VDM formal specification. The approach uses partition analysis of OCL expressions written as invariants and pre- and postconditions. The expressions are converted into disjunctive normal form to identify disjoint partitions for test generation. The work is one of the initial works to deal with test generation from static structure models and OCL.

Salas et al. [43,44] present a fault-based test case generation approach for OCL pre- and postconditions. The OCL expressions are mutated with faults and test cases are generated using partition analysis and constraint solving to kill these mutants. To automate the process, a prototype test case generator tool is developed that supports OCL.

Li et al. [45] present a test case generation approach from UML sequence diagrams and OCL expressions. The input sequence diagram is converted into a tree, which is used to extract the various condition predicates. The approach generates test data from the condition predicates by applying function minimization technique. The generated test cases successfully achieves *message path* and *constraint attribute* coverage of all the objects related to a message in the sequence diagram. The effectiveness of the approach is evaluated on an order ticketing system case study.

Bouquet et al. [36] present a functional testing approach that generate and manage test generated from UML/OCL models. The approach describes a MBT process from the functional requirements till the generation and execution of test case. Currently, three models including class diagram, object diagram, and state machine diagram are used for test generation. The approach describes modeling and automated test generation of a particular functionality of the StarUML project management framework.

Wießleder et al. [46,47] present an automatic test case generation approach for UML state machine, UML class diagram, and OCL expressions to satisfy boundary-based coverage criteria. OCL expressions on operations as pre- and postconditions and guard conditions of state machines are correlated to generate input partitions. From the input partitions, test suites are generated to satisfy boundary coverage criteria. To achieve boundary coverage, multiple coverage criterion including MC/DC is used. The authors use mutation testing to evaluate the effectiveness of the approach for the above coverage criteria and use part of a model of an elevator control example for explanation.

Noikajana et al. [48] present an approach for generating test cases for web services. The contracts are written in web service semantic language (WSDL-S) which is an extension of the web service description language, and OCL is used for describing services rules. From service rules specified in web service descriptions, pre- and postconditions of the various operations are extracted. A pairwise testing technique is used to generate test cases utilizing specific expressions of pre- and postcondition. To evaluate the effectiveness of the approach mutation testing is used on two web service examples. The generated test cases achieve desired coverage and kill all introduced mutants.

Brucker et al. [49] present an extension for specification-based testing. UML and OCL are used to represent specifications formally. In specifications, object graphs are described as states and state transitions are described using class model and state charts. The work uses deductive theorem prover for generating test cases from specifications written in OCL. The approach widely supports OCL language features, including recursive query operations. Equivalence relations are used for representing object identity instead of object graphs. The technique complies with HOL test generation system. The approach is demonstrated on various Linked list examples.

Cheon et al. [50] present an automated testing approach for Java programs that use OCL constraints as test oracles. The approach uses random algorithms for test generation. The work uses OCL constraints as test oracles by translating OCL into run-time checks using AspectJ.

Ascari et al. [51] present a mutation-based testing approach for class testing using OCL specifications and aspect-oriented programming. The paper focuses on specification faults by allowing validation of the implemented code according to the specifications. The approach is illustrated on a case study. A controlled experiment is also conducted to ensure the applicability of the proposed approach.

Jalila et al. [52,53] present an automated fault-based system specification testing framework. The specification is written in OCL. Based on OCL predicate-based fault classes, possible faults in the specification are anticipated and mutated in system under test. For the generation of test cases genetic algorithms with simulated annealing technique are used. To evaluate the generated data, a fitness function based on OCL predicate is defined. The branch coverage adequacy criterion is used for test generation. The experimental results indicate that the proposed methodology provides more coverage with minimum number of generated tests.

Ali et al. [16,54,55] present a search-based test data generation approach by solving OCL constraints. Authors formulate the test data generation approach as a search problem and uses various search heuristics to guide the search to solve OCL constraints. Solving of OCL constraints using search heuristics not only supports OCL collections, but also generates test data efficiently. To evaluate the search heuristics, multiple experiments on artificial and industrial case studies have been conducted for three search algorithms: Genetic Algorithms (GA), Evolutionary Algorithms (1 + 1(EA)), and Alternating variable algorithms (AVM). The empirical evaluations suggest that approach is significantly better than the existing OCL solvers. The approach is automated and fully supported by a tool.

Ali et al. [56] present a test data generation approach by generating boundary values from OCL constraints. The authors extended their previous works [16,54,55] by generating test data at the boundaries of each of the variable in a constraint. The authors discuss various cases involving different types of variable where authors rewrite the constraints for the generation of boundary values. For automated generation of boundary values ESOCL tool is created. Empirical evaluations on industrial case studies are conducted, which show that among the three search algorithms (AVM, (1 + 1) EA, and GA) the performance of AVM in terms of success rate is significantly higher than the other two in finding all the boundary values of constraints in fewer generations.

Chunhui et al. [57] present a use case modeling approach for system tests generation (UMTG) that automatically generates executable system test cases from use case specifications and a domain model. Authors combine natural language processing (NLP) with constraints for automated identification of test scenarios and test inputs. To extract behavioral information from use case specifications by means of NLP, their approach relies on use case specifications expressed in a restricted format. OCL is used to refine guard, pre-, and postconditions which are automatically identified by NLP. An algorithm is designed that builds path conditions that capture the constraints under which alternative flows are executed. The algorithm automatically identifies test inputs by solving such path conditions with the aid of an OCL constraint solver. The approach is demonstrated on industrial case study which shows that UMTG works well with use case specifications for an automotive sensor system.

As presented in Table 4, OCL constraints on the models need to be satisfied during MBT. The approaches in literature vary from converting OCL into some sort of formalism (e.g., VDM) and then applying the existing set of tools for data generation to more sophisticated strategies specifically targeting OCL constructs. A major challenge when converting OCL to the formal approach is state explosion problem, especially when dealing with collections (e.g., OCL Bag). On the other hand, the latter approaches, including search-based test data generation approaches that solve OCL constraints, show very promising results [16,25,52,54,56,58].

3.1.2 Model Verification and Validation

This section summarizes the existing literature on OCL-based verification and validation approaches for UML models. The use of OCL in model verification and validation ensures consistency between various UML models and their specification as per the requirements. Following, we present the OCL-based model validation and verification techniques.

Chiorean et al. [59] show the use of OCL in checking UML model consistency. The paper discusses a tool Object Constraint Language Environment (OCLE) that allows specification of OCL constraints, both at model and meta-model level. The OCLE tool uses the rules specified on the models to check for inconsistencies. The work is demonstrated on two examples, where the OCLE is used to specify rules related to UML. The models for the examples are then validated using OCLE. The work is an earlier work in the area and demonstrates the usefulness of OCL in identifying inconsistencies during modeling.

Richter et al. [60,61] present UML-based Specification Environment (USE) tool to validate models and OCL constraints. The authors formally

Table 4 Summary of Test Generation Approaches From OCL

Paper	Domain	Strategy Used	OCL Coverage	OCL Transformation/ Conversion	Case Study Type Description
Benattou et al. [18]	Test data generation	Partition analysis	Invariants, pre- and postconditions	VDM–DNF	Process scheduler
Salas et al. [43,44]	Test case generation	Partition analysis and constraint solving	Pre- and postconditions	CSP	Triangle example and grant object privilege
Li et al.[45]	Test case generation	Tree function minimization technique	Pre- and postconditions	—	Order ticketing system
Bouquet et al. [36]	Functional testing	Model-based testing	OCL expressions	—	Project management framework
Wießleder et al. [46,47]	Boundary coverage	Mutation testing	Invariants, pre- and postconditions	—	Elevator control
Noikajana et al. [48]	Test case generation	Pairwise testing	OCL expression	—	Rectangle type and increase date services
Brucker et al. [49]	Specification testing	Graph traversing	OCL expression	HOL	Linked list
Cheon et al. [50]	Java program testing	Assertions-random testing	Pre- and postconditions	Aspect J	Loyalty program
Ascari et al. [51]	Unit testing	Mutation-based testing	OCL expression	XMI	Collections

Jalila et al. [52,53]	Specification testing	Search-based testing	OCL expression	Boolean expressions	Patient monitoring and blood bank management system
Ali et al. [16,54,55]	Test data generation	Search-based software testing	OCL expression	—	Video conferencing system
Ali et al. [56]	Boundary values generation	Search-based software testing	OCL expression	Rewriting of constraints	Video conferencing system
Chunhui et al. [57]	Use case modeling for test generation	Natural language processing	OCL expression	Guards, pre- and postconditions	Automotive sensor system

define UML class diagram and provide precise definition of OCL language such that ambiguous interpretation for both the OCL and UML class diagram are avoided. On the basis of formal specifications, the tool was developed that support analysis, simulation, transformation, and validation of UML models with OCL constraints. The developed tool was used to validate well-formedness rules in UML class diagram.

Ziemann et al. [62] discuss an application of USE tool to validate OCL specification of industrial case study of advanced automatic train control system of Bay Area Rapid Transit (BART) system. The specifications of BART are validated by specifying using OCL constraints. The OCL specification expresses train speed, acceleration, and updated positions. The OCL specifications are validated with several test scenarios using USE tool by generating object diagram for a command file comprising of object creation and link creation instructions. The generated object diagram shows violation along with some inconsistencies for a particular constraint specified in OCL.

Richter et al. [63] present an aspect-oriented approach that helps in validating and testing of software program against the constraints specified on UML models. The focus of the approach is on defining monitors that observe the behavior of program and maps it to the model. For defining monitors aspect orientation is used that instruments the program without making any change in the application code. The monitoring approach provides mapping that ensures clear separation of abstraction level between the program behavior and the model behavior. The approach is applied on multiple examples.

Gogolla et al. [64–66] propose an automated snapshot generation-based validation approach for UML and OCL models. A snapshot represents system states that consist of objects having attributes', values, and links. A language ASSL (A Snapshot Sequence Language) is also proposed for constructing snapshots. The USE tool is extended to support ASSL and automated generation of consistent snapshots. The approach is also applied on multiple case studies and is successfully able to generate various snapshots.

Sohr et al. [67] proposed an approach to specify and validate authorization constraints. Authorization constraints are used to express high-level security policies for an organization. The paper focuses on nontemporal and history-based authorization constraints which are specified using OCL. The paper addresses the challenges for identification of conflicting and missing constraints. The paper also demonstrates how role-based access control (RBAC) constraints are validated using the USE tool.

Gogolla et al. [68] describe how USE tool can be utilized for ensuring consistency, independence, and checking of UML and OCL models using test scenarios. A test scenario used in the approach is a test case comprising of UML object or sequence diagram. To achieve consistency, a positive test case is created such that all invariants on model must hold. To achieve independence, a test case keeps UML models small such that no single invariant can be concluded from other stated invariants. For drawing conclusions (consequence) only basic properties are formulated as invariants. OCL plays key role in the approach since it is used for formulating constraints, reducing the test search space, and formulating search space properties.

Kuhlmann et al. [69–71] present a lightweight validation approach that uses model instances and relational logic for performing model validation. The existence or nonexistence of certain properties in model instances is used to draw conclusion about various model properties. The approach translates UML and OCL model to a relational logic and the USE tool is used for SAT solving. The approach uses bidirectional transformation in which UML class diagram and OCL concepts are transformed to and from relational logic. A uniform representation for OCL collection and string operations is also provided that helps in automated model validation [70]. Various OCL collections kinds such as set, bag, ordered set, sequence, and string-based operations are successfully represented in relational form to be used for the purpose of validation. The approach is implemented and applied on several examples.

Gogolla et al. [72] present an approach that performs automatic checking of features in UML and OCL model. The approach relies on USE tool and targets invariant independence. The approach is applied on a model transformation case study. The model transformation used as an example is a transformation from an entity relationship (ER) to relational database schema. The approach uses UML and OCL models for describing system structures. The approach is able to provide proof for the properties like model constraint independence and checking of redundancy absence in invariant sets.

Clavel et al. [73] present a rewriting-based automated UML class diagram validation tool. The tool validates the UML class diagram with respect to OCL constraints. The tool implementation is based on the equational specification of UML and OCL models which are written in Maude. Maude is a reflective programming language that implements membership equational logic and rewriting logic.

Chae et al. [74] propose an OCL-based approach for validating constraints written on class diagrams. The approach detect potential problem by performing analysis on constraint specified at class diagram. The approach is applied on four case studies related to information systems. The classes used in the case studies are classified as boundary classes, entity classes, and control classes to show the diversity. The approach is integrated in a UML modeling tool that generates warning at design time whenever a change made in analysis class.

Rull et al. [75] present a tool called AuRUS that analyze UML/OCL conceptual schemas to ensure their correctness. The tool instantiates a sample conceptual schema representing a particular situation and ensures that the required property in a schema does hold. The tool also provides explanation for unsatisfiability when a property is contradicting.

Czarnecki et al. [76] present an automated model verification technique for product lines. The technique uses feature-based model templates against the well formedness of OCL constraints. The methodology aims to ensure that no nonconforming template instance will be generated from a correct configuration. The OCL constraints are mapped to propositional formulas, which are solved by using SAT solver. The approach is implemented as a plug in to IBM Rational Software Modeler. The methodology ensures that no nonconforming template instance is generated from a correct configuration. OCL constraints are mapped to propositional formulas, which are solved by using SAT solver.

Cabot et al. [23,24,77] present an automated constraint programming methodology to verify UML models extended with OCL and operation contracts. Both the dynamic and static aspect of the system were successfully translated into constraint satisfaction problem (CSP) using UML to CSP tool and formally verified. Their methodology transforms a UML class diagram annotated with OCL constraints into a CSP. The resulting CSP is able to be verified against the original UML/OCL properties.

Demuth et al. [30] present use of Dresden OCL toolkit for model verification. Dresden toolkit is typically extended by the developer and designers to provide OCL support. The applications show how precise semantics are specified on both the model and the meta-models using OCL for verification. Authors have classified and discussed several OCL use cases supported by Dresden including model verification, testing design by contract, model transformation, code generation, etc.

Soeken et al. [78] present a Boolean satisfiability-based approach for verifying UML/OCL models. Authors describe how various states of UML model along with OCL constraint and respective verification task are encoded as SAT instance and automatically solved using off the shelf SAT solver. Furthermore, the authors perform an experiment to evaluate SAT-based UML/OCL model verification. The results show that SAT-based verification is faster when compared with other approaches.

Shaikh et al. [22] present a slicing technique that partitions the model into slices and verifies each slice independently. During slicing the property under verification is preserved. However, the approach is applicable only on class diagram which is annotated with unrestricted OCL constraints and other specific properties required to be verified.

Hilken et al. [20] present the idea of a base model that can be used in automatic verification. The base model represents core elements that can be used to express the important aspects of the system. Authors use transformations for expressing complex language constructs into relatively smaller one. This transformation enables the base model to be used with various verification engines.

Hilken et al. [21] present a comparison of two verification approaches for UML and OCL behavioral model and highlight their advantages and disadvantages. The approaches used for the comparison are filmstrip and unrolling. In filmstrip, UML/OCL descriptions which represent all the behavioral model elements are represented as static description and afterward are checked for interesting properties. In unrolling the dynamic behavior of the UML model is represented as skeleton and verification tasks are directly applied using Satisfiability Modulo Theory (SMT) and solving engines.

Przigoda et al. [79] present a methodology that translates UML behavioral and structural models enriched with OCL constraints into symbolic formulation. The authors also discuss how the symbolic formulation helps to perform various verification tasks such as consistency, reachability, etc. The authors proposed approach is almost fully automatable with limited human intervention. The presented framework used SMT solver to solve the symbolic formulation efficiently.

Anwar et al. [19] present a model verification technique for embedded systems. The authors use model-based software engineering and system verilog assertions (SVA) to handle the design verification aspect of embedded system. They have proposed an extension for system verilog as system verilog OCL (SVOCL) to represent the design verification requirements by means of SVA's. Transformation engine is also developed to ensure automation. The approach is evaluated on multiple case studies.

Fu et al. [80] present an approach that translates OCL invariants into web ontology language (OWL) axioms for inconsistency checking. The approach is evaluated using the TUCO prototype tool. The approach translates wide range of OCL expression into OWL axioms. The authors present the semantics and other preliminaries of OCL for proving correctness of the translation. To prove the correctness, induction on OCL structure is used. The TUCO tool successfully translates the OCL constraints and performed consistency checking on example study.

As presented in Table 5, OCL is extensively used for the verification and validation of systems and models. The USE tool [60] is widely used by researchers [61–66] for validating UML models. There are also numerous applications of model verification such as verifying configurations of the feature model with OCL constraints in product lines [76], design verification of embedded systems using SVA [19], transformation of OCL constraints to CSP [77], slicing of UML models for verification using OCL constraints [22], and translation of UML models into symbolic formulation using SMT solvers [79]. An important benefit of using OCL for verification and validation is tool support for automation. Most of the approaches related to model validation and verification convert OCL to some sort of formalism and a number of these only support a limited subset of OCL.

3.2 Experiences and Empirical Reports on OCL

This section summarizes the works that discuss empirical evaluation of OCL and report experiences of applying OCL in practice.

One of the initial works on empirical analysis of OCL is by Briand et al. [81,82]. The work focuses on evaluating the impact of OCL on UML-based development environment. The paper reports controlled experiment on two case studies. In the experiments final year undergraduate students and industry professionals were asked to develop UML models (UML class diagrams, state machines, and sequence diagrams) containing OCL to specify the system. The experiments were conducted to study model comprehension and maintainability. The results show that OCL was useful in specifying the modeling constraints; however, there was a learning curve when using OCL language for writing constraints specification on UML diagrams.

Correa et al. [83] present a controlled experiment that evaluates the impact of OCL expression structure on understandability of the OCL

Table 5 Summary of Verification and Validation Approaches

Paper	Domain	Strategy Used	OCL Transformation/ Conversion	Case Study Type Description	Tool Used
Chiorean et al. [18,59]	Model checking	Specification-based inconsistencies	—	Order system model, crash course model	OCLE
Richter et al. [60,61]	Model validation	Validation of well-formedness rule	—	Company model	USE
Ziemann et al. [62]	Constraints validation	Specification validation	—	Automated train control system	USE
Richter et al. [63]	Model validation	Aspect-oriented approach	—	Java application	—
Gogolla et al. [64–66]	Model validation	Snapshot generation	ASSL	Boss worker person	USE
Sohr et al. [67]	Constraints validation	Authorization constraints	—	Separation of duty, prerequisite role	USE
Gogolla et al. [68]	Model checking	Test scenarios	—	Model of train and wagon	USE
Kuhlmann et al. [69–71]	Model validation	Model instances and relational logic	Relational logic	UML class to relational model	SAT solver
Gogolla et al. [72]	Model checking	Invariants independence	—	ER to relational database schemata	—
Clavel et al. [73]	Model validation	Equational specification	Maude	Train wagon	—
Chae et al. [74]	Constraint validation	Structural constraint analysis	—	Analysis class model	OCLE

Continued

Table 5 Summary of Verification and Validation Approaches—cont'd

Paper	Domain	Strategy Used	OCL Transformation/ Conversion	Case Study Type Description	Tool Used
Rull et al. [75]	Model analysis	Conceptual schemas	—	Hiking events model	AuRUS
Czarnecki et al. [76]	Model verification	Feature-based models	Prepositional formulas	e-Commerce business model	SAT solver
Cabot et al. [23,24,77]	Model verification	Constraint programming	CSP	Paper researcher class model	UML to CSP
Demuth et al. [30]	Model verification	Dresden OCL toolkit	—	Plugin modeling language	Dresden OCL 2
Soeken et al. [78]	Model verification	Boolean satisfiability	Alloy	Person and company model	MFERT
Shaikh et al. [22]	Model verification	Partition analysis	—	Coach model	—
Hiken et al. [20]	Model verification	Base model	Constraint rewriting	Toll collect system	—
Hiken et al. [21]	Model verification	Filmstrip and unrolling	Symbolic formulation	Dinning philosopher model	SMT solver
Przigoda et al. [79]	Model verification	Symbolic formulation	Symbolic formulation	Traffic light model	SMT solver
Anwar et al. [19]	Model verification	System verilog assertion	SVOCL	Traffic light controller, arbiter, and car collision avoidance system	—
Fu et al. [80]	Inconsistency checking	Translation into OWL axioms	OWL axioms	University model	TUCO

constraints. The controlled experiment is conducted to evaluate the impact of five OCL smells on understandability. The experiment involved 23 professionals who were given OCL constraints with and without smells and were asked to answer a questionnaire evaluating the understandability of the constraints. The results show that the constraints which were refactored were significantly easier to understand.

Pandey et al. [84] highlight the evolution of OCL and discuss various areas of OCL usage. The authors discuss the usage of OCL to overcome the limitations of UML modeling in absence of well-define semantics and provide formal semantics of UML. OCL not only provides required degree of precision to UML models, but it also helps in performing verification and validation of UML models. The paper also highlights limitations of OCL that may have affected its applicability in industry.

Ali et al. [29] present their experiences of applying OCL on six industrial case studies. The case studies belong to diverse industrial domains including communication and control, energy equipment and services, Recycling, and Oil and Gas Production. In these case studies, OCL was applied to solve industrial problems including safety certification and automated product configuration. The result of the industrial case studies shows that a well-selected subset of OCL notations was sufficient for the selected problems for constraint evaluation, solving, and querying. It was observed that OCL constraint specification and enforcement works similarly at meta-levels of MOF. The paper also presents guidelines for practitioners that can help them to choose an appropriate use of OCL at different meta-levels.

In a recent work on empirical analysis of OCL, Yue et al. [85] compare OCL with Java for specifying constraints on models. The paper reports a controlled experiment to evaluate the applicability of OCL in the industrial settings. The control experiment compares OCL with Java and empirically answers the applicability challenges related to extensive training, slow learning curve, and significant effort to use the constraint language. The result shows that both OCL and Java are equally good in terms of learnability, however, OCL is better in modeling complicated constraints. Similarly, the constraints written in Java contain more errors as compared to OCL.

Ali et al. [86] present a controlled experiment to assess the OCL specification in industrial settings. The experiment involves modeling OCL specifications for Cancer Registry of Norway (CRN) by four industry participants and four researchers. On the basis of challenges in writing OCL specification authors have presents an easy to use OCL specification validation and evaluation tool (iOCL).

Table 6 Summary of Experiences and Empirical Reports on OCL

Paper	Domain	Strategy Used	Objective of the Study	Subjects Used
Briand et al. [81,82]	Empirical evaluation	Controlled experiment	Evaluating the impact of OCL on UML-based development environment	Two-case studies
Correa et al. [83]	Empirical evaluation	Controlled experiment	Evaluating the impact of OCL expression structure on understandability	23-professionals
Pandey et al. [84]	Experiences	Study	Areas of OCL usage	—
Ali et al. [29]	Experiences	OCL usage	Experiences of applying OCL on diverse domains	Six-industrial studies
Yue et al. [85]	Empirical evaluation	Controlled experiment	Comparison of OCL and Java for specifying constraints	29-graduate students
Ali et al. [86]	Controlled experiment	OCL modeling	Assessing OCL specification in real setting	Eight participants (four each from industry and researchers)

Though there are some works reporting empirical evaluations of using OCL in practice as shown in Table 6, however, there is still a room for further research in this area. Analysis of using OCL in large projects, issues in maintaining OCL specifications, comparative analysis of various tools supporting OCL, and further reports on successful industrial applications of OCL will be useful for both practitioners and researchers.

3.3 OCL Tools

As discussed earlier, there has been considerable research related to OCL. To make the language useful for the practitioners and researchers, tools supporting OCL are necessary to help the modeler specify, validate, and evaluate the constraints in an easy and user-friendly manner. This section describes in detail various tools that are developed to help the modelers in various stages of OCL specification.

USE tool [60,61] as discussed earlier is a widely used tool for validation of models and OCL constraints. USE allows textual and graphical modeling at both model and meta-model level. The goal of the USE tool is to support definition, validation and run-time checking of OCL invariants, and pre- and postconditions. USE tool allows writing of precise definition of OCL constraints such that ambiguous interpretation for both the OCL and UML diagrams could be avoided. The tool also supports analysis, simulation, transformation, and validation of UML models with OCL constraints.

Dresden OCL toolkit [87] is a mature, widely used set of libraries used by the practitioners to provide OCL support in their tools. As discussed earlier, Dresden OCL library provides support to parse and evaluate OCL constraints on various models including UML and EMF. It can be used as library or as an Eclipse plug-in project. Using Dresden OCL practitioners can specify constraints on both model and meta-model level and can even generate code. Dresden toolkit has been successfully used in model verification, testing design by contract, model transformation, code generation, and other projects [30].

Eclipse OCL [31] is an open-source implementation of the OMG OCL standard for EMF-based models including Ecore and UML on eclipse platform. The core component provides OCL integration by providing API support for parsing and evaluating OCL constraints. It also defines Ecore and UML implementation of OCL abstract syntax model and a visitor API for analyzing and transforming the abstract syntax tree model of OCL expressions. The extensibility API provided by eclipse OCL allows the developer to customize the parsing and evaluation environment of the parser.

OCLE [88] is a tool build to provide full OCL support to both UML model and meta-model. The tool not only provides support for checking well-formedness rules specified in OCL, but it also support compiling and debugging of OCL specifications. It generates Java programming language code to support dynamic checking. A graphical interface is also provided for modelers ease.

Dzidek et al. [89] discuss the lessons learned while developing a dynamic OCL constraint enforcement tool for Java language. The tool ocl2j automatically instruments OCL constraints represented as contracts in Java programs using aspect-oriented programming technique. The instrumentation performed by ocl2j is nonintrusive as aspect orientation helps to separate the assertions from the source code. The technique allows parallel development for Java as long as the class diagram and contract are stable. The tool successfully instrumented OCL constraints on Royal and Loyal case study. The authors also discuss various nontrivial issues related to instrumentation of @pre keyword and oclIsNew() expression that require significant work to address.

Hammad et al. [90] present an interactive tool iOCL, for specifying, validating, and evaluating OCL constraints. The tool helps the practitioners and researchers with the OCL syntax and presents only those details of OCL that are valid at a given step of constraint specification process. The goal of the tool is to reduce modeler effort requires for specifying constraints. Tool iOCL is provided as a web-based tool and is integrated with Eclipse OCL.

EsOCL [16,17] is a search-based constraint solver for OCL specifications. EsOCL uses evolutionary algorithms to generate data satisfying given constraints, based on heuristics defined for OCL constructs. EsOCL has been successfully used to generate test data for industrial scale software systems.

As shown in Table 7, there are some existing open-source and commercial tools such as IBM Rational Software Architect [32] and Papyrus [91] that provide modeling environment for the designers and practitioners to specify constraints on UML models using OCL. Addition to that Magic-Draw [37], Enterprise-Architect [38], and Argo-UML [92] are modeling tools that also provide facility of specifying and validating OCL constraints.

As highlighted by Briand et al. [81,82], there is a learning curve for practitioners in writing OCL constraints. The current set of tools can be significantly improved to ease writing of constraints. The tool presented in Ref. [90] is a step in the direction.

3.4 Extensions to OCL

While OCL has been applied successfully in a range of studies across various domains, the standard language is at times not sufficient to address certain domain-specific requirements, for example, requirement of modeling temporal constraints. Consequently, a number of extensions to the standard OCL have been proposed. The extensions bring new constructs language to OCL and extend its applicability to application domains not covered by the standard. In the following, we discuss some of the well-known extensions to OCL language. These include OCL extension specify temporal constraint OCL extension for temporal, spatial, and specifying actions.

Standard OCL language does not support specifying temporal constraints. A number of domains require such kind of constraint modeling,

Table 7 Summary of OCL Tools

Tool Name	Domain	Interface Support	OCL Support	Usage
USE tool [60,61]	Model validation	Textual and graphical modeling	Writing, validation, and run-time checking of OCL expression	Analysis, simulation, transformation, and validation of UML models
Dresden OCL toolkit [87]	Development library	Eclipse plug-in	Writing OCL constraints	Model verification, testing by contract, model transformation, code generation
Eclipse OCL [31]	Open-source OCL standard	API	Provide programmers classes and libraries	Parsing and evaluating OCL constraint
OCLE [88]	Model checking	Textual and graphical	Compiling and debugging of OCL specification	Static and dynamic checking of models, code generation
OCL2J [89]	OCL assertions	GUI	Instruments OCL in Java programs	Aspect-oriented programming
iOCL [90]	Writing and evaluation	Web-based graphical	Specifying validating and evaluating OCL constraints	Constraint specification
ESOCL [16,17]	OCL solver	Textual	Specifying, solving, and evaluating OCL constraint	Test data generation and constraint solving

specifically when modeling real-time embedded systems or modeling non-functional performance properties of systems. Following we discuss the various extensions to OCL dealing with modeling temporal constraints.

Ziaman et al. [93] extend OCL with elements of finite linear temporal logic and introduces past and future timing concepts as part of TOCL (temporal OCL) language. The semantics of the invariants are formally defined and constraints on the system structure and system behavior are specified as temporal invariants and pre- and postconditions of operations.

Bilal et al. [27] propose an extension to the OCL language to support pattern-based modeling of temporal constraints. A pattern provides the behavior that needs to be modeled along with a scope that limits the execution trace on which the pattern is to be applied. The proposed approach enables modeling of temporal constraints without relying on knowledge of formalisms that are typically found in temporal domains. The behavior to be tested is expressed as a temporal OCL expression which is then translated into regular expressions using formal semantics provided by the language extension. The approach is supported as an extension to Eclipse/MDT OCL plugin.

Another extension to support temporal expressions in OCL is proposed by Dou et al. [26]. The OCLR extension to OCL allows the designer to write temporal constraints based on the property specification patterns. The extension also adds support for referring to a specific event and temporal lapse between events. The syntax of the proposed extension is close to natural language to encourage wider use by industry practitioners. The results of using OCLR on industry case studies show the feasibility of the approach.

Bradfield et al. [94] extend the OCL for modeling temporal properties by providing a template-based extension using mu-calculus. The specifications are written at two levels of logic, with mu-calculus constraints specifying the high-level logic, whereas OCL is being used to specify domain-specific low-level logic.

An extension of OCL to support specification of integrity constraints to control topological relations of objects in spatial databases has been proposed by Pinet et al. [28]. The authors propose formalism based on OCL, referred as the Spatial OCL. While Spatial OCL can be used to specify integrity constraints, it cannot be used to define topological constraints involving objects with vague shapes, a common occurrence in spatial objects. To overcome the limitations of Pinet et al. [28], Bejaoui et al. [95] propose an extension to Spatial OCL. The proposed extension allows specification of constraints on regions with broad boundaries such as forests and lakes.

Robinson et al. [96] present a variant of OCL, OCLTM to represent properties derived from requirements to monitor them at runtime. OCLTM is proposed in the context of a goal-monitoring framework for requirements. The proposed extension supports pattern-based specification of constraints in addition to standard temporal operators and timeouts.

Actions in MDE represent units of fundamental executions that typically correspond to some processing. OCL, by definition, is a side effect-free language. However, researchers have also extended OCL to model actions. Kleppe et al. [97] introduce an extension to OCL to express dynamic requirements such as events or signals that have been received or will be received in future. Such a support is often required during business modeling, for example, to model stimulus and response rules. The authors introduce a new type of clause to OCL, the action clause which gives semantics to a model by presenting its dynamic requirements.

The ECL extension to OCL proposed by Deantoni et al. [98] provide extension to OCL for modeling concurrency and timed behavior. The authors extend OCL to allow explicit specification of events in a model and to support specifying behavior invariants that govern partial ordering of event occurrences in a model. The presented extension semantics are based on Clock Constraints Specification Language (CCSL), which is defined as part of UML profile for Modeling and Analysis of Real-time Embedded Systems (MARTE).

Table 8 summarizes the results of the survey for OCL extensions. As it can be observed, a number of approaches in the literature add new concepts and constructs to OCL. This is done primarily by adding new elements in the OCL meta-model or new constructs in the OCL grammar. A major challenge with such extensions is their alignment with OCL semantics. We did not find any study evaluating this aspect of OCL extensions. A number of such extensions have been proposed to support writing of temporal constraints in OCL. Similarly, researchers have extended OCL to allow specification of dynamic behavior using actions. Given the importance of specifying temporal constraints as highlighted by a number of works in literature, the constructs can be included as part of the standard OCL.

3.5 Other Works

There are a number of other research areas in software engineering where OCL is actively used. Following we discuss the works on formalizing OCL, use of OCL in the area of model transformations, and instance generation of meta-models.

3.5.1 Formalizing OCL

Since OCL is considered as a semiformal constraint specification language, a number of research works have focused on formalization of semantics of OCL. The formalization helps in automation and removes ambiguities in syntax and semantics of OCL.

Rickters et al. [99] highlight early efforts in formalizing the semantics for OCL. The paper describes and defines the syntax and semantics of OCL expressions and provides examples for their evaluations. The authors identify basic syntactic structures and assign semantics to them based on the set theory. The authors also provide solution to implicit flattening of complex

Table 8 Summary of Extension to Object Constraint Language

Paper	Target Feature	Proposed Constructs	Formalism Used	Evaluation Method	Compatibility With Standard OCL
Ziaman et al. [93]	Temporal constraints	Temporal invariants, pre- and postconditions	Liner temporal logic (LTL)	Academic example	No
Bilal et al. [27]	Temporal constraints	Pattern-based language, events	NA	Industry case study	Yes
Dou et al. [26]	Temporal constraints	Pattern-based language, events	NA	Industry case study	Partial
Bradfield et al. [94]	Temporal constraints	After, eventually pattern template	Observational mu-calculus	Academic example	No
Pinet et al. [28]	Spatial constraints for environmental information systems	Disjoint, contains, inside, equal, meet, cover, covered by, overlap	NA	Academic example	No
Bejaoui et al. [95]	Spatial constraints for environmental information systems	Weakly, fairly, strongly, and completely constructs	Region connection calculus	Case study	No
Robinson et al. [96]	Run-time goal monitoring from requirements	Pattern-based constraints, temporal operators, timeouts	NA	Case study	No
Kleppe et al. [97]	Business rule modeling	Actions	NA	Academic example	No
Deantoni et al. [98]	Causal and times relationships in models	Events (clocks)	Clock constraint specification language (CCL)	Academic example	No

results. The paper also discusses various issues resulting in an illegal expression evaluation and suggests various language improvements.

Brucker et al. [100] present a new formal model of OCL by embedding it into Isabelle/HOL language. The work provides automated reasoning support to OCL using a formal calculus to encode OCL semantics. The work also discusses an automated deduction technique based on derived rules from OCL semantics. The proposed methodology is applied on the application of test case generation for the triangle problem.

Roe et al. [101] present a mapping of UML models having OCL constraints to Object Z. The mapping uses the modeling constructs and generates Class skeleton in Object Z. The OCL constraints are also converted to Object Z schema. The authors also discuss the tool developed and the comparison of the proposed mapping with the existing approaches.

Kyas et al. [102] present a prototype tool that analyzes the syntax and semantics of OCL constraints together with UML model and translates them to the Prototype Verification System (PVS) theorem prover language. The UML class diagrams and state machines are translated to PVS. The translation process has successfully been applied on multiple industrial case studies.

Formalization of OCL has been an active area of research in which syntax and semantics of the OCL language are converted into a formal language for better reasoning. Formalization also helps in the extraction of illegal expressions. As presented in Table 9, the literature addresses numerous approaches [24,77,95] that focuses on the conversion of OCL semiformal syntax and

Table 9 Summary of Formalizing OCL Techniques

Paper	OCL Coverage	Strategy Used	Formal Model Used	Addition Support	Case Study Description
Richter et al. [99]	OCL expression	Formal syntax and semantic	NA	Implicit flattening of complex result	Rental system
Brucker et al. [100]	OCL expressions	Formal model	Isabelle/HOL	Automated reasoning and deduction	Triangle problem
Roe et al. [101]	OCL constraints	Mapping	Object Z	Generation of class skeleton	Person and bank accounts
Kyas et al. [102]	OCL constraints	Conversion	PVS theorem prover language	State machine and class diagram only	Sieve of Eratosthenes

semantics into more formal models. Formalizing OCL is still an open research direction to remove ambiguities and allow automated reasoning on OCL constraints. The major challenge in this regard is to introduce formalism without significantly complicating the syntax of the language.

3.5.2 Use of OCL in Model Transformations

Transformations are a fundamental part of MDE that deals with transforming models from one representation to another. Transformations take as input a source model that conforms to a source meta-model as input and generate a target model that conforms to a target meta-model based on a set of rules that map the source meta-model elements to target meta-model elements. The meta-models typically comprise of a large set of elements, which are restricted by a number of constraints usually written in OCL. In addition to constraints at meta-level, OCL is also used as a part of transformation language. In the following, we discuss the use of OCL in the area of model transformation.

Schürr et al. [103] describe the challenges associated with use of OCL as a graph transformation language. The OCL is compared with a graph transformation, PROGRESS, which is a path expression-based language. The PROGRESS language is similar to OCL in many ways, for example, operation manipulation in both works on class diagrams, both languages support collections and single object path expressions. However, in contrast with OCL, PROGRESS path expression supports functional abstraction and offers additional operators for conditions that are required for a graph transformation language.

Cariou et al. [104] present a method to verify transformation results with respect to the transformation specification. The work focuses on transforming contracts written in OCL for the verification of target transformation. A transformation contract represents expected transformation behavior using pre- and postcondition and invariants on transformation rules. A precondition of a transformation contract specifies state of a model before a transformation and postcondition contract specifies the state of a model after transformation. The proposed verification method is generic and can be applied on both manual and automatic transformations.

Cabot et al. [105] present an automated transformation approach from UML to SBVR (semantics of business vocabulary and rules) specification. The approach is able to extract OCL expressions used in the UML. The SBVR specification is used to describe business policies and rules in a language models that is closer to natural language. The proposed transformation aims to reduce the communication gap between modelers and business managers.

Bajwa et al. [106] present rule-based transformation approach to translate SBVR business rules into OCL. The automated transformation from SBVR rules to OCL constraints reduces the manual effort which is complex and time-consuming. The approach is implemented in SBVR2OCL prototype tool. The transformation is based on rules that are derived from the abstract syntax of SBVR and OCL. The approach is fully automated and requires business rules and target UML model as input.

González et al. [107] present a test data generation approach for model transformations by combining the partition and constraint analysis. For the generation of test data, the approach analyzes the OCL constraints in the source meta-model to fine-tune the partitions. The approach provides support of various OCL constructs and offers three different test model generation modes: a simple mode, multiple partition mode, and unique partition mode. A supporting tool is also implemented that can be used with other model generation approaches.

Jilani et al. [108] present a search-based test data generation approach for model transformation testing. The proposed approach uses search-based algorithms for test data generation. The approach is able to solve OCL constraints which are specified at meta-model level. The search algorithms used for generation are based on meta-heuristics (such as adoptive random testing). The approach also provides structural coverage of the transformation code and assures that generated test models achieve desired transformation code coverage. MOTTER tool is developed and successfully applied on Class to Relation database transformation example.

In the area of model transformations, Table 10 shows the approaches that used OCL as a transformation specification language as well as for verification and test generation of model transformations. A number of model transformation languages follow an OCL-like syntax for model query operations, which shows the acceptance of OCL by practitioners.

3.5.3 Instance Generation of Models

The OCL constraints at modeling language definition level (meta-model) typically define the well-formedness rules. These rules define the restrictions that the models developed in the modeling language should follow. A common task during model manipulation is automated generation of models from meta-models (i.e., instances of meta-models). This is typically referred to as instance generation. Following we discuss the

Table 10 Summary of OCL Techniques Used in Model Transformations

Paper	Use of OCL	Strategy Used	Support Provide by OCL	Addition Support	Case Study Description
Schürr et al. [103]	Graph transformation language	Comparison with PROGRESS	Collection, single object path expression, combination of expression	—	—
Cariou et al. [104]	Transformation contract	Verification of target model	Expected behavior pre- and postcondition, invariants on rules	Support both manual and automated transformation	Class to RDBMS transformation
Cabot et al. [105]	Constraints on models	UML2SBVR transformation	OCL expression on models	Business policies and rules	—
Bajwa et al. [106]	Transformation Target model	Mapping of SBVR2OCL	Mapping pf SBVR rules to OCL	Support Constraints on models	SBVR to OCL transformation
González et al. [107]	Source model constraints	Partition and constraint analysis	OCL expressions for test generation	Test generation for model transformation	Team submission
Jilani et al. [108]	Source model constraint and rules in transformation	Constraint solving using search-based algorithms	OCL expressions for transformation testing	Structural coverage of transformation code	Class to RDBMS transformation

works dealing with instance generation of models by resolving OCL constraints on meta-models.

Winkelmann et al. [109] present an instance generation approach. The approach is based on graph grammar and provides translation of restricted OCL constraints into graph grammar constraints. The approach automatically creates a graph grammar from a given meta-model. The grammar ensures typing and cardinality constraints. A restricted set of OCL constraints is also translated into the grammar that is checked during the process of instance generation. The approach is demonstrated on a state chart example.

Francisco et al. [110] present an approach for automated generation of test models following property-based testing. Property-based testing uses properties for specifying constraints written in OCL and generating tests based on it. The approach uses properties to generate test suites. For automated generation of test cases, QuickCheck tool is used. QuickCheck is an automatic property-based testing tool that allows test case generation, execution, and result analysis. The paper also shows advantages and disadvantages of the proposed approach as compared with the manual test suite generation.

Wu et al. [111] present a SMT-based approach for generating coverage-oriented meta-model instances. The approach uses two techniques for generating the instances. The first technique relies on a standard UML class diagram coverage criterion, while the second technique focuses on satisfying graph-based coverage criterion. Both the criteria defined in the techniques are translated to SMT formula, which are solved using an SMT solver.

Since OCL constraints allow specifying restrictions on UML models, these constraints need to be satisfied during instance generation. Table 11 summarizes the existing instance generation approaches [109–111] satisfy the OCL constraints during the instance generation process. Some of the works transform the applied OCL constraints into an intermediate model, e.g., graphs for better verification and generation of the correct instance. The examples of UML class diagram and state charts are typically used in the literature to demonstrate the results. The OCL constraint solving is a fundamental activity in such approaches. Although the challenge of meta-model instance generation from complex meta-models has not been completely resolved, the existing approaches have potential to be extended for this purpose. There is a vital need for a scalable and efficient approach for automated instance generation from models.

Table 11 Summary of Instance Generation Techniques

Paper	Use of OCL	Strategy Used	Addition Support	Case Study Description	Tool Support
Winkelmann et al. [109]	Transformation of OCL into graph grammar	Graph grammar	Generation of graph models	State chart model	—
Francisco et al. [110]	Properties written in OCL	Property-based testing	Automated generation of test suites	Process scheduler class model	QuickCheck
Wu et al. [111]	Constraints on model	SMT-based approach	UML class diagram coverage criteria and coverage-oriented meta-model instances	Department manager meta-model	SMT solver

3.5.4 Refactoring and Simplifying OCL Constraints

OCL has been introduced with the intention of being the standard language for specifying constraints on models. Ease of use and understandability are important factors in acceptance of the language by practitioners. While OCL aims to be an easier to use language than the other formalisms, at times OCL expressions and constraints can be quite complex and difficult to understand [112]. Consequently, there have been a number of efforts aimed at making OCL simpler to understand and apply.

Correa et al. [112] proposed a number of refactoring to improve understandability of OCL specifications. The work discusses a number of "OCL smells," similar to code smells and propose corresponding refactoring. An experimental study is presented to evaluate the impact of OCL smells and refactoring on the understandability of OCL expressions. The results of the study show that the presence of OCL smells has a negative impact on the understandability of OCL expressions. OCL specifications containing OCL smells take longer to understand and are less correct than specifications without them. The work provides useful guidelines for writing OCL constraints in more readable way.

Opoka et al. [113] present a set of OCL libraries, with the aim of simplifying the writing of OCL expressions. The libraries provide a set of useful and reusable OCL expressions. The developed libraries have been successfully used in sample projects demonstrating the feasibility of the approach. The libraries provide reusability, validation support, and documentation support.

Giese et al. [114] discuss the problem of redundancies in OCL specification. Such redundancies reduce the readability of the OCL specifications. The paper presents an approach to transform OCL expressions to simpler forms in the context of constraints expressing requirements. A prototype OCL simplifier is implemented to automate the simplification process.

Aspect orientation has emerged as widely used practice to avoid redundancies in software artifacts. Aspects allow the designers to separate core design elements from various crosscutting behaviors. The crosscutting behaviors are modeled separately in the form of aspects. Khan et al. [115] present an extension to OCL that bring the benefits of aspect orientation to constraint specification. The proposed approach allows the designer to specify core constraints separately from crosscutting constraints. The crosscutting constraints are written separately as constraint aspects, which are then woven in the core constraints using as aspect weaver. One benefit of the approach is that while at design time the crosscutting aspect is written

Table 12 Summary of Refactoring Techniques

Paper	Technique	Evaluation	Case Study Description
Correa et al. [112]	OCL smells	Experimental study	Bank system
Opoka et al. [113]	Extension of OCL	—	—
Giese et al. [114]	Transformation	Case study	Observer design pattern
Khan et al. [115]	Aspect orientation	Case study	EU-rental study

separately, the woven output is a standard OCL specification. This allows the approach to remain compatible with the existing tools and API developed for OCL. The results show that AspectOCL language can be effectively used to capture constraints in industry-scale case studies and can significantly reduce the constraint modeling effort.

Over the years, researchers and practitioners have faced challenges in applying OCL to particular domains due to a lack of constructs in the standard language. To overcome such challenges, new constructs have been added in the form of language extensions. Table 12 summarizes the refactoring techniques presented earlier. The works discussing code smells and libraries can be useful in making OCL more useable and easy to comprehend. However, significant empirical evaluations are further required to evaluate their effectiveness.

4. INSIGHTS ON ADVANCES IN OCL

OCL has been used in a number of industrial and academic projects. A large body of works in literature focuses on software verification and validation. The works range from test generation to model verification and validation using OCL.

Test generation from OCL specifications is significant for MBT. Recent trend in the area is the use of search-based strategies and constraint satisfaction for test generation. The search-based strategies have shown promising results with the works showing the scalability of the approach to industrial applications. The existing constraint satisfaction approaches either do not support complete OCL specification or they only provide a limited support (few constructs of OCL are supported). Even with a limited OCL support, the constraint solving approaches (based on SAT/SMT) require translation of OCL construct into intermediate languages, which make the process more complicated. Similarly, in CSP-based approaches, the OCL constructs are first translated to form a CSP problem and are then solved. The existing CSP solvers do not support a number of OCL constructs, such as collections, association, and three-order logic, which are commonly used in industry [16].

One benefit of the model verification through OCL is that it is automatable, and is well supported by UML modeling tools. Similar to verification, OCL has also been applied to validate models for their correct definition. Tools such as USE [60] are used by a large number of works [61,62] for validation of UML models using OCL. Researchers have extended the USE tool as per their requirements for model validation. For example, USE tool is extended for the generation of consistent snapshots [64,65], aspect-oriented-based constraint validation [63], and for the integration of role-based access control constraints [67]. The support provides by the USE tool helps the designers to achieve automated model validation using various techniques.

The OCL model validation has benefited the larger software industry and implementation of large systems such as advanced automated train control systems [62]. OCL is also used to perform analysis on UML class diagram through structural constraints [74], validating UML models transformed to relation logic and vice versa through equational specifications written in Maude [69–71], and AuRUS tool that analyzes UML/OCL conceptual schemas to ensure their correctness [75]. Model checking is another important aspect for which OCL has been used. The existing literature presents the use of model checking for various purposes, such as checking UML model consistency and identifying inconsistencies through OCLE tool [59], using USE tool [60] for ensuring consistency and independence, automated checking of UML and OCL model features in model transformation [72], and translation of OCL invariants into OWL axioms for inconsistency checking [80].

Though there are a few controlled experiments, empirical evaluations, and surveys conducted in the area, there is a significant room for further empirical investigation. The presented studies focus on writing specification and evaluating their impact on model comprehension, comparative analysis, maintainability, and applicability. Further empirical evaluations are required to evaluate use of OCL on large projects, scalable approaches for OCL constraint solving and instance generation from complex meta-models, and comparative tool analysis for OCL.

Even though there exists many commercial tools, such as Software Architect [32], Papyrus [91], Magic-Draw [37], Enterprise-Architect [38], Argo-UML [92], and Eclipse OCL [31] that help designers and practitioners to specify, validate, and evaluate OCL constraints, but still practitioner find it challenging to specify OCL constraints. Recently, a tool iOCL [90] was developed that guides the practitioners in writing constraints interactively. However, there is still room for improvement in tools that further ease the process of writing OCL specification and helps the modelers in validating and evaluating OCL constraints.

Another research area in the domain is the formalization of OCL as a language. A number of approaches in the literature focus on defining formal notations and removing semantic ambiguities of OCL. Formalization of OCL syntax into set theory, formal calculus, Object Z, CSP, PVS theorems provides formal reasoning and help in OCL constraints solving. But the problems in all such formalism, including partial OCL language support, computation expensive operation due to combinatorial explosion still require further investigation.

OCL has seen acceptance in industry and there have been some reports on successful usage of OCL in industrial and academic projects [29]. However, over the years, researchers have identified some limitations of the existing OCL standard. Often these limitations relate to a lack of support for constructs required to accomplish domain-specific tasks. Real-time and time-critical applications represent one such domain. One requirement of the domain is to model temporal constraints. The standard OCL specification does not provide constructs that can specify temporal properties and constraints. A number of researchers have addressed the problem and provided their own language extensions that allow designers to specify temporal constraints on system models. Similarly, other researchers have sought to extend the OCL language to incorporate new emerging practices, such as aspect orientation. The aspect-oriented extensions to OCL aim to bring the benefits of aspect orientation to constraint specification by allowing the designers to separate core constraints from crosscutting constraints. One fundamental challenge to wider adaption of such extensions, however, is the requirement of learning new nonstandard constructs on top of basic OCL constructs. This is further complicated by lack of industry-scale tool support for such extensions. Consequently, while the extensions may be successfully applied in academic studies and in the scope of the project where they are developed, wide scale adaption remains a challenge. On the other hand, extensions where the final output is always standard OCL [115] seems more feasible.

5. CONCLUSION AND FUTURE DIRECTIONS

MDE has seen a gradual increase in industry adaption since its inception. OCL is an OMG standard language for specifying constraints at model and meta-model levels.

With the increase in industrial usage of MDE, OCL as standard constraint language has also seen a rapid increase in use at various modeling levels. From the earlier definition of OCL as a part of UML, the OCL is now a standalone standard and has its applications beyond UML. For instance, OCL is being widely used for specifying language rules at meta-level while defining domain-specific languages and language profiles. The UML profile for MARTE uses the OCL constraint to specify language semantic rules. Similarly, the language rules Generic Modeling Environment (GME) are defined using OCL. Apart from its use at meta-level, it is also used as a constraint specification language for domain-specific languages and as a model query language. OCL has also received significant attraction from researchers. In this chapter, we reported a survey of various works in literature targeting OCL. The works were broadly classified into five categories: (i) verification and validation using OCL, (ii) experiences and empirical reports on OCL, (iii) OCL tools, (iv) OCL extensions, and (v) other works applying OCL in diverse areas. Among these areas verification and validation, using OCL has the largest number of research works. The tool support is there for basic tasks; however, there is a significant room for improvement in terms of their usability and scalability. However, there are certain work that report empirical studies on OCL, especially on evaluation of usability of OCL. However, this area still requires further studies on evaluating various aspects of OCL usage. Similarly, more industry experience reports on OCL will further support the industrial adoption of the language.

OCL has also recently seen an increased support by a number of industrial tools. The modern OCL tools provide sophisticated features for querying OCL models and evaluating constraints on the provided models. Additionally, the modern modeling tools, such as IBM Rational Software Architect, allow the engineers to specify OCL constraints and support model validation based on these constraints. These tools also allow writing constraints at meta-level (for example, while defining a profile) and enforcing the constraints at instance level. The support of tools has made writing and maintaining constraint specifications relatively less complicated than earlier. With the increase in tool support, OCL along with MDE has now a better chance to be used in more domains. For instance, we are already seeing the use of the OCL in domains other than software modeling (e.g., modeling business rules and modeling internet of things). However, the tool support of OCL still requires significant improvements. For instance, the current debugging support for OCL is very limited and is a major obstacle in large-scale adaption of OCL. This is a promising future research and development direction for the language. A provision of tool support with efficient debuggers for OCL will significantly increase the adoption of the language. Similarly, the provision of scalable OCL solvers and evaluators will also play a significant part in language adoption by industry.

Another criticism on OCL has been its complexity and the initial learning curve involved in learning and using the language constructs. There have been some attempts at proposing easier to use languages, including visual constraint modeling languages, to replace OCL, however, despite these attempts, OCL remains the only standard language for modeling constraints. Other formalisms for constraint specification purposes tend to be even more complex and difficult to comprehend for engineers who are not familiar with formal languages. OCL for now remains the simpler and often more practical alternative for constraint specification purposes.

The key challenge while writing constraints lies not only with complexity of OCL but also with complexity of constraint specification process itself. Identifying and writing constraints on models is a challenging task and requires thorough understanding of the domain and requirements, which is typically lacking.

Another evidence of increasing acceptance of OCL as a language is the number of languages that have been inspired by the syntax of OCL. A large number of model transformation languages use an OCL-like syntax, it allows easy mechanism for model query and manipulation as compared to programming and formal languages. For example, ATL has syntax that is very similar to OCL.

In our experience of using OCL for various industrial problems and our study of literature, we identify that during constraint specification only a small subset of OCL is required. A large part of OCL typically remains unused. For a novice engineer identifying the subset of OCL that will be used and focusing on learning that portion is not trivial. For instance, UML already allows selective usage of the language through language profiles. Provision of a similar mechanism in the language to define usage specific subset of OCL can greatly help in reducing the learning curve. Similarly, a number of researchers have proposed temporal extensions to OCL. Given the applicability of MDE in real-time embedded systems domain, perhaps a useful addition will be to provide a standard support for modeling time-related constraints in OCL. Such an extension will allow increased used in more domains and help improve the industry acceptance of the language. These can be interesting improvements in OCL as a language.

Overall, due to wide usage and support of modern tools, OCL as a language is poised to have wide spread acceptance in industry. The language has been applied successfully on various projects across multiple domains for different purposes. The purposes range from constraint specification at model level, language rule definition at meta-model level to model query and manipulation. None of the other proposed constraint specification language or formal languages enjoy other wide support from industry or standard organizations. Though there are some limitations in terms of tool support, the OCL is a pivotal part of MDE and will continue its growth with greater industry acceptance.

REFERENCES

[1] F. Terrier, S. Gérard, MDE Benefits for distributed, real time and embedded systems, in: B. Kleinjohann, L. Kleinjohann, R. Machado, C. Pereira, P.S. Thiagarajan (Eds.), From Model-Driven Design to Resource Management for Distributed Embedded Systems, vol. 225, Springer, USA, 2006, pp. 15–24.
[2] OMG. (2015). Unified Modeling Language Superstructure, Version 2.5, Object Management Group Available: http://www.omg.org/spec/UML/2.5.
[3] S. Friedenthal, A. Moore, R. Steiner, A Practical Guide to SysML: The Systems Modeling Language, Elsevier, 2008.
[4] C. Larman, Applying UML and Patterns: An Introduction to Object-Oriented Analysis and Design and Iterative Development, third ed., Prentice Hall, 2004.
[5] M.Z. Iqbal, A. Arcuri, L. Briand, Environment modeling with UML/MARTE to support black-box system testing for real-time embedded systems: methodology and industrial case studies, in: D. Petriu, N. Rouquette, Ø. Haugen (Eds.), Model Driven Engineering Languages and Systems, vol. 6394, Springer Berlin/Heidelberg, 2010, pp. 286–300.
[6] B. Douglass, Real-Time UML: Developing Efficient Objects for Embedded Systems, Addison-Wesley Longman Publishing Co., Inc., Boston, MA, USA, 1997.
[7] T. Yue, L. Briand, B. Selic, Q. Gan, Experiences with Model-based Product Line Engineering for Developing a Family of Integrated Control Systems: an Industrial Case Study, Simula Research Laboratory, Technical Report(2012–06), 2012.
[8] M. Usman, M.Z. Iqbal, M.U. Khan, A product-line model-driven engineering approach for generating feature-based mobile applications, J. Syst. Softw. vol. 123, (2017) 1–32.
[9] M. Usman, M.Z. Iqbal, M.U. Khan, in: A model-driven approach to generate mobile applications for multiple platforms, Software Engineering Conference (APSEC), 2014 21st Asia-Pacific, 2014, pp. 111–118.
[10] M.C. Jeruchim, P. Balaban, K.S. Shanmugan, Simulation of Communication Systems: Modeling, Methodology and Techniques, Springer Science & Business Media, 2006.
[11] S. Iftikhar, M.Z. Iqbal, M.U. Khan, W. Mahmood, in: An automated model based testing approach for platform games, Model Driven Engineering Languages and Systems (MODELS), ACM/IEEE 18th International Conference on 2015, 2015, pp. 426–435.

[12] B. Selic, The pragmatics of model-driven development, IEEE Softw. 20 (2003) 19–25.

[13] R. Binder, Testing Object-Oriented Systems: Models, Patterns, and Tools, Addison-Wesley Professional, 2000.

[14] OMG, Object Constraint Language Specification, Version 2.4, Object Management Group, 2016. http://www.omg.org/spec/OCL/2.4/.

[15] J. Warmer, A. Kleppe, The Object Constraint Language, Addison–Wesely, Reading, Mass., & Co, 1999.

[16] S. Ali, M. Iqbal, A. Arcuri, L. Briand, Generating Test Data from OCL Constraints with Search Techniques, IEEE Trans. Softw. Eng. 39 (2013) 26.

[17] S. Ali, M.Z. Iqbal, A. Arcuri, L. Briand, Solving OCL Constraints for Test Data Generation in Industrial Systems with Search Techniques, IEEE Trans. Softw. Eng. **39** (10) (2013) 1376–1402.

[18] M. Benattou, J.-M. Bruel, N. Hameurlain, in: Generating test data from OCL specification, Proc. ECOOP Workshop Integration and Transformation of UML Models, 2002.

[19] M.W. Anwar, M. Rashid, F. Azam, M. Kashif, Model-based design verification for embedded systems through SVOCL: an OCL extension for SystemVerilog, Des. Autom. Embed. Syst. 21 (2017) 1–36.

[20] F. Hilken, P. Niemann, R. Wille, M. Gogolla, Towards a base model for UML and OCL verification, in: Proceedings of the 11th Workshop on Model-Driven Engineering, Verification and Validation co-located with 17th International Conference on Model Driven Engineering Languages and Systems, MoDeVVa@MODELS 2014, Valencia, Spain, September 30, 2014. CEUR Workshop Proceedings 1235, CEUR-WS.org, 2014, pp. 59–68.

[21] F. Hilken, P. Niemann, M. Gogolla, R. Wille, Filmstripping and unrolling: a comparison of verification approaches for UML and OCL behavioral models, International Conference on Tests and Proofs, 2014, pp. 99–116.

[22] A. Shaikh, R. Clarisó, U.K. Wiil, N. Memon, in: Verification-driven slicing of UML/OCL models, Proceedings of the IEEE/ACM international conference on Automated software engineering, 2010, pp. 185–194.

[23] J. Cabot, R. Clarisó, D. Riera, in: Verification of UML/OCL class diagrams using constraint programming, Software Testing Verification and Validation Workshop, ICSTW'08. IEEE International Conference on 2008, 2008, pp. 73–80.

[24] J. Cabot, R. Clarisó, D. Riera, in: UMLtoCSP: a tool for the formal verification of UML/OCL models using constraint programming, Proceedings of the Twenty-Second IEEE/ACM International Conference on Automated Software Engineering, 2007, pp. 547–548.

[25] S. Ali, M.Z. Iqbal, A. Arcuri, in: Improved heuristics for solving OCL constraints using search algorithms, Proceeding of the Sixteen Annual Conference Companion on Genetic and Evolutionary Computation Conference Companion (GECCO), Vancouver,BC, Canada, 2014.

[26] W. Dou, D. Bianculli, L. Briand, in: OCLR: a more expressive, pattern-based temporal extension of OCL, European Conference on Modelling Foundations and Applications, 2014, pp. 51–66.

[27] B. Kanso, S. Taha, in: Temporal constraint support for OCL, International Conference on Software Language Engineering, 2012, pp. 83–103.

[28] F. Pinet, M. Duboisset, V. Soulignac, Using UML and OCL to maintain the consistency of spatial data in environmental information systems, Environ. Model. Software 22 (2007) 1217–1220.

[29] S. Ali, T. Yue, M.Z. Iqbal, R.K. Panesar-Walawege, in: Insights on the use of OCL in diverse industrial applications, International Conference on System Analysis and Modeling, 2014, pp. 223–238.

[30] B. Demuth, C. Wilke, in: Model and object verification by using dresden OCL, Proceedings of the Russian-German Workshop Innovation Information Technologies: Theory and Practice, Ufa, Russia, 2009, pp. 687–690.
[31] O. Eclipse, Project Team, Eclipse OCL Project. Eclipse Community, 2005. https://projects.eclipse.org/projects/modeling.mdt.ocl.
[32] IBM. Rational Software Architect. 2017, Available: https://www.ibm.com/developerworks/downloads/r/architect/.
[33] M. Richters, M. Gogolla, OCL: syntax, semantics, and tools, in: Object Modeling With the OCL, Springer, 2002, pp. 42–68.
[34] SmartState, SmartState-UML Statemachine Code Generation Tool, Available: http://www.smartstatestudio.com/, 2011.
[35] M.Z. Iqbal, A. Arcuri, L. Briand, Code Generation From UML/MARTE/OCL Environment Models to Support Automated System Testing of Real-Time Embedded Software, Simula Research Laboratory, Technical Report (2011-01), 2011.
[36] F. Bouquet, C. Grandpierre, B. Legeard, F. Peureux, in: A test generation solution to automate software testing, Proceedings of the 3rd International Workshop on Automation of Software Test, 2008, pp. 45–48.
[37] MagicDraw. 2017, Available: https://www.nomagic.com/products/magicdraw.
[38] Enterprise Architect. 2017, Available: http://sparxsystems.com/products/ea/.
[39] T.C.J. Warmer, Object Modeling With the OCL, 2002).
[40] J. Cabot, M. Gogolla, Object constraint language (OCL): a definitive guide, in: M. Bernardo, V. Cortellessa, A. Pierantonio (Eds.), Formal Methods for Model-Driven Engineering, Lecture Notes in Computer Science, **7320**, Springer, Berlin, Heidelberg, 2012, pp. 58–90.
[41] ATLAS Group, ATL: Atlas Transformation Language, 2007. http://www.eclipse.org/atl/.
[42] J. Oldevik, MOFScript User Guide, Version 0.6 (MOFScript v 1.1.11), 2006. http://www.uio.no/studier/emner/matnat/ifi/INF5120/v05/undervisningsmateriale/MOFScript-User-Guide.pdf.
[43] P.A.P. Salas, B.K. Aichernig, Automatic Test Case Generation for OCL: A Mutation Approach, UNU-IIST Report, 2005.
[44] B.K. Aichernig, P.A.P. Salas, in: Test case generation by OCL mutation and constraint solving, Fifth International Conference on Quality Software, 2005.(QSIC 2005), 2005, pp. 64–71.
[45] B.-L. Li, Z.-s. Li, L. Qing, Y.-H. Chen, in: Test case automate generation from UML sequence diagram and OCL expression, 2007 International Conference on Computational Intelligence and Security, 2007, pp. 1048–1052.
[46] S. Weißleder, D. Sokenou, in: Automatic test case generation from UML models and OCL expressions, Software Engineering (Workshops), 2008, pp. 423–426.
[47] S. Weißleder, B.-H. Schlingloff, in: Quality of automatically generated test cases based on OCL expressions, 2008 1st International Conference on Software Testing, Verification, and Validation, 2008, pp. 517–520.
[48] S. Noikajana, T. Suwannasart, in: An improved test case generation method for web service testing from WSDL-S and OCL with pair-wise testing technique, 33rd Annual IEEE International Computer Software and Applications Conference, 2009. COMPSAC'09, 2009, pp. 115–123.
[49] A.D. Brucker, M.P. Krieger, D. Longuet, B. Wolff, in: A specification-based test case generation method for UML/OCL, International Conference on Model Driven Engineering Languages and Systems, 2010, pp. 334–348.
[50] Y. Cheon, C. Avila, in: Automating Java program testing using OCL and AspectJ, 2010 Seventh International Conference on Information Technology: New Generations (ITNG), 2010, pp. 1020–1025.

[51] L.C. Ascari, S.R. Vergilio, in: Mutation testing based on OCL specifications and aspect oriented programming, 2010 XXIX International Conference of the Chilean Computer Science Society (SCCC), 2010, pp. 43–50.
[52] A. Jalila, D.J. Mala, M. Eswaran, Functional testing using OCL predicates to improve software quality, IJSSOE 5 (2015) 56–72.
[53] A. Jalila, D. Mala, M. Eswaran, Early identification of software defects using OCL predicates to improve software quality, J. Eng. Sci. Technol. 10 (2015) 307–321.
[54] S. Ali, M.Z. Iqbal, A. Arcuri, Empirically Evaluating Improved Heuristics for Test Data Generation From OCL Constraints Using Search Algorithms, Simula Research Laboratory 2012, 2012.
[55] S. Ali, M.Z. Iqbal, A. Arcuri, L. Briand, in: A search-based OCL constraint solver for model-based test data generation, 2011 11th International Conference on Quality Software (QSIC), 2011, pp. 41–50.
[56] S. Ali, T. Yue, X. Qiu, H. Lu, in: Generating boundary values from OCL constraints using constraints rewriting and search algorithms, 2016 IEEE Congress on Evolutionary Computation (CEC), 2016, pp. 379–386.
[57] C. Wang, F. Pastore, A. Goknil, L. Briand, Z. Iqbal, in: Automatic generation of system test cases from use case specifications, Proceedings of the 2015 International Symposium on Software Testing and Analysis, 2015, pp. 385–396.
[58] S. Ali, M.Z. Iqbal, M. Khalid, A. Arcuri, Improving the performance of OCL constraint solving with novel heuristics for logical operations: a search-based approach, Empir. Softw. Eng. 21 (6) (2016) 2459. https://doi.org/10.1007/s10664-015-9392-6.
[59] D. Chiorean, M. Paşca, A. Cârcu, C. Botiza, S. Moldovan, Ensuring UML models consistency using the OCL environment, Electron. Notes Theor. Comput. Sci. 102 (2004) 99–110.
[60] M. Richters, A Precise Approach to Validating UML Models and OCL Constraints, Citeseer, 2002.
[61] M. Richters, M. Gogolla, in: Validating UML models and OCL constraints, International Conference on the Unified Modeling Language, 2000, pp. 265–277.
[62] P. Ziemann, M. Gogolla, Validating OCL specifications with the use tool: an example based on the bart case study, Electron. Notes Theor. Comput. Sci. 80 (2003) 157–169.
[63] M. Richters, M. Gogolla, in: Aspect-oriented monitoring of UML and OCL constraints, AOSD Modeling With UML Workshop, 6th International Conference on the Unified Modeling Language (UML), San Francisco, USA, 2003.
[64] M. Gogolla, J. Bohling, M. Richters, Validating UML and OCL models in USE by automatic snapshot generation, Softw. Syst. Model. 4 (2005) 386–398.
[65] M. Gogolla, J. Bohling, M. Richters, Validation of UML and OCL models by automatic snapshot generation, International Conference on the Unified Modeling Language, 2003, pp. 265–279.
[66] M. Gogolla, F. Büttner, M. Richters, USE: a UML-based specification environment for validating UML and OCL, Sci. Comput. Program. 69 (2007) 27–34.
[67] K. Sohr, G.-J. Ahn, M. Gogolla, L. Migge, in: Specification and validation of authorisation constraints using UML and OCL, European Symposium on Research in Computer Security, 2005, pp. 64–79.
[68] M. Gogolla, M. Kuhlmann, L. Hamann, in: Consistency, independence and consequences in UML and OCL models, International Conference on Tests and Proofs, 2009, pp. 90–104.
[69] M. Kuhlmann, M. Gogolla, in: From UML and OCL to relational logic and back, International Conference on Model Driven Engineering Languages and Systems, 2012, pp. 415–431.
[70] M. Kuhlmann, M. Gogolla, in: Strengthening SAT-based validation of UML/OCL models by representing collections as relations, European Conference on Modelling Foundations and Applications, 2012, pp. 32–48.

[71] M. Kuhlmann, L. Hamann, M. Gogolla, in: Extensive validation of OCL models by integrating SAT solving into USE, International Conference on Modelling Techniques and Tools for Computer Performance Evaluation, 2011, pp. 290–306.
[72] M. Gogolla, F. Hilken, UML and OCL transformation model analysis: checking invariant independence, in: VOLT@ STAF, 2015, pp. 20–27. http://volt2015.big.tuwien.ac.at/data/submissions/paper_3.pdf.
[73] M. Clavel, M. Egea, in: ITP/OCL: a rewriting-based validation tool for UML + OCL static class diagrams, International Conference on Algebraic Methodology and Software Technology, 2006, pp. 368–373.
[74] H.S. Chae, K. Yeom, T.Y. Kim, Specifying and validating structural constraints of analysis class models using OCL, Inf. Softw. Technol. 50 (2008) 436–448.
[75] G. Rull, C. Farré, A. Queralt, E. Teniente, T. Urpí, AuRUS: explaining the validation of UML/OCL conceptual schemas, Softw. Syst. Model. 14 (2015) 953–980.
[76] K. Czarnecki, K. Pietroszek, in: Verifying feature based model templates against well-formedness OCL constraints, Proceedings of the 5th International Conference on Generative Programming and Component Engineering, 2006, pp. 211–220.
[77] J. Cabot, R. Clarisó, D. Riera, in: Verifying UML/OCL operation contracts, International Conference on Integrated Formal Methods, 2009, pp. 40–55.
[78] S. Flake, Enhancing the message concept of the object constraint language, in: SEKE, 2004, pp. 161–166. http://www.vldbarc.org/dblp/db/conf/seke/seke2004.html.
[79] N. Przigoda, M. Soeken, R. Wille, R. Drechsler, Verifying the structure and behavior in UML/OCL models using satisfiability solvers, IET Cyber Phys. Syst. Theor. Appl. 1 (2016) 49–59.
[80] C. Fu, D. Yang, X. Zhang, H. Hu, An approach to translating OCL invariants into OWL 2 DL axioms for checking inconsistency, Autom. Softw. Eng. 24 (2) (2017) 295. https://doi.org/10.1007/s10515-017-0210-9.
[81] L.C. Briand, Y. Labiche, H. Yan, M. Di Penta, in: A controlled experiment on the impact of the object constraint language in UML-based maintenance, Proceedings 20th IEEE International Conference on Software Maintenance, 2004, 2004, pp. 380–389.
[82] L.C. Briand, Y. Labiche, M. Di Penta, H. Yan-Bondoc, An experimental investigation of formality in UML-based development, IEEE Trans. Softw. Eng. 31 (2005) 833–849.
[83] A. Correa, C. Werner, M. Barros, in: An empirical study of the impact of OCL smells and refactorings on the understandability of OCL specifications, International Conference on Model Driven Engineering Languages and Systems, 2007, pp. 76–90.
[84] R. Pandey, Object constraint language (OCL): past, present and future, ACM SIGSOFT Softw. Eng. Notes 36 (2011) 1–4.
[85] T. Yue, S. Ali, Empirically evaluating OCL and Java for specifying constraints on UML models, Softw. Syst. Model. 15 (2016) 757–781.
[86] S. Ali, M. Hammad, H. Lu, J.F. Nygård, S. Wang, T. Yue, A Pilot Experiment to Assess Interactive OCL Specification in a Real Setting, Technical report (2017–01). Available at: https://www.simula.no/publications/pilot-experiment-assess-interactive-ocl-specification-real-setting.
[87] DresdenOCL, DresdenOCLToolkit, 2005.
[88] OCLE, Object Constraint Language Environment, Available, http://lci.cs.ubbcluj.ro/ocle/overview.htm.
[89] W.J. Dzidek, L.C. Briand, Y. Labiche, Lessons learned from developing a dynamic OCL constraint enforcement tool for Java, International Conference on Model Driven Engineering Languages and Systems, 2005, pp. 10–19.
[90] M. Hammad, T. Yue, S. Ali, S. Wang, iOCL: an interactive tool for specifying, validating and evaluating OCL constraints. In: ACM/IEEE 19th International Conference on Model Driven Engineering Languages and Systems (MODELS) Tool Demonstration Track, 2016, pp. 24–31.
[91] Papyrus Modeling Environment, 2017. Available: http://www.eclipse.org/papyrus/.

[92] L. Argoum, ArgoUML Tool, ArgoUML, 2017. Available at: http://argouml.tigris.org/.
[93] P. Ziemann, M. Gogolla, An extension of OCL with temporal logic, Critical Systems Development With UML–Proceedings of the UML, 2002, pp. 53–62.
[94] J. Bradfield, J.K. Filipe, P. Stevens, Enriching OCL using observational mu-calculus, International Conference on Fundamental Approaches to Software Engineering, 2002, pp. 203–217.
[95] L. Bejaoui, F. Pinet, M. Schneider, Y. Bédard, OCL for formal modelling of topological constraints involving regions with broad boundaries, GeoInformatica 14 (2010) 353–378.
[96] W. Robinson, Extended OCL for goal monitoring, Electronic Comm. EASST 9 (2007) 1–12.
[97] A. Kleppe, J. Warmer, in: Extending OCL to include actions, International Conference on the Unified Modeling Language, 2000, pp. 440–450.
[98] J. Deantoni, F. Mallet, ECL: The Event Constraint Language, an Extension of OCL With Events, INRIA, 2012.
[99] M. Richters, M. Gogolla, in: On formalizing the UML object constraint language OCL, International Conference on Conceptual Modeling, 1998, pp. 449–464.
[100] A.D. Brucker, B. Wolff, in: A proposal for a formal OCL semantics in Isabelle/HOL, International Conference on Theorem Proving in Higher Order Logics, 2002, pp. 99–114.
[101] D. Roe, K. Broda, A. Russo, Mapping UML Models Incorporating OCL Constraints Into Object-Z, Imperial College of Science, Technology and Medicine, Department of Computing, 2003.
[102] M. Kyas, H. Fecher, F.S. De Boer, J. Jacob, J. Hooman, M. Van Der Zwaag, et al., Formalizing UML models and OCL constraints in PVS, Electron. Notes Theor. Comput. Sci. 115 (2005) 39–47.
[103] A. Schürr, Adding graph transformation concepts to UML's constraint language OCL, Electron. Notes Theor. Comput. Sci. 44 (2001) 93–106.
[104] E. Cariou, N. Belloir, F. Barbier, N. Djemam, in: OCL contracts for the verification of model transformations, Proceedings of the Workshop the Pragmatics of OCL and Other Textual Specification Languages at MoDELS, 2009.
[105] J. Cabot, R. Pau, R. Raventós, From UML/OCL to SBVR specifications: a challenging transformation, Inf. Syst. 35 (2010) 417–440.
[106] I.S. Bajwa, M.G. Lee, in: Transformation rules for translating business rules to OCL constraints, European Conference on Modelling Foundations and Applications, 2011, pp. 132–143.
[107] C.A. González, J. Cabot, Test data generation for model transformations combining partition and constraint analysis, in: Theory and Practice of Model Transformations, Springer, 2014, pp. 25–41.
[108] A.A. Jilani, M.Z. Iqbal, M.U. Khan, A search based test data generation approach for model transformations, in: Theory and Practice of Model Transformations, Springer, 2014, pp. 17–24.
[109] J. Winkelmann, G. Taentzer, K. Ehrig, J.M. Küster, Translation of restricted OCL constraints into graph constraints for generating meta model instances by graph grammars, Electron. Notes Theor. Comput. Sci. 211 (2008) 159–170.
[110] M.A. Francisco, L.M. Castro, in: Automatic generation of test models and properties from UML models with OCL constraints, Proceedings of the 12th Workshop on OCL and Textual Modelling, 2012, pp. 49–54.
[111] H. Wu, An SMT-based approach for generating coverage oriented metamodel instances, Int. J. Inf. Syst. Model. Des. 7 (2016) 23–50.
[112] A. Correa, C. Werner, M. Barros, Refactoring to improve the understandability of specifications written in object constraint language, IET Softw. 3 (2009) 69–90.
[113] J. Chimiak-Opoka, in: OCLLib, OCLUnit, OCLDoc: pragmatic extensions for the object constraint language, International Conference on Model Driven Engineering Languages and Systems, 2009, pp. 665–669.

[114] M. Giese, D. Larsson, in: Simplifying transformations of OCL constraints, International Conference on Model Driven Engineering Languages and Systems, 2005, pp. 309–323.
[115] M.U. Khan, N. Arshad, M.Z. Iqbal, H. Umar, in: AspectOCL: extending OCL for crosscutting constraints, European Conference on Modelling Foundations and Applications, 2015, pp. 92–107.

ABOUT THE AUTHORS

Atif A. Jilani is currently an Assistant Professor at the Department of Computer Science, National University of Computer & Emerging Sciences (Fast-NU), Islamabad, Pakistan. He is also a research scientist at Software Quality Engineering and Testing (QUEST) Laboratory, Pakistan. He received his Master's degree in systems and software engineering from Mohammad Ali Jinnah University (M.A.J.U.), Islamabad campus, Pakistan in 2008. His research interests include model-driven engineering particularly model transformation, search-based software engineering, software testing, and empirical software engineering.

Muhammad Z. Iqbal is currently an Associate Professor at the Department of Computer Science, National University of Computer & Emerging Sciences (Fast-NU), Islamabad, Pakistan. He is also the chief scientist at Software Quality Engineering and Testing (QUEST) Laboratory and President of Pakistan Software Testing Board. He received his PhD degree in software engineering from University of Oslo, Norway in 2012. Before joining Fast-NU, he was a research fellow at Simula Research Laboratory, Norway. His research interests include model-driven engineering, mobile software engineering, software testing, and empirical software engineering. He has been involved in research projects in these areas since 2004.

Muhammad U. Khan is currently an Assistant Professor at the Department of Computer Science, National University of Computer & Emerging Sciences (Fast-NU), Islamabad, Pakistan. He is heading the Software Quality Engineering and Testing (QUEST) Laboratory and is a founding member of Pakistan Software Testing Board. He completed his PhD research work at INRIA, France and received his PhD degree in computer science from University of Nice, France in 2011. His research interests include model-driven engineering, empirical software engineering, aspect-oriented software engineering, model refactoring, and software testing.

Muhammad Usman is currently doing his PhD in Computer Science from National University of Computer and Emerging Sciences (Fast-NU), Islamabad, Pakistan. He is also a research scientist at Software Quality Engineering and Testing (QUEST) Laboratory, Pakistan. He received his Master's degree in systems and software engineering from Mohammad Ali Jinnah University (M.A.J.U.), Islamabad campus, Pakistan in 2009. His research interests include mobile software engineering, model-driven engineering, product-line engineering, and software testing.

CHAPTER FOUR

Advances in Techniques for Test Prioritization

Hadi Hemmati
University of Calgary, Calgary, AB, Canada

Contents

Abstract

With the increasing size of software systems and the continuous changes that are committed to the software's codebase, regression testing has become very expensive for real-world software applications. Test case prioritization is a classic solution in this context. Test case prioritization is the process of ranking existing test cases for execution with the goal of finding defects sooner. It is useful when the testing budget is limited and one needs to limit their test execution cost, by only running top *n* test cases, according to the testing budget. There are many heuristics and algorithms to rank test cases. In this chapter, we will see some of the most common test case prioritization techniques from software testing literature as well as trends and advances in this domain.

1. INTRODUCTION

Testing is one of the most practical methods of software quality assurance in industry [1,2]. The demand for testing is growing with the widespread application of Agile methodologies and the emphasis on continuous integration and

Advances in Computers, Volume 112
ISSN 0065-2458
https://doi.org/10.1016/bs.adcom.2017.12.004

delivery [3]. Such methodologies and approaches ideally suggest running all test cases of the system, after every modification to the code (i.e., after fixing a defect or adding a new feature to the software under test) to make sure that (a) the new changes are being tested with some newly developed test cases, as soon as possible, and (b) the new modifications do not break any of the already correct functionalities (regression testing [4]). However, frequent rerunning of all test cases on large-scale systems with many test suites is very expensive. It is often impossible to rerun all test cases after every commit to the code. To get a sense of how many test executions large software companies are dealing with, let's look at the code commits at Google in 2013–14 reported by Ankit Mehta's opening talk at Google Test Automation Conference in 2014 [5]. In her talk she explains that there were around 30,000 commits per day in early October 2014 in the Google codebase. This resulted in around 100 million test runs per day at Google. Newer statistic from a keynote speech by John Micco from Google at the International Conference in Software Testing, Verification, and Validation (ICST) in 2017 shows that at Google 4.2 million individual tests are running continuously. They run before and after code submission which results in 150 million test executions per day (on average 35 runs per test per day).

One strategy to tackle the problem of "too many test cases to rerun frequently" is scheduling the reexecution of test cases less frequently, e.g., once a day, as part of a nightly integration build, or before each major release [6]. There are at least two pitfalls of this practice: (1) you may still end up with larger-than-your-budget test suites to execute, especially if you opt for regular schedules such as nightly builds, and (2) by postponing the test executions you let the defects accumulate on top of each other, which makes the debugging process more difficult, whereas if you rerun your regression test suite after each change you will end up debugging a very small set of code changes [7]. Therefore, to follow continuous integration principles in large-scale systems, we need to choose a subset of all test cases to be executed in each build; or ideally, prioritize the test cases so that depending on the time constraints of the build, only the most important tests are executed.

The solution to the above problem is called "test case prioritization" (TCP). The concepts of selecting a subset of test cases or prioritizing them are not new in software testing. They are mostly used in the context of regression testing, where researchers have proposed several techniques to effectively minimize, select, and prioritize the existing test cases. The goal of test minimization is to eliminate the redundant test cases (e.g., those that verify the same functionality in a same way) from the test suite. Test case selection, on the other hand, selects a subset of test cases for execution, which are more relevant to the current (untested) change(s). Finally, test case prioritization techniques do not throw away any test case and only rank the existing test cases for execution, based on a heuristic [4]. In other words, when test cases are prioritized, one executes the test cases in the given order, until the testing budget is over. Thus, a TCP's target is to optimize the overall fault detection rate of the executed test cases for any given budget [8].

In this chapter, we specifically focus on the TCP techniques. The goal of this chapter is giving an introductory read on the advances in test prioritization. Thus we see both basic techniques and more advanced ones. However, we do not survey all existing techniques, in details (there are already many surveys and literature reviews in this domain, for interested readers [4,6,9–11]).

The rest of this chapter is organized as follows: in Section 2 TCP problem is introduced. Section 3 categorizes TCP studies from two perspectives (their input resources and optimization heuristics). Section 4 discusses different approaches for evaluating a TCP approach and Section 5 summarizes the most important trends and future directions in this context. Finally, Section 6 concludes this chapter.

2. TEST CASE PRIORITIZATION PROBLEM

Thomas et al. [12] define the problem of TCP as follows: "TCP is the problem of ordering the test cases within the test suite of a system under test (SUT), with the goal of maximizing some criteria, such as the fault detection rate of the test cases [13]. Should the execution of the test suite be interrupted or stopped for any reason, the more important test cases (with respect to the criteria) have been executed first." A more formal definition of TCP problem can be found in Ref. [14] as follows:

Definition 1. (*Test case prioritization*) Given: T, a test suite; PT, the set of permutations of T; and f, a performance function from PT to the real numbers. Find:

$$T' \in PT\ s.t.(\forall T'' \in PT)(T'' \neq T')[f(T') \geq f(T'')]$$

In this definition, PT is the set of all possible prioritizations of T and f is any function that determines the performance of a given prioritization. The definition of performance can vary, as developers will have different goals at different times [14].

In this paper, we break down the TCP problem to even smaller pieces. Essentially, each test prioritization technique needs to rank a set of test cases based on a performance function as explained earlier. However, it is also important to know what information and knowledge about test cases (and possibly the SUT) we have at our disposal. In other words, the type of input plays an important role in designing TCP algorithms. For instance, assume we have access to the code coverage information about each test case for its most recent execution. If we have such knowledge, we can incorporate that into the ranking algorithm. For example, one can simply sort test cases based on their most recent code coverage. Basically, the performance function here is code coverage and the hypothesis is that test cases with higher code coverage are more likely to detect faults. However, if we do not have access to code coverage at all, we cannot use this heuristic for our TCP algorithm and need to design a different TCP technique.

Researchers have proposed and evaluated many TCP techniques, based on a range of input data sources and prioritization algorithms; Yoo and Harman [4] and Kazmi et al. [9] provide two (relatively) up-to-date surveys of TCP techniques, which we suggest the referred readers to. In this paper, however, we will look at the TCP problem from a slightly different perspective and categorize many existing techniques (including most recent ones that are not listed in the above surveys) into different classes. We also provide a summary of trends and future direction in TCP study.

To categorize TCP techniques, we first need to generalize the steps that most TCP solutions take as follows:

1. Represent each test case by encoding the information you have from the test case that you are going to use for the prioritization. For example, encode a test case by its code coverage or historical failure record.
2. Define your performance function in a form of an objective to achieve, based on the information in hand (the encoded test cases). For example, your objective can be achieving a maximum statement coverage or a maximum diversity among test cases.
3. Apply an optimization algorithm (anything from a simple Greedy-based sorting to more sophisticated technique such as a Genetic Algorithm) to identify the most optimum ordering of test cases, with respect to the above objective.

Based on the above steps, we need to, at least, identify two factors for a TCP solution: (a) what are the input resources at our disposal for encoding test cases and setting up our objectives? and (b) what is the optimization strategy to achieve the best performance (most cost-effective approach)?

3. CATEGORIZING TCP TECHNIQUES

There are many TCP techniques available [4], where each technique may have access to different types of information and use different strategies to achieve the TCP's objective. In this section, we will categories TCP techniques from two perspectives: the "Input Resources for TCP" and the "Heuristics and Optimization Strategies for TCP."

3.1 Input Resources for TCP

In Ref. [8] the authors have summarized the most common input resources for a TCP technique. We provide and extended version of that category here, as follows:

- *Change information*: This source of information is quite common in the context of regression testing, where the overall idea of a TCP technique is prioritizing test cases that cover the changed part of the system over those that cover the same old code. The rationale behind this idea is that the defects that could be detected in the old code by the regression test suite are already detected; thus the limited testing budget should be assigned to the new/modified code.

  The main challenge here is to find out which test cases are affected (directly or indirectly) by each change, which is called change impact analysis [15,16]. Then the test cases can be represented using a simple Boolean variable that shows whether the test case has been in the affected subset or not. For instance, one can apply this idea on the method/function-level changes, which means that if Method A changes, all test cases that directly cover Method A are called "affected." In addition, any test case that covers any Method X that depends (directly or indirectly) on Method A will be denoted as "affected" and all "affected" test cases will be prioritized higher than the other test cases.

  Note that the overall ranking based on these Boolean values ("affected" or NOT) will not result in a total order. To make the ranking more precise, one can consider this information together with other sources of information for prioritization. It is also worth mentioning that the level of granularity may be different for change impact analysis. Examples are class, method, statement level. The finer grain analysis is more accurate but is also more costly.

  One of the caveats of applying this approach on large-scale codebases with frequent changes (commits) is that one may end up with very large set of affected test cases. For instance, in a recent study [17], Memon reports that at Google, with over 150 million test executions and 800K builds per day, change impact analysis is not feasible after each change (the cost were growing quadratically with two multiplicative linear factors: the code submission rate and the size of the test pool). Google's approach to solve this issue is not repeating the analysis after every change and postpone it to some milestone (every 45 min). However, the size of affected test suite may be still huge even after putting some milestone (e.g., one change alone affected over 150K test cases). Memon et al.'s improvement is, in addition, excluding affected tests that are very distant from the original code change (10 levels in the dependency graph).

- *Historical fault detection information*: Another quite common source of information for a TCP technique is the historical performance of the test cases. The overall idea here is that the test cases that detected defects in the past should be prioritized over those that have not detected any defects, so far. The rationale behind this idea is that test cases that (frequently) failed in the past, most probably, cover a challenging and complex part of the code. So it is wise to rerun them in the new built as well. This idea is the cornerstone of many defect prediction research studies, where the assumption is that the part of the code which used to have many defects is a risky part of the code and its chance to be defective in the new version is also high [18,19].

  An easy way for representing test cases using this information is again a Boolean variable to represent whether the test case has ever failed in the history of its execution or not [20,21]. A more complex representation of this information would use a weighted approach to combine failures in the past. In other words, rather than just identifying whether a test has ever failed or not, we would consider how many times it has failed, whether it has failed in more recent releases or older ones. In a recent study by Elbaum et al. [22], previous faults have been used as a basis of prioritization in the context of continuous integration at Google. Hemmati et al. [8] also have used a weighted history-based approach for prioritizing test cases. In their approach, the test cases that have failed more recently are given more priority.

  Noor and Hemmati [23] suggest that instead of looking for test cases that have failed in the past, we calculate the similarity between failing test cases in the past and the current set of test cases, to be prioritized. In their approach a probably of failure is assigned to each current test case by how close they are to the failing test case in the past and how recent was the failure.

A variation of this high-level principle of history-based prioritization may assign a higher rank to test cases that are similar to test cases that detected more severe faults [24]. To implement this, each test case needs to be represented by the severity of the defect(s) it has detected so far. It is quite common practice to use this information together with other sources in a TCP [25,26].

- *Code coverage data*: Another common resource that is being used in TCP techniques is the code coverage of each test case [4,14]. Code coverage can be measured in many different variations (e.g., statement, branch, method coverage). The coverage can be extracted by analyzing either the program execution (dynamically) or the test and source code (statically). Dynamic analysis is more accurate, but it requires a real execution of the test cases. However, executing test cases to calculate the coverage is not an option for TCP, due to the nature of the problem (the limited testing budget). Therefore, the dynamic code coverage data can only be used, if test case coverage data is already available from previous executions [8]. Note that this data may not be perfect because (a) the coverage data for the new test cases is not available yet, (b) the code changes between two releases may reduce the accuracy of such coverage data, and (c) in many scenarios, instrumenting the code for dynamic analysis and keeping all coverage data from previous versions are not practical.

  In the absence of execution information, coverage-based TCP techniques rank the test cases solely based on the static analysis of the test cases and/or the source code. For example, one can calculate method coverage of test cases by extracting the sequences of method calls in the source code for the given test cases by static analysis [27]. Of course, the availability of test scripts that execute specific method calls is a prerequisite here, which does not hold in some cases like manual test cases written in natural language. In addition, as explained, static analysis is not as precise as dynamically executing the system, when it comes to coverage calculations.

  There are at least two ways of performing static analysis for test case coverage calculation. The cheaper one, which is the least accurate, only analyzes the test script and extracts the first-level method call sequence. The more accurate and more expensive way, however, follows each first-level method call to the source code and extracts all the nested method call sequence from the code as well. As a simple example, assume our abstract test scripts for two test cases are as follows (note that each test script is represented by a vector of test case method calls and all other execution details are omitted):

$$\text{T1} : \langle \text{A, B} \rangle \text{ and } \text{T2} : \langle \text{C} \rangle$$

Using a first way of static analysis (only using the test code for coverage calculation) T1 is ranked higher than T2, given that it covers two source code methods A() and B(), whereas T2 only covers one, C(). However, assume the following scenario where A() and B() do not call any other methods in their body, but C() calls D() and E() and E() calls F(). Following the second way of static analysis (using the test and source code for coverage calculation) we should represent the T1 and T2 as follows:

$$\text{T1} : \langle \text{A}(), \text{B}() \rangle \text{ and } \text{T2} : \langle \text{C}(), \text{D}(), \text{E}(), \text{F}() \rangle$$

Thus T2 will be ranked higher, according to the represented code coverage data.

- *Specification or requirements*: In specification-based testing (aka model-based testing), each test case can be easily traced back to the specification models. Therefore, a typical model-based TCP technique has access to the specification models of the software under test. A typical scenario for model-based testing is specifying the software by a state machine and test cases by paths in the state machine. In such a scenario, a TCP technique would prioritize test cases by considering which paths in the model are executed by which test cases [28–30], which is called model coverage.

  There are also TCP techniques that prioritize test cases based on the software requirement artifacts [31–33]. Basically, if one knows what features of the software are tested by which test cases, one can prioritize test cases by "feature coverage." In other words, rather than focusing on the specification coverage we focus on the requirement coverage. The main challenge of all TCP techniques in this category is that they require extra information about the software, which is commonly not available (or up to date).
- *Test scripts*: Finally, there are a few TCP techniques that only look at the test scripts as a source of information. These TCP techniques are usually applicable in a wider range of domains, where the other mentioned sources of data may not be available. Given that the only available

resource in this category is a textual document (the test script), the proposed techniques are typically some sort of text analysis technique, which can become quite advanced.

For instance, Thomas et al. [12] derive a "topic model" from the test scripts, which approximates the features that each test case covers. Topic modeling is a statistical text mining technique that groups related words of several documents into multiple bags called topics. Each document will have a membership value assigned to each topic, based on how frequent the document discusses the topic. This idea has been used in this work to estimate how frequent each test case discusses (i.e., covers, in our context) each feature of the code.

There are also cases where simpler text analysis techniques have been used for TCP. For example, the test scripts are interpreted as strings of words [34,35], without any extracted knowledge attached to them. Given such data per test case, TCP techniques in this category may target different objectives such as diversifying test cases [34,35] or maximizing their coverage [12], which we explain in the next subsection.

A variation of "test scripts" (which are typically used for automated testing) are "test instructions" (which are typically used in manual testing). These instructions are designed based on test plans and are given to the manual testers to follow, typically on the graphical user interface (GUI) of the software. A test instruction is similar to test scripts because they are black box and have no access to source code, but they are different than test scripts because they are written in natural languages which makes it harder for an automated TCP technique to work on. Same as the test scripts, one can look at the test instruction as a document and apply text mining techniques to extract important features [8,36]. These ideas will be explained in more detail in the next subsection.

- *GUI and web events*: Event-driven software are quite common nowadays. According to Memon et al. [37] "such systems typically take sequences of events (e.g., messages, mouse-clicks) as input, change their state, and produce an output (e.g., events, system calls, text messages)." Examples include web applications, GUIs, mobile applications, and embedded software. Modeling test cases based on their sequence of events is a common approach in test generation and prioritization in this domain. For instance, Bryce et al. [38,39] used the notion of *t*-way event interaction [40] coverage as a heuristic for test prioritization, where the goal is covering all *t*-way event interactions, as soon as possible.

In another work, Sampath et al. prioritize web applications' test suites by test lengths, frequency of appearance of request sequences (which are the "events" in this context), and systematic coverage of parameter values and their interactions [41].

Note that availability of any of the above resources is very context dependent. However, in general, the type of testing heavily influences the available input resources. For instance, in the case of white-box unit testing, TCP techniques typically have access to both test and source codes, but in black-box testing the TCP techniques only have access to the test code and potentially specification or requirements information. In the context of regression testing, history of failure and change data may also be available and so on. Nevertheless, the take home message is that TCP technique can be categorized based on their available input resources, which defines how the test should be represented. The other perspective for categorizing TCP techniques is given test cases that are represented in a particular way (i.e., any representation that is applicable based on the above discussion on the available resources), what heuristic and algorithm is used by the TCP. In the next section, we will summarize some of these heuristics.

3.2 Heuristics and Optimization Strategies for TCP

The objective of most TCP techniques falls under one of these two heuristics: (a) maximizing some sort of coverage criterion and (b) maximizing diversity between test cases. There are several approaches to achieve these objectives. In the rest of this section, I will first explain the two main heuristics, in general, and then summarize some of the common strategies to realize those objectives.

3.2.1 TCP Heuristics

In this section, we will see two most common high-level heuristics that are used in TCP techniques.

- *Maximizing coverage.* Generally speaking, one can represent a test case as a set or sequence of some "features" or "aspects" of the SUT that are covered (i.e., being exercised, or verified, or just executed) during test execution. These "features" are defined based on the available "input resources for TCP," explained in the previous section. For instance, a test case can be represented by a set/sequence of code elements (e.g., statements or methods) it covers. Assume we represent test cases by sets of method names that they cover. Now we can create a total "feature set," which includes all method names that our test cases can possibly cover.

A "maximizing coverage" heuristic (or as it is usually called a "coverage-based" approach) suggests that a TCP technique must order test cases in a way that the test executions result in a high coverage of those features, as soon as possible. The intuition behind this strategy is that by maximizing "feature coverage," the test cases execute as much of the SUT as possible and thus increase their chances of fault detection.

Though the idea of "maximizing coverage" is mostly common for maximizing code coverage, as the example above, but its generic idea can be applicable in most situations, with respect to the available input resources and our testing perspective. Basically, given our choices of available input resources, what we consider as a feature defines what aspect of the software we want to test as soon as possible. Thus, all we need to do is to represent the test cases by their covered features and then try to order test cases so that higher coverage of those features is achieved as soon as possible.

Feature extraction: As mentioned, representing test cases by their covered features is the key in this heuristic. Features can be defined based on code, specification, requirements, GUI, etc. Sometime features are explicit and easy to map, e.g., when a feature is a code element (like statements or method calls), representing test cases by their covered code elements is straightforward. Sometimes the features are explicit but not easy to map to test cases. For instance, in case of requirements features, one might know exactly which requirements of the software are supposed to be tested, but there is no easy way to (automatically) map a test case to requirement features. Thus representing test cases by their covered requirement features is not as easy as representing them by their code-level features.

Finally, there are cases where the feature is implicit. In other words, the targeted feature is being estimated by some other surrogate metrics. For example, remember the case of input resources for manual testing, from the previous section? In that case, the authors' [42] targeted features are the GUI elements that are covered by each test case. However, all they have access to (input resources for TCP) is the test instructions. In such a case, they need to estimate the features covered by each test case using the information at their disposal. In that particular study, the authors used an NLP-based test mining approach to estimate the exercised GUI elements per test case and represented each test case using a vector of estimated GUI coverage info.

Obviously, the most accurate coverage-based TCP techniques are those that have an explicit access to the features and an automated precise way for collecting the feature coverage data from the test cases (e.g., a dynamic code coverage-based approach). But the overall idea of "maximizing coverage" can be exactly applied on any extracted (and estimated) feature set.

Note that there are several strategies to maximize coverage and we come back to that after explaining the other major heuristic for TCP, which is maximizing diversity.

- *Maximize diversity*. The overall idea behind this heuristic is allocating testing resources to a diverse set of "features." The intuition for this strategy is that set of test cases that are similar will likely detect the same faults, and therefore only one in the set needs to be executed. The rationale behind this idea is that if we are short in testing resources (which is the motivation for TCP) and we follow a "maximizing coverage" approach, we might end up having high coverage on some similar features, but no or very low coverage on some other features.

  As an example, assume that the software under test is a text editor application. The "features" are menu items in the GUI. You have three tests (T1–T3) to prioritize, where each test covers one or more menu items as follows:
  - T1 covers "open," "copy," "paste," and "save."
  - T2 covers "open," "cut," "paste," and "save as."
  - T3 covers "new document," and "close."

  In this example, we have eight features that can be covered by the test cases: "open," "copy," "cut," "paste," "save," "save as," "new document," and "close." Now assume that our coverage-based TCP already have picked T1, as the first test. So our coverage is 4/8=50%. To choose the second test a coverage-based heuristic does not have any preference between T2 and T3 since both cases will increase coverage to 6/8=75%. However, from the "maximize diversity" heuristic point of view, choosing T3 is more preferable, since T3 is quite different than T1, whereas T2 is quite similar to T1.

  It should be quite clear by now that the definition of "similarity" or "difference" between test cases is a key in designing a diversity-based heuristic. To be able to define "similarity" or "difference" between test cases, we start with representing (encoding) test cases as a set or sequence of "features." Exactly the same as the coverage-based approach. However, the diversity-based heuristic does not focus on maximizing the coverage of features. The new heuristic tries to rank test cases in a way that for any given budget the set of selected test cases executes/covers the most diverse features.

To measure diversity among a set of test cases, we need to define a "similarity" (or "distance") function that works on the features covered by the test cases. A common way to do this is representing the test cases by a vector of features they cover and then apply a standard distance function to calculate the distance between every two test cases. Then aggregate the paired distances (using an aggregation function) and use the final value as a diversity metric for that set of test cases.

There are several similarity/distance functions that can be used to determine similarity/distance among test cases. These functions will be covered in the next section.

3.2.2 Optimization Strategies

Optimization strategies can be the same or different for the two heuristics (maximizing coverage and diversity). In the rest of this section, I will first explain some of the common strategies for maximizing coverage and then explain optimization strategies for diversity-based approaches.

3.2.2.1 Maximizing Coverage

So far we have learned that prioritizing test cases starts with extracting features per test case based on the available resource and encoding each test case as a vector of features. Therefore, the goal of "maximizing coverage" as an optimization strategy is to rank test cases so that the highest "feature" coverage can be achieved with any given testing budget. There are several approaches to tackle this problem. Some are more common, but suboptimal, and some less intuitive but more effective. Let us start with the simplest approach. A Greedy algorithm:

- *Greedy approach for maximizing coverage*: The simplest coverage-based TCP approach that uses a Greedy algorithm works as follows:
  - Step 1: Identify a test case with the most number of covered "features" from the test suite (in case of ties use a tie breaker, e.g., random selection)
  - Step 2: Append the test case to your ranked list and remove it from the test suite
  - Step 3: If the test suite is not empty, back to step 1

Though this algorithm is pretty simple and practical, it has obvious shortcomings. The main issue with the simple Greedy is that it does not take into account the current coverage of features while selecting the next best. To understand this let us see the following example.

Assume we have a test suite of three test cases (T1, T2, and T3) and a source code class with five methods (M1 … M5). A feature in this example is a method. Therefore, each test case is represented by the methods it covers, as summarized in Table 1 (a checked cell (X) means the corresponding test case covers the indicated method).

The simple coverage-based Greedy algorithm selects T2 as the first choice, the test with the highest feature coverage (four out of five). Then the algorithm picks T1 as the second highest coverage (three out of five) and last choice is T3, resulting in ⟨T2, T1, T3⟩ ranked list. However, a quick investigation shows that in fact ⟨T2, T3, T1⟩ is a better choice, since after two test executions one can achieve 100% coverage, whereas the simple Greedy's output requires three test executions, two achieve 100% coverage (only 80% after two executions). This is simply due to the fact that the Greedy is not aware of the current covered features.

A variation of the Greedy-based approach modifies the algorithm by implementing a different way of coverage calculation per test case. It is usually called an "additional coverage" approach in the literature. The idea is that rather than calculating the raw coverage per test case, we assign an "additional coverage" value per test. The "additional coverage" tells you how much extra feature coverage we achieve if we append a selected test case to the so far ranked list. For instance, in the example above, T2's "additional coverage" is 80% because the ranked list is empty. But when T2 is added to the ranked list covering M1, M2, M3, or M5 does not add anything to the ranked list's coverage. Thus T1's "additional coverage" is zero and T3's "additional coverage" is 20% (going from four out of five to five out of five feature coverage).

Table 1 Example of Feature (Method) Coverage of Test Cases, Used for a Greedy-Based TCP

	T1	T2	T3
M1	X	X	
M2		X	
M3	X	X	
M4			X
M5	X	X	

- *Local vs global optima*: Both variations of the Greedy-based TCPs suffer from the same problem of "local optimum." According to Wikipedia [43], in the field of optimization, "a local optimum of an optimization problem is a solution that is optimal (either maximal or minimal) within a neighboring set of candidate solutions. This is in contrast to a global optimum, which is the optimal solution among all possible solutions, not just those in a particular neighborhood of values." Since a Greedy algorithm makes the locally optimal choice at each stage, in many problems, it does not produce an optimal solution [44].

  For instance, in the example below (Table 2), following an additional Greedy algorithm results in selecting T1 as the best choice. Then after T1, any order of T2, T3, and T4 is the same from the (extra) coverage point of view. So let's say we end up with the following ranked list: ⟨T1, T2, T3, T4⟩. In terms of covering the methods, we get 50% coverage, right away, by the first test case execution. After that we get ~16.7% (one out of six methods) extra coverage by each extra test case we run, until we reach 100% (with four test executions). Now look at the alternative ranked list of ⟨T2, T3, T4, T1⟩. We will have a lower coverage to start with here ~33.4% (two out of six methods), but then every extra test case brings in an additional here ~33.4% (two out of six). This will result in 100% coverage by executing the first three test cases (T2, T3, and T4).

  Depending on how one measures the effectiveness of a TCP with respect to coverage (we will talk about different options in the test prioritization evaluation section), the above example can show the suboptimality of the Greedy-based solutions. Basically, if we ignore the first quick start of the first ranked list (⟨T1, T2, T3, T4⟩) and look

Table 2 Example of Feature (Method) Coverage of Test Cases, Used to Show Local vs Global Solutions in Greedy-Based TCPs

	T1	T2	T3	T4
M1	X	X		
M2	X		X	
M3	X			X
M4		X		
M5			X	
M6				X

at the overall coverage or how fast we achieve the maximum coverage, the second ranked list (⟨T2, T3, T4, T1⟩) is a global optimum and the first ranked list is a local one.

- *Search-based coverage maximization*: To escape from the local optima, the alternative TCP optimization strategy is using global search techniques. The use of local and global search algorithms has become quite common in the automated software engineering domain [45]. The overall idea is to reformulate a software engineering task as an optimization problem and then automatically search for the best solution using a metaheuristic search algorithm. Metaheuristic search algorithms are general strategies that need to be adapted to the problem at hand.

  Many software engineering problems can be reformulated as optimization problems, for which search algorithms can be applied to solve them [46]. This has led to the development of a research area often referred to as Search-Based Software Engineering, for which several successful applications can be found in the literature, with a large representation from software testing [11,45].

  In our context, the problem in hand is maximizing the coverage of a list of test cases by reordering them in a way that high coverage achieves as soon as possible. Note that this problem, in general, is an NP-hard problem [47]. Therefore, using an exhaustive search for global optimum in most realistic problems is not an option, since the search space size for selecting the best order of n test cases is equal to the number of all permutations $= n!$.

  The space of all possible permutations of n test cases represents the search space. The order with the "best coverage" (fitness value) is called global optimum. A fitness function is defined to assign a fitness value to each permutation. A search algorithm can be run for an arbitrarily amount of time. The more time is used, the more elements of the search space can be evaluated. Unfortunately, in general, we will not know whether an element of the search space is a global optimum because such knowledge would require an evaluation of the entire search space. Therefore, stopping criteria need to be defined, as, for example, timeouts or fixed number of fitness evaluations.

  There are several types of global search algorithms that one can choose, e.g., Simulated Annealing (SA), Genetic Algorithms (GA), and Ant Colony Optimizations (ACO) [45]. On average, on all possible problems, all search algorithms perform equally, and this is theoretically proven in the famous No Free Lunch theorem [48]. Nevertheless, for specific classes of problems (e.g., software engineering problems), there can be significant differences among the performance of different search algorithms. Therefore, it is important to study and evaluate different search algorithms when there are a specific class of problems we want to solve, as, for example, software testing and its subproblems. This type of comparisons in software testing can be found, for example, in Refs. [4,28,49]. For the class of TCP problems, we need a global search algorithm like GA that does not stuck in local optima. On average, in most TCP scenarios in the literature, the global search-based approaches have shown to lead to better results than a Greedy-based solution or a local search-based approach such as Hill Climbing [50].

- *Fitness function and area under curve*: A fitness function in a search-based TCP technique can be designed in different ways. A typical approach for defining fitness values in coverage-based TCPs is using the area under a curve (AUC) concept. AUC simply means the area under the curve, where the curve represents the coverage per test case execution. As we saw in Table 2's example (local vs global search), the "Maximizing Coverage" strategy may be evaluated in different ways. One way to look at it is ordering the test cases so that we achieve 100% coverage (or whatever maximum we can achieve using our test suite) as soon as possible (with minimum test executions). Another way to look at it is having the highest coverage for any given subset of test case executions, following the ordered list. Essentially, this means that, since in practice we will end up running a subset of test cases and all test cases in that subset will be executed anyways, then we only care about the overall coverage but for any size subset. This latter evaluation method can be measured by AUC.

  To illustrate AUC measurement for TCP evaluation, one can plot the coverage of test cases on the y-axis, while the x-axis shows the number of test executions following the order of test cases given by a TCP technique. For instance, Fig. 1 compares the two orderings O1: $\langle$T1, T2, T3, T4$\rangle$ and O2: $\langle$T2, T3, T4, T1$\rangle$ from the previous example (Table 2).

  The actual AUC values for O1 and O2 in this example are AUC(O1) = 2.25 and AUC(2) = 2.33. So according to AUC O2 is a slightly better TCP technique. However, as we also mentioned earlier the illustrated AUCs shows that O2 achieves 100% coverage by three

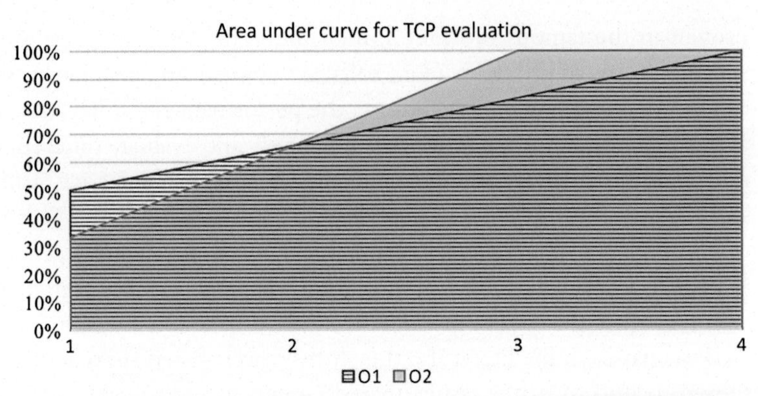

Fig. 1 Illustration of AUC measure for TCP evaluation in the example of Table 2.

test executions, whereas O1 requires four test cases for that level of coverage, which can be a more differentiating factor between O1 and O2 compared to the slight difference in their AUC. Nevertheless, AUC seems to be a reasonable measure to be used as fitness value in search-based TCPs.

So to summarize the search-based coverage maximization approach, we need a fitness function which typically is the AUC metric, and using a search-based algorithm like GA we explore different permutations to achieve higher AUC.

The exact way that the search space is explored is defined by the search algorithm. For instance, "GAs start off by an initial (typically random) population of individuals (each of them represents an ordered set of test cases, i.e., a candidate for an optimal solution) and improve the quality of individuals (in terms of the fitness value, i.e., AUC in this context) by iterating through generations. The transition from one generation to another consists of four main operations: selection, crossover, mutation, and sampling" [51]. There are several methods to implement each of these operations which their details are beyond the scope of this chapter. We encourage the interested readers to look at some of the related work such as Refs. [11,28,29,50,51].

Note that the AUC as defined above is just a heuristic to guide our search toward "better" solutions. It does not necessarily mean that a TCP that results in higher coverage-based AUC is "better." In the test prioritization evaluation section, we will talk more about what is the ideal ordering and how we can measure the "better," in this context.

3.2.2.2 Maximizing Diversity

As I explained in Section 3.2.1, this heuristic tries to rank test cases in a way that for any given budget the set of selected test cases executes/covers the most diverse features. To do so we need to follow three steps:

- Encode test cases
- Build distance/similarity matrix
- Maximize/minimize distances/similarities

The encoding of test cases as a vector of features has already been discussed in earlier sections. I also have explained the overall idea of diversity between test cases and the rationale behind it, in the maximize diversity section. In this section, I will explain the distance function and the optimization strategies for diversity-based TCPs, which is extracted from this thesis publication [50].

Once the test cases are encoded into vectors of features, they are given to a similarity/distance function. Note that it does not matter whether we define a distance function and maximize that or a similarity function and minimize it. Either cases can be transformed to the other one. Therefore, we just use distance function and maximization terminologies here.

A simple way of implementing the diversification concept is using a pairwise distance function that typically takes two sets/sequences of elements (we use "{ }" to represent sets and "⟨ ⟩" for representing sequences) and assigns a distance value to each pair. The results of measuring all these distance values are recorded in a "distance matrix" (in case of large test suites one can also replace this matrix generation phase with an on-the-fly distance calculation).

Note that the distance matrix can be an upper/lower triangular matrix, since the distance measure should be symmetric (the similarity between test case A and test case B is equal to the similarity between test case B and test case A). Therefore, we only need to store half of the matrix.

Given an encoding, one may use different set/sequence-based similarity functions [50,52]. The main difference between them is that set-based similarity measures, as opposed to sequence-based ones, do not take the order of features into account. For example, assume we represent test cases by the method names that are covered when running the test cases. Now assume test case T1 calls A() and then B(), whereas test case T2 calls B() and then A(). For a set-based distance functions T1 and T2 are identical. Next we will define some of the common set- and sequence-based distance functions.

- *Set-based distance functions*

  Set-based distance measures are widely used in data mining [50] to assess the distance of two objects described as multidimensional feature vectors, where the set is composed of the features' values [50]. In our case, each test case is also a vector of features. Each feature in the vector is taken from a limited alphabet of possible features. However, the vector size can be different since the length of test cases may vary.

  - *Basic counting*: This measure is a very basic function, which does not account for the features order in the test case [23]. The function requires each test to be represented as a set of unique features. That means that if for instance we represent test cases by the method names to cover, and the test covers a method twice, the set only indicates the covered method once. The basic counting function simply looks at the two sets and counts the overlaps. Number of overlapped features indicates similarity. Thus the raw distance is the number of non-common elements within the union of the two sets. For example, if T1:{a, b, c} and T2:{a, d}, then the raw distance value is 3 (for the following uncommon features "b," "c," and "d"). This will result in a distance metric that is between 0 (identical sets—remember that the order does not matter) and a MAX value (the size of the largest union of two sets in the test suite), representing the distance between possibly two large completely different test cases. To normalize the metric, one can divide the raw values to the MAX to get a metric in the [0,1] range.
  - *Hamming distance*: Hamming distance is one of the most used distance functions in the literature and is a basic edit distance [53–55]. The edit distance between two strings is defined as the minimum number of edit operations (insertions, deletions, and substitutions) needed to transform the first string into the second [53–55]. Hamming is only applicable on identical length strings and is equal to the number of substitutions required in one input to become the second one [53–55]. If all inputs are originally of identical length, the function can be used as a sequence-based measure. However, in realistic applications, test vectors usually have different lengths. Therefore, to obtain vectors of identical length, a binary string is produced to indicate which features, from the set of all possible features of the encoding, exist in the vector. This binary string, however, does not preserve the original order of elements in the test and therefore leads to a set-based distance function.

In our case, to use Hamming distance, each test case is represented as a binary string, where each bit corresponds to a unique feature among all possible features to cover by the test suite. A bit in the binary string is true only if the test case contains/covers the corresponding feature (e.g., the method call).

As an example, let's calculate the Hamming distance between T1:{M1, M2, M3} and T2:{M1, M4} from Table 2. The binary string representations of T1 and T2 are as follows: T1:⟨1,1,1,0,0,0⟩ and T2:⟨1,0,0,1,0,0⟩, where the 6 bits represent M1–M6 in the ascending order. Thus Hamming distance (T1, T2) = 3, since we need to modify at least 3 bits (#2, #3, and #4) T1 or T2 to make them identical.

- *Other set-based functions*: There are many other functions that can be used as a set-based function, e.g., Jaccard Index, Gower–Legendre (Dice), and Sokal–Sneath(Anti-Dice) measures [50] are all defined based in a generic similarity formula as below but with different weightings (w):

$$\text{similarity}(A, B) = \frac{|A \cap B|}{|A \cap B| + w(|A \cup B| - |A \cap B|)}$$

In the next section, we explain some of the common sequence-based distance functions.

- *Sequence-based distance functions*

  Sequence-based distance functions are more accurate in defining distances between two vectors of element because they do not typically exclude the duplicate elements and they keep the order of elements into account. Note that sequence-based functions are not necessarily better for diversity-based TCP. Depending on the context, a TCP may actually want to exclude duplicates and order of features.

  - *Edit distance function*: One of the most common sequence-based distance functions is edit distance. There are several implementations of edit distance. Levenshtein algorithm, which unlike Hamming distance is not limited to identical length sequences, is a well-known implementation of edit distance [55]. In a basic Levenshtein, each mismatch (substitutions) or gap (insertion/deletion) between two sequences of elements increases the distance by one unit.

For example, assume we show test cases by their sequence of method calls. Assume

$$T1:\langle M1, M2, M1, M3\rangle \text{ and}$$
$$T2:\langle M1, M4, M1\rangle$$

We can align the tests based on their common elements and then calculate the basic Levenshtein score as follows:

$$\mathbf{T1`:< M1, M2, M1, M3 > and}$$

$$| \quad | \quad | \quad |$$

$$\mathbf{T2`:< M1, M4, M1. ---- >}$$

Note that M2 and M4 are mismatches and M3 is a gap. So the distance is +1 for one mismatch and +1 for one missing element. That is +2. One can normalize it by the length of the longer test as well, which in this case is 4.

Therefore, normalized_Levenshtein (T1, T2) = 2/4 = 0.5

Other variations of Levenshtein may use different scores (operation weights) to matches, mismatches, and gaps. In addition, in some variations of the algorithm (mostly used in bioinformatics), there are different match scores (alphabet weight), based on the type of matches [55].

- *Other sequence-based functions*: There are many sequence-based distance functions that are defined and used in bioinformatics for matching and aligning DNA sequences [54]. They can be categorized in global (e.g., Needleman–Wunsch) or local (Smith–Waterman) alignments. Global alignment's idea is quite similar to edit distance's, but local alignment tries to only find partial matches. In the context of TCP, we may require to compare a short test case with a very long test that (almost) contains the short test case. In such cases, typical edit distance or global alignment measures show a high distance between the two test cases, but local alignment would show a low distance, which may be more appropriate in this example.

- *Maximizing distances*

In the last step of a diversity-based TCP technique, the distance matrix is given to an algorithm which maximizes the overall distance between all test cases for any given subset of the total test suite. In other words, we again use an aggregation measurement such as AUC, but this time rather than assigning the overall coverage for each *n* test case execution, we assign an aggregated distance value for the selected subset of test cases to be executed. There may be different ways to aggregate paired distance values to come up with one overall distance value per subset. One common way is averaging all the pairwise distances between all pairs of test cases in the selected set.

Note that the problem of prioritizing test cases to achieve such maximize diversity is again, in general, an NP-hard problem (traditional set cover) [47]. Therefore, using an exhaustive search in most realistic problems is not an option. Some of the other alternatives are: (1) Greedy, (2) Adaptive Random, (3) Clustering, and (4) Search-based algorithms.

- *Greedy based distance maximization*: There are several ways to implement a Greedy algorithm for distance maximization. I'll explain one option here. First, we choose a pair of test cases with maximum distance, among all pairs in the test suite. Then at each step, a test case that has the highest distance to the already selected test cases will be appended to the prioritized list. The distance of a test case (Tx) to a set of *n* test cases (Ti where i:1 … *n*) can be defined in different ways. For instance, we can take the average of all distance (Tx, Ti) for i:1 … *n*. Another options are taking Maximum or Minimum of all distance (Tx, Ti) for i:1 … *n*.
- *Adaptive random testing for distance maximization*: Adaptive random testing (ART) has been proposed as an extension to random testing [56]. Its main idea is that diversity among test cases should be rewarded, because failing test cases tend to be clustered in contiguous regions of the input domain. This has been shown to be true in empirical analyses regarding applications whose input data are of numerical type [57]. Therefore, ART is a candidate distance maximization strategy in our context as well. A basic ART algorithm described in Ref. [56] is quite similar to what was explained above for basic Greedy algorithm for distance maximization. There are different variations of ART [58] as well that can be applied on TCP problem.
- *Clustering-based distance maximization*: Clustering algorithms partition data points into groups, using a similarity/distance measure between pairs of data points and pairs of clusters, so that the points that belong to the same group are similar and those belonging to different groups are dissimilar. Though clustering techniques are not optimization techniques, the fact that clusters are formed based on the similarities/distances among data points makes these algorithms a potential solution for our maximization problem.

One way of applying clustering for prioritization is to start with selecting a test case (T1) in random. Then cluster the test cases into two partitions (C1 and C2). T1 will belong to one of the two clusters. Let's say T1 is in C1. Then we select a test case, T2 (either a random one or the centroid) from C2 to append to the ranked list. Then in each step, we split the biggest cluster into two and select a test case from the cluster that does not have a representative yet and append it to the ranked list, until we rank all test cases.

There are also many other ways that one can come up with, when applying clustering for prioritization. It is also possible to combine clustering with another maximization algorithm so that we first cluster the test cases and then order the clusters rather than ordering test cases. Finally, the tests within a cluster need to be ordered using other strategies.

- *Search-based distance maximization*: The overall idea here is quite similar to Search-based coverage maximization, explained earlier. Except that the fitness function should represent the aggregated measure based on the distance values. For instance, the average of all paired distance values for each set size, following the given ordering.

As an example, let's look at the Table 1 test cases. The encoded versions of those three test cases look like the following:

$$\begin{aligned}T1 &: \langle 1, 0, 1, 0, 1\rangle \\ T2 &: \langle 1, 1, 1, 0, 1\rangle \\ T3 &: \langle 0, 0, 0, 1, 0\rangle\end{aligned}$$

Now using a Hamming distance function, the distance matrix is:

	T1	T2	T3
T1	—	0.2	0.8
T2	—	—	1.0
T3	—	—	—

Finally, using a Greedy-based distance maximization, the first pair to select is T2 and T3 (with no particular preference among them) and then T1 is ranked as the last test to execute. To make a full ordering, one can choose the first test in random or choose the one among T2 and T3 that has higher coverage, which results in ⟨T2, T3, T1⟩ ordering.

4. TEST PRIORITIZATION EVALUATION

So far we have only focused on how to rank test cases based on available input resources. Given that all our techniques use heuristics (such as coverage or diversity of code, specification, history), the question is how good these heuristics are. In other words, assume we have the best optimization strategy and we rank test cases in an optimal way with respect to their let's say method coverage. Should we expect an optimal rank? How can we measure the quality of a ranked list? What is the ultimate goal of a TCP?

To answer these questions, we first need to (re)define our ideal ranking. We said earlier that there is no one way to define the best rank. The "best" might actually be different depending on the context, but overall, most people agree that the goal of a TCP is to rank test cases so that we find failing test cases as soon as possible. That means we need to prioritize failing test cases over nonfailing ones.

But this definition is not enough. First, we have to assume that each test case failure is mapped to one and only one defect. Second, we have to assume that all test failures are equally severe; otherwise we need to assign a weight per failure so that the failing tests with severe failure are ranked higher. Finally, we have to decide how to aggregate the results of several failing tests. Let me explain these in the following categories:

- *A test suite with one failing test case*: If a test suite contains only one failing test case, the goal of a TCP is putting that one test as high as possible in the ranked list. Ideally, as number one. If this is the case, the best evaluation measure for a prioritized list is the rank number of the failing test case in the list. For instance, if TCP1 prioritized the only failing test as #4 and TCP2 prioritized it as #20, TCP1 is a better prioritization technique. We can even say a TCP that ranks the failing test as #1 is the optimal solution. Note that this measure is also useful when the test suite has multiple failing tests, but as soon as a fault is detected, the code is fixed and the tests updated and reordered. Therefore, we always care about detecting the first fault.
- *A test suite with multiple failing test cases*: In a bit more generic setting, there might be more than one failing test cases per test suite. Note that we still assume that each test case detects maximum one defect and defects are equally severe. There are two different situations in this context, depending on whether the failures are unique or not.

- *Unique test case per defect*: If the test-to-failure mapping is unique, which means there are no two test cases that detect the same defect, what we need is to rank the failing test cases all up in the ranked list. A simple evaluation measure here would be the number of test cases required for detecting all defects. However, you may end up with scenarios where let's say TCP1 detects 8 defects out of 10 total detectable defects up front by just 8 test executions and then nothing for a while until it detects the rest by 40 test executions. However, TCP2 detects the first 8 by 10 test executions and all 10 defects by 15. Now depending on the context one may choose TCP1 or TCP2. If I only care about the first 80% of the defects or I have limited budget of let's say 10 test executions, then TCP1 is my choice. But if I care about all bugs and I have more testing budget, TCP2 may be my better choice.
- *Multiple tests per defect*: If the test case is not unique per failure, then the raw number of detected defects is not enough and we need to see how many unique defects have been defected so far. For instance, assume the first four test cases, ordered by TCP1, all fail but they all detect the same defect. However, among the first four test cases, ordered by TCP2, only two tests fail, but they detect separate defects. In this example, TCP2 is a better choice since we detect two defects rather than one (of TCP1 approach) by four test case executions.

- *Multiple faults per test case*: Finally, a test case alone can detect multiple faults. The best way to represent such cases is using a matrix, which is called fault matrix. It shows the test cases in rows/columns and unique faults in columns/rows (see Table 3).

A very common effectiveness metric that is used in most TCP literature is from this category and is called "Average Percentage of Fault-Detection" (APFD) [14]. APFD: APFD measures the average of the percentage of faults detected by a prioritized test suite. APFD is given by

$$\text{APFD} = 100^{*}\left(1 - \frac{TF_1 + TF_2 + \cdots + TF_m}{nm} + \frac{1}{2n}\right)$$

Table 3 Sample Fault Matrix

	T1	T2	T3	T4
Fault1	0	0	1	0
Fault2	1	0	1	0
Fault2	0	0	0	1

where n denotes the number of test cases, m is the number of faults, and TF_i is the number of tests which must be executed before fault i is detected. The overall idea behind APFD is calculating area under the curve where the x-axis is the number of test cases (or portion of test suite) executed in the given order and the y-axis is the number of unique faults detected by the test executions (quite similar to the explained concept of AUC for coverage-based optimization). Maximum APFD is when all defects are detected by the first test cases and minimum APFD is when no defect is detected by the first $n-1$ test cases.

- *APFD's weaknesses*: Though APFD is very common, but has its own limitations. Among them the most major one is when test suites are large with many test cases that do not detect any fault. This scenario, actually, happens to be a very common scenario given that test prioritization is more useful for large-scale test suites and typically these (regression) test suites have only a few failing test cases. The problem is that APFD cannot show the differences between two TCPs well, when the vast majority of test cases are not detecting any fault and the detecting tests are mostly ordered first. In other words, the gap between two curves (two different orderings) is in a small initial portion of the ordered test suite. Therefore, the gap between the overall AUCs for the two curves becomes insignificant.

  To overcome this issue, a simple approach is having a cutoff for the number of test cases in the suite that are being ranked. This way the very long tail of the curve will be excluded and the APFD results will be more normalized. Another option is transforming the raw APFDs by a logarithmic function to make the differences more visible.

- *Measures based on prediction effectiveness*: As we will discuss it in the next section, TCP problem can also be seen as a prediction problem, where the goal is predicting the failure likelihood of each test case in a test suite, in the new release, based on historical data. The prioritization then can be done by sorting the test cases based on such failure likelihood estimates. Most TCP techniques in this domain use historical data on some metrics (e.g., code coverage, historical failure, size) as a learning set and predict the failure of test cases based on the values of the metrics in this version of the software. Therefore, another category of evaluation metrics is standard prediction evaluation metrics such as precision, recall, F-measure, and accuracy [59].

- *The cost and severity factor*: The main two factors that are ignored in all above metrics are the cost of each test case execution and the severity of the detected faults per test case. These two factors (arguably) can be even more important that the raw number of detected faults is the basis for all the above metrics. The solution to consider these factors into account is actually quite simple. One can just use a weighted version of each metric. However, the challenge is finding the right weights. As an example, Elbaum et al. [24] proposed and studied one such variation of APFD which they call "cost-cognizant" metric APFDc.

  Regarding the cost measurement, there are several metrics introduced in the literature, e.g., test case/suite execution time, test case/suite size, defect discovery time, prioritization time [9] that can be used to quantify the cost measure.

  Severity can also be measured in different ways. In a study by Hettiarachchi [60], they used requirements modification status, complexity, security, and size of the software requirements as risk indicators and employed a fuzzy expert system to estimate the requirements risks.

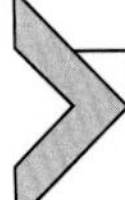

5. TRENDS AND FUTURE DIRECTIONS IN TCP TECHNIQUES

The advances in test prioritization in recent years are in three categories: (a) Scalable solutions, (b) Intelligent algorithms, and (c) New application domains.

(a) *Scalable solutions*

Looking at many earlier works in the domain of TCP, we realize the algorithms and heuristics proposed in those studies had only been evaluated on toy applications (very small size applications or lab-made programs). Therefore, scalability of the TCP techniques to real-world complex systems is definitely one of the target research directions. At least two trends are clear from the recent publications: Prioritizing system-level test suites and prioritizing test cases in the presence of frequent code changes.

- *System TCP*: In recent years, when researchers tried to apply the same old TCP techniques on real-world large-scale software systems, they have faced several scalability issues with respect to test execution cost. For example, Hemmati et al. [28] report that in the context of system-level testing of embedded systems, each test execution is very costly because there are hardware and network in the loop. This is very different situation that a typical unit-level testing scenario, where the external (e.g., network and database) access is mocked and each test execution is very fast. They report that even running 100 s of such test cases is not feasible for many teams, in their overnight regression test suite. The specific trend that this work has started is maximizing test suite's diversity rather than only focusing on its coverage. It is rather important for large-scale system where achieving very high coverage is very expensive; therefore, test cases prioritized by coverage-based approaches may fail to verify a wide-enough range of features of the software under test.

- *Continuous integration*: The other important trend in test case prioritization advances we observe today is dealing with frequent amount of changes a software undergoes. With the increasing interest in continuous integration, large software companies such as Google face scalability issues, even for running all unit test cases after each commit [17]. It is even expensive to just run the affected test cases by the code change. The new trends and advances in TCPs [17] suggest bounded change impact analysis based on effective evidence collected from earlier versions of the SUT.

Future directions (1): Based on the earlier discussion, it seems that one of the future directions on TCP research is conducting large-scale empirical studies with industrial software systems and evaluating the existing techniques in terms of their cost effectiveness in large and proposing modifications and relaxations to make the algorithms scale to the size and complexity of real-world systems.

(b) *Intelligent algorithms*

There is no shortage of new intelligent ideas proposed for TCP problem. Most techniques try to come up with a new metric for test prioritization that works better than the existing ones. Here we list some of the new and trendy heuristics for TCPs:

- *Search-based optimization*: As we saw earlier, the TCP problem can be formulated into an optimization problem and thus evolutionary search algorithms have been used in the literature to solve this problem. Genetic algorithm is one of the most common techniques in this domain. However, there are other search algorithms that have shown good results as well, for instance, Ant Colony Optimization [61], Particle Swarm, etc. [11].
- *Diversity-based (or similarity-based) TCP*: TCP techniques in this category [28,29,51,52,62] have got attention in academia in

recent years. The main idea as explained before is focusing on diversity of test cases rather than only coverage of some sort. This line of thought has also been very active in automated test generation literature. For example, Feldt et al. [63] studied diversity metrics and proposed new metrics that work on the diversity of the entire set rather than an aggregation of pairwise distances. In another work, Shi et al. [64] introduce an entropy-based metric for diversity measurement among test cases.

- *IR-based TCPs.* Information retrieval techniques have been used for extracting features from different resources to represent test cases. For instance, Kwon et al. [65] used an IR-based (TF-IDF) metric to extract features from code elements. Thomas et al. [12] used a topic modeling algorithm (LDA) to extract features from string literals and code comments and Hemmati et al. [8,36] used a topic modeling to extract features from textual test case instructions.
- *Historical failure*: Another category of TCP techniques that got attention in recent years is making the old-school "historical failure" metric more intelligent. As we saw earlier, the original algorithm simply orders test cases that fail in the past higher. Elbaum et al. [22] proposed a smarter and more scalable version of this idea by introducing a windowing mechanism to limit the considered history. They also look at how recent the failure was to rank the fault revealing test case. Noor and Hemmati also modified the basic algorithm by introducing the concept of "similarity to the failing test cases in the past" [23]. Basically, the idea is that many test cases are quite similar to older test cases (they may even be a slightly modified version of older test cases), but not exactly the same as those old test cases. In these scenarios, the test cases that are pretty similar (in terms of what behavior they test) to the old failing test cases should also be ranked higher. This makes the overall idea of historical failure metric more accurate and smooth (not a binary metric anymore).

There are also attempts to use multiple metrics together to get a better solution, overall. The main two strategies here are multiobjective search and machine learning-based approaches.

- *Multiobjective TCPs*: The use of multiobjective search algorithms on the TCP problem has been exercised in several studies [4,66]. One of the very common trends is adding the cost factor as a secondary criterion to an existing effectiveness criterion such as coverage or diversity [51,67]. Due to scalability issues of multiobjective searches [68], this strategy is typically used for only two to three objective optimizations.

- *Machine learning-based TCP*: The most straightforward way of using machine learning to combine heuristics for test prioritization is building prediction models (such as regression or Bayesian models) that accept a set of metrics (each based on one existing heuristic, e.g., coverage, historical failure detection, cost, diversity, change coverage) and predict the likelihood of "failure" for the test cases in the new release, based on historical data from previous releases. This idea falls under the category of history-based TCP approaches, since the learning dataset comes (typically) from the history. For example, Mirarab and Tahvildari [69] utilize Bayesian Networks (BNs) to incorporate source code changes, software fault-proneness, and test coverage data into a unified model. Noor and Hemmati [23] use a logistic regression model to aggregate metrics such as historical fault detection, code coverage, change coverage, and test case size for test failure prediction in the context of test prioritization.

  In another study [60], Hettiarachchi et al. used requirements "modification status," complexity, security, and size of the software requirements as risk indicators and employed a fuzzy expert system to estimate the requirements risks. The test would then be prioritized based on the estimated risk of the corresponding covered requirements.

Future directions (2): Based on the earlier discussions, proposing more intelligent and effective TCP heuristics are definitely one of the future directions in TCP. Domains such as search-based TCP and machine learning-based TCPs can be seen as the dominant players. However, use of information theory and diversity among test cases for TCP as well as use of NLP techniques is becoming attractive to researchers as well. One area that is not explored yet but has potentials is the use of Deep Learning for TCP. Deep Learning combined with NLP may be helpful to extract features from test cases that are not easily identifiable but yet very useful for prediction.

(c) *New application domains*

The last category of trends and future directions is about the application domain. Most earlier work on TCP studies was focused on test suites of simple C/C++/Java programs. Some of the new domains that have recently got special attention for test prioritization are product line software [66] and mobile applications [70].

- *Product line software*: In this context, typically a variability model (e.g., feature model) is used to prioritize test cases to cover software features. A feature model is a tree structure that captures the information of all the possible products of a software product line in terms of features and relationships among them [71]. The idea is to maximize the coverage of features across products that are being tested. Combinatorial testing [40] is also a useful approach that is being used in this domain. Note that most above discussed TCP techniques and approaches can be directly or indirectly applied in this domain. For example, Wang et al. [66] used a multiobjective TCP to maximize effectiveness (feature coverage) while minimizing the cost of test cases in a product line setting.
- *Mobile*: Another application domain that can be specifically benefitted from TCP is mobile application testing. One of the particular features of mobile applications that make them a perfect target for TCP is their battery limitation. Jabbarvand et al. proposed an energy-aware TCP approach that significantly reduces the number of tests needed to effectively test the energy properties of an Android app [70]. Basically, it is a coverage-based approach, but it relies on an energy-aware coverage criterion that indicates the degree to which energy-greedy segments of a program are tested. In another work, Zhai et al. [72] proposed a location-aware TCP. In their work a test case for a location-aware mobile app is a sequence of locations. They propose a suite of metrics that focus on the location of the device and points of interest (POI) for the application. For example, "sequence variance" measures the variance of a location sequence of a test case and "centroid distance" measures the distance from the centroid of a location sequence of a test case to the centroid of a set of POIs. The TCP tries to maximize the diversity of the locations while minimizing the relevance to the POIs.

Future directions (3): As soon as new application domains emerge, the TCP problem becomes relevant for them. Typically, as we saw for instance in the mobile app domain, the new TCPs will focus on features that are specific for such domains. Given the amount of interest in big data applications and cloud computing, we can predict that soon we will need TCP techniques that are focused for applications in these domains.

6. SUMMARY

TCP is one of the classic testing problems that have been studied in the literature for many years. There are several solutions for this problem depending on the available input resources for the prioritization approach and the heuristics employed. This chapter introduces some of these techniques and summarizes trends and future directions in the context of TCP.

REFERENCES

[1] G.J. Myers, C. Sandler, T. Badgett, The Art of Software Testing, John Wiley & Sons, 2011.
[2] J.A. Whittaker, What is software testing? And why is it so hard? IEEE Softw. 17 (1) (2000) 70–79.
[3] J. Humble, D. Farley, Continuous Delivery: Reliable Software Releases Through Build, Test, and Deployment Automation, Pearson Education, 2010.
[4] S. Yoo, M. Harman, Regression testing minimization, selection and prioritization: a survey, Softw. Test. Verification Reliab. 22 (2) (2010) 67–120.
[5] https://developers.google.com/google-test-automation-conference/2014/presentations. Accessed June 2017.
[6] E. Engström, P. Runeson, in: A qualitative survey of regression testing practices, International Conference on Product Focused Software Process Improvement, Springer Berlin Heidelberg, 2010.
[7] K. Beck, Test-Driven Development: By Example, Addison-Wesley Professional, 2003.
[8] H. Hemmati, Z. Fang, M.V. Mäntylä, B. Adams, Prioritizing manual test cases in rapid release environments, Softw. Test. Verif. Reliab. 27 (6) (2017) e1609. https://doi.org/10.1002/stvr.1609.
[9] R. Kazmi, D.N.A. Jawawi, R. Mohamad, I. Ghani, Effective regression test case selection: a systematic literature review, ACM Comput. Surv. 50 (2) (2017) 29:1–29:32.
[10] C. Catal, D. Mishra, Test case prioritization: a systematic mapping study, Softw. Q. J. 21 (3) (2013) 445–478.
[11] Z. Li, M. Harman, R.M. Hierons, Search algorithms for regression test case prioritization, IEEE Trans. Softw. Eng. 33 (4) (2007) 225–237.
[12] S.W. Thomas, H. Hemmati, A.E. Hassan, D. Blostein, Static test case prioritization using topic models, Empir. Softw. Eng. 19 (1) (2014) 182–212.
[13] W. Wong, J. Horgan, S. London, H. Agrawal, in: A study of effective regression testing in practice, International Symposium on Software Reliability Engineering, 1997, pp. 264–274.
[14] G. Rothermel, R. Untch, C. Chu, M.J. Harrold, Prioritizing test cases for regression testing, IEEE Trans. Softw. Eng. 27 (10) (2001) 929–948.
[15] A. Orso, T. Apiwattanapong, M.J. Harrold, Leveraging field data for impact analysis and regression testing, ACM SIGSOFT Softw. Eng. Notes ACM 28 (2003) 128–137.
[16] L.C. Briand, Y. Labiche, G. Soccar, in: Automating impact analysis and regression test selection based on UML designs, IEEE International Conference on Software Maintenance, 2002, pp. 252–261.
[17] A. Memon, in: Taming Google-Scale continuous testing, IEEE International Conference on Software Engineering: Software Engineering in Practice Track, 2017.
[18] M. D'Ambros, M. Lanza, R. Robbes, Evaluating defect prediction approaches: a benchmark and an extensive comparison, Empir. Softw. Eng. 17 (4–5) (2012) 531–577.

[19] K. Gao, T.M. Khoshgoftaar, H. Wang, N. Seliya, Choosing software metrics for defect prediction: an investigation on feature selection techniques, Softw. Pract. Exp. 41 (5) (2011) 579–606.
[20] J.M. Kim, A. Porter, in: A history-based test prioritization technique for regression testing in resource constrained environments, IEEE International Conference on Software Engineering, 2002, pp. 119–129.
[21] A.K. Onoma, W.T. Tsai, M. Poonawala, H. Suganuma, Regression testing in an industrial environment, Commun. ACM 41 (5) (1998) 81–86.
[22] S. Elbaum, G. Rothermel, J. Penix, in: Techniques for improving regression testing in continuous integration development environments, ACM SIGSOFT International Symposium on Foundations of Software Engineering, 2014, pp. 235–245.
[23] T.B. Noor, H. Hemmati, in: A similarity-based approach for test case prioritization using historical failure data, IEEE International Symposium on Software Reliability Engineering, 2015, pp. 58–68.
[24] S. Elbaum, A. Malishevsky, G. Rothermel, in: Incorporating varying test costs and fault severities into test case prioritization, IEEE International Conference on Software Engineering, 2001, pp. 329–338.
[25] S. Yoo, M. Harman, in: Pareto efficient multi-objective test case selection, ACM International Symposium on Software Testing and Analysis, 2007, pp. 140–150.
[26] M.G. Epitropakis, S. Yoo, M. Harman, E.K. Burke, in: Empirical evaluation of pareto efficient multi-objective regression test case prioritization, ACM International Symposium on Software Testing and Analysis, 2015, pp. 234–245.
[27] H. Mei, D. Hao, L. Zhang, J. Zhou, G.A. Rothermel, Static approach to prioritizing JUnit test cases, IEEE Trans. Softw. Eng. 38 (6) (2012) 1258–1275.
[28] H. Hemmati, A. Arcuri, L. Briand, Achieving scalable model-based testing through test case diversity, ACM Trans. Softw. Eng. Methodol. 22 (1) (2013) 42. Article 6.
[29] H. Hemmati, L. Briand, A. Arcuri, S. Ali, in: An enhanced test case selection approach for model-based testing: an industrial case study, ACM SIGSOFT International Symposium on Foundations of Software Engineering, 2010, pp. 267–276.
[30] Y. Chen, R.L. Probert, D.P. Sims, in: Specification-based regression test selection with risk analysis, Conference of the Center for Advanced Studies on Collaborative Research (CASCON), 2002, pp. 1–14.
[31] M.J. Arafeen, H. Do, in: Test case prioritization using requirements-based clustering, IEEE International Conference on Software Testing, Verification, and Validation, 2013, pp. 312–321.
[32] R. Krishnamoorthi, S.A. Sahaaya Arul Mary, Factor oriented requirement coverage based system test case prioritization of new and regression test cases, Inf. Softw. Technol. 51 (4) (2009) 799–808.
[33] H. Srikanth, L. Williams, J. Osborne, in: System test case prioritization of new and regression test cases, International Symposium on Empirical Software Engineering, 2005, pp. 64–73.
[34] Y. Ledru, A. Petrenko, S. Boroday, in: Using string distances for test case prioritization, International Conference on Automated Software Engineering, 2009, pp. 510–514.
[35] Y. Ledru, A. Petrenko, S. Boroday, N. Mandran, Prioritizing test cases with string distances, Autom. Softw. Eng. 19 (1) (2011) 65–95.
[36] H. Hemmati, Z. Fang, M.V. Mäntylä, B. Adams, in: Prioritizing manual test cases in traditional and rapid release environments, IEEE International Conference on Software Testing, Verification, and Validation, 2015, pp. 1–10.
[37] R.C. Bryce, S. Sampath, A.M. Memon, Developing a single model and test prioritization strategies for event-driven software, IEEE Trans. Softw. Eng. 37 (1) (2011) 48–64.

[38] R.C. Bryce, A.M. Memon, in: Test suite prioritization by interaction coverage, Workshop on Domain Specific Approaches to Software Test Automation: In Conjunction With ACM ESEC/FSE Joint Meeting, 2007.
[39] R.C. Bryce, C.J. Colbourn, Prioritized interaction testing for pair-wise coverage with seeding and constraints, Inf. Softw. Technol. 48 (10) (2006) 960–970.
[40] D.R. Kuhn, R.N. Kacker, Y. Lei, Introduction to Combinatorial Testing, CRC Press, 2013.
[41] S. Sampath, R.C. Bryce, G. Viswanath, V. Kandimalla, A.G. Koru, in: Prioritizing user-session-based test cases for web applications testing, IEEE International Conference on Software Testing, Verification, and Validation, 2008.
[42] F. Sharifi, H. Hemmati, Investigating NLP-based approaches for predicting manual test case failure, 11th IEEE Conference on Software Testing, Validation and Verification (ICST 2018), to appear.
[43] https://en.wikipedia.org/wiki/Local optimum. Accessed June 2017.
[44] https://en.wikipedia.org/wiki/Greedy_algorithm. Accessed June 2017.
[45] S. Ali, L. Briand, H. Hemmati, R.K. Panesar-Walawege, A systematic review of the application and empirical investigation of search-based test case generation, IEEE Trans. Softw. Eng. 36 (6) (2010) 742–762.
[46] M. Harman, B.F. Jones, Search-based software engineering, Inf. Softw. Technol. 43 (2001) 833–839.
[47] A.P. Mathur, Foundations of Software Testing, first ed., Addison-Wesley Professional, 2008.
[48] D. Wolpert, W.G. Macready, No free lunch theorems for optimization, IEEE Trans. Evol. Comput. 1 (1997) 67–82.
[49] Y. Chen, R.L. Probert, H. Ural, Regression test suite reduction based on SDL models of system requirements, J. Softw. Maint. Evol. Res. Pract. 21 (2009) 379–405.
[50] H. Hemmati, Similarity-Based Test Case Selection: Toward Scalable and Practical Model-Based Testing, PhD thesis, Simula Research Laboratory and Informatic Department, University of Oslo, 2011.
[51] D. Mondal, H. Hemmati, S. Durocher, in: Exploring test suite diversification and code coverage in multi-objective test case selection, IEEE International Conference on Software Testing, Verification and Validation, 2015.
[52] H. Hemmati, L. Briand, in: An industrial investigation of similarity measures for model-based test case selection, IEEE International Symposium on Software Reliability Engineering, 2010, pp. 141–150.
[53] G. Dong, J. Pei, Sequence Data Mining, Springer, 2007.
[54] R. Durbin, S.R. Eddy, A. Krogh, G. Mitchison, Biological Sequence Analysis: Probabilistic Models of Proteins and Nucleic Acids, Cambridge University Press, 1999.
[55] D. Gusfield, Algorithms on Strings, Trees and Sequences, Computer Science and Computational Biology, Cambridge University Press, 1997.
[56] T.Y. Chen, F.-C. Kuoa, R.G. Merkela, T.H. Tseb, Adaptive random testing: the ART of test case diversity, J. Syst. Softw. 83 (2010) 60–66.
[57] L.J. White, E.I. Cohen, A domain strategy for computer program testing, IEEE Trans. Softw. Eng. 6 (1980) 247–257.
[58] A. Arcuri, M.Z.I. Andrea, L. Briand, Random testing: theoretical results and practical implications, IEEE Trans. Softw. Eng. 38 (2) (2012) 258–277.
[59] I.H. Witten, E. Frank, M.A. Hall, C.J. Pal, Data Mining: Practical Machine Learning Tools and Techniques, Morgan Kaufmann, 2016.
[60] C. Hettiarachchi, H. Do, B. Choi, Risk-based test case prioritization using a fuzzy expert system, Inf. Softw. Technol. 69 (2016) 1–15.

[61] Y. Singh, A. Kaur, B. Suri, Test case prioritization using ant colony optimization, ACM SIGSOFT Softw. Eng. Notes 35 (4) (2010) 1–7.
[62] H. Hemmati, A. Arcuri, L. Briand, in: Empirical investigation of the effects of test suite properties on similarity-based test case selection, IEEE International Conference on Software Testing, Verification and Validation, 2011.
[63] R. Feldt, S. Poulding, D. Clark, S. Yoo, in: Test set diameter: quantifying the diversity of sets of test cases, IEEE International Conference on Software Testing, Verification and Validation, 2016.
[64] Q. Shi, Z. Chen, C. Fang, Y. Feng, B. Xu, Measuring the diversity of a test set with distance entropy, IEEE Trans. Reliab. 65 (1) (2016) 19–27.
[65] J.H. Kwon, I.Y. Ko, G. Rothermel, M. Staats, in: Test case prioritization based on information retrieval concepts, IEEE Asia-Pacific Software Engineering Conference, 2014.
[66] S. Wang, D. Buchmann, S. Ali, A. Gotlieb, D. Pradhan, M. Liaaen, in: Multi-objective test prioritization in software product line testing: an industrial case study, International Software Product Line Conference, 2014.
[67] S. Yoo, M. Harman, P. Tonella, A. Susi, in: Clustering test cases to achieve effective and scalable prioritization incorporating expert knowledge, ACM International Symposium on Software Testing and Analysis, 2009.
[68] S. Yoo, M. Harman, S. Ur, in: Highly scalable multi objective test suite minimization using graphics cards, International Symposium on Search Based Software Engineering, Springer Berlin Heidelberg, 2011.
[69] S. Mirarab, L. Tahvildari, in: A prioritization approach for software test cases based on Bayesian networks, International Conference on Fundamental Approaches to Software Engineering, Springer Berlin Heidelberg, 2007.
[70] R. Jabbarvand, A. Sadeghi, H. Bagheri, S. Malek, in: Energy-aware test-suite minimization for Android apps, ACM International Symposium on Software Testing and Analysis, 2016.
[71] A.B. Sánchez, S. Segura, A. Ruiz-Cortés, in: A comparison of test case prioritization criteria for software product lines, IEEE International conference on Software Testing, Verification and Validation, 2014.
[72] K. Zhai, B. Jiang, W.K. Chan, Prioritizing test cases for regression testing of location-based services: metrics, techniques, and case study, IEEE Trans. Serv. Comput. 7 (1) (2014) 54–67.

ABOUT THE AUTHOR

Dr. Hadi Hemmati is an assistant professor at the Department of Electrical and Computer Engineering, University of Calgary. Before joining University of Calgary, Dr. Hemmati was an assistant professor at the Department of Computer Science, University of Manitoba, Canada, and a postdoctoral fellow at University of Waterloo, and Queen's University. He received his PhD from Simula Research Laboratory, Norway. His main research interest is Automated Software Engineering with a focus on software testing and quality assurance. His research has a strong focus on empirically investigating software engineering practices in large-scale systems, using model-driven engineering and data science. He has/had industry research collaborations with several companies around the world.

CHAPTER FIVE

Data Warehouse Testing

Hajar Homayouni, Sudipto Ghosh, Indrakshi Ray
Colorado State University, Fort Collins, CO, United States

Contents

Abstract

Enterprises use data warehouses to accumulate data from multiple sources for data analysis and research. Since organizational decisions are often made based on the data stored in a data warehouse, all its components must be rigorously tested. Researchers have proposed a number of approaches and tools to test and evaluate different

Advances in Computers, Volume 112
ISSN 0065-2458
https://doi.org/10.1016/bs.adcom.2017.12.005

components of data warehouse systems. In this chapter, we present a comprehensive survey of data warehouse testing techniques. We define a classification framework that can categorize the existing testing approaches. We also discuss open problems and propose research directions.

1. INTRODUCTION

A data warehouse system gathers heterogeneous data from several sources and integrates them into a single data store [1]. Data warehouses are used for reporting and data analysis and form the core component of business intelligence (BI) [2]. The goal of data warehouses is to help researchers and data analyzers perform faster analysis and make better decisions [3]. Data warehousing also makes it possible to do data mining, which is the science of discovering patterns in the data for further decision-making, such as predictions or classifications [4]. Data warehouses often use large-scale (petabyte) data stores to keep archival as well as current data to enable data analyzers to find precise patterns based on long-term changes in the data.

Data warehouses are used in many application domains. A health data warehouse brings electronic health records from many hospitals into a single destination to help medical research on disease, drugs, and treatments. While each hospital focuses on transactions for current patients, the health data warehouse maintains historical data from multiple hospitals. This history often includes old patient records. The past records along with the new updates help medical researchers perform long-term data analysis. A weather data warehouse gathers observations from stations all around the world into a single data store to enable weather forecasting and climate change detection.

There are many components and processes involved in data warehousing. Fig. 1 shows the different components of a data warehouse system including (1) sources, (2) extract, transform, and load (ETL) process, (3) data warehouse, and (4) front-end applications.

The sources of a data warehousing system are data stores that provide data from various places. These sources store the data of entities belonging to one or more organizations. For example, the sources of a health data warehouse are obtained from multiple hospitals that are collaborating for medical research. There are different *models*, such as relational [5] or nonrelational [6] models, and *technologies*, such as database management systems

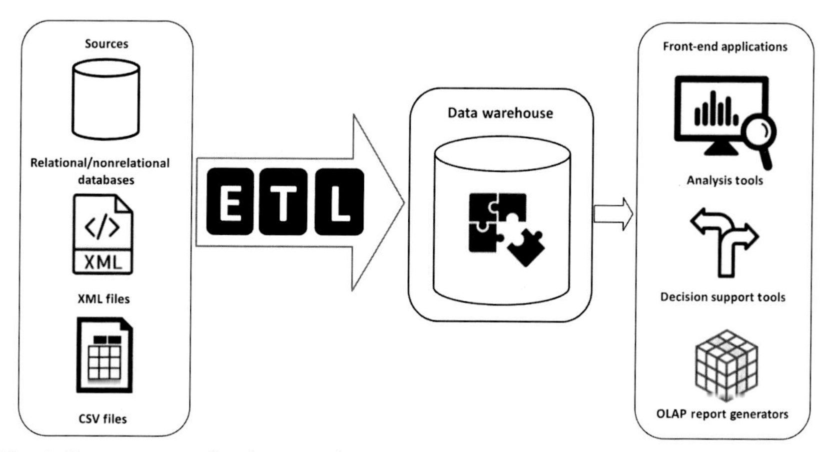

Fig. 1 Components of a data warehousing system.

(DBMSs) or extensible markup language (XML) or comma separated values (CSV) flat files that form the sources of data warehousing systems.

The ETL process selects data from the sources, resolves problems in the data, converts it into a common model appropriate for research and analysis, and writes it to the target data warehouse [1]. Among the four components presented in Fig. 1, the design and implementation of the ETL process require the largest effort in its development life cycle [3]. The ETL process presents many challenges, such as extracting data from multiple heterogeneous sources involving different models, detecting and fixing different types of errors in the data, and transforming the data into different formats that match the requirements of the target data warehouse.

The data warehouse keeps data gathered and integrated from different sources and stores a large number of records needed for long-term analysis. Implementations of data warehouses use various *data models*, such as dimensional or normalized models, and *technologies*, such as DBMS, data warehouse appliance (DWA), and cloud data warehouse appliance.

The front-end applications are in the form of desktop, web, and mobile applications that present business data with analysis to end users [3]. They include analysis and decision support tools and online analytical processing (OLAP) report generators. These applications make it easy for end users to construct complex queries to request information from data warehouses without requiring sophisticated programming skills.

Research is conducted and organizational decisions are made based on the data stored in a data warehouse [7]. For example, based on our health data warehouse, many critical studies such as the impacts of a specific medication on people from different groups are performed using patient, treatment, and medication data stored in the data warehouse. Thus, all the components of a data warehouse must be thoroughly tested using rigorous testing techniques. Although data warehouse design and implementation have received considerable attention in the literature, few systematic techniques have been developed for data warehouse testing [8].

There are a number of challenges in testing data warehouse systems:

- Heterogeneous sources and voluminous data involved in data warehousing make data warehouse testing harder than testing traditional software systems [7]. A comprehensive testing approach should take into account all possible sources and test inputs for adequate testing.
- Due to the confidentiality of data in the sources of data warehousing systems, testers typically do not have access to the real data. As a result, there is a requirement to create fake data with the relevant characteristics of real data that enables adequate testing.
- Most data warehouse testing approaches are created for a specific context and cannot be used for testing other data warehouse systems. This problem arises because testing approaches are typically designed based on business domain requirements and the data warehouse architecture. These testing approaches cannot be generalized and reused in projects with different domain requirements [9] and architectures.
- Data warehouse testing requirements are not formally specified, which makes them hard to verify. The tester needs to bridge the gap between the informal specifications and the formality required for verification and validation techniques [10].

ElGamal [9] presented several data warehouse testing approaches, and evaluated and compared them to highlight their limitations. The survey reported the comparison based on *what* (referring to the testing type), *where* (implying the data warehousing stage where the testing is applied), and *when* (stating whether the test takes place before or after data warehouse delivery) in a three-dimensional matrix. In this matrix, the rows represent the *where*-dimension that takes four values, namely, *sources to data store*, *data store to data warehouse*, *data warehouse to data mart* (a subset of data warehouse), and *data mart to front-end applications*. The columns of the matrix represent the *what*-dimension that takes three values, namely, *schema related tests*, *data related tests*, and *operational related tests*. The third dimension of the matrix is the *when*-dimension that takes two values, namely, *before system delivery* and *after system delivery*. The survey compared 10 data warehouse testing approaches and concluded that none of them addressed all the *what*, *where*, and *when* categories, and there are some test types that are not addressed by any of these approaches. In the matrix, the component being tested and the testing type are often described together which makes it hard to understand the comparison matrix. For example, the entries *data model* and *requirement testing* both fall under the *what* dimension of the matrix, which results in ambiguities in the interpretation of the matrix.

Gao et al. [11] compared contemporary data warehouse testing tools for data validation in terms of their operating environment, supported data sources, data validation checks, and applied case studies. This survey compared commercial and open source approaches that test the quality of the underlying data in the target data warehouse but did not consider approaches for testing the other data warehouse components that are shown in Fig. 1.

In this chapter, we present a comprehensive survey of existing testing and evaluation activities applied to the different components of data warehouses and discuss the specific challenges and open problems for each component. These approaches include both dynamic analysis as well as static evaluation and manual inspections. We provide a classification framework based on *what* is tested in terms of the data warehouse component to be verified, and *how* it is tested through categorizing the different testing and evaluation approaches. The survey is based on our direct experience with a health data warehouse, as well as from existing commercial and research efforts in developing data warehouse testing approaches. The rest of the chapter is organized as follows. Section 2 describes the components of a data warehouse. Section 3 presents a classification framework for testing data warehouse components. Sections 4–6 discuss existing approaches and their limitations for each testing activity. Finally, Section 7 concludes the chapter and outlines directions for future work.

2. DATA WAREHOUSE COMPONENTS

In this section, we describe the four components of a data warehousing system, which are (1) sources, (2) target data warehouse, (3) ETL process, and (4) front-end applications. We use the health data warehouse as a running example.

2.1 Sources and Target Data Warehouse

Sources in a data warehousing system store data belonging to one or more organizations for daily transactions or business purposes. The target data warehouse, on the other hand, stores large volumes of data for long-term analysis and mining purposes. Sources and target data warehouses can be designed and implemented using a variety of technologies including data models and data management systems.

A data model describes business terms and their relationships, often in a pictorial manner [12]. The following data models are typically used to design the source and target schemas:

- *Relational data model*: Such a model organizes data as collections of two-dimensional tables [5] with all the data represented in terms of tuples. The tables are *relations* of rows and columns, with a unique key for each row. Entity relationship (ER) diagrams [13] are generally used to design the relational data models.
- *Nonrelational data model*: Such a model organizes data without a structured mechanism to link data of different buckets (segments) [6]. These models use means other than the tables used in relational models. Instead, different data structures are used, such as graphs or documents. These models are typically used to organize extremely large datasets used for data mining because unlike the relational models, the nonrelational models do not have complex dependencies between their buckets.
- *Dimensional data model*: Such a model uses structures optimized for end-user queries and data warehousing tools. These structures include *fact* tables that keep measurements of a business process, and *dimension* tables that contain descriptive attributes [14]. Unlike relational models that minimize data redundancies and improve transaction processing, the dimensional model is intended to support and optimize queries. The dimensional models are more scalable than relational models because they eliminate the complex dependencies that exist between relational tables [15].

  The dimensional model can be represented by star or snowflake schemas [16] and is often used in designing data warehouses. These types of schemas are as follows:
  - *Star*: This type of schema has a fact table at the center. The table contains the keys to dimension tables. Each dimension includes a set of attributes and is represented via a one dimension table [17].
  - *Snowflake*: Unlike the star schema, the snowflake schema has normalized dimensions that are split into more than one dimension tables. The star schema is a special case of the snowflake schema with a single level hierarchy.

The sources and data warehouses use various data management systems to collect and organize their data. The following is a list of data management systems generally used to implement the source and target data stores.

- *Relational database management system (RDBMS)*: An RDBMS is based on the relational data model that allows linking of information from different *tables*. A table must contain what is called a key or index, and other tables may refer to that key to create a link between their data [6]. RDBMSs typically use Structured Query Language (SQL) [18] and are appropriate to manage structured data. RDBMSs are able to handle queries and transactions that ensure efficient, correct, and robust data processing even in the presence of failures.
- *Nonrelational database management system*: A nonrelational DBMS is based on a nonrelational data model. The most popular nonrelational database is Not Only SQL (NoSQL) [6], which has many forms, such as document-based, graph-based, and object-based. A nonrelational DBMS is typically used to store and manage large volumes of unstructured data.
- *Big data management system*: Management systems for big data need to store and process large volumes of both structured and unstructured data. They incorporate technologies that are suited to managing nontransactional forms of data. A big data management system seamlessly incorporates relational and nonrelational database management systems.
- *Data warehouse appliance (DWA)*: DWA was first proposed by Hinshaw [19] as an architecture suitable for data warehousing. DWAs are designed for high-speed analysis of large volumes of data. A DWA integrates database, server, storage, and analytics into an easy-to-manage system.
- *Cloud data warehouse appliance*: Cloud DWA is a data warehouse appliance that runs on a cloud computing platform. This appliance benefits from all the features provided by cloud computing, such as collecting and organizing all the data online, obtaining infinite computing resources on demand, and multiplexing workloads from different organizations [20].

Table 1 presents some of the available products used in managing the data in the sources and target data warehouses. The design and implementation of the databases in the sources are typically based on the organizational requirements, while those of the data warehouses are based on the requirements of data analyzers and researchers.

For example, the sources for a health data warehouse are databases in hospitals and clinic centers that keep patient, medication, and treatment information in several formats. Fig. 2 shows an example of possible sources in the health data warehouse. Hospital A uses a flat spreadsheet to keep records of

Table 1 Available Products for Managing Data in the Sources and Data Warehouses

Product Category	Examples
DBMS	Relational: MySQL [21], MS-SQL Server [22], PostgreSQL [23]
	Nonrelational: Accumulo [24], ArangoDB [25], MongoDB [26]
Big data management system	Apache Hadoop [27], Oracle [28]
Data warehouse appliance	IBM PureData System [29]
Cloud data warehouse	Google BigQuery [30], Amazon Redshift [31]

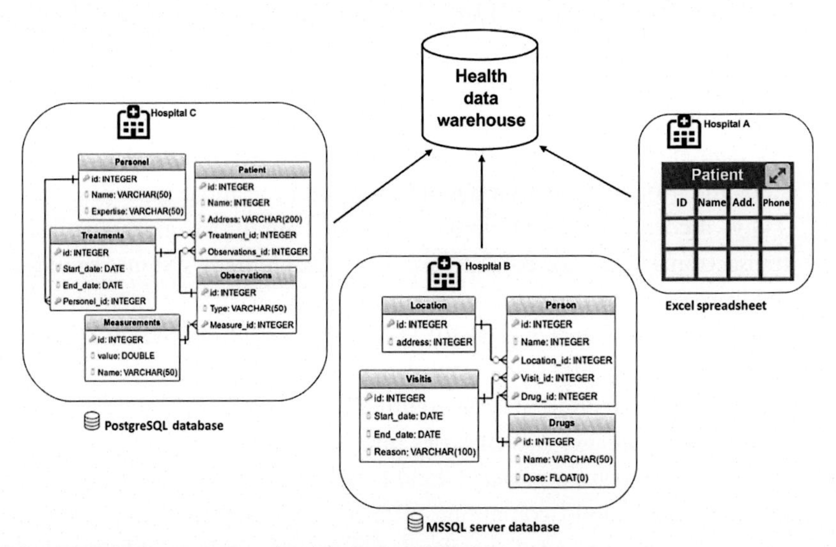

Fig. 2 Sample sources for a health data warehouse.

patient data. Hospital B uses an RDBMS for its data. Hospital C also uses an RDBMS but has a different schema than Hospital B. The data from different hospitals must be converted to a common model in the data warehouse.

The target data warehouse for health data may need to conform to a standard data model designed for electronic health records such as Observational Medical Outcomes Partnership (OMOP) Common Data Model (CDM) [32]. The OMOP CDM is a dimensional model that includes all the observational health data elements that are required for analysis use cases. The model supports the generation of reliable scientific evidence about disease, medications, and health outcomes.

2.2 Extract, Transform, and Load

The ETL process extracts data from sources, transforms it to a common model, and loads it to the target data warehouse. Fig. 3 shows the components involved in the ETL process, namely, extract, transform, and load.

1. *Extract*: This component retrieves data from heterogeneous sources that have different formats and converts the source data into a single format suitable for the transformation phase. Different procedural languages such as Transact-SQL or COBOL are required to query the source data. Most extraction approaches use Java Database Connectivity (JDBC) or Open Database Connectivity (ODBC) drivers to connect to sources that are in DBMS or flat file formats [33].

   Data extraction is performed in two phases. Full extraction is performed when the entire data is extracted for the first time. Incremental extraction happens when new or modified data are retrieved from the sources. Incremental extraction employs strategies such as log-based, trigger-based, or timestamp-based techniques to detect the newly added or modified data. In the log-based technique, the DBMS log files are used to find the newly added or modified data in the source databases. Trigger-based techniques create triggers on each source table to capture changed data. A trigger automatically executes when data is created or modified through a Data Manipulation Language (DML) event. Some database management systems use timestamp columns to specify the time and date that a given row was last modified. Using these columns, the timestamp-based technique can easily identify the latest data.

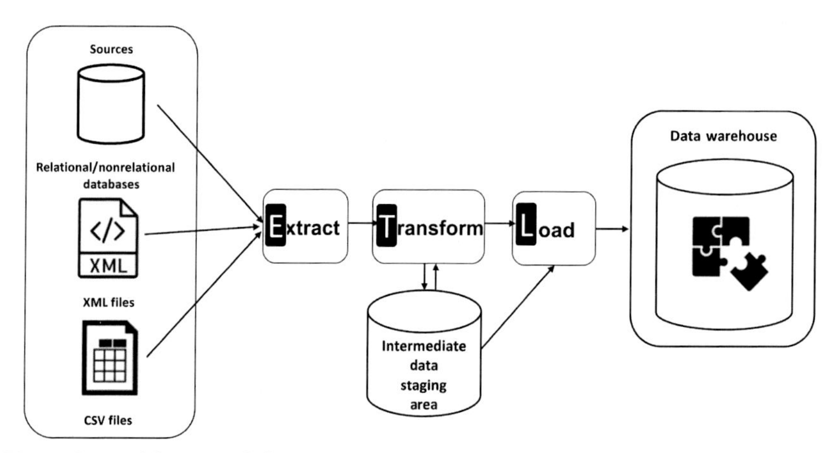

Fig. 3 General framework for ETL processes.

2. *Transform*: This component propagates data to an intermediate data staging area (DSA) where it is cleansed, reformatted, and integrated to suit the format of the model of a target data warehouse [3]. This component has two objectives.

   First, the transformation process cleans the data by identifying and fixing (or removing) the existing problems in the data and prepares the data for integration. The goal is to prevent the transformation of so-called dirty data [34, 35]. The data extracted from the sources is validated both syntactically and semantically to ensure that it is correct based on the source constraints. Data quality validation and data auditing approaches can be utilized in this step to detect the problems in the data. Data quality validation approaches apply quality rules to detect syntactic and semantic violations in the data. Data auditing approaches use statistical and database methods to detect anomalies and contradictions in the data [36]. In Table 2 we present some examples of data quality validation applied to data cleansing of patients in our health data warehouse.

   Second, it makes the data conform to the target format through the application of a set of transformation rules described in the source-to-target mapping documents provided by the data warehouse designers [33]. Table 3 presents examples of source-to-target mappings for generating a target table called *Patient* in our health data warehouse. The mappings include the names of the corresponding source and target tables, the source and target columns with their types, and selection conditions.
3. *Load*: This component writes the extracted and transformed data from the staging area to the target data warehouse [1]. The loading process varies widely based on the organizational requirements. Some data warehouses may overwrite existing data with new data on a daily, weekly, or

Table 2 Examples of Validation Applied to Data Cleansing

Validation	Example of a Violation
Incorrect value check	Birth_date=70045 is not a legal date format
Uniqueness violation check	Same SSN='123456789' presented for two people
Missing value check	Gender is null for some records
Wrong reference check	Referenced hospital=1002 does not exist
Value dependency violation check	Country='Germany' does not match zip code='77'

Table 3 Transforming Source Data to Generate Target Table *Patient*

Source			Target			
Table Name	**Column Name**	**Data Type**	**Table Name**	**Column Name**	**Data Type**	**Selection Condition**
PersonDim	PersonKey	Integer	Patient	Patient_id	Integer	Transform all the current patients
PersonDim	Name	String	Patient	Patient_name	String	Transform all the patients
PersonDim, AddressDim	AddressKey	Integer	Patient	Location_id	Integer	Transform all the patients with new addresses (after year 2000)
PersonDim, Concept	Sex	String	Patient	Gender	Integer	Transform all the patients with their sex using concept values female, male, other

monthly basis, while other data warehouses may keep the history of data by adding new data at regular intervals. The load component is often implemented using loading jobs that fully or incrementally transform data from DSA to the data warehouse. The full load transforms the entire data from the DSA, while the incremental load updates newly added or modified data to the data warehouse based on logs, triggers, or timestamps defined in the DSA.

The ETL components, namely, extract, transform, and load, are not independent tasks, and they need to be executed in the correct sequence for any given data. However, parallelization can be achieved if different components execute on distinct blocks of data. For example, in the incremental mode the different components can be executed simultaneously; the newly added data can be extracted from the sources while the previously extracted block of data is being transformed and loaded into the target data warehouse.

2.3 Front-End Applications

Front-end applications present the data to end users who perform analysis for the purpose of reporting, discovering patterns, predicting, or making complex decisions. These applications can be any of the following tools:

- *OLAP report generators*: These applications enable users and analysts to extract and access a wide variety of views of data for multidimensional analysis [37]. Unlike traditional relational reports that represent data in two-dimensional row and column format, OLAP report generators represent their aggregated data in a multidimensional structure called cube to facilitate the analysis of data from multiple perspectives [38]. OLAP supports complicated queries involving facts to be measured across different dimensions. For example, as Fig. 4 shows, an OLAP report can present a comparison of the number (fact) of cases reported for a disease (dimension) over years (dimension), in the same region (dimension).
- *Analysis and data mining*: These applications discover patterns in large datasets helping users and data analysts understand data to make better decisions [4]. These tools use various algorithms and techniques, such as classification and clustering, regression, neural networks, decision trees, nearest neighbor, and evolutionary algorithms for knowledge discovery from data warehouses. For example, clinical data mining techniques [39] are aimed at discovering knowledge from health data to extract valuable information, such as the probable causes of diseases, nature of progression, and drug effects.

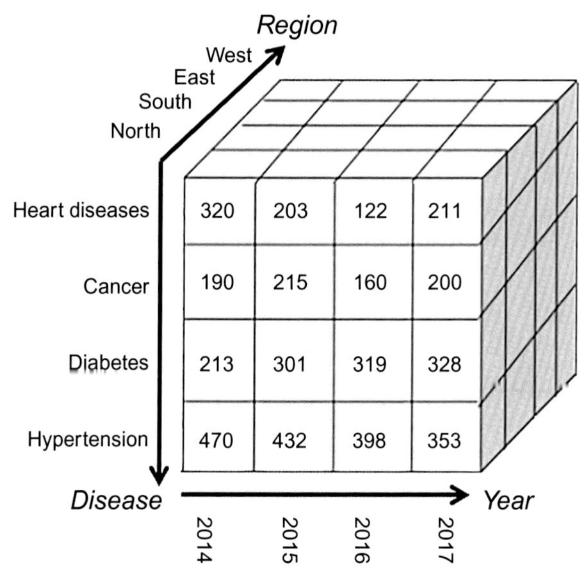

Fig. 4 OLAP cube example of the number of cases reported for diseases over time and regions.

- *Decision support*: These applications support the analysis involved in complex decision-making and problem solving processes [40] that involve sorting, ranking, or choosing from options. These tools typically use Artificial Intelligence techniques, such as knowledge base or machine learning to analyze the data. For example, a Clinical Decision Support [41] application provides alerts and reminders, clinical guidelines, patient data reports, and diagnostic support based on the clinical data.

3. TESTING DATA WAREHOUSE COMPONENTS

Systematic testing and evaluation techniques have been proposed by researchers and practitioners to verify each of the four components of a data warehouse to ensure that they perform as expected. We present a comprehensive survey by defining a classification framework for the testing and evaluation techniques applied to each of the four components.

Fig. 5 shows the classification framework for the techniques applicable to the sources, target data warehouse, ETL process, and front-end applications. The framework presents *what* is tested in terms of data warehouse components, and *how* they are tested. The following are the data warehouse components presented in the framework:

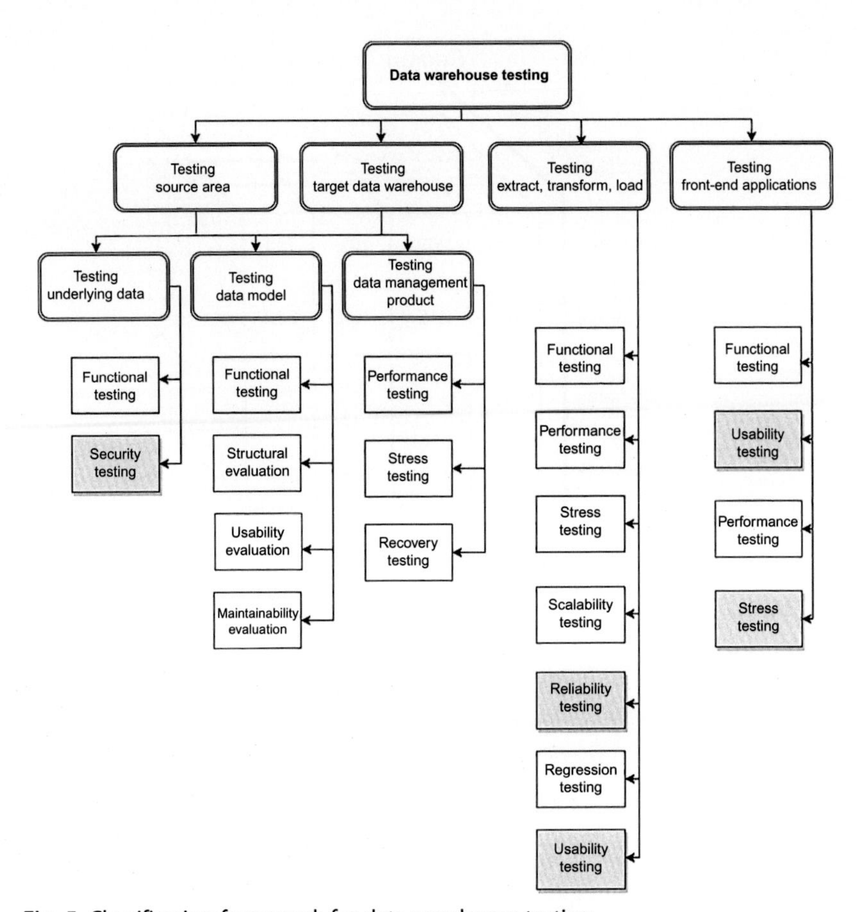

Fig. 5 Classification framework for data warehouse testing.

- *The sources and the target data warehouse* store data. As a result, the same types of testing and evaluation techniques apply to them. We consider three different aspects to classify the approaches used to test these two components; these are (1) testing the underlying data, (2) testing the data model, and (3) testing the product used to manage the data.
- *The ETL process* requires the largest effort in the data warehouse development life cycle [3]. As a result, most existing data warehouse testing and evaluation approaches focus on this process. Various functional and nonfunctional testing methods have been applied to test the ETL process because it directly affects the quality of data inside the data warehousing systems.
- *The front-end applications* in data warehousing systems provide an interface for users to help them interact with the back-end data store.

We categorize the existing testing and evaluation approaches as functional, structural, usability, maintainability, security, performance and stress, scalability, reliability, regression, and recovery testing. The shaded boxes represent the categories not covered by the existing testing approaches but that we feel are needed based on our experience with a real-world health data warehouse project.

Other researchers have also defined frameworks for testing techniques that are applicable to the different components of a data warehouse. Golfarelli and Rizzi [8] proposed a framework to describe and test the components in a data warehouse. They defined the data warehouse components as *schema*, *ETL*, *database*, and *front end applications*. However, the *schema* and *database* are not exactly data warehouse components. Instead they are features of the sources and the target data warehouses. The framework uses seven different testing categories (functional, usability, performance, stress, recovery, security, and regression) applicable to each of the data warehouse components. Some nonfunctional testing techniques such as those for assessing scalability, reliability, and maintainability are not included.

Mathen [1] surveyed the process of data warehouse testing considering two testing aspects, which were (1) testing underlying data and (2) testing the data warehouse components. The paper focused on two components in the data warehouse architecture, i.e., the ETL process and the client applications, and discussed testing strategies relevant to these components. Performance, scalability, and regression testing categories were introduced. Although testing the sources and the target data warehouse is critical to ensuring the quality of the entire data warehouse system, they were ignored in Mathen's testing framework. Moreover, other functional and non-functional aspects of testing data warehouse components, such as security, usability, reliability, recovery, and maintainability testing, and existing methods and tools for each testing type were not included.

In Sections 4–6, we describe the testing and evaluation activities necessary for each component in detail and present the challenges and open problems.

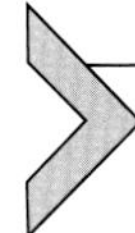

4. TESTING SOURCE AREA AND TARGET DATA WAREHOUSE

In this section, we target the locations that store the data in a data warehousing system, namely, the sources and the target data warehouse. If problems exist in the sources, they should be resolved before the data is extracted and loaded into a target where fault localization is much more expensive [7]. Fault localization is the process of finding the location of faults in a software system. Due to the fact that there are many components and processes involved in data warehousing systems, if the faulty data are propagated to the target data warehouse, finding the location of the original fault that caused subsequent error states will require a lot of effort. As a result, testing the source area is critical to ensuring the quality of data being propagated to the target data warehouse.

The quality of the target storage area is also important [42] because this is the place where the data analyzers and researchers apply their queries either directly or through the front-end applications. Any problem in the target data warehouse results in incorrect information. Thus, testing must ensure that the target meets the specifications and constraints defined for the data warehouse.

We considered three different aspects to test the source area and the target data warehouse. These are (1) testing the underlying data, (2) testing the data model, and (3) testing the data management product.

4.1 Testing Underlying Data

In this testing activity, the data stored in the sources and the target data warehouse is validated against organizational requirements, which are provided by domain experts in the form of a set of rules and definitions for valid data. If the underlying data fails to meet the requirements, any knowledge derived from the data warehouse will be incorrect [43].

We describe existing functional and security testing approaches based on testing the underlying data in data warehouses as well as propose approaches based on our experience to achieve high quality data in a health data warehouse.

4.1.1 Functional Testing of Underlying Data

Functional testing of the underlying data is a type of data quality testing that validates the data based on quality rules extracted from business requirements documents. The data quality test cases are defined as a set of queries that verify whether the data follows the syntactic and semantic rules. This testing activity uses domain-specific rules, which are a set of business rules that are internal to an organization.

Examples of the data elements that are verified using data quality tests are as follows:

- *Data type*: A data type is a classification of the data that defines the operations that can be performed on the data and the way the values of the data can be stored [44]. The data type can be numeric, text, Boolean, or date-time; these are defined in different ways in different languages.
- *Data constraint*: A constraint is a restriction that is placed on the data to define the values the data can take. Primary key, foreign key, and not-null constraints are typical examples.

Examples of semantic properties that we suggest are as follows:

- *Data plausibility*: A restriction that is placed on the data to limit the possible values it can take. For example, a US zip code can only take five digit values.
- *Logical constraint*: A restriction defined for the logical relations between data. For example, the *zip code=33293* does not match the *country=Germany*.

The data quality rules are not formally specified in the business requirements. The tester needs to bridge the gap between informal specifications and formal quality rules. Table 4 presents some examples of informally defined data quality rules for electronic health records [45]. Table 5 shows test cases defined as queries to verify the data quality rules presented in Table 4. Assume that after executing the test cases (queries), the test results are stored in a table called *tbl_test_results*. In this table, each record describes the failed assertion. The record includes the *test_id* that indicates the query number, *status* that takes as values *error* and *warning*, and *description* that contains a brief message about the failure. An empty table indicates that all the assertions passed.

Table 4 Data Quality Rules for Electronic Health Records

	Field	Data Quality Rule	Property
1	Weight	Should not be negative	Semantic (data plausibility)
2	Weight	Should be a numeric value	Syntactic (data type)
3	Sex	Should be *male* or *female* or *other*	Semantic (data plausibility)
4	Sex	Should not be null	Syntactic (data constraint)
5	Start_date, End_date	Start_date of patient visit should be before End_date	Semantic (logical constraint)
6	Start_date, End_date	Should be a date value	Syntactic (data type)

Table 5 Test Cases to Assess Electronic Health Records

	Query
1	INSERT INTO tbl_test_results (test_id, status, description) values (SELECT 1 AS test_id, 'error' AS status, 'weight is negative' AS description FROM tbl_patients WHERE weight<0)
2	INSERT INTO tbl_test_results (test_id, status, description) values (SELECT 2 AS test_id, 'error' AS status, 'weight is nonnumeric' AS description FROM tbl_patients WHERE weight.type<>DOUBLE OR weight.type<>INTEGER OR weight.type<>FLOAT)
3	INSERT INTO tbl_test_results (test_id, status, description) values (SELECT 3 AS test_id, 'error' AS status, 'Sex is invalid' AS description FROM tbl_patients WHERE !(Sex='Male' OR Sex='Female' OR Sex='Other'))
4	INSERT INTO tbl_test_results (test_id, status, description) values (SELECT 4 AS test_id, 'error' AS status, 'Sex is null' AS description FROM tbl_patients WHERE Sex=null)
5	INSERT INTO tbl_test_results (test_id, status, description) values (SELECT 5 AS test_id, 'error' AS status, 'start date is greater than end date' AS description FROM tbl_patients WHERE Start_date>End_date)
6	INSERT INTO tbl_test_results (test_id, status, description) values (SELECT 6 AS test_id, 'error' AS status, 'Invalid dates' AS description FROM tbl_patients WHERE Start_date.type<>Date OR End_date.type<>Date)

Data profiling [7] and data auditing [36] are statistical analysis tools that verify the data quality properties to assess the data and detect business rule violations, as well as anomalies and contradictions in the data. These tools are often used for testing the quality of data at the sources with the goal of rectifying data before it is loaded to the target data warehouse [33].

There exist data validation tools that perform data quality tests focusing on the target data. Data warehouse projects are typically designed for specific business domains and it is difficult to define a generalized data quality assurance model applicable to all data warehouse systems. As a result, the existing data quality testing tools are developed either for a specific domain or for applying basic data quality checks that are applicable to all domains. Other generalized tools let users define their desired data quality rules.

Achilles [46] proposed by the OHDSI community [47] is an example that generates specific data quality tests for the electronic health domain. This tool defines 172 data quality rules and verifies them using queries as test cases. The tool checks the data in health data warehouses to ensure consistency with the OMOP common data model. It also uses rules that check the semantics of health data to be plausible based on its rule set. Table 5 shows some examples.

Loshin [48] provided a data validation engine called GuardianIQ that does not define specific data quality rules but allows users to define and manage their own expectations as business rules for data quality at a high level in an editor. As a result, this tool can be used in any data warehousing project. The tool transforms declarative data quality rules into queries that measures data quality conformance with their expectations. Each data is tested against the query set and scored across multiple dimensions. The scores are used for the measurement of levels of data quality, which calculates to what extent the data matches the user's expectations.

Informatica Powercenter Data Validation [49] is another example of a tool that generates data quality tests and is generalized for use in any data warehouse project. It allows users to develop their business rules rapidly without having any knowledge of programming. The test cases, which are a set of queries, are generated from the user's business rules to be executed against the data warehouse under test.

Gao et al. [11] compare the existing data quality validation tools for general use in terms of the operation environment, supported DBMSs or products, data validation checks, and case studies. All the tools discussed in Gao et al.'s paper provide basic data quality validations, such as null value, data constraint, and data type checks. However, they do not assure the completeness of their data quality checks through well-defined test adequacy criteria. In software testing, a test adequacy criterion is a predicate that determines what properties of a software application must be exercised to constitute a complete test. We can define the test adequacy criteria for data quality tests as the number of columns, tables or constraints exercised by the quality tests. The set of test cases (queries) must contain tests to verify the properties of all the columns in all the tables of the sources or the target data warehouse.

Furthermore, the fault finding ability of the data quality tests is not evaluated in any of the surveyed approaches. We suggest that new research approaches be developed using mutation analysis techniques [50] to evaluate the ability of data quality tests to detect possible faults in the data. In these techniques, a number of faults are injected into the data to see how many of the faults are detected by the tests. Table 6 shows a number of sample faults to inject into the data to violate the data quality properties we defined in this section.

Table 6 Sample Faults Injected into Health Data for Mutation Analysis

Property	Fault Type
Data type	Store a string value in a numeric field
Data constraint	Copy a record to have duplicate values for a primary key field
Data plausibility	Store a negative value in a weight field
Logical constraint	Set a pregnancy status to *true* for a male patient

4.1.2 Security Testing of Underlying Data

Security testing of underlying data is the process of revealing possible flaws in the security mechanisms that protect the data in a data storage area. The security mechanisms must be built into the data warehousing systems. Otherwise, if access control is only built into the front-end applications but not into the data warehouse, a user may bypass access control by directly using SQL queries or reporting tools on the data warehouse [51].

Every source database may have its access privileges defined for its data based on organizational requirements. Data loaded to the target data warehouse is supposed to maintain the same security for the corresponding data in the sources, while enforcing additional policies based on the data warehouse requirements. For example, if the personal information of the patients in a hospital is protected via specific techniques such as by defining user profiles or database access control [8], the same protection must be applied for the patient data transformed to the target health data warehouse. Additional access polices may be defined on the target health data warehouse to authenticate medical researchers who want to analyze the patient data.

Security testing of the underlying data in a data warehouse involves a comparison of the access privileges defined for the target data with the ones defined for the corresponding source data to determine whether all the required protections are correctly observed in the data warehouse. For this purpose, we can define security tests by formulating queries that return defined permissions associated with the data in both the sources and the target data warehouse, and compare the permissions for equivalent data using either manual or automatic techniques.

4.2 Testing the Data Model

As the data model is the foundation for any database, it is critical to get the model right because a flawed model directly affects the quality of information. Data model tests ensure that the design of the model follows its standards both conceptually and logically and meets the organizational specifications. Documentation for the source and target model help equip testers with the required information for the systematic testing of data models.

4.2.1 Functional Evaluation of the Data Model

In this evaluation activity, the quality of the data model design is verified to be consistent with organizational requirements of the sources or the data warehouse. Some of the approaches are general enough to assess any data model (relational, nonrelational, or dimensional), while there exist other approaches that evaluate a specific data model.

Hoberman [12] created a data model scorecard to determine the quality of any data model design that can be applied to both the source area and the target data warehouse. The scorecard is an inspection checklist that includes a number of questions and the score for each question. The number in front of each question represents the score of the question assigned by Hoberman. The organization places a value between 0 and the corresponding score on each question to determine to what extent the model meets the functional requirements. The following is a description of each question related to the functional evaluation of data models and the corresponding scores:

1. Does the model capture the requirements (15)? This ensures that the data model represents the organizational requirements.
2. Is the model complete (15)? This ensures that both the data model and its metadata (data model descriptive information) are complete with respect to the requirements.
3. Does the model match its schema (10)? This ensures that the detail (conceptual, logical, or physical) and the perspective (relational, dimensional, or NoSQL) of the model matches its definition.
4. Is the model structurally correct (15)? This validates the design practices (such as primary key constraints) employed for building the data model.
5. Are the definitions appropriate (10)? This ensures that the definitions in the data model are correct, clear, and complete.
6. Is the model consistent with the enterprise (5)? This ensures that the set of terminology and rules in data model context can be comprehended by the organization.
7. Does the metadata match the data (10)? This ensures that the data model's description is consistent with the data model.

Golfarelli and Rizzi [8] proposed three types of tests on the conceptual and logical dimensional data model in a data warehouse:

- A *fact test* verifies whether or not the conceptual schema meets the preliminary workload requirements. The preliminary workload is a set of queries that business users intend to run against the target data warehouse. These queries help the data warehouse designers identify required facts, dimensions, and measurements in the dimensional data model [52]. For each workload, the fact test checks whether or not the required measures are included in the fact schema. This evaluation also measures the number of nonsupported workloads.
- A *conformity test* assesses how well the conformed dimensions are designed in a dimensional data model. Such a model includes fact tables that keep metrics of a business process, and dimension tables that contain descriptive attributes. A fact table contains the keys to the dimension tables. A conformed dimension is one that relates to more than one fact. These dimensions support the ability to integrate data from multiple business processes. The conformity test is carried out by measuring the sparseness of a bus matrix [53] that is a high-level abstraction of a dimensional data model. In this matrix, columns are the dimension tables, and rows are the fact tables (business processes). The matrix associates each fact with its dimensions. If there is a column in the matrix with more than one nonzero element, it shows the existence of a conformed dimension. If the bus matrix is dense (i.e., most of the elements are nonzero), it shows that there are dimensions that are associated with many facts, which indicates that the model includes overly generalized columns. For example, a *person* column refers to a wide variety of people, from employees to suppliers and customers while there is zero overlap between these populations. In this case, it is preferable to have a separate dimension for each population and associate them to the corresponding fact. On the other hand, if the bus matrix is sparse (i.e., most of the elements are zero), it shows that there is a few conformed dimension defined in the dimensional model, which indicates that the model includes overly detail columns. For example, each individual descriptive attribute is listed as a separate column. In this case, it is preferable to create a conformed dimension that is shared by multiple facts.
- A *star test* verifies whether or not a sample set of queries in the preliminary workload can be correctly formulated in SQL using the logical data model. The evaluation measures the number of nonsupported workloads.

The above functional evaluation activities are manually performed via inspections. There is a lack of automated techniques.

4.2.2 Structural Evaluation of the Data Model

This type of testing ensures that the data model is correctly implemented using the database schema. The database schema is assessed for possible flaws. MySQL schema validation plug-in performs general validation for relational data models [54]. It evaluates the internal structure of the database schema and performs the following checks:

1. Validate whether content that is not supposed to be empty is actually empty. The tool reports an error if any of the following empty content exists in the relational database schema:
   - A table without columns
   - A view without SQL code
   - A table/view not being referenced by at least one role
   - A user without privileges
   - A table/object that does not appear in any ER diagrams
2. Validate whether a table is correctly defined by checking the primary key and foreign key constraints in that table. The tool reports an error if any of the following incorrect definition exists in the relational database schema:
   - A table without primary key
   - A foreign key with a reference to a column with a different type
3. Validate whether there are duplications in the relational database objects. The tool reports an error if any of the following duplications exist in the relational database schema:
   - Duplications in object names
   - Duplications in roles or user names
   - Duplications in indexes
4. Validate whether there are inconsistencies in the column names and their types. The tool reports an error if the following inconsistency exists in the relational database schema:
   - Using the same column name for columns of different data types

The above approach targets the structural validation of the relational data schema but it does not apply to nonrelational and other data schema.

To assess the coverage of validation, we suggest using various structural metrics. These metrics are predicates that determine what properties of a schema must be exercised to constitute a thorough evaluation. The metrics are the number of views, routines, tables, columns, and structural constraints that are validated during the structural evaluations.

4.2.3 Usability Evaluation of the Data Model

Usability evaluation of a data model tests whether the data model is easy to read, understand, and use by the database and data warehouse designers. A data model is usually designed in a way to cover the requirements of database and data warehouse designers. There are many common data models designed for specific domains, such as health, banking, or business. The Hoberman scorecard [12] discussed in the functional evaluation of the data model also includes a number of questions and their scores for usability evaluation of any data model. The data warehouse designer places a value between 0 and the corresponding score on each question to determine to what extent the model meets the usability requirements. The following is a description of each question related to the usability evaluation of data models and the corresponding scores:

1. Does the model use generic structures, such as data element, entity, and relationship (10)? This ensures that the data model uses appropriate abstractions to be transferable to more generic domains. For example, instead of using phone number, fax number, or mobile number elements, an abstract structure contains phone and phone type which accommodates all situations.
2. Does the model meet naming standards (5)? This ensures that the terms and naming conventions used in the model follow the naming standards for data models. For example, inconsistent use of uppercase letter, lowercase letter, and underscore, such as in Last Name, FIRST NAME, and middle_name, indicates that naming standards are not being followed.
3. Is the model readable (5)? This ensures that the data model is easy to read and understand. For example, it is more readable to group the data elements that are conceptually related into one structure instead of scattering the elements over unrelated structures. For example, city, state, and postal code are grouped together.

The above approach involves human inspection, and there does not exist automated techniques for the usability testing of relational, nonrelational, and dimensional data models.

4.2.4 Maintainability Evaluation of the Data Model

Due to the evolving nature of data warehouse systems, it is important to use a data model design that can be improved during the data warehouse lifecycle. Maintainability assessments evaluate the quality of a source or target data model with respect to its ability to support changes during an evolution process [55].

Calero et al. [56] listed metrics for measuring the complexity of a data warehouse star design that can be used to determine the level of effort required to maintain it. The defined complexity metrics are for the table, star, and schema levels. The higher the values, the more complex is the design of the star model, and the harder it is to maintain the model. The metrics are as follows:

- *Table metrics*
  - Number of attributes of a table
  - Number of foreign keys of a table
- *Star metrics*
  - Number of dimension tables of a star schema
  - Number of tables of a star schema that correspond to the number of dimension tables added to the fact table
  - Number of attributes of dimension tables of a star schema
  - Number of attributes plus the number of foreign keys of a fact table
- *Schema metrics*
  - Number of fact tables of the star schema
  - Number of dimension tables of the star schema
  - Number of shared dimension tables that is the number of dimension tables shared for more than one star of the schema
  - Number of the fact tables plus the number of dimension tables of the star schema
  - Number of attributes of fact tables of the star schema
  - Number of attributes of dimension tables of the star schema

These metrics give an insight into the design complexity of the star data model, but there is no information in the Calero et al. paper on how to relate maintainability tests to these metrics. There is also a lack of work in developing metrics for other data models such as the snowflake model or relational data models.

4.3 Testing Data Management Product

Using the right product for data management is critical to the success of data warehouse systems. There are many categories of products used in data warehousing, such as DBMSs, big data management systems, data warehouse appliances, and cloud data warehouses that should be tested to ensure that it is the right technology for the organization. In the following sections, we describe the existing approaches for performance, stress, and recovery testing of the data management products.

4.3.1 Performance and Stress Testing of Data Management Product

Performance testing determines how a product performs in terms of responsiveness under a typical workload [57]. The performance of a product is typically measured in terms of response time. This testing activity evaluates whether or not a product meets the efficiency specifications claimed by the organizations.

Stress testing evaluates the responsiveness of a data management product using an extraordinarily large volume of data by measuring the response time of the product. The goal is to assess whether or not the product performs without failures when dealing with a database with a size significantly larger than expected [8].

Due to the fact that the demand for real-time data warehouses [3] and real-time analysis is increasing, performance and stress testing play a major role in data warehousing systems. Due to the growing nature of data warehousing systems, the data management product tolerance must be evaluated using unexpectedly large volumes of data. The product tolerance is the maximum volume of data the product can manage without failures and crashes. Comparing efficiency and tolerance characteristics of several data management products help data warehouse designers choose the appropriate technology for their requirements.

Performance tests are carried out on both real data or mock (fake) datasets with a size comparable with the average expected data volume [8]. However, stress tests are carried out on mock databases with a size significantly larger than the expected data volume. These testing activities are performed by applying different types of requests on the real or mock datasets. A number of queries are executed, and the responsiveness of the data management product is measured using standard database metrics. An important metric is the maximum query response time because query execution plays an important role in data warehouse performance measures. Both simple and multiple join queries are executed to validate the performance of queries on databases with different data volumes. Business users develop sample queries for performance testing with specified acceptable response times for each query [1].

Slutz [58] developed an automatic tool called Random Generation of SQL (RAGS) that stochastically generates a large number of SQL Data Manipulation Language (DML) queries that can be used to measure how efficiently a data management system responds to those queries. RAGS generates the SQL queries by parsing a stochastic tree and printing the query out. The parser stochastically generates the tree as it traverses the tree using database information (table names, column names, and column types). RAGS generates 833 SQL queries per second that are useful for performance and stress testing purposes.

Most performance and stress testing approaches in the literature focus on DBMSs [59], but there is a lack of work in performance testing of data warehouse appliances or cloud data warehouses.

4.3.2 Recovery Testing of Data Management Product

This testing activity verifies the degree to which a data management product recovers after critical events, such as power disconnection during an update, network fault, or hard disk failures [8].

As data management products are the key components of any data warehouse systems, they need to recover from abnormal terminations to ensure that they present correct data and that there are no data loss or duplications.

Gunawi et al. [60] proposed a testing framework to test the recovery of cloud-based data storage systems. The framework systematically pushes cloud storage systems into 40,000 unique failures instead of randomly pushing systems into multiple failures. They also extended the framework to evaluate the expected recovery behavior of cloud storage systems. They developed a logic language to help developers precisely specify recovery behavior.

Most data warehousing systems that use DBMSs or other transaction systems rely on the atomicity, consistency, isolation, and durability (ACID) properties [61] of database transactions to meet reliability requirements. Database transactions allow correct recovery from failures and keep a database consistent even after abnormal termination. Smith and Klingman [62] proposed a method for recovery testing of transaction systems that use ACID properties. Their method implements a recovery scenario to test the recovery of databases affected by the scenario. The scenario uses a two-phase transaction process that includes a number of service requests and is initiated by a client application. The scenario returns to the client application without completing the processing of transaction and verifies whether or not the database has correctly recovered. The database status is compared to the expected status identified by the scenario.

4.4 Summary

Table 7 summarizes the testing approaches that have been applied to the sources and the target data warehouse that we discussed in this section.

Table 7 Testing the Sources and the Target Data Warehouse

Test Category	Component	GuardianIQ [48]	Informatica [49]	Hoberman [12]	Golfarelli and Rizzi [8]	MySQL plug-in [54]	Calero et. al, [56]	Slutz [58]	Gunawi et al. [60]	Smith and Klingman [62]
Functional	Underlying data Data model Product	✓	✓	✓	✓					
Structural	Data model					✓				
Usability	Data model			✓						
Maintainability	Data model							✓		
Performance	Product							✓		
Stress	Product							✓		
Recovery	Product								✓	
Recovery	Product									✓
Security	Underlying data									

Gray shaded rows indicate that we did not find approaches or tools to support that kind of testing activity even though they are necessary for a real-world data-warehouse.

There are no methods proposed for the security testing of underlying data in data warehouse systems (as indicated by the gray shaded row in the table).

We have identified the following open problems in testing the sources and the target data warehouse.

- In the area of *functional testing of underlying data*, there is no systematic way to assure the completeness of the test cases written/generated by different data quality assurance tools. We suggest that new research approaches be developed using a test adequacy criterion, such as number of fields, tables, or constraints as properties that must be exercised to constitute a thorough test.
- Data quality rules are not formally specified in the business requirements for *the functional testing of the underlying data*. A tester needs to bridge the gap between informal specifications and formal quality rules.
- It is difficult to design a generalized *data quality test* applicable to all data warehouse systems because data warehouse projects are typically designed for specific business domains. There are a number of generalized tools that let users define their desired data quality rules.
- The fault finding ability of the *data quality tests* are not evaluated in the literature. One can use mutation analysis techniques to perform this evaluation.
- No approach has been proposed for *the security testing of underlying data*. One can compare the access privileges defined for the target data with the ones defined for the corresponding source data to ensure that all the required protections are correctly observed in the target data warehouse.
- There is a lack of automatic *functional evaluation* techniques for data models. The existing functional evaluation activities are manually performed through human inspections.
- There is a lack of *structural evaluation* techniques for nonrelational and dimensional schema. The existing approaches focus on the relational data schema.
- No formal technique has been proposed for *the usability testing of data models*. The proposed approaches are typically human inspections.
- In the area of *maintainability testing of data models*, a number of design complexity metrics have been proposed to get an insight into the capability of the data model to sustain changes. However, there is no information on how to design maintainability tests based on the metrics.
- The heterogeneous data involved in the data warehousing systems make *the performance and stress testing of data management products* difficult. Testers

must use large datasets in order to perform performance and stress tests. Generating this voluminous data that reflect the real characteristics of the data is an open problem in these testing activities.

- There is a lack of work in *performance and stress testing of data warehouse appliance and cloud data warehouses*. The proposed approaches in the literature typically focus on testing DBMSs.

5. TESTING ETL PROCESS

This testing activity verifies whether or not the ETL process extracts data from sources, transforms it into an appropriate form, and loads it to a target data warehouse in a correct and efficient way. As the ETL process directly affects the quality of data transformed to a data warehouse [9], it has been the main focus of most data warehouse testing techniques [3]. In this section, we describe existing functional, performance, scalability, reliability, regression, and usability testing approaches as well as propose a new approach based on our experience in testing the ETL process in a health data warehouse [63].

5.1 Functional Testing of ETL Process

Functional testing of ETL process ensures that any changes in the source systems are captured correctly and propagated completely into the target data warehouse [3]. Two types of testing have been used for evaluating the functionality of ETL process, namely, data quality and balancing tests.

5.1.1 Data Quality Tests

This testing activity verifies whether or not the data loaded into a data warehouse through the ETL process is consistent with the target data model and the organizational requirements. Data quality testing focuses on the quality assessment of the data stored in a target data warehouse. Data quality tests are defined based on a set of quality rules provided by domain experts. These rules are based on both domain and target data model specifications to validate the syntax and semantics of data stored in a data warehouse. For example, in our health data warehouse project, we use data quality rules from six clinical research networks, such as Achilles [46] and PEDSnet [64] to write test cases as queries to test the data quality. Achilles and PEDSnet define a number of rules to assess the quality of electronic health records, and report errors and warnings based on the data. These quality rules are defined and periodically updated in a manner to fit the use and

Table 8 Examples of Achilles Data Quality Rules

Rule_id	Data Quality Rule	Status	Description
19	Year of birth should not be prior to 1800	Warning	Checks whether or not year of birth is less than 1800
32	Percentage of patients with no visits should not exceed a threshold value	Notification	Checks whether or not the percentage of patients that have no visit records is greater than 5

need of the health data users. Achilles defines its data quality rules as SQL queries while PEDSnet uses R. Table 8 shows examples of two data quality rules that are validated in Achilles.

Note that the tools described in Section 4.1.1 to test the quality of the underlying data in a data warehousing system can also be used to execute data quality tests for the ETL process. The difference is that in the context of ETL testing, the tools have a different purpose, which is to test any time data is added or modified through the ETL process.

5.1.2 Balancing Tests

Balancing tests ensure that the data obtained from the source databases is not lost or incorrectly modified by the ETL process. In this testing activity, data in the source and target data warehouse are analyzed and differences are reported.

The balancing approach called *Sampling* [65] uses source-to-target mapping documents to extract data from both the source and target tables and store them in two spreadsheets. Then it uses the *Stare and Compare* technique to manually verify data and determine differences through viewing or *eyeballing* the data. Since this task can involve the comparison of billions of records, most of the time, a few number of the entire set of records are verified through this approach.

IBM QuerySurge [65] is a commercial tool that was built specifically to automate the balancing tests through query wizards. The tool implements a method for fast comparison of validation query results written by testers [66]. The query wizards implement an interface to make sure that minimal effort and no programming skills are required for developing balancing tests and obtaining results. The tool compares data based on column, table, and record count properties. Testers select the tables and columns to be compared in the wizard. The problem with this tool is that it only compares data that is not modified during the ETL transformation, which is claimed to be 80% of data. However, the goal of ETL testing should also be to validate data that has been reformatted and modified through the ETL process.

Another method is *Minus Queries* [65] in which the difference between the source and target is determined by subtracting the target data from the source data to show existence of unbalanced data. The problem with this method is the potential for false positives. For example, as many data warehouses keep historical data, there may be duplicate records in the target data warehouse corresponding to the same entity and the result might report an error based on the differences in number of records in the source and the target data warehouse. However, these duplications are actually allowed in the target data warehouse.

We propose to identify discrepancies that may arise between the source and the target data due to an incorrect transformation process. Based on these discrepancies we define a set of properties, namely, completeness, consistency, and syntactic validity.

Completeness ensures that all the relevant source records get transformed to the target records. Consistency and syntactic validity ensure correctness of the transformation of the attributes. Consistency ensures that the semantics of the various attributes are preserved in the transformation process. Syntactic validity ensures that no problems occur due to the differences in the syntax between the source and the target data.

In our project, we generated balancing tests to compare the data in the health data warehouse, which uses a dimensional database on Google BigQuery, with the corresponding data in sources, which use dimensional patient databases of two hospitals [63].

The source-to-target mappings available in the ETL transformation specifications provide the necessary information to identify corresponding tables and attributes in the sources and target data warehouse and assist in developing an appropriate testing strategy [7]. The ETL transformations include one-to-one, many-to-one, and many-to-many mappings. We used the mapping documents from the health data warehouse to extract corresponding source and target tables and attributes, both for modified and nonmodified data. Then we generated a set of test assertions as queries to compare the source and target data verifying the proposed properties.

The fault finding ability of the balancing tests are not evaluated in any of the surveyed approaches. We proposed to use the mutation analysis technique [50] to evaluate the ability of our balancing tests in detecting possible

Table 9 Mutation Operators Used to Inject Faults in Health Data

Operator	Description
AR	Add random record
DR	Delete random record
MNF	Modify numeric field
MSF	Modify string field
MNF_{min}	Modify min of numeric field
MNF_{max}	Modify max of numeric field
MSF_{length}	Modify string field length
MF_{null}	Modify field to null

faults in the data. Table 9 shows mutation operators we proposed to inject faults into the data to assess the effectiveness of the balancing tests.

Due to the voluminous data involved in data warehousing, a comprehensive functional test of ETL must consider all possible inputs for an adequate testing. However, there is a limitation in defining adequate test inputs in the literature. As it is impossible to test the functionality of the ETL process with all possible test inputs, one can use systematic input space partitioning [67] techniques to generate test data. Input space partitioning is a software testing technique that groups the input data into partitions of equivalent data called *equivalent classes*. Test inputs can be derived from each partition. A comprehensive test must generate at least one input for each partition.

Moreover, ETL testers typically do not have access to real data because of the confidentiality of data in the sources, and they need to generate mock data that correctly represents the characteristics of the real data. There are a number of mock data generator tools, such as Mockaroo [68] and Databene Benerator [69] that randomly generate test data. However, testers must generate data in a systematic manner, such as through input space partitioning techniques to cover data from all equivalent classes.

5.2 Performance, Stress, and Scalability Testing of ETL Process

Performance tests assess whether or not the entire ETL process is performed within the agreed time frames by the organizations [70]. The goal of performance testing of ETL is to assess ETL processing time under typical workloads that are comparable with the average expected data volume [42].

Stress tests also assess the ETL processing time but under a workload which is significantly larger than the expected data volume. The goal of stress testing of ETL is to assess ETL tolerance by verifying whether or not it crashes or fails when dealing with an extraordinarily large volume of data.

Scalability testing of ETL is performed to assess the process in terms of its capability to sustain further growth in data warehouse workload and organizational requirements [1, 70]. The goal of scalability testing is to ensure that the ETL process meets future needs of the organization. Mathen [1] stated that this growth mostly includes an increase in the volume of data to be processed through the ETL. As the data warehouse workload grows, the organizations expect ETL to sustain extract, transform, and load times. Mathen [1] proposed an approach to test the scalability of ETL by executing ETL loads with different volumes of data and comparing the times used to complete those loads.

In all of these testing activities, the processing time of ETL is evaluated when a specific amount of data is extracted, transformed, and loaded into the data warehouse [8]. The goal is to determine any potential weaknesses in ETL design and implementation, such as reading some files multiple times or using unnecessary intermediate files or storage [1]. The initial extract and load process, along with the incremental update process must be evaluated through these testing activities.

Wyatt et al. [71] introduced two primary ways to measure the performance of ETL, i.e., time-based and workload-based. In the time-based method, they check if the ETL process was completed in a specific time frame. In the workload-based approach they test the ETL process using a known size of data as test data, and measure the time to execute the workload. Higher performing ETL processes will transfer the same volume of data faster. The two approaches can be blended to check if the ETL process was completed in a specific time frame using a specific size of data.

The above testing approaches test the entire ETL process under different workload conditions. However, the tests should also focus on the extraction, transformation, and load components separately and validate each component under specific workloads. For example, the performance testing of the extraction component verifies whether or not a typical sized data can be extracted from the sources in an expected time frame. Applying the tests separately on the constituent ETL components and procedures helps localize existing issues in the ETL design and implementation, and determine the areas of weaknesses. The weaknesses can be addressed by using an alternative technology, language, algorithm, or intermediate files. For example, consider the performance issue in the *Load* component of ETL, which incorrectly uses the full mode and loads the entire data every time instead of loading only the new added or modified data. In this case, the execution time of the *Load* component is considerably longer than the *Extract* and *Transform* components. This problem can be localized if we apply the tests on the individual components instead of on the entire ETL process.

5.3 Reliability Testing of ETL Process

This type of testing ensures the correctness of the ETL process under both normal and failure conditions [71]. Normal conditions represent situations in which there are no external disturbances or unexpected terminations in the ETL process. To validate the reliability of the ETL process under normal conditions, we want to make sure that given the same set of inputs and initial states, two runs of ETL will produce the same results. We can compare the two result sets using properties such as completeness, consistency, and validity.

Failure conditions represent abnormal termination of ETL as a result of loss of connection to a database or network, power failure, or a user terminating the ETL process. In such cases the process should be able to either complete the task later or restore the process to its starting point. To test the reliability of the ETL process under abnormal conditions, we can simulate the failure conditions and compare the results from a failure run with the results of a successful run to check if the results are correct and complete. For example, we should check that no records were loaded twice to the target data warehouse as a result of the failure condition followed by rerunning the ETL process.

Most ETL implementations indirectly demonstrate reliability features by relying on the ACID properties of DBMSs [71]. If DBMSs are used in a data warehousing system to implement the sources or the target data warehouse components, these components recover from problems in the Extract, Transform, or Load processes. However, there are data management products other than DBMSs used in the data warehousing systems that do not support ACID properties (e.g., Google Cloud Bigtable [72] that is a NoSQL data management product). In such cases, reliability needs to be addressed separately and appropriate test cases must be designed.

Note that balancing tests may be performed to compare the two result datasets in both normal and abnormal conditions to verify the proposed properties in addition to other reliability tests.

5.4 Regression Testing of ETL Process

Regression tests check if the system still functions correctly after a modification. This testing phase is important for ETL because of its evolving nature [8]. With every new data warehouse release, the ETL process needs to evolve to enable the extraction of data from new sources for the new applications. The goal of regression testing of ETL is to ensure that the enhancements and modifications to the ETL modules do not introduce new faults [73]. If a new program is added to the ETL, interactions between the new and old programs should be tested.

Manjunath et al. [74] automated regression testing of ETL to save effort and resources with a reduction of 84% in regression test time. They used Informatica [49] to automatically generate test cases in SQL format, execute test cases, and compare the results of the source and target data. However, their approach uses the *retest all* [75] strategy, which reruns the entire set of test cases for regression testing of the ETL process. Instead, they could use *regression test selection* [75] techniques to run a subset of test cases to test only the parts of ETL that are affected by project changes. These techniques classify the set of test cases into retestable and reusable tests for regression testing purposes in order to save testing cost and time. A retestable test case tests the modified parts of the ETL process and needs to be rerun for the safety of regression testing. A reusable test case tests the unmodified parts of the ETL process and does not need to be rerun, while it is still valid [76].

Mathen [1] proposed to perform regression testing by storing test inputs and their results as expected outputs from successful runs of ETL. One can use the same test inputs to compare the regression test results with the previous results instead of generating a new set of test inputs for every regression test [1].

5.5 Usability Testing of ETL Process

The ETL process consists of various components, modules, databases, and intermediate files with different technologies, DBMSs, and languages that require many prerequisites and settings to be executed on different platforms. ETL is not a one-time process; it needs to be executed frequently or any time data is added or modified in the sources. As a result, it is important to execute the entire process with configurations that are easy to set up and modify.

Usability testing of ETL process assesses whether or not the ETL process is easy to use by the data warehouse implementer. This testing activity determines how easy it is to configure and execute ETL in a data warehouse project.

We suggest to assess the usability of the ETL process by measuring the manual effort involved in configuring ETL in terms of time. The configuration effort includes (1) providing connection information to the sources, data staging area, or target data warehouse, (2) installing prerequisite packages, (3) preprocessing of data before starting the ETL process, and (4) human interference to execute jobs that run each of the extract, transform, and load components. We can also do a survey with different users that gives us more information about the difficulty level.

5.6 Summary

Table 10 summarizes the existing approaches to test different aspects of the ETL process. As can be seen from the table, scalability, reliability, and usability testing were not reported in the literature even though they are critical for a comprehensive testing of the process. We identified the following open problems in ETL testing. We summarize areas and ideas for future investigation.

- In *the functional testing of ETL*, there is not a systematic way to assure the completeness of the test cases written/generated as a set of queries. As with the functional testing of underlying data, we can use appropriate test adequacy criteria to evaluate and create a thorough test.
- There is a lack of systematic techniques to generate mock test inputs for *the functional testing of the ETL process*. We can use input space partitioning techniques to generate test data for all the equivalent classes of data. Current tools generate random test data with not much similarity with the characteristics of real data.
- The fault finding ability of *the balancing tests* is not evaluated in the surveyed approaches. We can use mutation analysis for this evaluation.
- In *the performance, stress, and scalability testing of ETL*, the existing approaches test the entire ETL process under different workloads. We can apply tests to the individual components of ETL to determine the areas of weaknesses.

Table 10 Testing Extract, Transform, and Load (ETL)

Testing Category	GuardianIQ [48]	Informatica [49]	QuerySurge [65]	Wyatt et al. [71]	Mathen [1]	Manjunath et al. [74]
Functional	✓	✓	✓			
Performance				✓		
Stress						
Scalability					✓	
Reliability						
Regression						✓
Usability						

Gray shaded rows indicate that we did not find approaches or tools to support that kind of testing activity even though they are necessary for a real-world data-warehouse.

- The heterogeneous data involved in the data warehousing systems make *the performance, stress, and scalability testing of the ETL process* difficult. Testers must use large heterogeneous datasets in order to perform tests.
- The existing ETL implementations rely on ACID properties of transaction systems, and ignore *the reliability testing of the ETL process*. We can perform the balancing tests proposed in Section 5.1.2 to compare the results of the ETL process under normal conditions with the ones under abnormal conditions to verify the properties, namely, completeness, consistency, and syntactic validity.
- No approach has been proposed to *test the usability of the ETL process*. We define this testing activity as the process of determining whether or not the ETL process is easy to use by the data warehouse implementer. One can test the usability of the ETL process by assessing the manual effort involved in configuring ETL that is measured in terms of time.

6. TESTING FRONT-END APPLICATIONS

Front-end applications in data warehousing are used by data analyzers and researchers to perform various types of analysis on data and generate reports. Thus, it is important to test these applications to make sure the data are correctly, effectively, and efficiently presented to the users.

6.1 Functional Testing of Front-End Applications

This testing activity ensures that the data is correctly selected and displayed by the applications to the end users. The goal of testing the functionality of the front-end applications is to recognize whether the analysis or end result in a report is incorrect, and whether the cause of the problem is the front-end application rather than the other components or processes in the data warehouse.

Golfarelli and Rizzi [8] compared the results of analyses queries displayed by the application with the results obtained by executing the same queries directly (i.e., without using the application as an interface) on the target data warehouse. They suggested two different ways to create test cases as queries for functionality testing. In a black-box approach, test cases are a set of queries based on user requirements. In a white-box approach, the test cases are determined by defining appropriate coverage criteria for the dimensional data. For example, test cases are created to test all the facts, dimensions, and attributes of the dimensional data.

The approaches proposed by Golfarelli and Rizzi are promising. In our project, we have used Achilles, which is a front-end application that performs quality assurance and analysis checks on health data warehouses in the OMOP [77] data model. The queries in Achilles are executable on OMOP.

The functional testing of the front-end applications must consider all possible test inputs for adequate testing. As it is impossible to test the functionality of the front-end applications with all possible test inputs, we can use systematic input space partitioning [67] techniques to generate test data.

6.2 Usability Testing of Front-End Applications

Two different aspects of usability of front-end applications need to be evaluated during usability testing, namely, ease of configuring and understandability.

First, we must ensure that the front-end application can be easily configured to be connected to the data warehouse. The technologies used in the front-end application should be compatible with the ones used in the data warehouse; otherwise, it will require several intermediate tools and configurators to use the data warehouse as the application's back-end. For example, if an application uses JDBC drivers to connect to a data warehouse, and the technology used to implement the data warehouse does not support JDBC drivers, it will be difficult to connect the front-end apps to the back-end data warehouse. We may need to reimplement parts of the application that set up connections to the data warehouse or change the query languages that are used. We suggest evaluating this usability characteristic by measuring the time and effort required to configure the front-end application and connect it to the target data warehouse.

Second, we must ensure that the front-end applications are understandable by the end users [70], and the reports are represented and described in a way that avoids ambiguities about the meaning of the data. Existing approaches to evaluate the usability of generic software systems [78] can be used to test this aspect of front-end applications. These evaluations involve a number of end users to verify the application. Several instruments are utilized to gather feedback from users on the application being tested, such as paper prototypes [79], and pretest and posttest questionnaires.

6.3 Performance and Stress Testing of Front-End Applications

Performance testing evaluates the response time of front-end applications under typical workloads, while stress testing evaluates whether the application performs without failures under significantly heavy workloads. The workloads are identified in terms of number of concurrent users, data volumes, and number of queries. The tests provide various types of workloads to the front-end applications to evaluate the application response time.

Filho et al. [80] introduced the OLAP Benchmark for Analysis Services (OBAS) that assesses the performance of OLAP analysis services responsible for the analytical process of queries. The benchmark uses a workload-based evaluation that processes a variable number of concurrent executions using variable-sized dimensional datasets. It uses the Multidimensional Expressions (MDX) [81] query language to perform queries over multidimensional data cubes.

Bai [82] presented a performance testing approach that assesses the performance of reporting systems built using the SQL Server Analysis Services (SSAS) technology. The tool uses the MDX query language to simulate user requests under various cube loads. Metrics such as average query response time and number of queries answered are defined to measure the performance of these types of reporting services.

The above two tools compare the performance of analysis services such as SSAS or Pentaho Mondrian. However, the tools do not compare analysis services that support communication interfaces other than XML for Analysis (XMLA), such as OLAP4J.

6.4 Summary

Table 11 summarizes existing approaches that test front-end applications in data warehousing systems. Although it is important to assess usability, none of the approaches addressed usability testing. Stress testing of the front-end applications is not reported in the surveyed approaches. Below

Table 11 Testing Front-End Applications

Test Category	Golfarelli and Rizzi [8]	Filho et al. [80]	Bai [82]
Functional	✓		
Usability			
Performance		✓	✓
Stress			

Gray shaded rows indicate that we did not find approaches or tools to support that kind of testing activity even though they are necessary for a real-world data-warehouse.

is a summary of the open problems in testing front-end applications, and ideas for future investigation.

- Existing approaches proposed for *functionality testing of the front-end applications* compare the results of queries in the application with the ones obtained by directly executing the same queries on the target data warehouse.
- *Functional testing of the front-end applications* must consider all types of test inputs. We can use input space partitioning techniques to generate test data for this testing activity.
- To support *usability testing of front-end applications*, we define a new aspect of testing to assess how easy it is to configure the application. We can test this aspect of usability by measuring the manual effort involved in configuring the front-end application in terms of time.
- The voluminous data involved in the data warehousing systems makes *performance and stress testing of the front-end applications* difficult. Testers must use large datasets in order to perform realistic tests.

7. CONCLUSION

In this chapter, we described the challenges, approaches, and open problems in the area of testing data warehouse components. We described the components of a data warehouse using examples from a real-world health data warehouse project. We provided a classification framework that takes into account *what* component of a data warehouse was tested, and *how* the component was tested using various functional and nonfunctional testing and evaluation activities. We surveyed existing approaches to test and evaluate each component. Most of the approaches that we surveyed adapted traditional testing and evaluation approaches to the area of data warehouse testing. We identified gaps in the literature and proposed directions for further research. We observed that the following testing categories are open research areas.

- *Security testing* of the underlying data in the source and target components
- *Reliability testing* of the ETL process
- *Usability testing* of the ETL process
- *Usability testing* of the front-end applications
- *Stress testing* of the front-end applications

Future research needs to focus on filling the above gaps for comprehensively testing data warehouses. Moreover, the following techniques need to be developed or improved in all the testing activities in order to enhance the overall verification and validation of the data warehousing systems.

Test automation needs to improve to decrease the manual effort involved in data warehouse testing by providing effective test automation tools. The data involved in data warehouse testing are rapidly growing. This makes it impossible to efficiently test data warehouses while relying on manual activities. Existing testing approaches require a lot of human effort in writing test cases, executing tests, and reporting results. This makes it difficult to run tests repeatedly and consistently. Repeatability is a critical requirement of data warehouse testing because we need to execute the tests whenever data are added or modified in a data warehouse. The approaches previously discussed in this chapter are based on statistical analysis, manual inspections, or semiautomated testing tools that still need manual effort for generating test input values and assertions. Existing approaches to software test automation can be utilized to fully automate the tasks involved in data warehouse testing. However, automatic test assertion generation (oracle problem) is an open problem for software systems in general [83] because the expected test outputs for all possible test inputs are typically not formally specified. Testers often manually identify the expected outputs using informal specifications or their knowledge of the problem domain. The same problem exists for automatically generating test assertions for testing the data warehouse components. If the expected outputs are not fully specified in the source-to-target transformation rules or in data warehouse documentation, it will be difficult to automatically generate test assertions. Future research needs to fill the gap between informal specifications and formally specified outputs to automatically generate test assertions.

Like other generic software systems, data warehouse projects need to implement agile development processes [84], which help produce results faster for end users and adapt the data warehouse to ever-changing user requirements [85]. The biggest challenge for testing an agile data warehouse project is that the data warehouse components are always changing. As a result, testing needs to adapt as part of the development process. The design and execution of these tests often take time that agile projects typically do not have. The correct use of regression test selection techniques [75] can considerably reduce the time and costs involved in the iterative testing of agile data warehousing systems. These techniques help reduce costs by selecting and executing only a subset of tests to verify the parts of the data warehouse that are affected by the changes. At the same time, testers need to take into account trade-offs between the cost of selecting and executing tests, and the fault detection ability of the executed tests [75].

There will be a growing demand for real-time analysis and data requests [3]. The data warehouse testing speed needs to increase. Most of the existing functional and nonfunctional testing activities rely on testing the entire source, target, or intermediate datasets. As the data are incrementally extracted, transformed, and loaded into the target data warehouse, tests need to be applied to only the newly added or modified data in order to increase the speed of testing. Testing with the entire data should be applied only in the initial step where the entire data are extracted from the sources, transformed, and loaded to the target data warehouse for the first time.

Most of the existing tools rely on using real data as test inputs, while testers typically do not have access to the real data because of privacy and confidentiality concerns. Systematic test input generation techniques for software systems can be used in future studies to generate mock data with the characteristic of the real data and with the goal of adequately testing the data warehouse components. For example, we proposed to use random mock data generation tools that populate a database with randomly generated data while obeying data types and data constraints. Genetic and other heuristic algorithms have been used in automatic test input generation for generic software systems [86] with the goal of maximizing test coverage. The same idea can be utilized in data warehouse testing to generate test data for testing different components with the goal of maximizing test coverage for the component under test.

Identifying a test adequacy criterion helps testers evaluate their set of tests and improve the tests to cover uncovered parts of the component under test. Determining adequate test coverage is a limitation of current testing approaches. Further research needs to define test adequacy criteria to assess the completeness of test cases written or generated for different testing purposes. For example, test adequacy criteria for testing the underlying data can be defined as the number of tables, columns, and constraints that are covered during a test activity. The adequacy criteria can also be defined as the number of data quality rules that are verified by the tests. Test adequacy criteria for white-box testing of the ETL process can be defined as the number of statements or branches of the ETL code that are executed during the ETL tests.

The fault finding abilities of existing testing approaches need to be evaluated. Mutation analysis techniques [50] can be used in future studies to evaluate the number of injected faults that are detected using the written/generated test cases. These techniques systematically seed a number of faults that simulate real faults in the program under test. The techniques execute the tests and determine whether or not the tests can detect the injected faults. Faults can be injected into both the code and the data in a data warehousing system. Different functional and nonfunctional tests are supposed to fail as a result of the injected faults. For example, balancing tests should result in failures because of imbalances caused by the seeded data faults. Functional testing of front-end application must fail due to the incorrect data that is reported in the final reports and analysis. A fault in the ETL code may result in the creation of unnecessary intermediate files during the ETL process and cause the performance tests to fail. Using these techniques help testers evaluate test cases and improve them to detect undetected faults.

Due to the widespread use of confidential data in data warehousing systems, security is a major concern. Security testing of all the components of data warehouses must play an important role in data warehouse testing. There are different potential security challenges in data warehousing systems that need to be addressed in future studies.

First, there are many technologies involved in the data warehousing implementations. Different DBMSs, data warehouse products, and cloud systems are being used to store and manage data of the sources, the intermediate DSA, and the target data warehouse. Correctly transforming data access control and user privileges from one technology to the other is a significant challenge in the security of the data warehousing systems. Future research in security testing needs to develop techniques that compare the privileges defined in the sources and the ones defined in the target of a transformation to ensure that all the privileges are correctly transformed without losing any information.

Second, due to the large number of interactive processes and distributed components involved in data warehousing systems, especially those containing sensitive data, there are many potential security attacks [87], such as man-in-the-middle, data modification, eavesdropping, or denial-of-service. The goal of such attacks may be to read confidential data, modify the data that results in misleading or incorrect information in the final reports, or disrupting any service provided by the data warehousing system. Some of the consequences can be detected using the previously discussed functional and nonfunctional testing approaches. For example, if the data were modified through an attack, balancing tests (Section 5.1.2) can detect the faulty data in the target data warehouse. However, comprehensive security techniques can prevent these types of attacks before the problem is propagated to the target data warehouse where error detection and fault localization is much more expensive. Security testing should detect vulnerabilities in the code, hardware, protocol implementations, and database access controls in a data warehousing project and report them to the data warehouse developers to avoid the exploitation of those vulnerabilities.

An alternative to data warehouse testing is to develop the data warehouse in a way that proves that the data warehouse implements the specifications and it will not fail under any circumstances. This approach is called *correct by construction* [88] in software engineering context. It guarantees the correct construction of software, and thus, does not require testing. Data warehousing systems include different distributed components, programs, and processes on various platforms that are implemented using different technologies. The large number of factors that affect data warehousing systems at run-time makes it practically impossible to prove that a data warehouse meets its specifications under all circumstances. Testing must be performed to validate the data warehousing systems in different situations. As a result, data warehouse testing is likely to be an active research field in the near future.

REFERENCES

[1] M.P. Mathen, Data warehouse testing, Infosys DeveloperIQ Magazine (2010) 1–8.
[2] N. Dedic, C. Stanier, An evaluation of the challenges of multilingualism in data warehouse development, in: 18th International Conference on Enterprise Information Systems, Rome, Italy, ISBN: 978-989-758-187-8, 2017, pp. 196–206.
[3] V. Rainardi, Building a Data Warehouse with Examples in SQL Server, first ed., Apress, 2008, ISBN: 1590599314, 9781590599310, pp. 477–489.
[4] M.J. Berry, G. Linoff, Data Mining Techniques: For Marketing, Sales, and Customer Support, second, John Wiley & Sons, Inc., 1997. ISBN: 978-0-471-17980-1
[5] A.V. Aho, J.D. Ullman, Foundations of Computer Science, C Edition, W. H. Freeman, 1994, ISBN: 978-0-7167-8284-1.
[6] J. Han, E. Haihong, G. Le, J. Du, Survey on NoSQL database, in: 6th International Conference on Pervasive Computing and Applications, 2011, pp. 363–366.
[7] D. Vucevic, W. Yaddow, Testing the Data Warehouse Practicum: Assuring Data Content, Data Structures and Quality, Trafford Publishing, 2012. ISBN: 978-1-4669-4356-8.
[8] M. Golfarelli, S. Rizzi, A comprehensive approach to data warehouse testing, in: 12th ACM International Workshop on Data Warehousing and OLAP, New York, NY, USA, ISBN: 978-1-60558-801-8, 2009, pp. 17–24.
[9] N. ElGamal, A. ElBastawissy, G. Galal-Edeen, Data warehouse testing, in: The Joint EDBT/ICDT Workshops, New York, USA, ISBN: 978-1-4503-1599-9, 2013, pp. 1–8.
[10] H.M. Sneed, Testing a datawarehouse—an industrial challenge, in: Academic Industrial Conference—Practice And Research Techniques, 2006, pp. 203–210.
[11] J. Gao, C. Xie, C. Tao, Big data validation and quality assurance—issuses, challenges, and needs, in: IEEE Symposium on Service-Oriented System Engineering, 2016, pp. 433–441.
[12] S. Hoberman, Data Model Scorecard: Applying the Industry Standard on Data Model Quality, first ed., Technics Publications, 2015, ISBN: 978-1-63462-082-6.
[13] Q. Li, Y.-L. Chen, Entity-relationship diagram, in: Modeling and Analysis of Enterprise and Information Systems, Springer, Berlin, Heidelberg, ISBN: 978-3-540-89555-8, 978-3-540-89556-5, 2009, pp. 125–139.
[14] M. Varga, On the differences of relational and dimensional data model, in: 12th International Conference on Information and Intelligent Systems, 2001, pp. 245–251.

[15] R. Kimball, M. Ross, W. Thornthwaite, J. Mundy, B. Becker, The Data Warehouse Lifecycle Toolkit, second ed., Wiley, 2008, ISBN: 978-0-470-14977-5.
[16] V. Gopalkrishnan, Q. Li, K. Karlapalem, Star/snow-flake schema driven object-relational data warehouse design and query processing strategies, in: DataWarehousing and Knowledge Discovery, Springer, Berlin, Heidelberg, 1999, pp. 11–22.
[17] Deleted in review.
[18] A. Askoolum, Structured query language, in: System Building With APL + WinJohn Wiley & Sons, Ltd, ISBN: 978-0-470-03434-7, 2006, pp. 447–477.
[19] F. Hinshaw, Data warehouse appliances: driving the business intelligence revolution, DM Review Magazine (2004) 30–34.
[20] M. Armbrust, A. Fox, R. Griffith, A.D. Joseph, R. Katz, A. Konwinski, G. Lee, D. Patterson, A. Rabkin, I. Stoica, M. Zaharia, A view of cloud computing, Commun. ACM 53 (4) (2010) 50–58. ISSN: 0001-0782.
[21] MySQL, https://www.mysql.com (accessed 02.05.17).
[22] Microsoft SQL Server, https://www.microsoft.com/sql-server (accessed 02.05.17).
[23] PostgreSQL, https://www.postgresql.org (accessed 02.05.17).
[24] Apache Accumulo, https://accumulo.apache.org (accessed 02.05.17).
[25] ArangoDB: Highly Available Multi-model NoSQL Database, https://www.arangodb.com (accessed 02.05.17).
[26] MongoDB, https://www.mongodb.com (accessed 02.05.17).
[27] Apache Hadoop, https://hadoop.apache.org (accessed 02.05.17).
[28] Oracle: Integrated Cloud Applications and Platform Services, https://www.oracle.com (accessed 02.05.17).
[29] IBM PureApplication, http://www-03.ibm.com/software/products/pureapplication (accessed 02.05.17).
[30] BigQuery: Analytics Data Warehouse, https://cloud.google.com/bigquery (accessed 02.05.17).
[31] Amazon Redshift | Data Warehouse Solution, https://aws.amazon.com/redshift (accessed 02.05.17).
[32] Observational Medical Outcomes Partnership, http://omop.org (accessed 14.04.17).
[33] R. Kimball, J. Caserta, The Data Warehouse ETL Toolkit: Practical Techniques for Extracting, Cleaning, Conforming, and Delivering Data, first ed., Wiley, 2004. ISBN: 978-0-7645-6757-5.
[34] J. Barateiro, H. Galhardas, A survey of data quality tools, Datenbank Spektrum 14 (2005) 15–21.
[35] M.L. Lee, H. Lu, T.W. Ling, Y.T. Ko, Cleansing data for mining and warehousing, in: 10th International Conference on Database and Expert Systems Applications, Springer, Berlin, Heidelberg, 1999, pp. 751–760.
[36] S. Chaudhuri, U. Dayal, An overview of data warehousing and OLAP technology, ACM SIGMOD Record 26 (1) (1997) 65–74. ISSN: 0163-5808.
[37] G. Colliat, OLAP, relational, and multidimensional database systems, ACM SIGMOD Record 25 (3) (1996) 64–69, ISSN: 0163-5808.
[38] Crystal Reports: Formatting Multidimensional Reporting Against OLAP Data, http://www.informit.com/articles/article.aspx?p=1249227 (accessed 10.10.17).
[39] J. Iavindrasana, G. Cohen, A. Depeursinge, H. Muller, R. Meyer, A. Geissbuhler, Clinical data mining: a review, Yearb. Med. Inform. 48 (1) (2009) 121–133. ISSN: 0943-4747. https://imia.schattauer.de/contents/archive/issue/2347/issue/special/manuscript/11863/show.html.
[40] J.P. Shim, M. Warkentin, J.F. Courtney, D.J. Power, R. Sharda, C. Carlsson, Past, present, and future of decision support technology, Decis. Support. Syst. 33 (2) (2002) 111–126, ISSN: 0167-9236.

[41] E.S. Berner, Clinical Decision Support Systems: Theory and Practice, second ed., Springer, 2006, ISBN: 978-0-387-33914-6.
[42] M. Golfarelli, S. Rizzi, Data warehouse testing: a prototype-based methodology, Inf. Softw. Technol. 53 (11) (2011) 1183–1198, ISSN: 0950-5849.
[43] M.P. Neely, Data Quality Tools for Data Warehousing—A Small Sample Survey, State University of New York at Albany, 1998.
[44] M.A. Weiss, Data Structures & Algorithm Analysis in C++, fourth, Pearson, 2013. ISBN: 978-0-13-284737-7.
[45] M.G. Kahn, T.J. Callahan, J. Barnard, A.E. Bauck, J. Brown, B.N. Davidson, H. Estiri, C. Goerg, E. Holve, S.G. Johnson, S.-T. Liaw, M. Hamilton-Lopez, D. Meeker, T.C. Ong, P. Ryan, N. Shang, N.G. Weiskopf, C. Weng, M.N. Zozus, L. Schilling, A harmonized data quality assessment terminology and framework for the secondary use of electronic health record data, EGEMS (Washington, DC) 4 (1) (2016) 1244.
[46] OHDSI/Achilles, https://github.com/OHDSI/Achilles (accessed 11.05.17).
[47] G. Hripcsak, J.D. Duke, N.H. Shah, C.G. Reich, V. Huser, M.J. Schuemie, M.A. Suchard, R.W. Park, I.C.K. Wong, P.R. Rijnbeek, J. van der Lei, N. Pratt, G.N. Norén, Y.-C. Li, P.E. Stang, D. Madigan, P.B. Ryang, Observational Health Data Sciences and Informatics (OHDSI): opportunities for observational Researchers, Stud. Health Technol. Inform. 216 (2015) 574–578, ISSN: 0926-9630.
[48] D. Loshin, Rule-based data quality, in: 11th ACM International Conference on Information and Knowledge Management, New York, NY, USA, ISBN: 978-1-58113-492-6, 2002, pp. 614–616.
[49] Informatica, https://www.informatica.com/ (accessed 14.04.17).
[50] Y. Jia, M. Harman, An analysis and survey of the development of mutation testing, IEEE Trans. Softw. Eng. 37 (5) (2011) 649–678. ISSN: 0098-5589.
[51] K.B. Edwards, G. Lumpkin, Security and the Data Warehouse, Tech. rep., Oracle, 2005.
[52] M.K. Mohania, A.M. Tjoa, Data Warehousing and Knowledge Discovery: 12th International Conference, DaWaK, Springer, 2010, ISBN: 978-3-642-15104-0.
[53] M. Golfarelli, S. Rizzi, Data warehouse design: modern principles and methodologies, first, McGraw-Hill, Inc., 2009, ISBN: 978-0-07-161039-1
[54] MySQL: MySQL Workbench Manual: 9.2.3 Schema Validation Plugins, https://dev.mysql.com/doc/workbench/en/wb-validation-plugins.html (accessed 14.04.17).
[55] G. Papastefanatos, P. Vassiliadis, A. Simitsis, Y. Vassiliou, Design metrics for data warehouse evolution, in: International Conference on Conceptual Modeling, Springer, Berlin, Heidelberg, 2008, pp. 440–454.
[56] C. Calero, M. Piattini, C. Pascual, M.A. Serrano, Towards data warehouse quality metrics, in: The International Workshop on Design and Management of Data Warehouses Interlaken, Switzerland, 2001, pp. 1–10.
[57] B. Erinle, Performance testing with JMeter 2.9, first ed., Packt Publishing Ltd, 2013, ISBN: 978-1-78216-584-2, 978-1-78216-585-9.
[58] D.R. Slutz, Massive Stochastic Testing of SQL, in: 24th International Conference on Very Large Data Bases, San Francisco, CA, USA, ISBN: 978-1-55860-566-4, 1998, pp. 618–622.
[59] D. Chays, Y. Deng, P.G. Frankl, S. Dan, F.I. Vokolos, E.J. Weyuker, An agenda for testing relational database applications, Softw. Test. Verif. Rel. 14 (1) (2004) 17–44, ISSN: 0960-0833.
[60] H.S. Gunawi, T. Do, P. Joshi, P. Alvaro, J.M. Hellerstein, A.C. Arpaci-Dusseau, R.H. Arpaci-Dusseau, K. Sen, D. Borthakur, FATE and DESTINI: a framework for cloud recovery testing, in: 8th USENIX Conference on Networked Systems Design and Implementation, 2011, pp. 238–252.
[61] T. Haerder, A. Reuter, Principles of transaction-oriented database recovery, ACM Comput. Surv. 15 (4) (1983) 287–317, ISSN: 0360-0300.

[62] B. Smith, V.J. Klingman. Method and Apparatus for Testing Recovery Scenarios in Global Transaction Processing Systems, US Patent 707/999.202, 1997.
[63] H. Homayouni, An Approach for Testing the Extract-Transform-Load Process in Data Warehouse Systems (Master's thesis), Department of Computer Science, Colorado State University, 2018, Available at http://www.cs.colostate.edu/etl/papers/Thesis.pdf.
[64] PEDSnet, https://pedsnet.org/ (accessed 11.05.17).
[65] QuerySurge: Big Data Testing, ETL Testing & Data Warehouse Testing, http://www.querysurge.com/ (accessed 20.09.17).
[66] M. Marin, A data-agnostic approach to automatic testing of multi-dimensional databases, in: 7th International Conference on Software Testing, Verification and Validation 2014, pp. 133–142.
[67] I. Burnstein, Practical Software Testing: A Process-Oriented Approach, Springer, 2003, ISBN: 978-0-387-95131-7.
[68] Mockaroo - Random Data Generator|CSV/JSON/SQL/Excel, https://mockaroo.com/ (accessed 15.05.17).
[69] Databene Benerator, http://databene.org/databene-benerator/ (accessed 15.05.17).
[70] J. Theobald, Strategies for testing data warehouse applications, Inf. Manage. 17 (6) (2007) 20.
[71] L. Wyatt, B. Caufield, D. Pol, Principles for an ETL benchmark, in: Performance Evaluation and Benchmarking. Springer, Berlin, Heidelberg, 2009, pp. 183–198.
[72] Google Cloud Bigtable, https://cloud.google.com/bigtable/ (accessed 11.05.17).
[73] L.T. Moss, S. Atre, Business Intelligence Roadmap: The Complete Project Lifecycle for Decision-Support Applications, Addison-Wesley Professional, 2003. ISBN: 978-0-201-78420-6.
[74] T.N. Manjunath, R.S. Hegad, H.K. Yogish, R.A. Archana, I.M. Umesh, A case study on regression test automation for data warehouse quality assurance, Int. J. Inf. Technol. Knowl. Manage. 5 (2) (2012) 239–243.
[75] T.L. Graves, M.J. Harrold, J.-M. Kim, A. Porter, G. Rothermel, An empirical study of regression test selection techniques, ACM Trans. Softw. Eng. Methodol. 10 (2) (2001) 184–208. ISSN: 1049-331X.
[76] L.C. Briand, Y. Labiche, G. Soccar, Automating impact analysis and regression test selection based on UML designs, in: International Conference on Software Maintenance2002, pp. 252–261.
[77] Observational Medical Outcomes Partnership Common Data Model (OMOP CDM), https://www.ohdsi.org/data-standardization/the-common-data-model, (accessed 20.09.17).
[78] J.S. Dumas, J.C. Redish, A Practical Guide to Usability Testing, Rev Sub Edition, Intellect Ltd, 1999. ISBN: 978-1-84150-020-1.
[79] C. Snyder, Paper Prototyping: The Fast and Easy Way to Design and Refine User Interfaces, first ed., Morgan Kaufmann, 1877.
[80] B.E.M. de Albuquerque Filho, T.L.L. Siqueira, V.C. Times, OBAS: an OLAP benchmark for analysis services., J. Inf. Data Manag. 4 (3) (2013) 390, ISSN: 21787107.
[81] B.-K. Park, H. Han, I.-Y. Song, XML-OLAP: a multidimensional analysis framework for XML warehouses, in: Data Warehousing and Knowledge DiscoverySpringer, Berlin, Heidelberg, 2005, pp. 32–42.
[82] X. Bai, Testing the performance of an SSAS cube using VSTS, in: 7th International Conference on Information Technology: New Generations, Las Vegas, NV, USA, 2010, pp. 986–991.
[83] E.T. Barr, M. Harman, P. McMinn, M. Shahbaz, S. Yoo, The oracle problem in software testing: a survey, IEEE Trans. Softw. Eng. 41 (5) (2015) 507–525.
[84] D. Greer, Y. Hamon, Agile software development, Softw. Pract. Exp. 41 (9) (2011) 943–944, ISSN: 1097-024X.

[85] L. Corr, J. Stagnitto, Agile Data Warehouse Design: Collaborative Dimensional Modeling, From Whiteboard to Star Schema, third ed., DecisionOne Press, 2011, ISBN: 978-0-9568172-0-4.
[86] R.P. Pargas, M.J. Harrold, R.R. Peck, Test-data generation using genetic algorithms, Softw. Test. Verif. Rel. 9 (1999) 263–282.
[87] H.C.A. Van Tilborg, S. Jajodia, Encyclopedia of Cryptography and Security, second ed., Springer, 2011, ISBN: 978-1-4419-5905-8.
[88] L.P. Carloni, K.L. McMillan, A. Saldanha, A.L. Sangiovanni-Vincentelli, A methodology for correct-by-construction latency insensitive design, in: IEEE/ACM International Conference on Computer-Aided Design. Digest of Technical Papers (Cat. No. 99CH37051)1999, pp. 309–315.

ABOUT THE AUTHORS

Hajar Homayouni is a Ph.D. student in Computer Science at Colorado State University. She received the Master's degree in Computer Science from Colorado State University in 2018. Her research interests are in software testing, data warehousing, Extract-Transform-Load design and testing, data quality testing, and machine learning.

Sudipto Ghosh is an Associate Professor in the Computer Science Department at Colorado State University. He received the Ph.D. degree in Computer Science from Purdue University, USA, in 2000. His teaching and research interests include model-based software development and software testing. He is on the editorial boards of IEEE Transactions on Reliability, Information and Software Technology, Journal of Software Testing, Verification and Reliability, and Software Quality Journal. He was a Program co-chair of ICST 2010 and DSA 2017. He was a general co-chair of MODELS 2009 and Modularity 2015. He is a Senior Member of the IEEE and a member of the ACM.

Indrakshi Ray is a Professor in the Computer Science Department at Colorado State University. She has been a visiting faculty at Air Force Research Laboratory, Naval Research Laboratory, and at INRIA, Rocquencourt, France. She obtained her Ph.D. in Information Technology from George Mason University. Dr. Ray's research interests include security and privacy, database systems, and formal methods for software assurance. She has published over a hundred technical papers in refereed journals and conference proceedings with the support from agencies including Air Force Research Laboratory, Air Force Office of Scientific Research, National Institute of Health, National Institute of Standards and Technology, National Science Foundation, and the United States Department of Agriculture. She is on the editorial board of IEEE Transactions on Dependable and Secure Computing, IEEE Transactions on Services Computing, and Computer Standards and Interfaces. She has been a guest editor of ACM Transactions of Information Systems Security and Journal of Digital Library. She was the Program Chair of ACM SACMAT 2006, Program Co-Chair for ICISS 2013, CSS 2013, IFIP DBSec 2003, and General Chair of SACMAT 2008. She has served on the program committees of various conferences including ACM SACMAT, DBSec, EDBT, ESORICS, ICDE, and VLDB. She is a senior member of the IEEE and a member of the ACM.

CHAPTER SIX

Mutation Testing Advances: An Analysis and Survey

Mike Papadakis*, Marinos Kintis*, Jie Zhang‡, Yue Jia†, Yves Le Traon*, Mark Harman†

*Luxembourg University, Luxembourg, Luxembourg
†University College London, London, United Kingdom
‡Peking University, Beijing, China

Contents

Advances in Computers, Volume 112
ISSN 0065-2458
https://doi.org/10.1016/bs.adcom.2018.03.015

Abstract

Mutation testing realizes the idea of using artificial defects to support testing activities. Mutation is typically used as a way to evaluate the adequacy of test suites, to guide the generation of test cases, and to support experimentation. Mutation has reached a maturity phase and gradually gains popularity both in academia and in industry. This chapter presents a survey of recent advances, over the past decade, related to the fundamental problems of mutation testing and sets out the challenges and open problems for the future development of the method. It also collects advices on best practices related to the use of mutation in empirical studies of software testing. Thus, giving the reader a "mini-handbook"-style roadmap for the application of mutation testing as experimental methodology.

1. INTRODUCTION

How can we generate test cases that reveal faults? How confident are we with our test suite? Mutation analysis answers these questions by checking the ability of our tests to reveal some artificial defects. In case our tests fail to reveal the artificial defects, we should not be confident on our testing and should improve our test suites. Mutation realizes this idea and measures the confidence inspired by our testing. This method has reached a maturity phase and gradually gains popularity both in academia and in industry [1]. Figs. 1 and 2 record the current trends on the number of publications according to the data we collected (will be presented later). As demonstrated by these figures, the number of scientific publications relying on mutation analysis is continuously increasing, demonstrated in Fig. 1, and numerous of these contributions appear in the major software engineering venues, shown in Fig. 2. Therefore, mutation testing can be considered as a mainstream line of research.

The underlying idea of mutation is to force developers to design tests that explore system behaviors that reveal the introduced defects. The diverge types of defects that one can use, leads to test cases with different properties

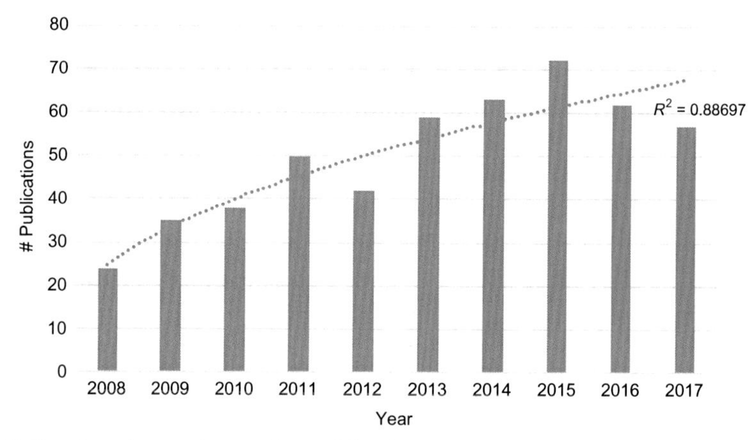

Fig. 1 Number of mutation testing publications per year (years: 2008–2017).

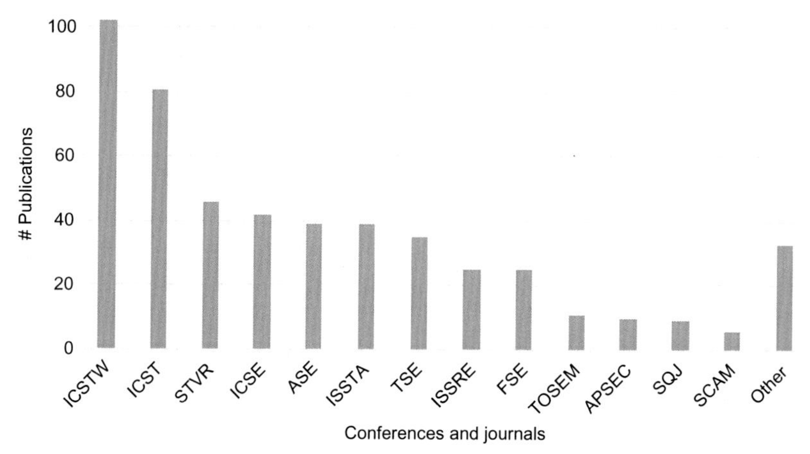

Fig. 2 Number of mutation testing publications per scientific venue.

as these are designed to reflect common programming mistakes, internal boundary conditions, hazardous programming constructs, and emulate test objectives of other structural test criteria. Generally, the method is flexible enough that it can be adapted to check almost everything that developers want to check, i.e., by formulating appropriate defects.

The flexibility of mutation testing is one of the key characteristics that makes it popular and generally applicable. Thus, mutation has been used for almost all forms of testing. Its primary application level is unit testing but several advances have been made in order to support other levels, i.e., specification [2], design [3], integration [4], and system levels [5]. The method has been applied on the most popular programming languages such as C [6],

C++ [7], C# [8], Java [9], JavaScript [10], Ruby [11], including specification [2] and modeling languages [12]. It has also been adapted for the most popular programming paradigms such as object-oriented [13], functional [14], aspect-oriented, and declarative-oriented [15,16] programming.

In the majority of the research projects, mutation was used as an indicator of test effectiveness. However, recently researchers have also focused on a different aspect: the exploration of the mutant behaviors. Thus, instead of exploring the behavior space of the program under analysis the interest shifts to the behavior of the mutants. In this line of research, mutants are used to identify important behavior differences that can be either functional or nonfunctional. By realizing this idea, one can form methods that assist activities outside software testing. Examples of this line of research are methods that automatically localize faults [17], automatically repair software [18], and automatically improve programs' nonfunctional properties such as security [19], memory consumption [20], and execution speed [20,21].

Mutation analysis originated in early work in the 1970s [22–24], but has a long and continuous history of development improvement, with particularly important advances in breakthroughs in the last decade, which constitute the focus of this survey. Previous surveys can be traced back to the work of DeMillo [25] (in 1989), which summarized the early research achievements. Other surveys are due to Offutt and Untch [26] (in 2000) and Jia and Harman [27] (in 2011). Offutt and Untch summarized the history of the technique and listed the main problems and solutions at that time. Subsequently, Jia and Harman comprehensively surveyed research achievements up to the year 2009.

There is also a number of specialized surveys on specific problems of mutation testing. These are a survey on the equivalent mutant problem by Madeyski et al. [28] (in 2014), a systematic mapping of mutation-based test generation by Souza et al. [29] (in 2014), a survey on model-based mutation testing by Belli et al. [30] (in 2016), and a systematic review on search-based mutation testing by Silva et al. [31] (in 2017). However, none of these covers the whole spectrum of advances from 2009. During these years there are many new developments, applications, techniques, and advances in mutation testing theory and practice as witnessed by the number of papers we analyze (more than 400 papers). These form the focus of the present chapter.

Mutation testing is also increasingly used as a foundational experimental methodology in comparing testing techniques (whether or not these techniques are directly relevant to mutation testing itself). The past decade has also witnessed an increasing focus on the methodological soundness and threats to validity that accrue from such use of mutation testing and the experimental methodology for wider empirical studies of software testing. Therefore, the present chapter also collects together advices on best practices, giving the reader a "mini-handbook"-style roadmap for the application of mutation testing as an experimental methodology (in Section 9).

The present chapter surveys the advances related to mutation testing, i.e., using mutation analysis to detect faults. Thus, its focus is the techniques and studies that are related to mutation-guided test process. The goal is to provide a concise and easy to understand view of the advances that have been realized and how they can be used. To achieve this, we categorize and present the surveyed advances according to the stages of the mutation testing process that they apply to. In other words, we use the mutation testing process steps as a map for detailing the related advances. We believe that such an attempt will help readers, especially those new to mutation testing, understand everything they need in order to build modern mutation testing tools, understand the main challenges in the area, and perform effective testing.

The survey was performed by collecting and analyzing papers published in the last 10 years (2008–2017) in leading software engineering venues. This affords our survey a 2-year overlap in the period covered with the previous survey of Jia and Harman [27]. We adopted this approach to ensure that there is no chance that the paper could "slip between the cracks" of the two surveys. Publication dating practices can differ between publishers, and there is a potential time lag between official publication date and the appearance of a paper, further compounded by the blurred distinction between online availability and formal publication date. Allowing this overlap lends our survey a coherent, decade-wide, time window, and also aims to ensure that mutation testing advances do not go uncovered due to such publication date uncertainties.

Thus, we selected papers published in the ICSE, SigSoft FSE, ASE, ISSTA, ICST, ICST Workshops (ICSTW), ISSRE, SCAM, and APSEC conferences. We also collected papers published in the TOSEM, TSE, STVR, and SQJ journals and formed our set of surveyed papers. We augmented this set with additional papers based on our knowledge. Overall, we selected a set of 502 papers, which fall within five generic categories, those that deal with the code-based mutation testing problems (186 papers), those concerning model-based mutation testing (40 papers), those that tackle problems unrelated to mutation testing problems (25 papers), those that describe mutation testing tools (34 papers), and those that use mutation testing only to perform test assessment (217 papers). In an attempt to provide a complete view of the fundamental advances in the area we also refer to some of the publications that were surveyed by the two previous surveys on the topic, i.e., the surveys of Offutt and Untch [26] and Jia and Harman [27], which have not been obviated by the recent research.

The rest of the chapter is organized as follows: Section 2 presents the main concepts that are used across the chapter. Sections 3 and 4, respectively, motivate the use of mutation testing and discuss its relation with real faults. Next, the regular and other code-based advances are detailed in Sections 5 and 6. Applications of mutation testing to other artifacts than code and a short description of mutation testing tools are presented in Sections 7 and 8. Sections 9 and 10 present issues and best practices for using mutation testing in experimental studies. Finally, Section 11 concludes the chapter and outlines future research directions.

2. BACKGROUND

Mutation analysis refers to the process of automatically mutating the program syntax with the aim of producing semantic program variants, i.e., generating artificial defects. These programs with defects (variants) are called *mutants*. Some mutants are syntactically illegal, e.g., cannot compile, named "*stillborn*" mutants, and have to be removed. Mutation testing refers to the process of using mutation analysis to support testing by quantifying the test suite strengths. In the testing context, mutants form the objectives of the test process. Thus, test cases that are capable of distinguishing the behaviors of the mutant programs from those of the original program fulfill the test objectives. When a test distinguishes the behavior of a mutant (from that of the original program) we say that the mutant is "*killed*" or "*detected*"; in a different case, we say that the mutant is "*live*."

Depending on what we define as program behavior we can have different *mutant-killing conditions*. Typically, what we monitor are all the observable program outputs against each running test: everything that the program prints to the standard/error outputs or is asserted by the program assertions. A mutant is said to be killed *weakly* [32], if the program state immediately after the execution of the mutant differs from the one that corresponds to the original program. We can also place the program state comparison at a later point after the execution of a mutant. This variation is called *firm mutation* [32]. A mutant is *strongly killed* if the original program and the mutant exhibit some observable difference in their outputs. Thus, we have variants of mutation, called weak, firm, and strong. Overall, for weak/firm mutation, the condition of killing a mutant is that the program state has to be changed, while the changed state does not necessarily need to propagate to the output (as required by strong mutation). Therefore, weak mutation is expected to be less effective than firm mutation, which is in turn less effective than strong mutation. However, due to failed error propagation (subsequent computations may mask the state differences introduce by the mutants) there is no formal subsumption relation between any of the variants [32].

Mutants are generated by altering the syntax of the program. Thus, we have syntactic transformation rules, called "*mutant operators*," that define how to introduce syntactic changes to the program. For instance, an arithmetic mutant operator alters the arithmetic programming language operator, changing + to −, for example. A basic set of mutant operators, which is usually considered as a minimum standard for mutation testing [33] is the five-operator set proposed by Offutt et al. [34]. This set includes the relational (denoted as ROR), logical (denoted as LCR), arithmetic (denoted as AOR), absolute (denoted as ABS), and unary insertion (denoted as UOI) operators. Table 1 summarizes these operators.

Defining mutant operators is somehow easy, we only need to define some syntactic alterations. However, defining useful operators is generally challenging. Previous research followed the path of defining a large set (almost exhaustive) of mutant operators based on the grammar of the language. Then, based on empirical studies, researchers tend to select subsets of them in order to improve the applicability and scalability of the method. Of course both the definition of operators and mutant selection (selection of

Table 1 The Popular Five-Operator Set (Proposed by Offutt et al. [34])

Names	Description	Specific Mutation Operator
ABS	Absolute value insertion	{(*e*, 0), (*e*,`abs` (*e*)), (*e*,`-abs` (*e*))}
AOR	Arithmetic operator replacement	{((a *op* b), *a*), ((a *op* b), b), (*x*, *y*)|*x*, *y* ∈ {+,-,*,/,%}∧ *x*≠*y*}
LCR	Logical connector replacement	{((a *op* b), *a*), ((a *op* b), *b*), ((a *op* b), `false`), ((a *op* b), `true`), (*x*, *y*)|*x*, *y* ∈ {&, | , ^, &&, ||}∧ *x*≠*y*}
ROR	Relational operator replacement	{((a *op* b), `false`), ((a *op* b), `true`), (*x*, *y*)|*x*, *y* ∈ {>,>=, <,<=,==,!=} ∧ *x*≠*y*}
UOI	Unary operator insertion	{(*cond*,! *cond*), (*v*,-*v*), (*v*, ∼*v*), (*v*,--*v*), (*v*, *v*--), (*v*,++ *v*), (*v*, *v*++)}

representative subsets) form the two sides of the same coin. Since all possible operators are enormous if not infinite, the definition of small sets of them can be viewed as a subset selection (among all possible definitions). Here, we refer to mutant reduction as the process of selecting subsets of operators over a given sets of them. We discuss studies that define mutant operators in Section 5.1.1 and mutant reduction in Section 5.1.2.

Based on the chosen set of mutant operators, we generate a set of mutant instances that we use to perform our analysis. Thus, our test objectives are to kill the mutants (design test cases that kill all the mutants). We can define as "*mutation score*" or *mutation coverage* the ratio of mutants that are killed by our test cases. In essence, the mutation score denotes the degree of achievement of our test cases in fulfilling the test objectives. Thus, mutation score can be used as an adequacy metric [32].

Adequacy criteria are defined as predicates defining the objectives of testing [32]. Goodenough and Gerhart [35] argue that criteria capture what properties of a program must be exercised to constitute a thorough test, i.e., one whose successful execution implies no errors in a tested program. Therefore, the use of mutation testing as a test criterion has the following three advantages [6]: to point out the elements that should be exercised when designing tests, to provide criteria for terminating testing (when coverage is attained), and to quantify test suite thoroughness (establish confidence).

In practice using mutation score as adequacy measure, implicitly assumes that all mutants are of equal value. Unfortunately, this is not true [36]. In practice, some mutants are *equivalent*, i.e., they form functionally equivalent versions of the original program, while some others are *redundant*, i.e., they are not contributing to the test process as they are killed whenever other mutants are killed. Redundant mutants are of various forms. We have the *duplicated mutants*, mutants that are equivalent between them but not with the original program [37]. We also have the *subsumed mutants* [38] (also called *joint mutants* [39]), i.e., mutants that are jointly killed when other mutants are killed [36,39].

The problem with subsumed mutants is that they do not contribute to the test assessment process because they are killed when other mutants are also killed. This means that eliminating these mutants does not affect the selection/generation of test cases but the computation of the mutation score. Thus, the metric is inflated and becomes hard to interpret. As the distribution of mutants tend to be unpredictable and the identification of mutant equivalences and redundancies is an undecidable problem [37], it is hard to judge the test suite strengths based on the mutation score. In other words, the accuracy of the score metric is questionable. We will discuss this issue in depth in Section 9.

3. WHAT IS SPECIAL ABOUT MUTATION TESTING

Mutation testing principles and concepts share on a long heritage with more general scientific investigation, essentially drawing on the common sense of "trial and error" (that predates civilisation), and also resting on the foundations of inferential statistics and Popperian science [40].

One of the fundamental problems in software testing is the inability to know practically or theoretically when one has tested sufficiently. Practitioners often demand of researchers a method to determine when testing should cease. Unfortunately, this revolves around the question of what is intended by sufficiency; if we are to test in order to be sufficient to demonstrate the absence of bugs, then we are forced against the impossibility of exhaustive testing.

Faced with this challenge, much literature has centered on concepts of coverage, which measure the degree of test effort, that each coverage technique's advocates hope will be correlated with test achievement. For instance Table 2 reports on the main studies on this subject. There has been much empirical evidence concerning whether coverage is correlated with faults revelation [41–43], a problem that remains an important subject of study to the present day [6,41]. However, even setting aside the concern of whether such correlation exists, nonmutation-based forms of coverage suffer from a more foundational problem; they are essentially unfalsifiable (with respect to the goal of fault revelation), in the Popperian sense of science [40].

Mutation testing is important because it provides a mechanism by which assertions concerning test effectiveness become falsifiable; failure to detect certain kinds of mutants suggest failure to detect certain kinds of faults. Alternative coverage criteria can only be falsified in the sense of stating that should some desired coverage item remain uncovered, then claims to test effectiveness remain "false." Unfortunately, it is not usually possible to cover every desired coverage item, it is typically undecidable whether this criterion has been achieved in any case [59]. Coverage of all mutants is also undecidable [37], but mutation testing forms a direct link between faults and test achievements, allowing more scientific (in the sense intended by Popper) statements of test achievement than other less-fault-orientated approaches.

Mutation testing also draws on a rich seam of intellectual thought that is currently becoming more popular in other aspects of science and engineering. Such counterfactual reasoning can even be found beyond science, in the humanities, where historians use it to explore what would have happened had certain historical events failed to occur. This is a useful intellectual tool

Table 2 Summary of the Main Studies Concerned With the Relationship of Test Criteria and Faults

Author(s) [Reference]	Year	Test Criterion	Summary of Primary Scientific Findings
Frankl and Weiss [44,45]	1991, 1993	Branch, all-uses	All-uses relates with test effectiveness, while branch does not.
Offutt et al. [46]	1996	All-uses, mutation	Both all-uses and mutation are effective but mutation reveals more faults.
Frankl et al. [47]	1997	All-uses, mutation	Test effectiveness (for both all-uses and mutation) is increasing at higher coverage levels. Mutation performs better.
Frankl and Iakounenko [48]	1998	All-uses, branch	Test effectiveness increases rapidly at higher levels of coverage (for both all-uses and branch). Both criteria have similar test effectiveness.
Briand and Pfahl [49]	2000	Block, c-uses, p-uses, branch	There is no relation (independent of test suite size) between any of the four criteria and effectiveness.
Chen et al. [50]	2001	Block	Coverage can be used for predicting the software failures in operation.
Andrews et al. [42]	2006	Block, c-uses, p-uses, branch	Block, c-uses, p-uses, and branch coverage criteria correlate with test effectiveness.
Namin and Andrews [51]	2009	Block, c-uses, p-uses, branch	Both test suite size and coverage influence (independently) the test effectiveness.
Li et al. [52]	2009	Prime path, branch, all-uses, mutation	Mutation testing finds more faults than prime path, branch and all-uses.

Papadakis and Malevris [53]	2010	Mutant sampling, first- and second-order mutation	First-order mutation is more effective than second order and mutant sampling. There are significantly less equivalent second-order mutants than first-order ones.
Ciupa et al. [54]	2011	Random testing	Random testing is effective and has predictable performance.
Kakarla et al. [55]	2011	mutation	Mutation-based experiments are vulnerable to threats caused by the choice of mutant operators, test suite size, and programming language.
Wei et al. [56]	2012	Branch	Branch coverage has a weak correlates with test effectiveness.
Baker and Habli [57]	2013	Statement, branch, MC/DC, mutation, code review	Mutation testing helps improving the test suites of two safety-critical systems by identifying shortfalls where traditional structural criteria and manual review failed.
Hassan and Andrews [58]	2013	Multi-Point Stride, data-flow, branch	Def-uses is (strongly) correlated with test effectiveness and has almost the same prediction power as branch coverage. Multi-Point Stride provides better prediction of effectiveness than branch coverage.
Gligoric et al. [59,60]	2013, 2015	AIMP, DBB, branch, IMP, PCC, statement	There is a correlation between coverage and test effectiveness. Branch coverage is the best measure for predicting the quality of test suites.
Inozemtseva and Holmes [61]	2014	Statement, branch, modified condition	There is a correlation between coverage and test effectiveness when ignoring the influence of test suite size. This is low when test size is controlled.

Continued

Table 2 Summary of the Main Studies Concerned With the Relationship of Test Criteria and Faults—cont'd

Author(s) [Reference]	Year	Test Criterion	Summary of Primary Scientific Findings
Just et al. [43]	2014	Statement, mutation	Both mutation and statement coverage correlate with fault detection, with mutants having higher correlation.
Gopinath et al. [62]	2014	Statement, branch, block, path	There is a correlation between coverage and test effectiveness. Statement coverage predicts best the quality of test suites.
Ahmed et al. [63]	2016	Statement, mutation	There is a weak correlation between coverage and number of bug-fixes
Ramler et al. [64]	2017	Strong mutation	Mutation testing provides valuable guidance toward improving the test suites of a safety-critical industrial software system
Chekam et al. [6]	2017	Statement, branch, weak & strong mutation	There is a strong connection between coverage attainment and fault revelation for strong mutation but weak for statement, branch and weak mutation. Fault revelation improves significantly at higher coverage levels.
Papadakis et al. [41]	2018	Mutation	Mutation score and test suite size correlate with fault detection rates, but often the individual (and joint) correlations are weak. Test suites of very high mutation score levels enjoy significant benefits over those with lower score levels.

to help increase understanding and analysis of the importance of these events the influence (or forward dependence [65] as we might think of it within the more familiar software engineering setting. In software engineering, counterfactual reasoning plays a role in causal impact analysis [66], allowing software engineers to discover the impact of a particular event, thereby partly obviating the problem "correlation is not causation."

In mutation testing, we create a "counterfactual version of the program" (a mutant) that represents what the program would have looked like had it contained a specific chosen fault, thereby allowing us to investigate what would have happened if the test approach encounter a program containing such a fault. Causal impact analysis relies on recent development in statistics In more traditional statistical analysis, mutation testing also finds a strong resonance with the foundations of sampling and inferential frequentist statistical analysis. A statistician might, for instance, seek to estimate the number of fish in a pool by catching a set of fish, marking these, and returning them to the pool, subsequently checking how many marked fish are present in a random sample. Of course such an approach measures not only the number of fish in the pool, but also the effectiveness of the recatching approach used in resampling. In a similar way, creating sets of mutant programs (marking fish) and then applying the test technique (resampling) can also be used to estimate the number of faults in the program, albeit confounded by concurrently measuring the effectiveness of the testing technique.

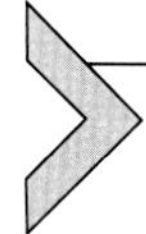

4. THE RELATIONS BETWEEN MUTANTS AND FAULT REVELATION

The underlying idea of mutation is simple; we form defects and we ask testers to design test cases that reveal them. Naturally, readers may ask why such an approach is effective. Literature answers this question in the following ways:

- First (theoretically), by revealing the formed defects we can demonstrate that these specific types of defects are not present in the program under analysis [67]. In such a case we assume that the formed mutants represent the fault types that we are interested in. In a broader perspective, the mutants used are the potential faults that testers target and thus, they are in a sense equivalent to real faults.
- Second (theoretically and practically), when test cases reveal simple defects such as mutants (defects that are the result of simple syntactic alterations), they are actually powerful enough to reveal more complex defects. In such a case, we assume that test cases that reveal the used types of defects also

reveal more complex types (multiple instances of the types we used) [68]. Thus, mutation helps revealing a broader class of faults (than those used) composed of the simple and complex types of faults that were used.

- Third (practically), when we design test cases that kill mutants we are actually writing powerful test cases. This is because we are checking whether the defects we are using can trigger a failure at every location (or related ones) we are applying them to. In such a case, we assume that test cases that are capable of revealing mutants are also capable of revealing other types of faults (different from the mutants). This is because mutants require checking whether test cases are capable of propagating corrupted program states to the observable program output (asserted by the test cases). Thus, potential faulty states (related to the mutants) have good chances to became observable [6].

The aforementioned points motivated researchers to study and set the foundation of mutation testing.

The first point has been shown theoretically based on the foundations of fault-based testing [67,69]. The practical importance of this assertion is that in case we form the employed mutants as common programming mistakes, then we can be confident that testers will find them. Thus, we can check against the most frequent faults. This premise becomes more important when we consider the competent programmer hypothesis [24], which states that developers produce programs that are nearly correct, i.e., they require a few syntactic changes to reach the correct version. This hypothesis implies that if we form mutants by making few simple syntactic changes we can represent the class of frequent faults (made by "competent programmers"). A recent study by Gopinath et al. [70] has shown that defects mined from repositories involve three to four tokens to be fixed, confirming to some extent the hypothesis. Generally, the recent studies have not consider this subject and thus, further details about the competent programmer hypothesis can be found in the surveys of Offutt and Untch [26] and Jia and Harman [27].

The second point is also known as the mutant coupling effect [68]. According to Offutt [68] the mutant coupling effect states "Complex mutants are coupled to simple mutants in such a way that a test data set that detects all simple mutants in a program will detect a large percentage of the complex mutants." Therefore, by designing test cases that reveal almost all the mutants used, we expect a much larger set of complex mutants to be also revealed. This premise has been studied both theoretically and practically (details can be found in the surveys of Offutt and Untch [26] and Jia and Harman [27]). Recent studies on the subject only investigated (empirically) the relationship between simple and complex mutants. For example the study of

Gopinath et al. [71] demonstrate that many higher order mutants are semantically different from the simple first-order ones they are composed of. However, the studies of Langdon et al. [72] and Papadakis and Malevris [53] show that test cases kill a much larger ratio of complex (higher order) mutants than simple ones (first-order ones) indicating that higher order mutants are of relatively lower strength than the first-order ones.

The third point is a realization of the requirement that the mutants must influence the observable output of the program (the test oracle). To understand the importance of this point we need to consider the so-called RIPR model (reachability, infection, propagation, revealability) [32]. The RIPR model states that in order to reveal a fault, test cases must: (a) reach the faulty locations (reachability), (b) cause a corruption (infection) to the program state (infection), (c) propagate the corruption to the program output (propagation), and (d) cause a failure, i.e., make the corrupted state observable to the user, be asserted by the test cases (revealability). Therefore, when designing test cases to kill mutants, we check the sensitivity of erroneous program states to be observable. Empirical evidence by Chekam et al. [6] has shown that this propagation requirement makes mutation strong and distinguishes it from weak mutation and other structural test criteria. In particular the same study demonstrates that mutant propagation is responsible for the revelation of 36% of the faults that can be captured by strong mutation.

Overall, the fundamental premise of mutation testing can be summarized as "if the software contains a fault, there will usually be a set of mutants that can only be killed by a test case that also detects that fault" [73]. This premise has been empirically investigated by the study of Chekam et al. [6] which demonstrated a strong connection between killing mutants and fault revelation. Similarly, the studies of Baker and Habli [57], Ramler et al. [64], and Ahmed et al. [74] have shown that mutation testing provides valuable guidance toward improving existing test suites.

A last reason that makes mutation strong is the fact that its test objectives are the formed defects. Thus, depending on the design of these defects, several test criteria can be emulated. Therefore, mutation testing has been found to be a strong criterion that subsumes, or probably subsumes[a] almost all other test criteria [32]. Thus, previous research has shown that strong mutation probably subsumes weak mutation [75], all data-flow criteria [47,55], logical criteria [76], and branch and statement criteria [52]. These studies suggest that a small set of mutant operators can often result in a set of test cases that is as strong as the ones resulting from other criteria.

[a] Subsumption is not guaranteed but it is probable to happen [32].

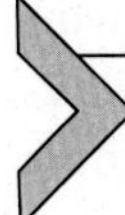

5. THE REGULAR CODE-BASED MUTATION TESTING PROCESS

This section details the code-based mutation testing advances. We categorize and present the surveyed advances according to the stages of the mutation testing process that they apply to. But before we begin the presentation we have to answer the following question: what is the mutation testing process? Fig. 3 presents a detailed view of the modern mutation testing process. This process forms an extension of the one proposed by Offutt and Untch [26]. The extension is based on the latest advances in the area. The important point here is that the process keeps the steps that are inherently manual outside the main loop of the testing process (steps in bold). The remaining steps can be sufficiently automated (although manual analysis may still be necessary).

Overall, the process goes as follows: first, we select a set of mutants that we want to apply (Step 1, detailed in Section 5.1) and instantiate them by forming actual executable programs (Step 2, detailed in Section 5.2). Next, we need to remove some problematic mutants, such as equivalent, i.e., mutants that are semantically equivalent to the original program despite being syntactically different, and redundant mutants, i.e., mutants that are semantically different to the original program but are subsumed by others (Step 3, detailed in Section 5.3).

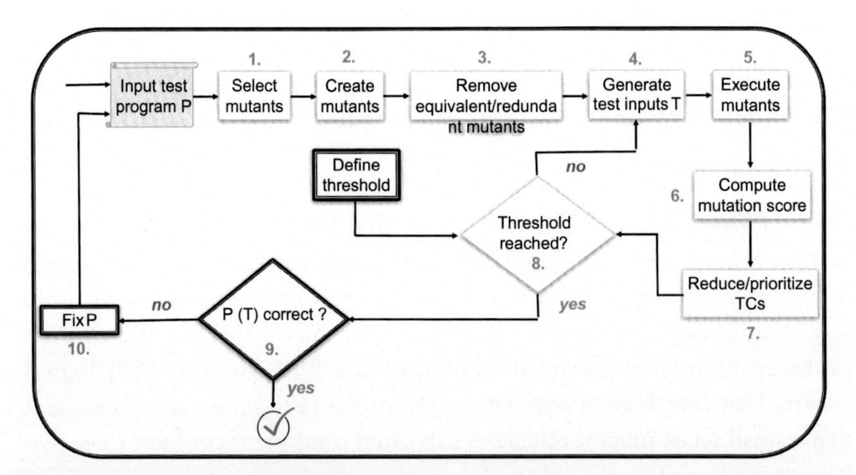

Fig. 3 Modern mutation testing process. The process forms an adaptation of the Offutt's and Untch's proposition [26] based on the latest advances in the area. *Bold boxes* represent steps where human intervention is mandatory.

Once we form our set of mutants we can generate our mutation-based test suite, execute it against our mutants and determine a score. In this step, we design (either manually or automatically) test cases that have the potential to kill our mutants (Step 4, detailed in Section 5.4) and execute them with all the mutants (Step 5, detailed in Section 5.5) to determine how well they scored (Step 6, detailed in Section 5.6). Subsequently, we perform test suite reduction by removing potentially ineffective test cases from our test suite. At this stage, we can also perform test case prioritization so that we order the most effective test cases first (Step 7, detailed in Section 5.7). The Steps 4–7 are repeated until the process results in a mutation score that is acceptable (Step 8, detailed in Section 5.8).

The last part of the process is when the user is asked to assess whether the results of the test executions were the expected ones (Step 9, detailed in Section 5.9). This step regards the so-called test oracle creation problem that involves the tester to assert the behavior of the test execution. In case faults are found then developers need to fix the problems and relaunch the process (Step 10, detailed in Section 5.10) until we reach an acceptable score level and cannot find any faults.

5.1 Mutant Selection (Step 1)

Mutation testing requires selecting a set of mutant operators based on which the whole process is applied. Thus, we need to define the specific syntactic transformations (mutant operators) that introduce mutants. Since defining mutant operators requires the analysis of the targeted language these may result in an enormous number of mutants. Therefore, in practice it might be important to select representative subsets of them. Section 5.1.1 refers to works that define mutant operators, while Section 5.1.2 refers to strategies that select subsets of mutants from a given set of mutants (aiming at reducing the cost of the process).

5.1.1 Mutant Operators

A large amount of work has focused on designing mutant operators that target different (categories of) programming languages, applications, types of defects, programming elements, and others.

5.1.1.1 Operators for Specific Programming Languages

Anbalagan and Xie [77] proposed a mutant generation framework for *AspectJ*, an aspect-oriented programming language. This framework uses two mutant operators; *pointcut strengthening* and *pointcut weakening*, which are used to increase or reduce the number of joint points that a pointcut matches.

Derezinska and Kowalski [8] introduced six object-oriented mutant operators designed for the intermediate code that is derived from compiled *C#* programs. Their work also revealed that mutants on the intermediate language level are more efficient than the high-level source code-level mutants.

Estero-Botaro et al. [78] defined 26 mutant operators for *WS-BPEL*—the Web Services Business Process Execution Language. Later on, they further quantitatively evaluated these operators regarding the number of stillborn and equivalent mutants each operator generates [79]. On the same topic, Boubeta-Puig et al. [80] conducted a quantitative comparison between the operators for *WS-BPEL* and those for other languages. The results indicate that many of the *WS-BPEL* operators are different due to the lack of common features with other languages (e.g., functions and arrays).

Mirshokraie et al. [10,81] proposed a set of *JavaScript* operators. These are designed to capture common mistakes in *JavaScript* (such as changing the *setTimeout* function, removing the *this* keyword, and replacing *undefined* with *null*). Experimental results indicate the efficiency of these operators in generating nonequivalent mutants.

Delgado-Pérez [7,82] conducted an evaluation of class-level mutant operators for *C++*. Based on the results, they propose a C++ mutation tool, MuCPP, which generates mutants by traversing the abstract syntax tree of each translation unit with the Clang API.

5.1.1.2 Operators for Specific Categories of Programming Languages

Derezinska and Kowalski [8] explored and designed the mutant operators for *object-oriented programs* through *C#* programs. They advocated that traditional mutant operators are not sufficient for revealing object-oriented flaws. Hu et al. [83] studied in depth the equivalent mutants generated by object-oriented class-level mutant operators and revealed differences between class-level and statement-level mutation: statement-level mutants are more easy to be killed by test cases.

Ferrari et al. [84] focused on *aspect-oriented programs*. They design a set of operators based on the aspect-oriented fault types. Similarly to the work of Anbalagan and Xie [77], this work uses *AspectJ* as a representative of aspect-oriented programs. Except for pointcut-related operators, operators for general declarations, advice definitions, and implementations are also adopted.

Bottaci [85] introduced an mutation analysis approach for dynamically typed languages based on the theory that the mutants generated by modifying types are very easily killed and should be avoided. Gligoric et al. [86] mentioned that "almost all mutation tools have been developed for statically typed languages," and thus proposed *SMutant*, a mutation tool that postpones mutation until execution and applies mutation testing dynamically instead of statically. In this way, the tool is able to capture type information of *dynamic languages* during execution.

5.1.1.3 Operators for Specific Categories of Applications

Alberto et al. [87] investigated the mutation testing approach for *formal models*. In particular, they introduced an approach to apply mutation testing to *Circus* specifications as well as an extensive study of mutant operators. Praphamontripong et al. [88,89] and Mirshokraie et al. [10] designed and studied mutant operators for web applications. Example operators are link/field/transition replacement and deletion. In a follow up study, Praphamontripong and Offutt [90] refined the initial set of operators (they exclude three operators), i.e., operators proposed in [89], by considering the redundancy among the mutants they introduce.

Deng et al. [91,92] defined mutant operators specific for the characteristics of *Android apps*, such as the event handle and the activity lifecycle mutant operators. Usaola et al. [93] introduced an abstract specification for defining and implementing operators for context-aware, mobile applications. Similarly, Linares-Vasquez et al. [94] introduced 38 mutation operators for Android apps. These operators were systematically derived by manually analysis types of Android faults. Oliveira et al. [95] proposed 18 *GUI-level* mutant operators. Also focused on GUI mutants, Lelli et al. [96] specially designed mutation operators based on the created fault model. Abraham and Erwig [97] proposed operators for *spreadsheets*. Dan and Hierons [98] introduced how to generate mutants aiming at floating-point comparison problems. Jagannath et al. [99] introduced how to generate mutants for actor systems.

Maezawa et al. [100] proposed a mutation-based method for validating Asynchronous JavaScript and XML (Ajax) applications. The approach is based on delay-inducing mutant operators that attempt to uncover potential delay-dependent faults. The experimental study suggests that by killing these mutants, actual errors can be revealed.

The study of Xie et al. [101] describes a mutation-based approach to analyze and improve parameterized unit tests (PUTs). The authors propose appropriate mutant operators that alter the effectiveness of the PUT test by varying the strength of its assumptions and assertions.

5.1.1.4 Operators for Specific Categories of Bugs

Brown et al. [102] proposed a technique to mine mutation operators from source code repositories. The intuition of this work is that by making mutants syntactically similar to real faults one can get semantically similar mutants.

Loise et al. [19] uses mutation testing to tackle security issues. They proposed 15 security-aware mutant operators for Java. Nanavati et al. [103,104] realized that few operators are able to simulate memory faults. They proposed nine memory mutant operators targeting common memory faults. Garvin and Cohen [105] focus on feature interaction faults. An exploratory study was conducted on the real faults from two open source projects and mutants are proposed to mimic interaction faults based on the study's results. Al-Hajjaji et al. [106] specially focus on variability-based faults, such as feature interaction faults, feature independency faults, and insufficient faults. Based on real variability-related faults, they derive a set of mutant operators for simulating them.

Additionally, other studies focus on the design of mutant operators for different levels or different programming elements. Mateo et al. [5] defined system-level mutant operators. Applying mutation at the system level faces two problems: first, mutating one part of the system can lead to an anomalous state in another part, thus comparing program behavior is not a trivial task; and, second, mutation's execution cost. To resolve these problems, the authors turn to weak mutation by introducing *flexible weak mutation* for the system level. The proposed approach is embedded in a mutation tool named Bacterio. Delamaro et al. [107] designed three new deleting mutant operators that delete variables, operators, and constants.

Additional categories of operators are due to Belli et al. [108], who proposed mutant operators for go-back functions (which cancel recent user actions or system operations), including basic mutant operators (i.e., transaction, state, and marking mutant operators), stack mutant operators (i.e., write replacement and read replacement), and high order mutant operators.

Gopinath and Walkingshaw [109] proposed operators targeting type annotations. This line of work aims at evaluating the appropriateness of type annotations. Jabbarvand and Malek [110] introduced an energy-aware framework for Android application. In this work, a set of 50 mutant operators mimicking energy defects was introduced. Arcaini et al. [111] proposed operators targeting regular expressions. These aim at assisting the generation of tests based on a fault model involving the potential mistakes one could made with regex.

5.1.2 Mutant Reduction Strategies

Mutant reduction strategies aim at selecting representative subsets from given sets of mutants. The practical reason for that is simply to reduce the application cost of mutation (since all the costly parts depend on the number of mutants).

Perhaps the simpler way of reducing the number of mutants is to randomly pick them. This approach can be surprisingly effective and achieve reasonably good trade-offs. Papadakis and Malevris [53] report that randomly selecting 10%, 20%, 30%, 40%, 50%, and 60% of the mutants results in a fault loss of approximately 26%, 16%, 13%, 10%, 7%, and 6%, respectively. Zhang et al. [112] reports that by killing randomly selected sets of mutants, composed of more than 50% of the initial set, results in killing more than 99% of all the mutants. Recently, Gopinath et al. [113] used large open source programs and found that a small constant number of randomly selected mutants is capable of providing statistically similar results to those obtained when using all mutants. Also, they found that this sample is independent of the program size and the similarity between mutants.

An alternative way of selecting mutants is based on their types. The underlying idea is that certain types of mutants may be more important than others and may result in more representative subsets than random sampling. Namin et al. [114] used a statistical analysis procedure to identify a small set of operators that sufficiently predicts the mutation score with high accuracy. Their results showed that it is possible to reduce the number of mutants by approximately 93%. This is potentially better than the mutant set of the previous studies, i.e., the five-operator set of Offutt et al. [34]. Investigating ways to discover relatively good trade-offs between cost and effectiveness, Delgado et al. [82] studied a selective approach that significantly reduce the number of mutants with a minimum loss of effectiveness for C++ programs. Delamaro et al. [107,115] experimented with mutants that involve deletion operations (delete statements or part of it) and found that they form a cost-effective alternative to other operators (and selective mutation strategies) as it was found that they produce significantly less equivalent mutants. In a later study, Durelli et al. [116] studied whether the manual analysis involved in the identification of deletion equivalent mutants differs from that of other mutants and found no significant differences. The same study also reports that relational operators require more analysis in order to asses their equivalence.

A later study of Yao et al. [117] analyzed the mutants produced by the five-operator set of Offutt et al. [34] and found that equivalent and stubborn mutants are highly unevenly distributed. Thus, they proposed dropping the ABS class and a subclass of the UOI operators (postincrement and decrement) to reduce the number of equivalent mutants and to improve the accuracy of the mutation score. Zhang et al. [118,119] conducted an empirical study regarding the scalability of selective mutation and found that it scales well for programs involving up to 16,000 lines of code. To further improve scalability, Zhang et al. [120] demonstrated that the use of random mutant selection with 5% of the mutants (among the selective mutant operators) is sufficient for predicting the mutation score (of the selective mutants) with high accuracy.

A comparison between random mutant selection and selective mutation was performed by Zhang et al. [112]. In this study, it was found that there are no significant differences between the two approaches. Later, Gopinath et al. [121] reached a similar conclusion by performing a theoretical and empirical analysis of the two approaches. The same study also concludes that the maximum possible improvement over random sampling is 13%.

Overall, as we discuss in Section 9, all these studies were based on the traditional mutation scores (using all mutants) that is vulnerable to the "subsumed mutant threat" [36]. This issue motivated the study of Kurtz et al. [122], which found that mutant reduction approaches (selective mutation and random sampling) perform poorly when evaluated against subsuming mutants.

Both random sampling and selective mutation are common strategies in the literature. Gligoric et al. [123] applied selective mutation to concurrent programs and Kaminski and Ammann [124] to logic expressions. Papadakis and Traon [17,125] adapt both of them for the context of fault localization and report that both random sampling and selective mutation that use more than 20% of all the mutants are capable of achieving almost the same results with the whole set of mutants.

Another line of research aiming at reducing the number of mutants is based on the notion of higher order mutants. In this case, mutants are composed by combining two or more mutants at the same time. Polo et al. [126] analyzed three strategies to combine mutants and found that they can achieve significant cost reductions without any effectiveness loss. Later, studies showed that relatively good trade-offs between cost and effectiveness can be achieved by forming higher order combination strategies [39,53,127]. In particular, Papadakis and Malevris [53] found that second-order strategies can achieve a reduction of 80%–90% of the equivalent mutants, with approximately 10% or less of test effectiveness loss. Similar results are reported in the studies of Kintis et al. [39], Madeyski et al. [28], and Mateo et al. [128], who found that second-order strategies are significantly more efficient than the first-order ones. Taking advantage of these benefits and ameliorate test effectiveness losses, Parsai et al. [129] built a prediction model that estimates the first-order mutation score given the achieved higher order mutation score.

Other attempts to perform mutant reduction are based on the mutants' location. Just et al. [130] used the location of the mutants on the program abstract syntax tree to model and predict the utility of mutants. Sun et al. [131] explored the program path space and selected mutants that are as diverse as possible with respect to the paths covering them. Gong et al. [132] selected mutants that structurally dominate the others (covering them results in covering all the others). This work aims at weak mutation and attempts to statically identify dominance relations between the mutants. Similarly, Iida and Takada [133] identify conditional expressions that describe the mutant-killing conditions, which are used for identifying some redundant mutants. Pattrick et al. [134] proposed an approach that identifies hard-to-kill mutants using symbolic execution. The underlying idea here is that mutants with little effect on the output are harder to kill. To determine the effect of the mutants on the program output, Pattrick et al. suggests calculating the range of values (on the numerical output expressions) that differ when mutants are killed. This method was latter refined by Pattrick et al. [135] by considering the semantics (in addition to numeric ones) of Boolean variables, strings, and composite objects.

Another attempt to reduce the number of mutants is to rank them according to their importance. After doing so, testers can analyze only the number of mutants they can handle based on the available time and budget by starting from the higher ranked ones. The idea is that this way, testers will customize their analysis using the most important mutants. In view of this, Sridharan et al. [136] used a Bayesian approach that prioritizes the selection of mutant operators that are more informative (based on the set of the already analyzed mutants). Along the same lines, Namin et al. [137] introduced MuRanker an approach that predicts the difficulty and complexity of the mutants. This prediction is based on a distance function that combines three elements; the differences that mutants introduce on the control-flow-graph representation (Hamming distance between the graphs), on the Jimple representation (Hamming distance between the Jimple codes), and on the code coverage differences produced by a given set of test cases (Hamming distance between the traces of the programs). All these together allow testers to prioritize toward the most difficult to kill mutants.

Mirshokraie et al. [10,81] used static and dynamic analysis to identify the program parts that are likely to either be faulty or to influence the program output. Execution traces are used in order to identify the functions that play an important role on the application behavior. Among those, the proposed approach then mutates selectively: (a) variables that have a significant impact on the function's outcome and (b) the branch statements that are complex. The variables with significant impact on the outcome of the functions are identified using the usage frequency and dynamic invariants (extracted from the execution traces), while complex statements are identified using cyclomatic complexity.

Anbalagan and Xie [77] proposed reducing mutants, in the context of pointcut testing of AspectJ programs, by measuring the lexical distance between the original and the mutated pointcuts (represented as strings). Their results showed that this approach is effective in producing pointcuts that are both of appropriate strength and similar to those of the original program.

Finally, mutant reduction based on historical data has also been attempted. Nam et al. [138] generated calibrated mutants, i.e., mutants that are similar to the past defects of a project (using the project's fix patterns), and compared them with randomly selected ones. Their results showed that randomly selected mutants perform similarly to the calibrated ones. Inozemtseva et al. [139] proposed reducing the number of mutants by mutating the code files that contained many faults in the past.

5.2 Mutant Creation (Step 2)

This stage involves the instantiation of the selected mutants as actual executables. The easiest way to implement this stage is to form a separate source file for each considered mutant. This approach imposes high cost as it requires approximately 3 s (on average) to compile a single mutant of a large project [37]. Therefore, researchers have suggested several techniques to tackle this problem.

The most commonly used technique realizes the idea of meta-mutation, also known as mutant schemata [140], which encodes all mutants in a single file. This is achieved by parameterizing the execution of the mutants [141]. The original proposition of mutant schemata involved the replacement of every pair of operands that participate in an operation with a call to a meta-function that functions as the operand [140]. The meta-functions are controlled through global parameters. This technique has been adopted by several tools and researchers, Papadakis and Malevris [141] use special meta-functions to monitor the mutant execution and control the mutant application. Wang et al. [142] and Tokumoto et al. [143] use meta-functions that fork new processes. Bardin et al. [144] and Marcozzi et al. [145,146] instrument the program with meta-functions that do not alter the program state, called labels, to record the result of mutant execution at the point of mutation and apply weak mutation.

Another approach involves bytecode manipulation [9,147]. Instead of compiling the mutants, these approaches aim at generating mutants by manipulating directly the bytecode. Coles et al. adopt such an approach for mutating Java bytecode [147]. Derezinska and Kowalski [8] and Hariri et al. [148] adopt the same approach for mutating the Common Intermediate Language of .NET and LLVM Bitcode, respectively.

Other approaches involve the use of interpreted systems, such as the ones used by symbolic evaluation engines [149]. A possible realization of this attempt is to harness the Java virtual machines in order to control and introduce the mutants [150]. Finally, Devroey et al. [151] suggested encoding all mutants as a product line. The mutants can then be introduced as features of the system under test.

5.3 Statically Eliminating Equivalent and Redundant Mutants (Step 3)

This step involves the identification of problematic mutants before their execution. This is a process that is typically performed statically. The idea is that some equivalent mutants, i.e., mutants that are semantically equivalent to the original program despite being syntactically different, and some redundant mutants, i.e., mutants that are semantically different to the original program but are subsumed by others, can be identified and removed prior to the costly test execution phase. By removing these "useless" types of mutants we gain two important benefits: first, we reduce the effort required to perform mutation and, second, we improve the accuracy of the mutation score measurement. Unfortunately, having too many "useless" mutants obscures the mutation testing score measurement by either overestimating or underestimating the level of coverage achieved. This last point is particularly important as it is linked to the decision of when to stop the testing process, i.e., Step 8 (Section 5.8).

5.3.1 Identifying Equivalent Mutants

Detecting equivalent mutants is a well-known undecidable problem [28]. This means that it is unrealistic to form an automated technique that will identify all the equivalent mutants. The best we can do is to form heuristics that can remove most of these mutants. One such effective heuristic relies on compiler optimization techniques [37,152]. The idea is that code optimizations transform the syntactically different versions (mutants) to the optimized version. Therefore, semantically equivalent mutants are transformed to the same optimized version. This approach is called Trivial Compiler Optimization

(TCE) and works by declaring equivalences only for the mutants that their compiled object code is identical to the compiled object code of the original program. Empirical results suggest TCE is surprisingly effective, being able to identify at least 30% of all the equivalent mutants.

Other techniques that aim at identifying equivalent mutants are of Kintis and Malevris [153–155] who observed that equivalent mutants have specific data-flow patterns which form data-flow anomalies. Thus, by using static data-flow analysis we can eliminate a large portion of equivalent mutants. This category of techniques includes the use of program verification techniques, such as value analysis and weakest precondition calculus. Program verification is used to detect mutants that are unreachable or mutants that cannot be infected [144,145].

A different attempt to solve the same problem is based on identifying killable mutants. This has been attempted using (static) symbolic execution [149,156]. Such attempts aim at executing mutants symbolically in order to identify whether these can be killable with symbolic input data. Other approaches leverage software clones to tackle this issue [157]. Since software clones behave similarly, their (non)equivalent mutants tend to be the same. Therefore, likely killable mutants can be identified by projecting the mutants of one clone to the other [157].

Literature includes additional techniques for the identification of equivalent mutants using dynamic analysis. These require test case execution and, thus, are detailed in the Step 6 (Section 5.6).

5.3.2 Identifying Redundant Mutants

Redundant mutants, i.e., mutants that are killed when other mutants are killed, inflate the mutation score with the unfortunate result of skewing the measurement. Thus, it is likely that testers will not be able to interpret the score well and end up wasting resources or performing testing of lower quality than intended [152]. To this end, several researchers have proposed ways to statically reduce redundancies.

Researchers have identified redundancies between the mutants produced by the mutant operators. The initial attempts can be found in the studies of Foster [158], Howden [159], and Tai [160,161] which claimed that every relational expression should only be tested to satisfy the $>$, $==$, and $<$ conditions. Tai [160,161] also argued that compound predicates involving n conditional AND/OR operators should be tested with $n + 2(2 * n + 3)$ conditions. Along the same lines, Papadakis and Malevris [149,162] suggested inferring the mutant infection conditions (using symbolic execution) of the mutants produced by all operators and simplify them in order to reduce the effort required to generate mutation-based test cases. This resulted in restricted versions for the Logical, Relational, and Unary operators.

More recently, Kaminski et al. [76,163] analyzed the fault hierarchy of the mutants produced by the relational operators and showed that only three instances are necessary. Just et al. [164] used a similar analysis and identified some redundancies between the mutants produced by the logical mutant operators. In a subsequent work, Just et al. [165] showed that the unary operator can be also improved. Putting all these three cases together, i.e., Logical, Relational, and Unary operators, results in significant gains in both required runtime execution (runtime reductions of approximately 20%) and mutation score accuracy (avoiding mutation score overestimation which can be as high as 10%). Along the same lines, Fernandes et al. [166] proposed 37 rules that can help avoiding the introduction of redundant mutants. Their results showed that these rules can reduce (on average) 13% of the total number of mutants.

All these approaches are based on a "local" form analysis, which is at the predicate level (designed for the weak mutation). Thus, applying them on strong mutation may not hold due to: (a) error propagation that might prohibit killing the selected mutants [141] and (b) multiple executions of the mutated statements caused by programming constructs such as loops and recursion [167,168]. Empirical evidence by Lindström and Márki [168] confirms the above problem and shows that there is a potential loss on the mutation score precision of 8%, at most.

Recently, Trivial Compiler Optimization (TCE) [37,152] has been suggested as a way to reduce the adverse effects of this problem. Similar to the equivalent mutant identification, TCE identifies duplicate mutant instances by comparing the compiled object code of the mutants. Empirical results have shown that TCE identifies (on average) 21% and 5.4% of C and Java mutants [152]. The benefits of using TCE is that it is conservative as all declared redundancies are guaranteed and deals with strong mutation.

Finally, Kurtz et al. [169] attempts to identify nonredundant mutants using (static) symbolic execution. The idea is that by performing differential symbolic execution between the mutants it is possible to identify such redundancies.

5.4 Mutation-Based Test Generation (Step 4)

According to the RIPR model [32], in order to kill a mutant we need test cases that reach the mutant, cause an infection on the program state, manifest the infection to the program output at an observable to the user point (asserted by the test cases). Formulating these conditions as requirements we can drive the test generation process. Currently, there are three main families of approaches aiming at tackling this problem named as (static) constraint-based test generation [149,170], search-based test generation [171,172], and concolic/dynamic symbolic execution [141,172]. Additional details regarding the automatic test generation and mutation-based test generation can be found in the surveys of Anand et al. [173] and Souza et al. [29].

5.4.1 Static Constraint-Based Test Generation

Constraint-based methods turn each one of the RIPR conditions into a constraint and build a constraint system that is passed to a constraint solver. Thus, the mutant-killing problem is converted to a constraint satisfaction problem [170]. Wotawa et al. [174] and Nica [175] proposed formulating the original and mutant programs (one pair at a time) as a constraint system and use solvers to search for a solution that makes the two programs differ by at least one output value. Kurtz et al. [169] adopted the same strategy in order to identify subsuming mutants.

Papadakis and Malevris [149,176] suggested formulating the RIPR conditions under selected paths in order to simplify the constraint formulation and resolution process. A usual problem of path selection is the infeasible path problem, i.e., paths that do not represent valid execution paths, which is heuristically alleviated using an efficient path selection strategy [149,176].

Other attempts are due to Papadakis and Malevris [177] and Holling et al. [156] who used out of the box symbolic execution engines (JPF-SE [178] and KLEE [179], respectively) to generate mutation-based test cases. These approaches instrument the original program with mutant-killing conditions that the symbolic execution engine is asked to cover (transforms the mutant-killing problem to code reachability problem). Riener et al. [180] suggested using bounded model-checking techniques to search for solutions (counter examples) that expose the studied mutants.

5.4.2 Concolic/Dynamic Symbolic Execution Test Generation

To overcome the potential weaknesses of the static methods, researchers proposed dynamic techniques such as concolic/dynamic symbolic execution. Similar to the static methods, the objective is to formulate the RIPR conditions. However, dynamic techniques approximate the symbolic constraints based on the actual program execution. Therefore, there is a need to embed the mutant-killing conditions within the executable program and guide test generation toward these conditions.

The first approach that uses concolic/dynamic symbolic execution is that of Papadakis et al. [177,181] that targets weak mutation. The main idea of this approach is to embed the mutant infection conditions within the schematic functions that are produced by the mutant schemata technique (described earlier in Section 5.2). This way, all the mutants are encoded into one executable program along with their killing conditions (mutant infection conditions). Subsequently, by using a concolic/dynamic symbolic execution tool we can directly produce test cases by targeting the mutant infection conditions. Similarly, Zhang et al. [182] and Bardin et al. [144,183] use annotations to embed the mutant infection conditions within the program under analysis. Along the same lines, Jamrozik et al. [184] augment the path conditions with additional constraints, similar to mutant infection conditions, to target mutants.

All the earlier approaches are actually performing some form of weak mutation as they produce test cases by targeting mutants' reachability and infection conditions. Performing weak mutation often results in tests that can strongly kill many mutants [32]. However, these tests often only kill (strongly) trivial mutants which usually fail to reveal faults [6]. To improve test generation, there is a need to formulate the mutant propagation condition on top of the reachability and infection conditions. This is complex as it involves the formulation of the two executions (the one of the original programs and the one of the mutants) along all possible execution paths. Therefore, researchers try to heuristically approximate this condition through search. Papadakis and Malevris [141] search the path space between the mutation point until the program output. This helps finding inputs that satisfy the propagation condition. Finally, another approach proposed by Harman et al. [172] searches the program input space using a constrained search engine (reachability and infection conditions are augmented with an extra conjunct to additional constraints).

5.4.3 Search-Based Test Generation

Other dynamic test generation techniques use search-based optimization algorithms to generate mutant-killing test cases. The idea realized by this class of methods is to formulate and search the program input domain under the guidance of a fitness function. The primary concern for these approaches is to define a fitness function that is capable of capturing the RIPR conditions and effectively identify test inputs that satisfy these conditions.

There are many different search-based optimization algorithms to choose from, but in the case of mutation, the most commonly used ones are the hill climbing [172,185] and genetic algorithms. As mentioned earlier, the main concern of these methods is the formulation of the fitness function. For instance, Ayari et al. [186] formulates the fitness as mutant reachability (distance from covering mutants) and Papadakis et al. [177,181] formulates the fitness as fulfillment of mutant infection conditions (distance from infecting mutants).

As with the already-presented techniques, formulating the propagation condition in the fitness function is not straightforward and thus, it is approximated by formulating indirect objectives. Fraser and Zeller [171,187] measure the mutants' impact (the number of statements with changed coverage, between a mutant and the original programs, along the test execution) to form the propagation condition. Papadakis and Malevris [188–190] measure the distance to reach specific program points which when impacted (covered by the original program execution and not by the mutant execution or vice versa) result in mutant killing. These are determined based on the mutants that have been killed by the past executions.

Patrick et al. [191] proposed a technique to evolve input parameter subdomains based on their effectiveness in killing mutants. The experimental evaluation of this approach suggests that it can find optimized subdomains whose test cases are capable of killing more mutants than test cases selected from random subdomains.

The most recent approaches try to formulate the mutant propagation condition by measuring the disagreement between the test traces of the original program and the mutants. Fraser and Arcuri [192] count the number of executed predicates that differ while Souza et al. [185] measure the differences in the branch distances between the test executions.

5.5 Mutant Execution (Step 5)

Perhaps the most expensive stage of mutation testing is the mutant execution step. This step involves the execution of test cases with the candidate test cases. Thus, given a program with n mutants and a test suite that contains m tests, we have to perform $n \times m$ program executions at maximum. For instance, consider a case where we have 100 mutants and a test suite that requires 10 s to execute for the original program. In this case, we expect that our analysis will complete in 1000 s. This makes the process time-consuming (since a large number of mutants is typically involved) and, thus, limits the scalability of the method.

Tests	Mutants			
	m_1	m_2	m_3	m_4
t_1	✓			✓
t_2	✓			✓
t_3			✓	✓
t_4			✓	
t_5			✓	✓

Fig. 4 Example mutant matrix.

To reduce this overhead, several optimizations have been proposed. We identify two main scenarios where the optimizations may appear. The first one, which we refer to as "Scenario A," regards the computation of mutation score, while the second one, which we refer to as "Scenario B," regards the computation of a mutant matrix (a matrix that involves the test execution results of all tests with all the mutants; an example appears in Fig. 4). This mutant matrix is used by many techniques such as the mutation-based fault localization [17], the oracle construction [171], test suite reduction [193], and prioritization [194].

The difference between the aforementioned scenarios is that when computing the mutation score (Scenario A) we only need to execute a mutant until it is killed. Therefore, we do not need to reexecute the mutants that have already been killed by a test case with other test cases. This simple approach achieves major execution savings. However, it does not apply on the second scenario where we need to execute all mutants with all test cases.

To illustrate the difference, consider the example mutant matrix of Fig. 4. To construct this mutant matrix (Scenario B), we need 20 executions (four mutants executed with the five tests). A naive approach for computing the mutation score (Scenario A) that does the same will also require 20 executions. However, mutation score calculation requires computing the number of mutants that are killed. Therefore, once a mutant is killed we do not need to reexecute it. In the above example, the tester will make four executions for t_1 (mutants m_1, m_2, m_3, and m_4), and he will determine that m_1 and m_4 are killed. Then, he will execute all the live mutants with t_2 (2 executions, mutants m_2 and m_3), and he will determine that none of them is killed. Then, he will execute the same mutants with t_3 (two executions, mutants m_2 and m_3) and will determine that m_3 is killed. For the last couple of test cases t_4 and t_5 he will execute only the mutant m_2. The sum of these executions is 10 which is greatly reduced compared to the initial requirement of 20 executions.

In the above analysis, we implicitly consider that there is an order of the test cases we are using. Therefore, by using different orders we can reduce further the number of test executions. In the example of Fig. 4, if we execute t_1 and t_3 first and then t_2, t_4, and t_5, we can reduce the number of test executions to 9. Zhang et al. [195] realized this idea using test case prioritization techniques. Similarly, Just et al. [196] proposed using the fastest test cases first and Zhu et al. [197] selected pairs of mutants and test cases to run together based on the similarity of mutants and test cases (identified by data-compression techniques). Of course, these two approaches only apply to Scenario A.

Regarding both Scenarios A and B, there are several optimizations that try to avoid executing mutants that have no chance of being killed by the candidate test cases. Thus, mutants that are not reachable by any test should not be executed as there is no chance of killing them. This is one of the initial test execution optimizations that has been adopted in the Proteum mutation testing tool [198]. This tool records the execution trace of the original program and executes only the mutants that are reachable by the employed tests. In practice, this optimization achieves major speed-ups compared to the execution of all tests (in both considered scenarios).

Papadakis and Malevris [177,181] observed that it is possible to record with one execution all the mutants that can be infected by a test. This was implemented by embedding the mutant infection conditions within the schematic functions that are produced by the mutant schemata technique (described in Section 5.2). Thus, instead of executing every mutant with every test, it is possible to execute all mutants at once, by monitoring the coverage of the infection conditions. Durelli et al. [150] suggested harnessing the Java virtual machine instead of schematic functions in order to record the mutant infection conditions. Both these methods resulted in major speed-ups (up to five times).

When performing strong mutation many mutants are not killed despite being covered by test cases, simply because the mutant execution did not infect the program state. Therefore, there is no reason to strongly execute mutants that are not reached and infected by the candidate test cases. Based on this observation, Papadakis and Malevris [141] proposed to strongly execute only the mutants that are reached and infect the program state at the mutant expression point. Along the same lines, Kim et al. [199] proposed optimizing test executions by avoiding redundant executions identified using statement-level weak mutation. Both the studies of Papadakis and Malevris [141] and Kim et al. [199] resulted in major execution savings. More recently, Just et al. [200] reported that these approaches reduce the mutant execution time by 40%. A further extension of this approach is to consider mutant infection at the mutant statement point (instead of mutant expression) [142].

Other test execution advances include heuristics related to the identification of infinite loops caused by mutants. Such infinite loops are frequent and greatly affect mutant execution time. However, since determining whether a test execution can terminate or not is an undecidable problem heuristic solutions are needed. The most frequent practice adopted by mutation testing tools to terminate test execution is with predefined execution time thresholds, e.g., if it exceeds three times the original program execution time. Mateo et al. [201,202] proposed recording program execution and determine whether potential infinite loops are encountered by measuring the number of encountered iterations.

All the aforementioned works aim at removing redundant executions. Another way to reduce the required effort is to take advantage of common execution parts (between the original program and the mutants). Thus, instead of executing every mutant from the input point to the mutation point, we can have a shared execution for these parts and then a different execution for the rest (from the mutation point to the program output). Such an approach is known as split-stream execution [203]. The separation of the execution is usually performed using a fork mechanism [143]. Empirical results suggest that such an approach can substantially improve the mutant execution process [142,143]. It is noted that these approaches are orthogonal to those that are based on mutant infection. Therefore, a combination of them can further improve the process and substantially enhance the scalability of the method, as demonstrated by Wang et al. [142].

An alternative way of speeding-up mutation testing is by leveraging parallel processing. This is an old idea that has not been investigated much. There are many tools supporting parallel execution of mutants, such as [147,204,205], but they do not report any results or specific advances. Mateo et al. [206] report results from five algorithms and shows that the mutant execution cost is reduced proportionally to the number of nodes that one is using.

Finally, there are also approaches tackling the problem in specialized cases. For instance, Gligoric et al. [207,208] suggest a method for the efficient state-space exploration of multithreaded programs. This work involves optimization techniques and heuristics that achieve substantial mutant execution savings. In the context of fault localization, Gong et al. [209] proposed a dynamic strategy that avoids executing mutants that do not contribute to the computation of mutant suspiciousness and achieves 32.4%–87% cost reductions. In the case of regression testing, Zhang et al. [210] identify the mutants that are affected by the program changes (made during regression) and executes only those in order to compute the mutation score. The affected mutants are identified with a form of slicing (dependencies between mutants and program changes). Wright et al. [211] uses mutant schemata and parallelization to optimize the test of relational database schemas. Zhou and Frankl [212] proposed a technique called *inferential checking* that determines whether mutants of database updating statements (INSERT, DELETE, and UPDATE) can be killed by observing the state change they induce.

5.6 Mutation Score Calculation and Refinement (Step 6)

The mutant execution aims at determining which mutants are killed and which are not. By calculating this number, we can compute the mutation score that represents the level of the test thoroughness achieved. Determining whether a test execution resulted in killing a mutant requires observing and comparing the program outputs. Thus, depending on what we define as a program output we can have different killing conditions. Usually, what constitutes the program output is determined by the level of granularity that the testing is applied to. Usually in unit testing the program output is defined as the observable (public access) return values, object states (or global variables), exceptions that were thrown (or segmentation faults), and program crashes. In system level, program output constitutes everything that the program prints to the standard/error outputs, such as messages printed on the monitor, behavior of user interfaces, messages sent to other systems, and data stored (in files, databases, etc.).

In the case of nondeterministic systems, it is necessary to define mutant-killing conditions based on a form of oracle that models the behavior of the obtained outputs. Patrick et al. [213] use pseudo-oracles to test stochastic software. Rutherford et al. [214] use discrete-event simulations (executable specifications) to define assertions and sanity checks that model how "reasonable" are the test execution results (distribution topology, communication failure, and timing) of distributed systems.

Observing and comparing the program outputs often requires a test driver that it is program specific and, thus, researchers usually approximate program outputs by observing a subset of it, usually defined by the test assertions (and program crashes). Alternative techniques involve the use of stubs, oracle data, log messages, and internal program states that will be detailed later on in Section 5.9. Mateo et al. [128] proposed *flexible weak mutation*, an approach for system-level mutation testing that considers mutants as killed when they result in corrupted object states. Object states are checked after the execution of every method call. Wu et al. [104] record execution paths and determine whether mutants cause any deviations from the original program's ones (execution of different paths).

Computing the mutation score requires the removal of equivalent mutants. As already discussed in Section 5.3, identifying equivalent mutants is a manual task that is partially addressed through static heuristics. Since the problem is important, there are some attempts to approximate the mutation score using dynamic heuristics. The idea is that mutants that are not killed by the tests but are capable of causing differences on the program state are likely to be killable [215]. This idea was initially introduced by Grun et al. [215] and, later, studied by the works of Schuler and colleagues [216–218]. Overall, these studies examined several heuristics that measure different types of impact (breaking program invariants, changed return values, altered control-flow and data-flow) and showed that measuring whether mutants cause deviations on the program execution forms the best option.

The use of mutants' impact provides opportunities to define mutant selection and classification strategies. Schwarz et al. [219] defined a mutant selection strategy by selecting a small set of mutants with high impact and diverge locations (all over the codebase). Mutant classification provides opportunities to achieve good trade-offs between effectiveness and efficiency. Papadakis and Traon [220,221] defined such strategies and found that mutant classification is beneficial when low-quality test suites are used.

Other attempts to refine and approximate the mutation score with the use of mutant classification are due to Kintis et al. [222,223]. Kintis et al. observed that killable mutants are likely to compose a higher order mutant that behaves differently than the first-order ones that it is composed of. Based on this observation, a mutant classification strategy that identifies 81% of the killable mutants with a precision of 71% was proposed.

5.7 Reduce/Prioritize Test Cases (Step 7)

This step involves the test suite reduction and/or test suite prioritization. Test reduction refers to the process of removing test cases that are somehow redundant, i.e., test cases that when removed from the test suites do not change the mutation score. Test prioritization refers to the process of ordering test cases in such a way that mutants are killed as early as possible.

Mutation-based test suite reduction has been suggested by Usaola et al. [224], using a greedy algorithm. The idea is to iteratively select the test cases that kill the maximum number of mutants that were not killed by the previously selected test cases. Hao et al. [225] used mutation to estimate a confidence level for the fault detection loss experienced due to the reduced test suites. Therefore, users can minimize their test suites using structural criteria and get an estimation of the potential fault detection capability loss based on the mutants.

Shi et al. [193] used mutants to reduce test suites and measured different trade-offs between reduced test suites and fault detection loss when reduced test suites kill fewer mutants than the original (nonreduced) ones. Similar to this work, Alipour et al. [226] proposed reducing (simplifying) individual tests rather than removing some of them and measured the trade-offs between reductions and fault detection loss.

Regarding test case prioritization, Lou et al. [194] studied two prioritization schemes; one based on the number of mutants killed and one based on the distribution of the killed mutants and found that prioritizing based on the number of killed mutants performs best. Nguyen et al. [227] proposed ordering first the test cases that kill the most mutants in order to support the audit testing of web service compositions. In their work, they considered only a subset of mutants, which is the ones that do not violate the explicit contract with the service under analysis.

5.8 Confidence Inspired by Mutation Score (Step 8)

The mutation testing process stops when mutation score reaches a user-specified threshold. In theory, this threshold reflects the level of confidence that developers have on the testing performed. Unfortunately, there are very few studies related to this subject, i.e., measuring the relationship between mutation score and fault revelation. Along these lines, Li et al. [52] experimented with mutation-adequate test suites (test suites that kill all killable mutants) and showed that these tests reveal more faults than the ones of structural testing criteria. This result is in line with the results of older studies that showed the superiority of mutation testing over other structural test criteria [27,55].

The most recent studies on the subject are those of Papadakis et al. [41] and Chekam et al. [6] that studied the fault revelation ability of mutation testing. The most important finding of the studies is that the "relationship between strong mutation and fault revelation exhibits a form of threshold behavior" [6] and that "achieving higher mutation scores improves significantly the fault detection." This means that there is a strong connection between mutation score and fault revelation only at higher mutation score levels (above a specific threshold). However, below that level, the mutation score is completely disconnected from fault revelation. In practice, this means that inadequate test suites that fail to reach relatively high mutation scores are vulnerable to noise effects and testers should not be confident on their testing (based on them). Perhaps more importantly, the same study shows that strong mutation-adequate test suites are capable of revealing at least 90% of the program faults [6].

The study of Tengeri et al. [228] suggests that mutation testing forms a good indicator of the expected number of defects in a system (number of real faults reported after the release of the system). Since these defects are those missed by the testing process they can be viewed as quality indicators of the test suite thoroughness.

Generally, it is important to consider the role of equivalent and redundant mutants when studying the relationship between mutation score and fault revelation. In practice, the existence of both equivalent and redundant mutants makes the evaluation of the exact mutation score value obscure, with the unfortunate effect of overestimating or underestimating the true score [36,152,229]. In particular, equivalent mutants tend to reduce the true mutation score, while redundant mutants have mixed effects. Therefore, reliably studying the mutant-fault relation requires, to some extend, adequate solutions for these problems.

Overall, we know very little regarding this fundamental aspect of software testing (confidence inspired by mutation score). Studies increasing our understanding on this respect are important and should form one of the main subjects addressed by future research. Similarly, studies addressing the equivalent and redundant mutant problems are also key to this problem.

5.9 Test Oracles (Step 9)

Once we create test inputs and reach the desire level of mutation score, we need to check whether the program under test behaves as expected. Additionally, we need to equip our tests with test oracles that assert the desired behavior for future use. This is the phase where we actually find faults (when the program does not behave as expected). Unfortunately, in the absence of formal specifications this task is carried out manually.

One of the first attempts to automate this process is due to Fraser and Zeller [171,187] who used mutants to guide oracle assertion creation. The idea is to check (and assert) the part of the program output that is responsible for killing the mutants. Fraser and Zeller [171,187] formulate this as a search problem and devised an automated approach that generates test assertions. Testers are then asked to validate these assertions. The same method has also been extended to identify relevant pre- and postconditions suitable for parameterized tests [101,230]. In the same vein, Knauth et al. [231] evaluated the quality of contracts (written in the Java Modeling Language) by mutating them.

The use of mutants in creating test oracles has been a common practice in automated test generation tools. Evosuite [232] adopts this practice for generating test oracles for Java programs. Yoshida et al. [233,234] use the same method to support test generation for C/C++ programs. Jahangirova et al. [235,236] use mutants to detect relevant observed state differences and abstract them into test oracles.

Mutants have also been used to drive the creation of oracle data (a set of variables that should be monitored during testing) [237,238]. In this work, internal program variables are monitored and ranked according to their ability to kill mutants. Similar to the work of Fraser and Zeller, mutants assist the creation and minimization of the oracle data. Jahangirova et al. [239] use test generation and mutation testing to assess and improve oracles (code assertions). Additional details regarding test oracles can be found in the survey of Barr et al. [240].

5.10 Debugging (Step 10)

Research on mutation-based debugging has followed two main directions, namely, fault localization and fault fixing. The former refers to the problem of locating the code areas that are responsible for a given failure while the later to the problem of automatically repairing the fault using the available test suite.

5.10.1 Mutation-Based Fault Localization

Mutation-based fault localization was introduce by Papadakis and Le Traon [17,241] with their work on the Metallaxis method. The underlying idea of Metallaxis is that mutants killed mostly by failing tests have a connection (interaction) with the program defects that caused the program failures. Thus, mutants killed mostly by failing tests provide indications regarding the faulty program locations. Empirical results on this approach demonstrated that mutation-based fault localization is significantly superior to other types of fault localization techniques, such as spectrum-based fault localization [17,242].

Metallaxis was later extended to support mutant reduction techniques, such as selective mutation [125] and has been released as an automated tool called Proteum/FL [204]. The idea of Metallaxis was later extended by Moon et al. [243], who introduce the MUSE method. MUSE works by checking whether mutants turn the failing test cases into passing or not. The difference from Metallaxis is that MUSE does not consider the mutants that are killed (have different outputs from the original program) by failing test cases but still they are not passing.

Other mutation-based fault localization techniques are those of Zhang et al. [244] who studied fault localization on the context of evolving programs and localized suspicious program edits. Hong et al. [245,246] extended MUSE for multilingual programs. Empirical results demonstrated that mutation-based techniques identify the faulty program locations (and edits) as the most suspicious statements. Another work of this type is that of Murtaza et al. [247,248]. In this work, it was observed that the test execution traces produced by mutants and faults are similar. Musco et al. [249] used mutants to approximate a causal graph. This approach realizes the idea of tracking causality in call graphs by exploring the test paths that lead to killing mutants.

5.10.2 Mutation-Based Fault Repair and Other Debugging Activities

Mutation has been used to support program repair activities by Debroy and Wong [250,251]. These works observe that many faults are fixed by simple syntactic transformations. Therefore, since mutants are simple syntactic transformations, they form potential patch candidates. The advantage of this technique is that it is simple and can be completely automated by a mutation testing tool.

Generally, automated fault repair is a large field of research with many specialized applications. Most of the approaches use genetic programming or constraint-based techniques to select and check whether special types of mutants can fix the underlying faults.

One of the first attempts in the area is due to Weimer et al. [18,252], who used genetic programming with statement deletion, statement insertion, statement replacement, and crossover mutant operators in order to support the automated bug fixing. Empirical results demonstrated that this approach can be particularly effective [253]. Later, Weimer et al. [254] leveraged mutation testing advances in order to improve the performance of fault fixing. In this work, the duality between mutation testing and fault repair is detailed along with potential opportunities for cross fertilization between the approaches. Other fault repair approaches combining search and mutation testing are by Tan and Roychoudhury [255], who introduce a regression repair technique that searches for mutants (using eight types of mutant operators) that fix regression faults.

There are many automated fault repair techniques that use some form of syntactic transformations (can be viewed as specialized higher order mutants) in order perform program repair. For instance, Long and Rinard [256] and Kim et al. [257] use syntactic patterns to perform program repair. However, as these approaches fall outside the scope of the present paper, the interested reader is redirected to a specialized survey on this subject [258].

Other debugging techniques relying on a form of mutation testing is Angelic debugging [259]. The idea of angelic debugging is to perform a form of data state mutation in order to correct the program execution. The idea is to identify the set of values that can substitute the program state (at runtime) and results in a form of "correct" execution.

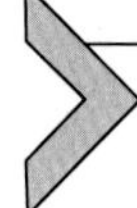

6. ALTERNATIVE CODE-BASED MUTATION TESTING ADVANCES

This section details code-based testing advances that do not conform to the mutation testing process, as depicted in Fig. 3. These advances include predictions of the mutation score, gamification of mutation testing, the use of search algorithms, and diversity-aware testing techniques.

Mutation testing requires performing the mutant execution step which is expensive even when using the test execution optimizations that were discussed in Section 5.5. To deal with this problem, many researchers have proposed the use of alternative proxies to measure the fault revealing potential of test cases. Gligoric et al. [59,60] found that branch coverage measurements are strongly correlated with mutation scores. Therefore, they argued that branch coverage could be used as an alternative to real faults and mutation faults. Later, Gopinath et al. [62] conducted a large empirical study and found that statement coverage scores correlate strongly with mutation scores. Unfortunately, there are also studies demonstrating that coverage does not correlate well with mutant detection [61].

Zhang et al. [260] built a classification model that predicts the mutants that are killed without executing them. The model relies on a number of features related to mutants, tests, and coverage measures and predicts the mutant execution results with a relatively good precision (with over 0.85 precision and recall).

Designing mutation-based tests is a tedious and potentially boring task that most developers try to avoid. As testing requires the involvement of developers, their motivation is crucial. In view of this, the studies of Rojas and Fraser [261] and Rojas et al. [262] suggested making these activities entertaining by gamefying mutation testing. The game includes two main roles: the attacker and the defender. The former aims at creating subtle non-equivalent mutants and the latter at creating test cases to kill these mutants. Overall, this approach helps educating and motivating developers. It can also help crowdsourcing complex tasks, such as test generation and adequacy evaluation [262].

Search-based mutation testing forms an alternative approach to traditional mutation testing. Instead of selecting mutants from a predefined set of operators, it uses meta-heuristic search techniques to evolve and optimize the generation of higher order mutants [263]. The idea here is to seek tailored mutants that fit to the particular goals of the testers. Thus, search-based techniques can be employed in order to search the space of all possible mutants for those that are subtle (mutants that are killed by only few test cases) [264], representative of all mutants [38], and realistic (both semantically and syntactically close to the original program) [72].

There is a number of approaches that utilize mutant optimization: Jia and Harman study ways to combine first-order mutants so that they produce subtle faults [38,264] for C programs. Harman et al. [265] report that by using search it is possible to improve test effectiveness (between 5.6% and 12%) while enjoying 15% improved efficiency (in Java programs). Langdon et al. [72] used grammar-based, biobjective, strongly typed genetic programming to form realistic mutants for C programs. Along the same lines, Omar et al. [15] experiment with Java and AspectJ mutants. Their results demonstrate that it is possible to generate subtle higher order mutants [266,267]. Omar et al. [267,268] experimented with different ways of combining first-order mutants in order to improve the efficiency of the approach. They found that a form of local search performs best. Wu et al. [269] use higher order mutation to genetically improve the nonfunctional properties of the program under test. Their approach yields time and memory performance improvements of approximately 18% and 20%, respectively.

Finally, Shin et al. [270,271] proposed using mutants as a test suite diversity measure and defined a mutation-based diversity test criterion. The idea of this approach is to construct test cases that can distinguish every mutant from every other mutant. Thus, instead of trying to distinguish the behavior of the mutants from that of the original program, they proposed to distinguish the behavior of every mutant from all the others.

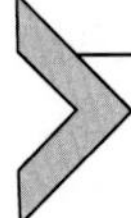

7. ADVANCES BEYOND CODE-BASED MUTATION TESTING

This section presents approaches that use mutants to support software engineering activities other than code-based testing. We first outline those belonging to the model-based testing, and continue with those related to security testing and, finally, other applications and testing approaches.

7.1 Model-Based Testing

An important line of mutation-based research regards its application to model artifacts. Older approaches called it specification-based mutation but the newer ones refer to it as model-based mutation. Here, we briefly discuss these approaches. For a detailed description and discussion on this subject, we point the reader to the specialized survey of Belli et al. [30].

Henard et al. [272] propose the use of mutation to test software product lines. The variability of software product lines and configurable systems is compactly represented by feature models. Therefore, the study of Henard et al. introduces mutant operators that mutate feature models (and their constraints). The idea is to transform the feature model into an equivalent logic formula, which is mutated using logical operators (using a tool called MutaLog [273]). Subsequently, the effectiveness of the selected configurations (to detect conformance faults of feature models) is evaluated based on their ability to detect mutants. In a later study, the feature model mutants were used to assist the automatic repair of feature models [274]. This is achieved by iteratively mutating the model under analysis until it reaches the desirable state.

An advantage of using feature model mutants is that by targeting them, it is possible to generate (automatically) a small set of test configurations. Henard et al. [275] applied this idea using search-based optimization methods in order to minimize the number of selected configurations and maximize the number of killed mutants. Filho et al. [276,277] used a multiobjective optimization approach to achieve several trade-offs between the number of selected configurations and number of killed mutants, their diversity, etc.

Generating feature model mutants with the aim of selecting test configurations has also been attempted by Arcaini et al. [278]. The difference of this method from that of Henard et al. is that mutants are introduced directly on the feature model under test (instead of the logic formula). The study also reports results from a test configuration generation technique that attempts to kill these mutants. The same approach was then used to evaluate the conformance of feature models to a software product line [279]. This approach also attempts to automatically detect and remove conformance faults from the feature model, similar to Henard et al. [274]. The difference is that since mutants are applied directly on the feature model the resulting models are expected to be easy to understand.

The study of Trakhtenbrot [280] focuses of testing statechart-based models for reactive systems. This approach is concerned with specific semantics of statechart models that are not aligned with the model's implementation. These semantics are the "zero-time" abstraction and "maximal parallelism," which are the subjects of mutation. Considering the conformance relation of action systems, Aichernig and Jöbstl [281] proposed a technique for encoding the semantics (of action systems) as constrains to be incorporated in the test conformance relations. These relations form the mutant killable conditions. Similarly, Aichernig et al. [282,283] developed a mutation-based test generation technique for UML state machines.

Devroey et al. [12] introduced the notion of featured model-based mutation analysis, a flexible formalism based on featured transition systems, which enable the optimized generation, configuration, and execution of mutants. The main idea behind this approach is to represent the model mutants as products of a software product line [151]. Based on this idea, the authors demonstrate that the technique can speed-up mutant execution up to 1000 times when compared to other behavioral model mutation approaches. Similarly, Belli and Beyazit [284] propose a mutant generation technique that attempts to limit the introduction of equivalent and duplicated mutants. The same approach aims at optimizing the test case execution by avoiding the comparison of the mutants with the original models.

El-Fakih et al. [285] present a mutation-based test case generation technique for extended finite state machines (EFSMs) that evaluates whether the EFSM under test conforms to user defined faults. As part of the technique, the EFSM under test is mutated and test cases able to kill the generated mutants are generated. Another technique, proposed by Su et al. [286], utilizes mutated GUI models for test case generation of Android applications. This particular approach mutates an autogenerated stochastic GUI model of the application, represented as a FSM, in order to search for better models that will result in different, and potentially better, event sequences compared to the original model. Finally, Aichernig et al. [287] combine property-based testing with model-based mutation testing to generate efficiently test suites that target specific coverage criteria based on EFSMs.

As for code-based mutation, detecting equivalent mutants at the model level is a tricky problem. When considering behavioral models such as automata, this problem can be formulated as a language equivalence problem. Indeed, if two automata accept the same language, then their traces are the same and no test case can distinguish them. Language equivalence is P-SPACE complete but efficient algorithms exist. Devroey et al. [3] compared one of such algorithms with two sorts of simulations: one that is completely random and one that exploits syntactic differences between the models to direct trace generation to infected states. Biased simulations proved to be efficient for strong mutation cased on large models while the exact approach was more interesting for weak mutation.

Belli et al. [108] present a mutation-based technique to test "go-back" functions modeled by pushdown automata. This approach uses mutant operators that affect the transitions, state, and stack of pushdown automata. Aichernig and Lorber [288,289] propose a model-based mutation testing technique for timed automata that tackles the state-space explosion problem caused when unfolding timed automata. The method improves the test execution using the unfolded structure of the original specifications. Larsen et al. [290] build on this work and further improve its efficiency. Zhou et al. [291] present a specification-based mutation approach to test safety-critical systems. This method defines mutant operators for the input output symbolic transition system modeling language and introduce a test case generation technique to create test cases based on these mutants. Adra et al. [292] study the application of mutation to agent-based systems. This approach defines mutant operators to address the properties of this type of systems.

Stephan et al. [293,294] present a technique that compares model-clone detection techniques for Simulink models using mutation. This approach introduces a set of structural mutant operators designed to compare model-clone detectors. The design of the operators was based on the authors' observations of potential model edit operations in publicly available models. In an extension of this work, Stephan et al. [295] present a taxonomy of Simulink mutant operators that represent realistic edit scenarios when modeling.

Although this taxonomy was created with the comparison of model-clone detectors in mind, the authors suggest that it can represent Simulink model mutations in general. Later, Pill et al. [296] developed a mutation testing framework for Simulink models, named SIMULTATE.

Testing model transformations using mutation has been attempted by Khan and Hassine [297]. In this approach the authors introduce specific mutant operators for the Atlas Transformation Language. Later, Troya et al. [298] focus on the same subject and presented an extensive set of mutant operators that uses both first and higher order mutation transformations [299]. Another study that tests model transformations using mutation is due to Aranega et al. [300], who focuses on how to support the generation of mutation-adequate test cases for checking model transformations. In this case, test cases are test models. The intuition behind the approach is that building a test model that is able to kill alive mutants from scratch is difficult, thus, the approach attempts to provide guidance to select some of the already available test models (test cases) to modify to kill the alive mutants.

Bartel et al. [301] focus on testing dynamically adaptive systems. These systems are governed by adaptation policies that incorporate how and when the system will adapt. The approach focuses on testing whether these adaptation policies are correctly implemented using mutation. Thus, a set of mutant operators is defined using a meta-model that represent the policy formalisms. The approach also suggests a specialization of these mutant operators for the case of action-based adaptation policies.

Mutation testing has also been applied on NuSMV models. Arcaini et al. [302] create mutants of such models and checks whether the NuSMV model advisor (an automatic static model review tool) can statically detect these mutants. Mutant operators for UML domain models have been defined by Kaplan et al. [303]. This study aims at generating test cases based on information provided by the domain model, expressed as a UML class diagram with invariants, and the use case model of the application under test. Fraser and Wotawa [304] present another model-based mutation approach aiming at determining property violations of a model. The approach relies on the notion of property relevance which relates test cases to model properties, in an attempt to connect the failing of a test case with a violation of a property.

The application of mutation testing at system requirements that are expressed in a natural language has also been attempted. Trakhtenbrot [305] introduced a semantic mutation approach that introduces mutants related to the intended meaning of the requirements (requirements expressed by predifined patterns) by altering the pattern of the requirements. This enables the use of the mutants for test generation and test evaluation.

Mutation testing has also been applied on Alloy models. Sullivan et al. [306] performed mutation testing in declarative programming paradigm (Alloy language) to support test case generation and showed that it is robust at revealing real faults.

Finally, mutation has been applied on aspect-oriented programs by mutating state models. Xu et al. [307] propose two strategies for generating tests from such models. The first one leverages structural information from the state model to generate test cases, whereas the second one is based on counterexamples generated by model-checking (counterexamples that form illegal sequences of events in the original model). The study of Lindström et al. [308] introduces a mutation-based approach to test aspect-oriented models. The approach proposes a set of mutant operators targeting specific features of aspect-oriented modeling. Abstract tests created to kill the generated mutants evaluate the modeling of cross-cutting concerns and the weaving process, as well.

7.2 Security Testing

This section presents mutation-based approaches related to security. More precisely, applications on testing security policies [309–313], regression testing of security policies [314] and testing security protocols [315], are shortly described.

Testing security policies using mutation has been suggested by Mouelhi et al. [310]. In this study, Mouelhi et al. propose a meta-model that captures different rule-based security policy formalisms. This meta-model forms the subject for mutation which is performed using a set of (proposed) generic operators that can simulate faults in the various instantiations of the model. Along the same lines, Mouelhi et al. [312] present another mutation-based technique that automatically transforms functional test cases into security test cases (test the security policy). In this approach, mutation is used for two purposes: to identify the subset of the functional test cases that are impacted by the security policy and to relate this subset's functional test cases to specific security policy rules.

Dadeau et al. [315,316] propose a mutation-based test generation and evaluation technique that validates an implementation of a security protocol that is written in the High-Level Security Protocol Language. These approaches propose mutant operators that introduce leaks in the security protocols and creates abstract test cases for HLPSL models by targeting/killing mutants.

Other attempts to test security policies are due to Bertolino et al. [311] who propose a fault model for history-based security policies. This study aims at policies written in the PolPA language and proposes modification rules that attempt to simulate faults that can occur in the implementation of the policy decision point (PDP) and target only the static behavior of the PDP. Elrakaiby et al. [309] and Nguyen et al. [313] attempt to test the obligation policy enforcement and delegation policies. These goals are achieved by using mutant operators that introduce changes in key elements of the obligation policy management and delegation features.

Finally, Hwang et al. [314] investigate test selection techniques for regression testing of security policies. They proposed three techniques toward this goal, one of which is based on mutation analysis. This particular technique first uses mutation analysis to correlate policy rules and tests cases and, subsequently, it applies test selection. The test selection is performed by selecting test cases that are correlated with rules involved with syntactic changes between the original policy and its mutants.

7.3 Supporting Adaptive Random Testing, Boundary Value Analysis, and Combinatorial Interaction Testing

Mutation testing has been used to support or extend several not mutation-based testing methods. Thus, it has been used to support adaptive random testing, boundary value analysis, and combinatorial interaction testing.

In the context of combinatorial interaction testing, Papadakis et al. [317] proposed mutating the constraints between the program input parameters. Thus, instead of selecting input combinations that satisfy the input constraints only, the authors proposed selecting the combinations that make the mutated constraints invalid. The underlying idea of this approach is that the difference between the original and the mutant constraints define some form of "boundary" conditions that trigger faults. Empirical results with faulty applications demonstrated that mutants have a stronger correlation with faults than the input parameter combinations.

Zhang et al. [318] proposed a mutation-based extension to boundary value analysis. The approach mutates some predicates of a given path condition in order to define boundary values. Similarly to Papadakis et al. [317], these values are the solutions that satisfy the path condition and at the same time differentiate the original predicate from its mutants. The authors also propose a way to generate test cases that cover these boundary conditions based on constrained combinatorial interaction testing.

Patrick and Jia [319,320] proposed a technique, named Kernel Density, to support adaptive random testing. This technique guides the test selection process based on the killed mutants. Thus, tests killing new mutants are considered to be more distant than those killing the same ones. Empirical results show that test cases selected by this approach kill more mutants than the ones selected by adaptive random testing.

7.4 Other Mutation-Based Applications

This section describes approaches tackling general software engineering problems. These include program analysis, software verification, code clones, defect prediction, and regression testing.

Mutation analysis has been used to automatically detect loop invariants by mutating postcondition clauses [321]. In such a way, many invariant candidates are generated and invalid invariants are discarded based on appropriate counterexamples. The study of Galeotti et al. [321] describes several ways to mutate the post conditions, as well as, ways to eliminate some trivial cases. Similarly, Andrés et al. [322] propose an automated framework, named PASTE, that uses mutants to evaluate the fault revealing ability of system invariants (generated from specifications) in the context of passive testing of stochastic timed systems. The approach evaluates the strengths of the invariants (and prioritizes them) based on the number of killed mutants. Subsequent works detail (its mutation module, its mutant operators and the algorithms it incorporates) extend and evaluate the framework further [323,324].

Pankumhang et al. [325] propose a code instrumentation technique, named iterative instrumentation, for measuring code coverage when testing time-sensitive systems. The approach is based on weak mutation analysis and instruments the program by inserting exit statements at the instrumentation points considered.

Other applications of mutation testing include its use for software verification. Groce et al. [326] used mutants to make developers familiar with software verification. The idea is to focus on incorrect programs (mutants) in order to understand when and how the verification process fails (by observing failures to detect problems caused by mutants).

Using mutation analysis to create software clones forms another application of mutation. Roy and Cordy [294] introduce several mutant operators that model typical copy/paste activities of developers and create clones based on the application of these operators. The same study also uses these clones to evaluate different clone detection techniques. Along these lines, the work of Svajlenko et al. [327] presents mutant operators that create fork constructs in order to assist the study and analysis of code similarity.

More recently, Bower et al. [328] proposed using mutation to assist the prediction of software defects. This technique combines traditional source code metrics with a number of mutation analysis metrics to built defect classifiers. The mutation analysis metrics that were used are classified into static, e.g., the number of mutants a mutant operator generated, and dynamic ones, e.g., the number of mutants killed. The study concludes that mutation-based metrics significantly improve the performance of defect prediction and that the best results are obtained by using a combination of static and dynamic metrics.

In the context of regression testing, Zhang et al. [329] used special forms of mutants to improve the fault detection ability of regression test suites. The approach mutates both the old and the new versions of the program under test and executes them with the available test suites. The detected differences between the two versions are considered as problems.

Di Nardo et al. [330] present a mutation-based technique to automatically generate faulty input data within complex data structures from existing field data. The approach uses six generic mutant operators that mutate the field data and guide their selection (using a data model). Results from an industrial case study show that it performs better, in terms of code coverage, than the manual testing performed by domain experts.

As discussed earlier, the existence of equivalent mutants constitutes one of the major costs of mutation's application. However, many researchers have started viewing this as an advantage in certain cases. Arcaini et al. [331,332] seek to find opportunities to improve the quality of the artifact under consideration. For example, they suggest that equivalent mutants can be used for improving code readability and for refactoring purposes. It is suggested that benefits from equivalent mutants may arise on all software artifacts where mutation applies. For instance, in feature models it is possible to detect dead and false optional features and redundant constraints. A similar approach is that of Baudry et al. [333], who suggested that equivalent mutants can be seen as diverse program versions. Therefore, by generating equivalent program versions, one can produce multiple diverse program variants (which can support security purposes, such as moving target defense).

Lisper et al. [334] introduced the concept of targeted mutation, which aims at nonfunctional properties. The idea underlying this approach is to introduce mutants that are relevant to a targeted nonfunctional property and use them as guides for generating and augmenting test suites. These test suites can then be used for estimating the worst-case execution time.

Finally, another mutation-based testing technique refers to testing relational database schemas. Wright et al. [211] investigate ways to make the application of mutation to database schemas more efficient. To this end, the authors propose and evaluate four cost-reduction approaches that leverage mutant schemata and parallelization. The results of the empirical study conducted suggest that the mutation analysis time can be reduced by the approaches proposed but they also indicate that their performance can be influenced by the underlining database management system (DBMS).

8. TOOLS FOR MUTATION TESTING

One important factor for the successful application of mutation is the availability of automated frameworks that support its application steps. This section discusses the tools that were introduced or were used in the studies we surveyed. Table 3 outlines the corresponding tools along with the year of their creation, their application artifact, and a concise description of key characteristics.

As it can be seen from the table, our analysis concluded in 76 tools, most of which where introduced between 2008 and 2017, that apply mutation to different software artifacts. By closely examining the table, it becomes obvious that there is an increasing growth in mutation testing tools with the creation of approximately 10 tools per year. Most of these tools target the implementation level languages but there are also tools that target specification languages and models.

At the implementation level, the mutation testing tools target mostly the C and Java programming languages. Most of the tools focus on the support of traditional, method-level mutant operators, and strong mutation, with few tools supporting object-oriented operators and weak or higher order mutation. Additionally, there have been various tools proposed that apply mutation to dynamically typed programming languages and concurrency-related aspects.

For the noncode-based tools, there have been proposed various tools for many model notations, including Extended and Timed Finite State Machines, Simulink models, Feature Models, etc., that automate the application of mutation. Furthermore, automated frameworks for mutating security policies and protocols have also been introduced.

9. MUTATION-BASED TEST ASSESSMENT: USE AND THREATS TO VALIDITY

Mutation testing is a popular technique for assessing the fault revealing potential of test suites. Much work on empirical software engineering relies

Table 3 Mutation Testing Tools

Name and Ref	Year	Application	Description
`mutate` [335]	N/A	C	Supports method-level mutant operators
`Jester` [336]	2001	Java	Supports source-code-level (src-level) mutant generation
`Proteum` [337]	2001	C	Supports an extensive set of method-level mutant operators and *interface mutation* (intermethod-level operators)
`mutgen` [338,339]	2003	C	Supports method-level mutant operators
`muJava` [9,340,341]	2004	Java	Implements src-level mutant generation and supports method-level and object-oriented (OO) mutant operators
`ByteME` [342]	2006	Java	Implements bytecode-level mutant generation and supports method-level and object-oriented (OO) mutant operators
`SQLMutation` [343]	2006	SQL	Supports traditional and SQL-specific mutant operators for SQL queries
`Jumble` [344]	2007	Java	Implements bytecode-level mutant generation and supports method-level mutant operators
`ESTP` [345]	2008	C	Supports 20 traditional C mutant operators
Not named [346]	2008	Sulu	Supports method-level mutant operators (drawn from the study of Andrews et al. [339])
`Milu` [347]	2008	C	Supports method-level mutant operators and higher order mutation
Not named [304]	2008	NuSMV models	Supports specification-based mutation (drawn from the study of Black et al. [2])

Continued

Table 3 Mutation Testing Tools—cont'd

Name and Ref	Year	Application	Description
Not named [348]	2008	Code generated from Simulink models	Seeds faults into the implementations generated from Simulink models
Not named [310]	2008	Security policies	Supports security-policy-access-control meta-model mutation (applied on policies defined in various notations (e.g., RBAC and OrBAC)
Not named [349]	2008	LOTOS specifications	Supports mutation testing for LOTOS specifications
Not named [77]	2008	AspectJ	Supports mutant operators for the creation of *pointcut* mutants that vary the strength of the corresponding pointcut in terms of the number of joint points it matches
`Javalanche` [205]	2009	Java	Implements bytecode-level mutant generation and supports method-level mutant operators and mutant classification based on mutants' impact
`JDama` [350,351]	2009	SQL/JDBC	Implements bytecode-level mutant generation and supports SQL-related operators and weak mutation
`AjMutator` [352,353]	2009	AspectJ	Supports mutant operators for AspectJ Pointcut Descriptors (PCDs) [84] and automated equivalent mutant detection
`GAmera` [354]	2009	WS-BPEL	Supports mutation testing for WS-BPEL composition
Not named [124]	2009	boolean logic	Supports mutant operators for possible DNF logic faults
`PASTE` [322–324]	2009	TFSM	Supports passive testing of systems presenting stochastic-time information using mutant operators specific to timed finite state machines (TFSM)

Not named [355]	2009	Z	Supports mutant operators for Z specifications
Not named [356]	2009	GCC-XML	Supports a subset of mutant operators proposed by Ellims et al. [357]
Not named [358]	2009	LUSTRE / SCADE	Supports mutant operators for LUSTRE/SCADE programs
Not named [231]	2009	Java	Supports mutant operators that follow the fault classification of Durães and Madeira [359]
PIT [147]	2010	Java	Implements bytecode-level mutant generation and supports method-level mutant operators
MutMut [208]	2010	Java	Supports concurrency-related mutant operators
GenMutants [182]	2010	.Net	Supports method-level mutant operators
Judy [360]	2010	Java	Implements src-level mutant generation and supports method-level and OO mutant operators
webMuJava [88]	2010	HTML/JSP	Supports specific mutant operators for web components written in HTML and JSP languages
Bacterio [5]	2010	Java	Supports method-level mutant operators for system-level testing using flexible weak mutation
Not named [141,177,181,188,190]	2010		Supports method-level mutant operators
Major [361]	2011	Java	Supports method-level mutant operators
Para μ [362]	2011	Java	Supports OO and concurrency-related mutant operators and higher order mutation

Continued

Table 3 Mutation Testing Tools—cont'd

Name and Ref	Year	Application	Description
`ILMutator` [8]	2011	C#	Implements CLI-level mutant generation and supports method-level and OO mutant operators
`SMutant` [86]	2011	Smalltalk	Supports traditional, method-level mutant operators in a dynamically typed language
`MuBPEL` [363]	2011	WS-BPEL	N/A
`jMuHLPSL` [315]	2011	HLPSL	Supports mutant operators that introduce leaks in security protocols
Not named [364]	2011	SPADE	Mutates the flow pattern description of input and output streams and the SPADE code of components
Not named [365]	2011	Aglets	Supports mutant operators specific to mobile agent systems that affect the movement, communication, run method, creation, event listeners, and agent proxy of an agent
Not named [366]	2011	Java	Supports method-level mutant operators based on the selective mutation approach and higher order mutation
`SMT-C` [367]	2012	C	Supports the semantic-related and method-level mutant operators
`mutant (muRuby)` [11,368]	2012	Ruby	Supports Ruby-specific mutant operators
Not named [309]	2012	Obligation policies	Supports mutant operators specific to obligation policy enforcement
Not named [149]	2012		Supports traditional, method-level mutant operators

CCMUTATOR [369]	2013	C/C++	Supports concurrency-related mutant operators, higher order mutation, and target applications written using the PThreads and C++11 concurrency constructs
Comutation [123]	2013	Java	Supports concurrency-related mutant operators [370]
SchemaAnalyst [371]	2013	SQL	Supports mutant operators related to relational schema integrity constraints, applied to multiple database management systems
XACMUT [372]	2013	XACML	Supports mutant operators targeting XACML 2.0 security policies
Mutandis [10,81]	2013	JavaScript	Supports JavaScript-specific mutant operators
Not named [373]	2013	Web service compositions	Supports two types of mutant operators for web service compositions: one that is internal to the service and one that models inconsistencies across different services of the composition
Not named [313]	2013	Security policies	Supports mutant operators specific to delegation policies based on a formal analysis of key delegation features
Not named [272]	2013	Feature models	Supports mutant operators for mutating feature models
MutPy[374]	2014	Python	Implements traditional and python-specific mutation operators
MuCheck [14]	2014	Haskell	Supports mutant operators targeting functional constructs and higher order mutation
HOMAJ [375]	2014	AspectJ/Java	Supports higher order mutation
Not named [376]	2014	HTML/CSS	Supports mutant operators that seed presentation defects to web pages

Continued

Table 3 Mutation Testing Tools—cont'd

Name and Ref	Year	Application	Description
Not named [377]	2014	EFSM	Supports mutants that introduce single transfer faults (STFs) and double transfer faults (DTFs) to extended finite state machines (EFSM) models
Not named [378]	2014	Data-flow languages	Supports two mutant operators that model common mistakes when creating power plant control programs
MutaLog [273]	2014	Logic mutation	Supports mutant operators for mutating logic expressions
`REDECHECK` [379]	2015	HTML/CSS	Supports mutant operators for layout defects in responsive web sites
Not named [380]	2015	Spreadsheets	Supports mutant operators for spreadsheets (*spreadsheet mutation*)
Not named [381]	2015	FSM	Supports mutant operators for FSM specifications (based on the studies of Fabbri et al. [382] and Petrenko et al. [383])
Not named [384]	2015	Component-level sequence and state diagrams	Supports architecture- and design-level mutant operators
Not named [385]	2015	HTML/JavaScript	Supports mutant operators for the Model-View-Controller frameworks of web application development
Not named [103,104]	2015	C	Supports memory-related mutant operators that model memory faults and control flow deviation as a mutant-killing condition
Not named [91]	2015	Android apps	Supports android-specific mutant operators, affecting intents, events, activity lifecycle, and XML files; and the method-level mutant operators of `muJava`
`MoMut` [386]	2015	UML models	Supports model-based mutation testing for UML state charts, class diagrams, and instance diagrams

MμVM [143]	2016	C	Implements bitcode-level mutant generation and supports higher order mutation
Not named [387]	2016	FBD	Supports mutant operators for FBD language
Not named [388]	2016	Simulink	Supports mutant operators that model common Simulink fault patterns
Not named [233]	2016	C++	Supports mutant operators similar to the ones of PIT for the Java language
Not named [148]	2016	C	Implements LLVM-level mutant generation and supports method-level mutant operators typically used in other tools, e.g., Milu
Not named [36,389]	2016	C	Supports traditional, method-level mutant operators
Not named [242]	2016		Supports traditional, method-level mutant operators
Vibes [12,390]	2016	Transition systems, Statechart models	Implements featured model-based mutation analysis and supports the Fabbri et al. [382] operator set (both first order and higher order)
μDroid [110]	2017	Android apps	Implements energy-aware mutation operators derived from a specific energy defect model
MDroid+[94]	2017	Android apps	Implements mutation operators to test Android applications based on a specifically designed Android fault model derived from manual analysis of various software artifacts
Not named [102]	2017	Source code	Extracts mutation operators from changes made in the development history of projects in an attempt to produce more "realistic" mutants
LittleDarwin [391]	2017	Java	Supports method-level mutation operators, higher order mutation, mutant sampling, and disjoint/subsuming mutant analysis

Continued

Table 3 Mutation Testing Tools—cont'd

Name and Ref	Year	Application	Description
MuCPP [7]	2017	C++	Implements class-level, object-oriented mutation operators for C++ programs
MutRex [111]	2017	Regular expressions	Implements mutation operators based on a specific fault model for regular expressions
BacterioWeb [93]	2017	Android apps	Implements mutation operators for Android applications
Not named [6]	2017	C	Supports method-level mutant operators
Not named [142]	2017	C	Implements method-level mutation operators and the AccMut approach [142] to reduce the cost of mutant execution
Not named [19]	2017	Java	Implements security-aware mutation operators

on the use of artificial faults (mutants or manually seeded faults). Researchers employ mutants to perform controlled experiments and assess the relative strengths of test strategies. We call this practice as *mutation-based test assessment.*

A typical mutation-based test assessment scenario arises when we want to determine whether one method, say Method-1, is more effective than another one, say Method-2. For instance, suppose that the methods to compare are a random test generation (Method-1) and a search-based test generation (Method-2). In this case, our objective is to check whether one of them has a higher fault revealing ability within a given amount of time. This is assessed by counting the number of killed mutants, i.e., the technique that kills the highest number of mutants is the winning one. Of course, in this particular case, the approaches are stochastic and thus, the experiment needs to be repeated multiple times and assessed by a statistical test, but in every case the technique that kills statistically significant more mutants is the winning one.

This is an intuitive choice made by many empirical studies. However, is it safe to conclude that Method-1 which kills more mutants than Method-2 is better? Actually, it is hard to draw any such conclusion unless we carefully consider and control a number of parameters. As we shall discuss in this section, there are influential factors lying at the heart of mutation-based test assessment that can hamper our ability to assess the fault revealing potential of the techniques.

9.1 The Use of Mutation in Empirical Studies

Using mutants as an effectiveness metric is a common practice, which previous research suggests that is adopted by more than a quarter of software testing controlled experiments [36]. To demonstrate the importance and popularity of this practice, we collected the papers that conduct empirical studies and use mutation testing as an assessment method and we analyze them to identify: (1) how often mutation testing is used as an assessment method; (2) the types of assessment that are used; (3) the tools that are frequently used; and (4) the languages mutation is applied to.

The results of our analysis show that, in most cases, mutation was applied to evaluate test techniques. Thus, mutants are used as proxies for real faults and the mutation score is used as an indicator of real fault detection. Fig. 5 presents the distribution of these studies in a yearly basis, including the overall growth trend. It is clear from the figure that the application of mutation testing, to experimental studies has been steadily increasing over the last 10 years.

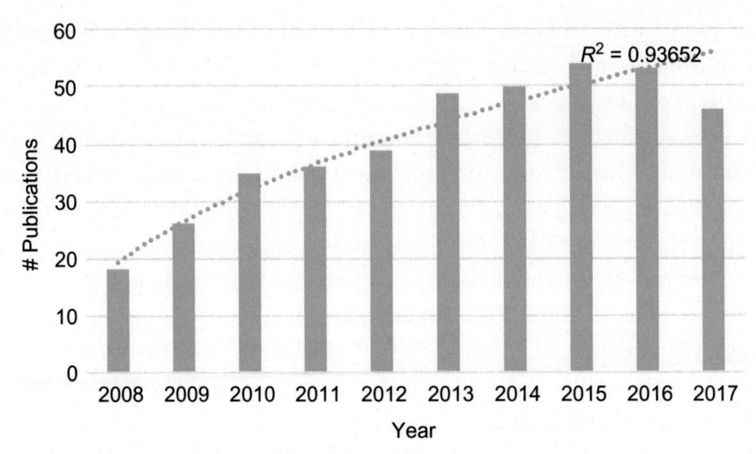

Fig. 5 Number of empirical studies using mutation testing as an assessment method.

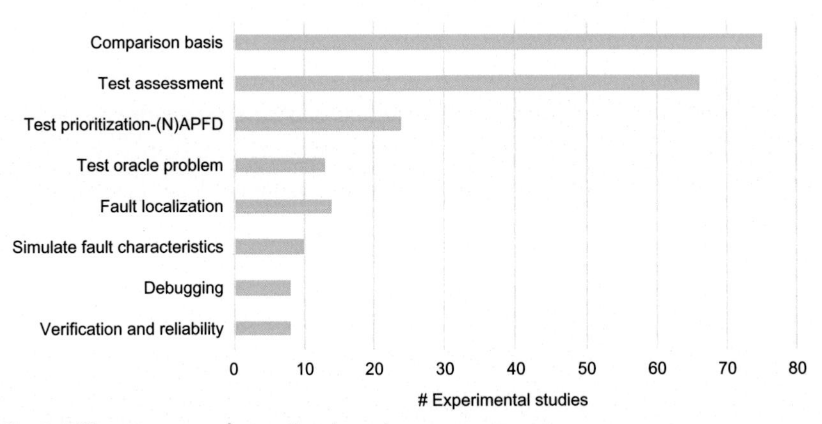

Fig. 6 Different types of mutation-based test assessment.

It is important to note that many of the analyzed studies are strictly not concerned with mutation testing; their objectives do not include mutation-related software engineering problems. Rather, mutation is a mechanism to validate the study and not the subject of the study.

Overall, from the papers we surveyed, we identified 190 papers falling into this category. Taking these findings into account, we can conclude that an increasing number of scientific results rely on mutation.

Fig. 6 presents the types of mutation-based test assessment. As can be seen from the figure, mutation's primary use in experimental studies is for comparing test techniques, *Comparison Basis*. The second largest category refers to *Test Assessment* and includes the test effectiveness of single techniques (without comparison), i.e., a test technique kills this number of

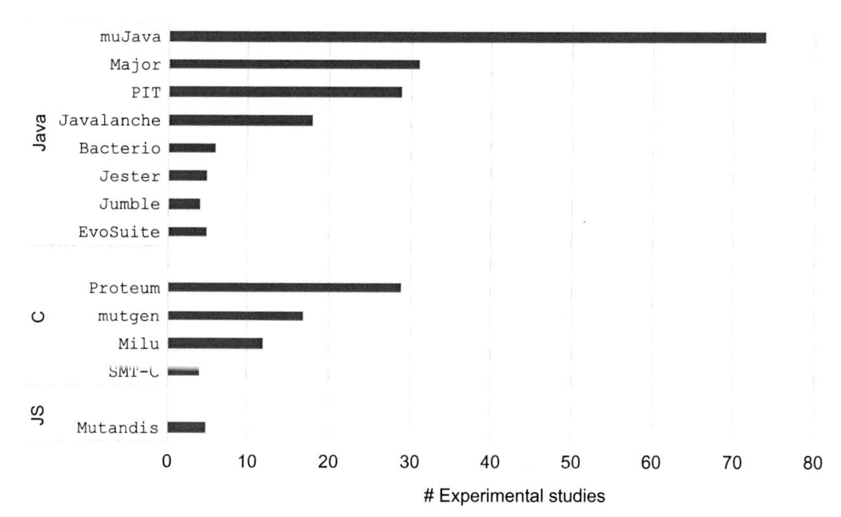

Fig. 7 Distribution of mutation testing tools in experimental studies.

mutants. The other categories involve assessments for test prioritization (*Test Prioritization-(N)APFD*),[b] test oracle (*Test Oracle Problem*), localization of faults (*Fault Localization*), verification techniques (*Verification and Reliability*), debugging (*Debugging*), and other techniques.

Fig. 7 depicts the mutation testing tools that were used in the experimental studies along with the number of studies using these tools. It is noted that the figure includes only the most frequently used tools.[c] As can be seen from the figure, muJava [9] and Proteum [198] are the most frequently used tools in experimental studies (operate on Java and C, respectively). Other frequently used Java mutation systems are PIT [147], Major [361], and Javalanche [205]. For C, the mutgen framework, used in the study of Andrews et al. [339] and Milu [347] are some of the most frequently used tools.

Fig. 8 depicts the most frequently used languages in experimental studies. It can be seen that mutation is mostly applied at the code/implementation level with the respective test subjects implemented in Java and C. Other commonly used programming languages that are used in mutation experiments include JavaScript and AspectJ. Finally, mutation has also been applied to other test subjects such as security protocols [316], feature models [272] and regular expressions [111], which are distributed in the remaining categories of the figure.

---

[b] Test prioritization techniques are typically assessed based on Average Percentage Faults Detected (APFD) or Normalized APFD (NAPFD) [392].

[c] The figure does not include tools that were used in fewer than four publications.

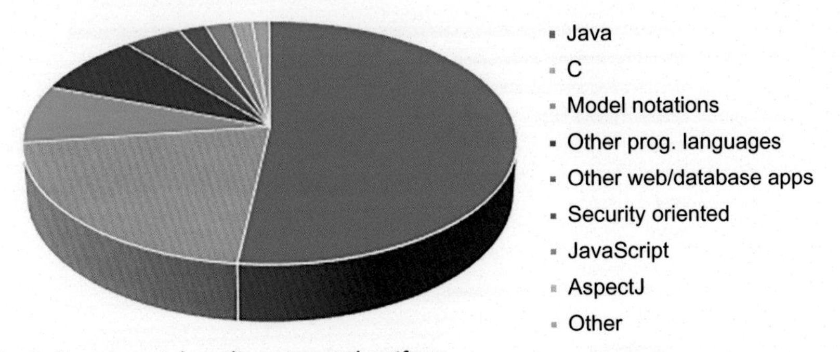

Fig. 8 Experimental studies: targeted artifacts.

Since mutation testing is increasingly used in experimental studies, potential issues with this practice can have serious implications on many research studies. Although there is some empirical evidence suggesting that mutants behave like real faults [41–43], these are only preliminary results, contradicted by other studies [63,393], and can be questioned if not suitable experimental care is taken. Unfortunately, recent research has shown that mutation testing is vulnerable to a number of confounding factors, such as those discussed in this section, that researchers should be aware and cater for. The influence of these factors can be severe and lead to questioning many findings of empirical research. In the remainder of this section, we discuss the influential factors that can bias the results of mutation-based test assessment and how we can mitigate them.

9.2 Programming Language and Mutant Operators

One of the main factors influencing the effectiveness of mutation-based test assessment is the programming language and the mutants that are used in the experimental study [33,393,394]. Applying mutation testing requires defining mutants based on the language's constructs. Thus, it is likely that we are not able to use the same mutants for different languages. Additionally, languages following different typing disciplines and programming paradigms may also require different sets of mutant operators. Certain types of mutants might be more effective in one paradigm than another. For instance, strongly typed languages may produce fewer mutants than weakly typed ones [152]. Similarly, object oriented code tends to have simpler method functionality, but more complex interactions among the methods (or classes) than imperative (procedural) languages [395]. Therefore, mutants encoding intramethod faults [9] might not be effective. Another example is the Java language runtime checking, which may result in mutants that are easier to kill than those in C.

Namin and Kakarla [393] demonstrated that the correlation between mutants and faults differs significantly across different programming languages. Kintis et al. [152] and Baker and Habli [57] also report significant differences between mutants of different languages, in particular between C—Java and C—Ada, respectively. Similarly, there are significant difference across different types of mutant operators, as reported by Namin and Kakarla [393]. Kurtz et al. [122] showed that it is hard to select mutant operator sets that perform similarly well on different programs. Therefore, when using mutation testing, *it is important to carefully select mutant operators that are appropriate to the programming language studied.*

9.3 Subsumed Mutant Threat

A major concern when using mutation testing is related to the "quality" of the employed mutants. In case the mutants we are using are trivial, then we only measure the ability of test suites to cover some parts of the code, instead of their ability to uncover faults at these parts. This problem is called the "subsumed mutant threat" [36]. The problem becomes particularly important when we have large number of redundant mutants. Unfortunately, when assessing testing methods, one test technique might achieve a significant advantage over another by killing redundant than nonredundant mutants. This would be a case of inadequately scientific methodology leading to possible incorrect scientific conclusions.

Recent research has shown that redundant mutants tend to skew the mutation score measurement leading to serious threats to the validity of empirical research. Andrews et al. [42] noted that the difficulty of revealing faults and killing mutants may influence the experimental results. Thus, they pointed out that it may be important to filter out the subset of trivial mutants in order to set a representative relation between mutants and real faults. Visser [396] suggested controlling for mutants' reachability in order to identify mutants that are hard to kill (hard to infect and propagate). Papadakis et al. [36] used the notion of mutant subsumption, demonstrating empirically that there is a very good chance (estimated to be more than 60% for arbitrary experiments) to compromise scientific conclusions, due to this subsumed mutant thread [36]. Similarly, Kurtz et al. [122] replicated previous studies on selective mutation and found that they perform well when considering redundant mutants but perform poorly when discarding them.

There are many studies advocating some form of "refined" mutation score for mitigating the problems caused by mutant redundancies. The first study attempting to address this problem was that of Kintis et al. [39] who suggested using disjoint mutants, i.e., a small representative subset of all mutants in order to remove all the mutant redundancies from the set of mutants that is used for test assessment. Consider that we have a set of *N* mutants, a representative subset, say *D*, means that any test suite that kills this subset of mutants also kills the *N* mutants. No redundancy between the mutants of *D* means that it is not safe to remove any mutant from this set because in this case we fail to kill all the *N* mutants.

Computing the true disjoint mutant set is impossible and thus, in the context of controlled experiments, it is approximated by a test suite. This dynamic approximation of the disjoint mutants can be computed using the Algorithm 1 [36,167]. In this algorithm, the live and duplicate mutants are removed first (from *S*, lines 2 and 3). Then, the mutant that is joint (subsuming) with the highest number of live mutants is selected (lines 8–15). This is the mutant that it is killed by test cases, which manage to collaterally kill the highest number of other mutants. This mutant is then added to the disjoint set *D* (line 16) and the joint mutants are removed from *S* (line 17). This process is repeated until *S* is empty. Finally, the set of disjoint mutants, *D*, is returned.

Other studies of redundancy reduction include that of Kaminski et al. [76,163] and Just et al. [164], which suggested removing some instances of the relational and logical mutant operators, in order to improve the accuracy of the mutation score. Ammann et al. [397] introduced the notion of minimal mutants (smallest possible set of mutants)[d] and Kurtz et al. [398] suggested selecting the minimal sets of mutants using mutant subsumption graphs. In another study, Kurtz et al. [169] proposed using symbolic execution to approximate subsuming mutants. Papadakis et al. [37] and Kintis et al. [152] suggested using compiler optimizations to remove duplicated mutants (a special form of redundant mutants) as a way to strengthen experimental rigor.

Overall, all these studies found that a large percentage of mutants are redundant, indicating potential inflation problems for the studies that have not take account of redundancy. Kintis et al. [39] report that disjoint mutants were approximately 9% of all the mutants, Ammann et al. [397] that minimal mutants were 1.2%, and Kurtz et al. [398] that they were 4%. These results

[d] The difference between the notions of minimal and disjoint mutants is that minimal mutants is the smallest possible representative set of mutants while the disjoint ones is a set that has no redundancies (perhaps not the minimal one) [33,39].

ALGORITHM 1 Disjoint Mutants

```
Input: A set S of mutants
Input: A set T of test cases
Input: A matrix M of size |T|×|S| such as M_ij = 1 if test_i kills mutant_j
Output: The disjoint mutant set D from S
D = ∅
                                             /* Remove live mutants */
S = S \{m ∈ S | ∀i ∈ 1..|T|, M_ij ≠ 1}
                                             /* Remove duplicate mutants */
S = S \{m ∈ S | ∃m'∈ S | ∀i ∈ 1..|T|, M_ij(m) = M_ij(m')}
while (|S| > 0) do
    maxJoint = 0
    jointMut = null
    maxMutDisjoint = null
                                             /* Select the most disjoint mutant */
    foreach (m ∈ S) do
        sub_m = {m' ∈ S|∀i ∈ 1..|T|,(M_ij(m) = 1) ⇒ (M_ij(m') = 1)}
        if (|sub_m| > maxJoint) then
            maxJoint = |sub_m|
            maxMutDisjoint = m
            jointMut = sub_m
        end
    end
                                             /* Add the most disjoint mutant to D */
    D = D ∪ {maxMutDisjoint}
                                             /* Remove the joint mutants from the remaining */
    S = S \jointMut
end
return D
```

motivated the work of Papadakis et al. [36] that found that redundant mutants can bias experimental results (approximately 60% for arbitrary experiments). Therefore, researchers should *identify and discard as many subsumed mutants as possible before conducting any test assessment.*

9.4 Test Suite Strength and Size

Failure to account for test suite strength can also adversely affect the scientific findings of empirical studies. Chekam et al. [6], studied the relation between faults and mutants, report that low-strength test suites are vulnerable to noise effects "two studies with below-threshold coverage may yield different findings, even when the experimenters follow identical experimental procedures." Thus, their study concluded that test suite strength plays a central role when conducting an experiment.

In particular, the study of Chekam et al. [6] showed that there is no practical difference between test criteria when relatively low-strength test suites are used. By contradiction, higher-strength test suites yield larger differences for test criteria. This is particularly important, because it indicates that empirical studies need to improve the strength of their test suites before conducting the experiment. Unfortunately, the mutation strength (over other test techniques) is only observable using strong test suites. For instance, one might conclude that a test technique or criterion is ineffective (compared to another), while in fact it is not, simply because the superiority of this criterion is only observable using stronger test suites.

Recent studies have also identified that test suite size (number of test cases) introduces another confounding factor that should be brought under experimental control [61]. Going a step further, Namin and Kakarla [393] observed that by using different test suite sizes, an experimenter will observe different correlations between faults and mutants. Since test suite size can be considered as a proxy measure of test suite's strength, this finding reconfirms the findings of Chekam et al. [6], suggesting experiments should consider both test suite size and test suite strength.

Another source of variation of empirical results is due to selection of candidate test cases. Consider the case where we want to compare two test techniques. In this case, we need two test suites, one that simulates the result of the first technique and one that simulates the result of the second technique. This is usually performed by randomly sampling of test cases from a test pool or by randomly generating test suites. Thus, two sets of test cases are to be compared. A problem typically arises in this scenario is that these two test suites need to adequately simulate the results of the techniques. Therefore, the two sets need to be free from redundant test cases [33,167,397], which might otherwise inadvertently bias experimental results. An easy way to ameliorate this problem is to select test cases that only increase coverage, while discarding all the others [33,42].

Another concern derives from the test case selection. As this may be stochastic, it is likely that different selections of test cases (at random) may result in different results; perhaps very different results if the experimenter happens to be unlucky. To reduce this problem, researchers usually select multiple sets of test cases and perform an inferential statistical analysis on the set of results as a whole [399,400]. Delamaro and Offutt [401] investigated the influence of using multiple sets of test cases (selected randomly) and found that "averaging over multiple programs was effective in reducing the variance in the mutation scores introduced by specific tests." Therefore, they found that in case it is too expensive to perform multiple repeated experiments a single test set (per program) over a relatively large number of subject can be enough to provide accurate average values.

Overall, researchers are advised to *carefully select their test suites*. Depending on the evaluation scenario it might be important to *control for test suite size* and *reduce redundant test cases*.

9.5 Mutation Testing Tools

An often-ignored parameter that can also inadvertently bias experimental findings relates to the choice of mutation testing tool. Mutation testing tools implement different operators and have different implementation details, most of which can influence the experimental outcome [394]. As already explained, the choice of mutant operators affects significantly the results of an experiment. However, different implementations of the same operators are likely to produce different mutants and merely provide divergent results.

The studies of Kintis et al. [394,402] and Gopinath et al. [403] demonstrate that there is a large degree of disagreement between the judgements made by the most popular Java mutation testing tools. The studies of Kintis et al. [394] and Marki and Lindström [404] cross evaluate the Java mutation testing tools and identify specific implementation weaknesses. This motivated the work of Laurent et al. [33,167], who compared Java mutation testing tools and implemented one (called PIT_{RV} [147]) that is "at least as strong as the mutants of all the other tools together." Unfortunately, these studies are only concerned with Java so there is no clear evidence concerning the C mutation testing tools (or tools for other less widely studied languages). Taken together, these results suggest that the *choice of a mutation tool need to be carefully introduced and justified* in best practice empirical studies.

9.6 Clean Program Assumption

Mutation-based test assessment can be viewed as a simulation that involves two "roles"; the faults role (played by the mutants) and the "oracle" role (played by the original program). By aligning this simulation to the reality, we can say that developers produce the faulty programs (simulated by the mutants) which they test using a test oracle (simulated by the original program).

Naturally, testers apply their tools and techniques on the mutant program versions, check whether they can find any unexpected behavior, as defined by the test oracle and report on any bugs found.

As intuitive as this seems, the practice of test assessment is performed differently. It is a common practice to apply tools and test techniques on the original program and then check their fault revealing power by executing tests on the mutants. This practice may be less time-consuming but it makes an implicit assumption that coverage measurements (or the application of test techniques) on the original program are representative (or very similar) of those on the mutant programs. This assumption is called the "clean program assumption" (CPA) [6]. The assumption can be problematic since test suites are assessed on the mutant program versions instead of the original program from which (and for which) they were applied.

Unfortunately, Chekam et al. [6] demonstrated that the CPA does not hold and therefore cannot be relied upon. The study also showed that CPA has the potential of changing the outcome of empirical studies if not brought under experimental control. Overall, the Chekam et al. revisited previous empirical questions concerning the usefulness of test adequacy criteria, using a robust methodology that accounts for CPA and showed that mutation testing outperform statement and branch coverage for real fault revelation. These results suggest that experiments dealing with the real fault revelation question, should *report on the CPA*. If it is not possible to take CPA into account (potentially due to execution cost), researchers are advised to report the amount of time required by the performed study.

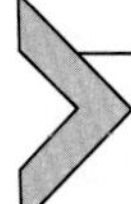

10. A SEVEN-POINT CHECK LIST OF BEST PRACTICES ON USING MUTATION TESTING IN CONTROLLED EXPERIMENTS

The fundamental experimental factors surveyed in Section 9 highlight the many pitfalls that can compromise or even invalidate the scientific findings and conclusions of a controlled experiment that uses mutation testing. It can be a daunting challenge for experimenters and researchers to be sure they have catered for all of the potential threats to validity that have accrued over four decades of literature recording the development of mutation testing.

Therefore, to address this challenge, in this section, we provide a simple seven-step checklist that aims to give experimenters the confidence that they are compliant with best practice reporting of results. Ensuring that all seven steps are met is relatively straightforward, because it simply involves explaining and justifying choices that may affect conclusion validity. Nevertheless, experimenters who follow these seven steps help other researchers replicate and investigate, properly, the influence of such potentially confounding factors, thereby contributing to the overall experimental robustness of their study.

1. *Mutant selection*: Explain the choice of mutant operators. One of the most important things that experimenters need to explain is the appropriateness of the chosen mutant operators with respect to the programming language used.
2. *Mutation testing tool*: Justify the choice of mutation testing tool. The choice of mutation testing tool needs to be made carefully as at the current state, mutation testing tools differ significantly [394,402]. To support the reproducibility and comprehension of the experimental results, researchers should also clearly describe the exact version of the employed mutation testing tool. If the used tool is not a publicly available, researchers should list the exact transformation rules (mutant instances supported by each operator [394,402]) that are supported by the mutant operators selected. Unfortunately, our survey found that more than a quarter of the empirical studies does not report such details. The objective is to provide readers with the low-level details that might vary from one study to another, so that these can be accounted for in subsequent studies.
3. *Mutant redundancy*: Justify the steps taken to control mutant redundancy. As we discussed in Section 9.3, mutant redundancy may have a large impact on the validity of the assessment. Therefore, it is important to explain how mutant redundancy is handled (perhaps in the threats to validity section). Where possible, experimenters are advised to additionally use techniques such as TCE [152] to remove the duplicate mutants (in case the interest is on the achieved score of a technique), or a dynamic approximation of the disjoint mutation score [33,36] (in case the interest is on comparing test techniques). As already discussed, the approximation of the disjoint mutants can be made by using Algorithm 1. In case these techniques are expensive, researchers are advised to clarify this and contrast their findings on a (small) sample of cases where mutant redundancy is controlled.
4. *Test suite choice and size*: Explain the choice of test suite and any steps taken to account for the effects of test suite size, where appropriate. Ideally, an experimenter would like to have large, diverse (i.e., mutants are killed by multiple test cases), and high-strength (i.e., killing the

majority of the mutants) test suites. As such test suites are rare in most of the open source projects, researchers are advised to demonstrate and contrast their findings with a (small) sample of subjects with strong and diverse test suites (perhaps in addition to the chosen subjects). Alternatively, experimenters may consider using automated tools to augment their test suites. Overall, the objective is to allow other researchers to create a similar test suite and/or to experiment with different choices of suite and measure the effects of such choices.

5. *Clean program assumption*: Explain what the study relies on the CPA assumption. Ideally, where possible, the CPA should not be relied upon; testing should be applied to the faulty programs (instead of the clean, nonfaulty ones). If this is not possible (potentially due to execution cost or lack of resources), researchers are advised to note the reliance on the CPA. Its effects may be small in some cases, justifying reliance on this assumption. Either way, explicitly stating whether or not it is relied upon will aid clarity and facilitate subsequent studies.
6. *Multiple experimental repetitions*: Clarify the number of experimental repetitions. Ideally, when techniques make stochastic choices they should be assessed by multiple experimental repetitions [400,405]. In practice, this might not be possible due to the required execution time or other constraints. In this case, researchers have to choose between experiments with many subjects but few repetitions or experiments with few subjects and many repetitions; research suggests that it is preferable to choose the second option [401]. Of course, this choice needs to be clarified according to the specific context and goals of the study.
7. *Presentation of the results*: Clarify the granularity level of the empirical results. Many empirical studies compute mutation scores over the whole subject projects they are using (one score per project). Since, this practice may not generalize to other granularity levels[e] (such as unit level) [167], researchers should report and explain the suitability of the chosen granularity level at the given application context.

11. CONCLUSION AND FUTURE DIRECTIONS

This chapter surveyed the recent trends and advances on mutation testing. It offers a concise description of the mutation testing problems, methods, applications, and best practices for applying mutation testing

[e] Two methods can have a similar number of mutants killed on a project (overall number), but quite different numbers, of mutants killed, on the individual units of the project.

(either as a test technique or as an experimental methodology). Based on the data we collected, we demonstrate that there is a growing interest in the subject. Interestingly, even 8 years after the first observation of this trend, by Jia and Harman [27], the interest in the field is still increasing markedly.

The interest in the field is related to both fundamental research advances and practical applications such as tool support and use in controlled experiments. Our analysis shows that many tools and techniques have been introduced these last 10 years. Many of these advances are already widely used by researchers. At the same time major companies report that they experiment with mutation in order to include in their practices. Hopefully, practitioners will soon use mutation as well. All these observations may be seen as evidence supporting the claim that mutation testing is reaching a state of maturity. In summary, the research interest in mutation testing is divided into the following general categories:

- Solutions to the problems of mutation analysis (fundamental advances of mutation).
- Mutation applied on new languages and artifacts. New mutation testing tools also appear.
- Use of mutants as a means to support other software engineering activities (e.g., fault localization [17]).
- Use of mutation testing advances to support controlled experiments.

Recent work in the area focuses on building scalable and practical tools that can push mutation testing toward industry and everyday use. The rest of this section is dedicated to summarize the mutation testing open problems, barriers, and areas that we believe will attract attention in the near future.

11.1 Open Problems

One of the main open problems of mutation regards the detection of the equivalent and redundant mutants. As we already discussed, there are many techniques tackling this problem, either directly or indirectly, but unfortunately the problem remains largely unresolved. Overall, the current research results show that only few of the mutants produced (approximately 5%) is practically useful. The rest is noise to the process with severe consequences [36].

Overall, mutation testing requires models that will guide the mutations toward small semantic deviations that are in a sense disjoint, instead of blind syntactical mutations. Unfortunately, there is no clear theory or consensus on which types and instances of mutants we should use. Some initial results indicate that almost all the mutant operators are of some value. The fact that most of the existing tools are limited to a small number of mutant operators is restrictive and to some extend arbitrary. Thus, in future, mutation may be tailored toward few diverse and "useful" mutants that bring value to the tester (regardless of the operators used) [41].

The lack of clear theory on which mutants are of some value has restricted most of the previous research on first-order mutation. Higher order mutation appears to have similar characteristics with the first-order mutation as it produces subtle mutants. Of course the great majority of them are redundant, but in theory a smart mutant selection process can identify them. Therefore, future mutation may identify ways to generate and use those valuable higher order mutants.

Another important aspect concerns the automatic mutation-based generation of test cases and test oracles. Although the last 10 years there are major advances on this area of research, the problem remains. Most of the automated approaches fail to kill a substantial number of mutants and recent empirical evaluations show that automatic test generation techniques fail to cover most of the critical program areas. Therefore, there is little work on improving test suites using mutants. Perhaps this is attributed to the lack of understanding and modeling of the error propagation. Recent research has shown that failed mutant propagation is the basic ingredient that makes mutation testing powerful [6]. Much work remains to be done until we can automatically produce high quality test cases through high quality mutants.

Although researchers have identified mutation as a strong test criterion, there is neither clear understanding nor much empirical evidence concerning whether and when mutants are correlated with real faults. What types of faults are not captured by simple or complex mutants? What percentage of future regression errors can we capture with mutations? When is it appropriate to stop the testing process? How should we integrate mutation testing into our development process? Of course these questions need to be answered under the light of specific development paradigms and application domains. These are open questions, hopefully to be answered by future research.

Model-based mutation is one of the areas that has not been researched much (compared to code-based mutation) over the last years. Despite this, we see a growing interest toward this direction. There is a recent dedicated survey on this subject [30] and multiple high profile publications over the last couple of years. Additionally, very recently efficient and scalable tools have been built, e.g., the VIBeS tool [12], which hopefully will push the research in this area further.

Finally, there are many new areas of research that can benefit from the use of mutants. The current trend is to explore the behavior space of mutants, instead of the original program, to identify several interesting aspects, either functional or nonfunctional. Thus, the conformance of models, the generation and improvement of models, the improvements of program security, and debugging activities are only a few examples where mutants have been shown to be spectacularly useful and effective. Future research is heading toward this line of research with many new and exciting applications of mutation analysis.

REFERENCES

[1] J. Offutt, A mutation carol: past, present and future, Inf. Softw. Technol. 53 (10) (2011) 1098–1107, https://doi.org/10.1016/j.infsof.2011.03.007.

[2] P.E. Black, V. Okun, Y. Yesha, Mutation operators for specifications, in: The Fifteenth IEEE International Conference on Automated Software Engineering, ASE 2000, Grenoble, France, September 11–15, 2000, 2000, p. 81, https://doi.org/10.1109/ASE.2000.873653.

[3] X. Devroey, G. Perrouin, M. Papadakis, A. Legay, P. Schobbens, P. Heymans, Automata language equivalence vs. simulations for model-based mutant equivalence: an empirical evaluation, in: 2017 IEEE International Conference on Software Testing, Verification and Validation, ICST 2017, Tokyo, Japan, March 13–17, 2017, 2017, pp. 424–429, https://doi.org/10.1109/ICST.2017.46.

[4] M.E. Delamaro, J.C. Maldonado, A.P. Mathur, Interface mutation: an approach for integration testing, IEEE Trans. Softw. Eng. 27 (3) (2001) 228–247, https://doi.org/10.1109/32.910859.

[5] P.R. Mateo, M.P. Usaola, J. Offutt, Mutation at system and functional levels, in: Third International Conference on Software Testing, Verification and Validation, ICST 2010, Paris, France, April 7–9, 2010, Workshops Proceedings, 2010, pp. 110–119, https://doi.org/10.1109/ICSTW.2010.18.

[6] T.T. Chekam, M. Papadakis, Y.L. Traon, M. Harman, An empirical study on mutation, statement and branch coverage fault revelation that avoids the unreliable clean program assumption, in: Proceedings of the 39th International Conference on Software Engineering, ICSE 2017, Buenos Aires, Argentina, May 20–28, 2017, 2017, pp, 597–608. http://dl.acm.org/citation.cfm?id=3097440.

[7] P. Delgado-Pérez, I. Medina-Bulo, F. Palomo-Lozano, A. García-Domínguez, J.J. Domínguez-Jiménez, Assessment of class mutation operators for C++ with the MuCPP mutation system, Inf. Softw. Technol. 81 (2017) 169–184, https://doi.org/10.1016/j.infsof.2016.07.002.

[8] A. Derezinska, K. Kowalski, Object-oriented mutation applied in common intermediate language programs originated from C#, in: Fourth IEEE International Conference on Software Testing, Verification and Validation, ICST 2012, Berlin, Germany, March 21–25, 2011, Workshop Proceedings, 2011, pp. 342–350, https://doi.org/10.1109/ICSTW.2011.54.

[9] Y. Ma, J. Offutt, Y.R. Kwon, MuJava: an automated class mutation system, Softw. Test. Verif. Reliab. 15 (2) (2005) 97–133, https://doi.org/10.1002/stvr.308.

[10] S. Mirshokraie, A. Mesbah, K. Pattabiraman, Efficient JavaScript mutation testing, in: Sixth IEEE International Conference on Software Testing, Verification and

Validation, ICST 2013, Luxembourg, Luxembourg, March 18–22, 2013, 2013, pp. 74–83, https://doi.org/10.1109/ICST.2013.23.

[11] N. Li, M. West, A. Escalona, V.H.S. Durelli, Mutation testing in practice using Ruby, in: Eighth IEEE International Conference on Software Testing, Verification and Validation, ICST 2015 Workshops, Graz, Austria, April 13–17, 2015, 2015, pp. 1–6, https://doi.org/10.1109/ICSTW.2015.7107453.

[12] X. Devroey, G. Perrouin, M. Papadakis, A. Legay, P. Schobbens, P. Heymans, Featured model-based mutation analysis, in: Proceedings of the 38th International Conference on Software Engineering, ICSE 2016, Austin, TX, USA, May 14–22, 2016, 2016, pp. 655–666, https://doi.org/10.1145/2884781.2884821.

[13] Y. Ma, Y.R. Kwon, J. Offutt, Inter-class mutation operators for Java, in: 13th International Symposium on Software Reliability Engineering (ISSRE 2002), November 12–15, 2002, Annapolis, MD, USA, 2002, pp. 352–366, https://doi.org/10.1109/ISSRE.2002.1173287.

[14] D. Le, M.A. Alipour, R. Gopinath, A. Groce, MuCheck: an extensible tool for mutation testing of haskell programs, in: International Symposium on Software Testing and Analysis, ISSTA '14, San Jose, CA, USA, July 21–26, 2014, 2014, pp. 429–432, https://doi.org/10.1145/2610384.2628052.

[15] E. Omar, S. Ghosh, An exploratory study of higher order mutation testing in aspect-oriented programming, in: 23rd IEEE International Symposium on Software Reliability Engineering, ISSRE 2012, Dallas, TX, USA, November 27–30, 2012, 2012, pp. 1–10, https://doi.org/10.1109/ISSRE.2012.6.

[16] J. Tuya, M.J.S. Cabal, C. de la Riva, Mutating database queries, Inf. Softw. Technol. 49 (4) (2007) 398–417, https://doi.org/10.1016/j.infsof.2006.06.009.

[17] M. Papadakis, Y.L. Traon, Metallaxis-FL: mutation-based fault localization, Softw. Test. Verif. Reliab. 25 (5–7) (2015) 605–628, https://doi.org/10.1002/stvr.1509.

[18] C. Le Goues, T. Nguyen, S. Forrest, W. Weimer, GenProg: a generic method for automatic software repair, IEEE Trans. Softw. Eng. 38 (1) (2012) 54–72, https://doi.org/10.1109/TSE.2011.104.

[19] T. Loise, X. Devroey, G. Perrouin, M. Papadakis, P. Heymans, Towards security-aware mutation testing, in: 2017 IEEE International Conference on Software Testing, Verification and Validation Workshops, ICST Workshops 2017, Tokyo, Japan, March 13–17, 2017, 2017, pp. 97–102, https://doi.org/10.1109/ICSTW.2017.24.

[20] Y. Jia, F. Wu, M. Harman, J. Krinke, Genetic improvement using higher order mutation, in: Genetic and Evolutionary Computation Conference, GECCO 2015, Madrid, Spain, July 11–15, 2015, Companion Material Proceedings, 2015, pp. 803–804, https://doi.org/10.1145/2739482.2768417.

[21] W.B. Langdon, B.Y.H. Lam, M. Modat, J. Petke, M. Harman, Genetic improvement of GPU software, Genet. Program. Evolvable Mach. 18 (1) (2017) 5–44, https://doi.org/10.1007/s10710-016-9273-9.

[22] R.G. Hamlet, Testing programs with the aid of a compiler, IEEE Trans, Softw. Eng. 3 (4) (1977) 279–290.

[23] R.A. DeMillo, R.J. Lipton, F.G. Sayward, Program mutation: a new approach to program testing, in: Infotech State of the Art Report, Software Testing, vol, 2, 1979, pp. 107–126.

[24] R.A. DeMillo, R.J. Lipton, F.G. Sayward, Hints on test data selection: help for the practicing programmer, IEEE Comput. 11 (4) (1978) 34–41, https://doi.org/10.1109/C-M.1978.218136.

[25] R.A. DeMillo, Test adequacy and program mutation, in: Proceedings of the 11th International Conference on Software Engineering, Pittsburg, PA, USA, May 15–18, 1989, 1989, pp. 355–356, https://doi.org/10.1145/74587.74634.

[26] A.J. Offutt, R. Untch, Mutation 2000: uniting the orthogonal, in: W,E. Wong (Ed.), Mutation 2000, Kluwer, San Jose, CA, USA, 2001, ISBN: 0-7923-7323-5, pp. 45–55.

[27] Y. Jia, M. Harman, An analysis and survey of the development of mutation testing, IEEE Trans. Softw. Eng. 37 (5) (2011) 649–678, https://doi.org/10.1109/TSE.2010.62.

[28] L. Madeyski, W. Orzeszyna, R. Torkar, M. Jozala, Overcoming the equivalent mutant problem: a systematic literature review and a comparative experiment of second order mutation, IEEE Trans. Softw. Eng. 40 (1) (2014) 23–42, https://doi.org/10.1109/TSE.2013.44.

[29] F.C. Souza, M. Papadakis, V.H.S. Durelli, M.E. Delamaro, Test data generation techniques for mutation testing: a systematic mapping, in: Proceedings of the 11th ESELAW, 2014, pp, 1–14.

[30] F. Belli, C.J. Budnik, A. Hollmann, T. Tuglular, W.E. Wong, Model-based mutation testing—approach and case studies, Sci. Comput. Program. 120 (2016) 25–48, https://doi.org/10.1016/j.scico.2016.01.003.

[31] R.A. Silva, S. do Rocio Senger de Souza, P.S.L. de Souza, A systematic review on search based mutation testing, Inf. Softw. Technol. 81 (2017) 19–35, https://doi.org/10.1016/j.infsof.2016.01.017.

[32] P. Ammann, J. Offutt, Introduction to Software Testing, second ed,, Cambridge University Press, 2016. ISBN: 978-1-107-17201-2.

[33] T. Laurent, M. Papadakis, M. Kintis, C. Henard, Y.L. Traon, A. Ventresque, Assessing and improving the mutation testing practice of PIT, in: 2017 IEEE International Conference on Software Testing, Verification and Validation (ICST), 2017, pp. 430–435, https://doi.org/10.1109/ICST.2017.47.

[34] A.J. Offutt, A. Lee, G. Rothermel, R.H. Untch, C. Zapf, An experimental determination of sufficient mutant operators, ACM Trans. Softw. Eng. Methodol. 5 (2) (1996) 99–118, https://doi.org/10.1145/227607.227610.

[35] J.B. Goodenough, S.L. Gerhart, Toward a theory of test data selection, IEEE Trans. Softw. Eng. 1 (2) (1975) 156–173, https://doi.org/10.1109/TSE.1975.6312836.

[36] M. Papadakis, C. Henard, M. Harman, Y. Jia, Y.L. Traon, Threats to the validity of mutation-based test assessment, in: Proceedings of the 25th International Symposium on Software Testing and Analysis, ISSTA 2016, Saarbrücken, Germany, July 18–20, 2016, 2016, pp. 354–365, https://doi.org/10.1145/2931037.2931040.

[37] M. Papadakis, Y. Jia, M. Harman, Y.L. Traon, Trivial compiler equivalence: a large scale empirical study of a simple, fast and effective equivalent mutant detection technique, in: 37th IEEE/ACM International Conference on Software Engineering, ICSE 2015, Florence, Italy, May 16–24, 2015, vol. 1, 2015, pp. 936–946, https://doi.org/10.1109/ICSE.2015.103.

[38] Y. Jia, M. Harman, Higher order mutation testing, Inf. Softw. Technol. 51 (10) (2009) 1379–1393, https://doi.org/10.1016/j.infsof.2009.04.016.

[39] M. Kintis, M. Papadakis, N. Malevris, Evaluating mutation testing alternatives: a collateral experiment, in: 17th Asia Pacific Software Engineering Conference, APSEC 2010, Sydney, Australia, November 30–December 3, 2010, 2010, pp. 300–309, https://doi.org/10.1109/APSEC.2010.42.

[40] Karl Popper, https://en.wikiquote.org/wiki/Karl_Popper (accessed 06.10.2017).

[41] M. Papadakis, D. Shin, S. Yoo, D. Bae, Are mutation scores correlated with real fault detection? A large scale empirical study on the relationship between mutants and real faults, in: Proceedings of the 40th International Conference on Software Engineering, ICSE 2018, Gothenburg, Sweden, May 27–3 June, 2018, 2018,

[42] J.H. Andrews, L.C. Briand, Y. Labiche, A.S. Namin, Using mutation analysis for assessing and comparing testing coverage criteria, IEEE Trans. Softw. Eng. 32 (8) (2006) 608–624, https://doi.org/10.1109/TSE.2006.83.

[43] R. Just, D. Jalali, L. Inozemtseva, M.D. Ernst, R. Holmes, G. Fraser, Are mutants a valid substitute for real faults in software testing? in: Proceedings of the 22nd ACM SIGSOFT International Symposium on Foundations of Software Engineering (FSE-22), Hong Kong, China, November 16–22, 2014, 2014, pp, 654–665, https://doi.org/10.1145/2635868.2635929.

[44] P.G. Frankl, S.N. Weiss, An experimental comparison of the effectiveness of the all-uses and all-edges adequacy criteria, in: Symposium on Testing, Analysis, and Verification, 1991, pp. 154–164, https://doi.org/10.1145/120807.120821.

[45] P.G. Frankl, S.N. Weiss, An experimental comparison of the effectiveness of branch testing and data flow testing, IEEE Trans. Softw. Eng. 19 (8) (1993) 774–787, https://doi.org/10.1109/32.238581.

[46] A.J. Offutt, J. Pan, K. Tewary, T. Zhang, An experimental evaluation of data flow and mutation testing, Softw. Pract. Exp. 26 (2) (1996) 165–176, https://doi.org/10.1002/(SICI)1097-024X(199602)26:2⟨165::AID-SPE5⟩3.0.CO;2-K.

[47] P.G. Frankl, S.N. Weiss, C. Hu, All-uses vs mutation testing: an experimental comparison of effectiveness, J. Syst. Softw. 38 (3) (1997) 235–253, https://doi.org/10.1016/S0164-1212(96)00154-9.

[48] P.G. Frankl, O. Iakounenko, Further empirical studies of test effectiveness, in: SIGSOFT 1998, Proceedings of the ACM SIGSOFT International Symposium on Foundations of Software Engineering, Lake Buena Vista, FL, USA, November 3–5, 1998, 1998, pp. 153–162, https://doi.org/10.1145/288195.288298.

[49] L.C. Briand, D. Pfahl, Using simulation for assessing the real impact of test-coverage on defect-coverage, IEEE Trans. Reliab. 49 (1) (2000) 60–70, https://doi.org/10.1109/24.855537.

[50] M. Chen, M.R. Lyu, W.E. Wong, Effect of code coverage on software reliability measurement, IEEE Trans. Reliab. 50 (2) (2001) 165–170, https://doi.org/10.1109/24.963124.

[51] A.S. Namin, J.H. Andrews, The influence of size and coverage on test suite effectiveness, in: Proceedings of the Eighteenth International Symposium on Software Testing and Analysis, ISSTA 2009, Chicago, IL, USA, July 19–23, 2009, 2009, pp. 57–68, https://doi.org/10.1145/1572272.1572280.

[52] N. Li, U. Praphamontripong, J. Offutt, An experimental comparison of four unit test criteria: mutation, edge-pair, all-uses and prime path coverage, in: Second International Conference on Software Testing Verification and Validation, ICST 2009, Denver, CO, USA, April 1–4, 2009, Workshops Proceedings, 2009, pp. 220–229, https://doi.org/10.1109/ICSTW.2009.30.

[53] M. Papadakis, N. Malevris, An empirical evaluation of the first and second order mutation testing strategies, in: Third International Conference on Software Testing, Verification and Validation, ICST 2010, Paris, France, April 7–9, 2010, Workshops Proceedings, 2010, pp. 90–99, https://doi.org/10.1109/ICSTW.2010.50.

[54] I. Ciupa, A. Pretschner, M. Oriol, A. Leitner, B. Meyer, On the number and nature of faults found by random testing, Softw. Test. Verif. Reliab. 21 (1) (2011) 3–28, https://doi.org/10.1002/stvr.415.

[55] S. Kakarla, S. Momotaz, A.S. Namin, An evaluation of mutation and data-flow testing: a meta-analysis, in: Fourth IEEE International Conference on Software Testing, Verification and Validation, ICST 2012, Berlin, Germany, March 21–25, 2011, Workshop Proceedings, 2011, pp. 366–375, https://doi.org/10.1109/ICSTW.2011.51.

[56] Y. Wei, B. Meyer, M. Oriol, Is branch coverage a good measure of testing effectiveness? in: Springer, Berlin, Heidelberg, 2012, ISBN: 978-3-642-25231-0, pp. 194–212, https://doi.org/10.1007/978-3-642-25231-0_5.

[57] R. Baker, I. Habli, An empirical evaluation of mutation testing for improving the test quality of safety-critical software, IEEE Trans. Softw. Eng. 39 (6) (2013) 787–805, https://doi.org/10.1109/TSE.2012.56.

[58] M.M. Hassan, J.H. Andrews, Comparing multi-point stride coverage and dataflow coverage, in: 35th International Conference on Software Engineering, ICSE 2013, San Francisco, CA, USA, May 18–26, 2013, 2013, pp. 172–181, https://doi.org/10.1109/ICSE.2013.6606563.

[59] M. Gligoric, A. Groce, C. Zhang, R. Sharma, M.A. Alipour, D. Marinov, Comparing non-adequate test suites using coverage criteria, in: International Symposium on Software Testing and Analysis, ISSTA '13, Lugano, Switzerland, July 15–20, 2013, 2013, pp. 302–313, https://doi.org/10.1145/2483760.2483769.

[60] M. Gligoric, A. Groce, C. Zhang, R. Sharma, M.A. Alipour, D. Marinov, Guidelines for coverage-based comparisons of non-adequate test suites, ACM Trans. Softw. Eng. Methodol. 24 (4) (2015) 22:1–22:33, https://doi.org/10.1145/2660767.

[61] L. Inozemtseva, R. Holmes, Coverage is not strongly correlated with test suite effectiveness, in: 36th International Conference on Software Engineering, ICSE '14, Hyderabad, India, May 31 June 7, 2014, 2014, pp. 435–445, https://doi.org/10.1145/2568225.2568271.

[62] R. Gopinath, C. Jensen, A. Groce, Code coverage for suite evaluation by developers, in: 36th International Conference on Software Engineering, ICSE '14, Hyderabad, India, May 31–June 7, 2014, 2014, pp. 72–82, https://doi.org/10.1145/2568225.2568278.

[63] I. Ahmed, R. Gopinath, C. Brindescu, A. Groce, C. Jensen, Can testedness be effectively measured? in: Proceedings of the 24th ACM SIGSOFT International Symposium on Foundations of Software Engineering, FSE 2016, Seattle, WA, USA, November 13–18, 2016, 2016, pp, 547–558, https://doi.org/10.1145/2950290.2950324.

[64] R. Ramler, T. Wetzlmaier, C. Klammer, An empirical study on the application of mutation testing for a safety-critical industrial software system, in: Proceedings of the Symposium on Applied Computing, SAC 2017, Marrakech, Morocco, April 3–7, 2017, 2017, pp. 1401–1408, https://doi.org/10.1145/3019612.3019830.

[65] D. Binkley, S. Danicic, T. Gyimóthy, M. Harman, A. Kiss, L. Ouarbya, Formalizing executable dynamic and forward slicing, in: 4th International Workshop on Source Code Analysis and Manipulation (SCAM 04), 2004, pp, 43–52. Los Alamitos, CA, USA.

[66] K.H. Brodersen, F. Gallusser, J. Koehler, N. Remy, S.L. Scott, Inferring causal impact using Bayesian structural time-series models, Ann, Appl. Stat. 9 (2015) 247–274.

[67] L.J. Morell, A theory of fault-based testing, IEEE Trans. Softw. Eng. 16 (8) (1990) 844–857, https://doi.org/10.1109/32.57623.

[68] A.J. Offutt, Investigations of the software testing coupling effect, ACM Trans. Softw. Eng. Methodol. 1 (1) (1992) 5–20, https://doi.org/10.1145/125489.125473.

[69] J. Voas, G. McGraw, Software Fault Injection: Inoculating Programs Against Errors, John Wiley & Sons, 1997, ISBN: 0-471-18381-4.

[70] R. Gopinath, C. Jensen, A. Groce, Mutations: how close are they to real faults? in: 25th IEEE International Symposium on Software Reliability Engineering, ISSRE 2014, Naples, Italy, November 3–6, 2014, 2014, pp, 189–200, https://doi.org/10.1109/ISSRE.2014.40.

[71] R. Gopinath, C. Jensen, A. Groce, The theory of composite faults, in: 2017 IEEE International Conference on Software Testing, Verification and Validation (ICST), 2017, pp. 47–57, https://doi.org/10.1109/ICST.2017.12.

[72] W.B. Langdon, M. Harman, Y. Jia, Efficient multi-objective higher order mutation testing with genetic programming, J. Syst. Softw. 83 (12) (2010) 2416–2430, https://doi.org/10.1016/j.jss.2010.07.027.

[73] R. Geist, A.J. Offutt, F.C. Harris Jr, Estimation and enhancement of real-time software reliability through mutation analysis, IEEE Trans. Comput. 41 (5) (1992) 550–558, https://doi.org/10.1109/12.142681.

[74] I. Ahmed, C. Jensen, A. Groce, P.E. McKenney, Applying mutation analysis on kernel test suites: an experience report, in: 2017 IEEE International Conference on Software Testing, Verification and Validation Workshops (ICSTW), 2017, pp. 110–115, https://doi.org/10.1109/ICSTW.2017.26.
[75] A.J. Offutt, S.D. Lee, An empirical evaluation of weak mutation, IEEE Trans. Softw. Eng. 20 (5) (1994) 337–344, https://doi.org/10.1109/32.286422.
[76] G. Kaminski, P. Ammann, J. Offutt, Improving logic-based testing, J. Syst. Softw. 86 (8) (2013) 2002–2012, https://doi.org/10.1016/j.jss.2012.08.024.
[77] P. Anbalagan, T. Xie, Automated generation of pointcut mutants for testing pointcuts in AspectJ programs, in: 19th International Symposium on Software Reliability Engineering (ISSRE 2008), November 11–14, 2008, Seattle/Redmond, WA, USA, 2008, pp. 239–248, https://doi.org/10.1109/ISSRE.2008.58.
[78] A. Estero-Botaro, F. Palomo-Lozano, I. Medina-Bulo, Mutation operators for WS-BPEL 2.0, in: 21th International Conference on Software & Systems Engineering and their Applications, 2008,
[79] A. Estero-Botaro, F. Palomo-Lozano, I. Medina-Bulo, Quantitative evaluation of mutation operators for WS-BPEL compositions, in: Third International Conference on Software Testing, Verification and Validation, ICST 2010, Paris, France, April 7–9, 2010, Workshops Proceedings, 2010, pp. 142–150, https://doi.org/10.1109/ICSTW.2010.36.
[80] J. Boubeta-Puig, I. Medina-Bulo, A. García-Domínguez, Analogies and differences between mutation operators for WS-BPEL 2.0 and other languages, in: Fourth IEEE International Conference on Software Testing, Verification and Validation, ICST 2012, Berlin, Germany, March 21–25, 2011, Workshop Proceedings, 2011, pp. 398–407, https://doi.org/10.1109/ICSTW.2011.52.
[81] S. Mirshokraie, A. Mesbah, K. Pattabiraman, Guided mutation testing for JavaScript web applications, IEEE Trans. Softw. Eng. 41 (5) (2015) 429–444, https://doi.org/10.1109/TSE.2014.2371458.
[82] P. Delgado-Pérez, S. Segura, I. Medina-Bulo, Assessment of C++ object-oriented mutation operators: a selective mutation approach, Softw. Test. Verif. Reliab. 27 (4–5) (2017) e1630. ISSN: 1099-1689, https://doi.org/10.1002/stvr.1630.
[83] J. Hu, N. Li, J. Offutt, An analysis of OO mutation operators, in: Fourth IEEE International Conference on Software Testing, Verification and Validation, ICST 2012, Berlin, Germany, March 21–25, 2011, Workshop Proceedings, 2011, pp. 334–341, https://doi.org/10.1109/ICSTW.2011.47.
[84] F.C. Ferrari, J.C. Maldonado, A. Rashid, Mutation testing for aspect-oriented programs, in: First International Conference on Software Testing, Verification, and Validation, ICST 2008, Lillehammer, Norway, April 9–11, 2008, 2008, pp. 52–61, https://doi.org/10.1109/ICST.2008.37.
[85] L. Bottaci, Type sensitive application of mutation operators for dynamically typed programs, in: Third International Conference on Software Testing, Verification and Validation, ICST 2010, Paris, France, April 7–9, 2010, Workshops Proceedings, 2010, pp. 126–131, https://doi.org/10.1109/ICSTW.2010.56.
[86] M. Gligoric, S. Badame, R. Johnson, SMutant: a tool for type-sensitive mutation testing in a dynamic language, in: SIGSOFT/FSE'11 19th ACM SIGSOFT Symposium on the Foundations of Software Engineering (FSE-19) and ESEC'11: 13th European Software Engineering Conference (ESEC-13), Szeged, Hungary, September 5–9, 2011, 2011, pp. 424–427, https://doi.org/10.1145/2025113.2025181.
[87] A.D.B. Alberto, A. Cavalcanti, M. Gaudel, A. Simão, Formal mutation testing for Circus, Inf. Softw. Technol. 81 (2017) 131–153, https://doi.org/10.1016/j.infsof.2016.04.003.
[88] U. Praphamontripong, J. Offutt, Applying mutation testing to web applications, in: Third International Conference on Software Testing, Verification and Validation,

ICST 2010, Paris, France, April 7–9, 2010, Workshops Proceedings, 2010, pp. 132–141, https://doi.org/10.1109/ICSTW.2010.38.

[89] U. Praphamontripong, J. Offutt, L. Deng, J. Gu, An experimental evaluation of web mutation operators, in: Ninth IEEE International Conference on Software Testing, Verification and Validation Workshops, ICST Workshops 2016, Chicago, IL, USA, April 11–15, 2016, 2016, pp. 102–111, https://doi.org/10.1109/ICSTW.2016.17.

[90] U. Praphamontripong, J. Offutt, Finding redundancy in web mutation operators, in: 2017 IEEE International Conference on Software Testing, Verification and Validation Workshops (ICSTW), 2017, pp. 134–142, https://doi.org/10.1109/ICSTW.2017.30.

[91] L. Deng, N. Mirzaei, P. Ammann, J. Offutt, Towards mutation analysis of Android apps, in: Eighth IEEE International Conference on Software Testing, Verification and Validation, ICST 2015 Workshops, Graz, Austria, April 13–17, 2015, 2015, pp. 1–10, https://doi.org/10.1109/ICSTW.2015.7107450.

[92] L. Deng, J. Offutt, P. Ammann, N. Mirzaei, Mutation operators for testing Android apps, Inf. Softw. Technol. 81 (2017) 154–168, https://doi.org/10.1016/j.infsof.2016.04.012.

[93] M.P. Usaola, G. Rojas, I. Rodríguez, S. Hernández, An architecture for the development of mutation operators, in: 2017 IEEE International Conference on Software Testing, Verification and Validation Workshops (ICSTW), 2017, pp. 143–148, https://doi.org/10.1109/ICSTW.2017.31.

[94] M. Linares-Vásquez, G. Bavota, M. Tufano, K. Moran, M. Di Penta, C. Vendome, C. Bernal-Cárdenas, D. Poshyvanyk, Enabling mutation testing for Android apps, in: Proceedings of the 2017 11th Joint Meeting on Foundations of Software Engineering, ACM, New York, NY, USA, 2017, ISBN: 978-1-4503-5105-8, pp. 233–244, https://doi.org/10.1145/3106237.3106275.

[95] R.A.P. Oliveira, E. Alégroth, Z. Gao, A.M. Memon, Definition and evaluation of mutation operators for GUI-level mutation analysis, in: Eighth IEEE International Conference on Software Testing, Verification and Validation, ICST 2015 Workshops, Graz, Austria, April 13–17, 2015, 2015, pp. 1–10, https://doi.org/10.1109/ICSTW.2015.7107457.

[96] V. Lelli, A. Blouin, B. Baudry, Classifying and qualifying GUI defects, CoRR (2017). http://arxiv.org/abs/1703.09567.

[97] R. Abraham, M. Erwig, Mutation Operators for Spreadsheets, IEEE Trans. Softw. Eng. 35 (1) (2009) 94–108, https://doi.org/10.1109/TSE.2008.73.

[98] H. Dan, R.M. Hierons, Semantic mutation analysis of floating-point comparison, in: Fifth IEEE International Conference on Software Testing, Verification and Validation, ICST 2012, Montreal, QC, Canada, April 17–21, 2012, 2012, pp. 290–299, https://doi.org/10.1109/ICST.2012.109.

[99] V. Jagannath, M. Gligoric, S. Lauterburg, D. Marinov, G. Agha, Mutation operators for actor systems, in: Third International Conference on Software Testing, Verification and Validation, ICST 2010, Paris, France, April 7–9, 2010, Workshops Proceedings, 2010, pp. 157–162, https://doi.org/10.1109/ICSTW.2010.6.

[100] Y. Maezawa, K. Nishiura, H. Washizaki, S. Honiden, Validating ajax applications using a delay-based mutation technique, in: ACM/IEEE International Conference on Automated Software Engineering, ASE '14, Vasteras, Sweden, September 15–19, 2014, 2014, pp. 491–502, https://doi.org/10.1145/2642937.2642996.

[101] T. Xie, N. Tillmann, J. de Halleux, W. Schulte, Mutation analysis of parameterized unit tests, in: Second International Conference on Software Testing Verification and Validation, ICST 2009, Denver, CO, USA, April 1–4, 2009, Workshops Proceedings, 2009, pp. 177–181, https://doi.org/10.1109/ICSTW.2009.43.

[102] D.B. Brown, M. Vaughn, B. Liblit, T. Reps, The care and feeding of wild-caught mutants, in: Proceedings of the 2017 11th Joint Meeting on Foundations of Software

Engineering, ACM, New York, NY, USA, 2017, ISBN: 978-1-4503-5105-8, pp. 511–522, https://doi.org/10.1145/3106237.3106280.
[103] J. Nanavati, F. Wu, M. Harman, Y. Jia, J. Krinke, Mutation testing of memory-related operators, in: Eighth IEEE International Conference on Software Testing, Verification and Validation, ICST 2015 Workshops, Graz, Austria, April 13–17, 2015, 2015, pp. 1–10, https://doi.org/10.1109/ICSTW.2015.7107449.
[104] F. Wu, J. Nanavati, M. Harman, Y. Jia, J. Krinke, Memory mutation testing, Inf. Softw. Technol. 81 (2017) 97–111, https://doi.org/10.1016/j.infsof.2016.03.002.
[105] B.J. Garvin, M.B. Cohen, Feature interaction faults revisited: an exploratory study, in: IEEE 22nd International Symposium on Software Reliability Engineering, ISSRE 2011, Hiroshima, Japan, November 29–December 2, 2011, 2011, pp. 90–99, https://doi.org/10.1109/ISSRE.2011.25.
[106] M. Al-Hajjaji, F. Benduhn, T. Thüm, T. Leich, G. Saake, Mutation operators for preprocessor-based variability, in: Proceedings of the Tenth International Workshop on Variability Modelling of Software-intensive Systems, Salvador, Brazil, January 27–29, 2016, 2016, pp. 81–88, https://doi.org/10.1145/2866614.2866626.
[107] M.E. Delamaro, J. Offutt, P. Ammann, Designing deletion mutation operators, in: Seventh IEEE International Conference on Software Testing, Verification and Validation, ICST 2014, March 31 2014–April 4, 2014, Cleveland, OH, USA, 2014, pp. 11–20, https://doi.org/10.1109/ICST.2014.12.
[108] F. Belli, M. Beyazit, T. Takagi, Z. Furukawa, Mutation testing of "Go-Back" functions based on pushdown automata, in: Fourth IEEE International Conference on Software Testing, Verification and Validation, ICST 2011, Berlin, Germany, March 21–25, 2011, 2011, pp. 249–258, https://doi.org/10.1109/ICST.2011.30.
[109] R. Gopinath, E. Walkingshaw, How good are your types? Using mutation analysis to evaluate the effectiveness of type annotations, in: 2017 IEEE International Conference on Software Testing, Verification and Validation Workshops (ICSTW), 2017, pp. 122–127, https://doi.org/10.1109/ICSTW.2017.28.
[110] R. Jabbarvand, S. Malek, muDroid: an energy-aware mutation testing framework for Android, in: Proceedings of the 2017 11th Joint Meeting on Foundations of Software Engineering, ACM, New York, NY, USA, 2017, ISBN: 978-1-4503-5105-8, pp. 208–219, https://doi.org/10.1145/3106237.3106244.
[111] P. Arcaini, A. Gargantini, E. Riccobene, MutRex: a mutation-based generator of fault detecting strings for regular expressions, in: 2017 IEEE International Conference on Software Testing, Verification and Validation Workshops (ICSTW), 2017, pp. 87–96, https://doi.org/10.1109/ICSTW.2017.23.
[112] L. Zhang, S. Hou, J. Hu, T. Xie, H. Mei, Is operator-based mutant selection superior to random mutant selection? in: Proceedings of the 32nd ACM/IEEE International Conference on Software Engineering, vol, 1, ICSE 2010, Cape Town, South Africa, May 1–8, 2010, 2010, pp. 435–444, https://doi.org/10.1145/1806799.1806863.
[113] R. Gopinath, A. Alipour, I. Ahmed, C. Jensen, A. Groce, How hard does mutation analysis have to be, anyway? in: 26th IEEE International Symposium on Software Reliability Engineering, ISSRE 2015, Gaithersbury, MD, USA, November 2–5, 2015, 2015, pp, 216–227, https://doi.org/10.1109/ISSRE.2015.7381815.
[114] A.S. Namin, J.H. Andrews, D.J. Murdoch, Sufficient mutation operators for measuring test effectiveness, in: 30th International Conference on Software Engineering (ICSE 2008), Leipzig, Germany, May 10–18, 2008, 2008, pp. 351–360, https://doi.org/10.1145/1368088.1368136.
[115] M.E. Delamaro, L. Deng, V.H.S. Durelli, N. Li, J. Offutt, Experimental evaluation of SDL and one-op mutation for C, in: Seventh IEEE International Conference on Software Testing, Verification and Validation, ICST 2014, March 31 2014–April 4, 2014, Cleveland, OH, USA, 2014, pp. 203–212, https://doi.org/10.1109/ICST.2014.33.

[116] V.H.S. Durelli, N.M.D. Souza, M.E. Delamaro, Are deletion mutants easier to identify manually? in: 2017 IEEE International Conference on Software Testing, Verification and Validation Workshops (ICSTW), 2017, pp, 149–158, https://doi.org/10.1109/ICSTW.2017.32.
[117] X. Yao, M. Harman, Y. Jia, A study of equivalent and stubborn mutation operators using human analysis of equivalence, in: 36th International Conference on Software Engineering, ICSE '14, Hyderabad, India, May 31–June 7, 2014, 2014, pp. 919–930, https://doi.org/10.1145/2568225.2568265.
[118] J. Zhang, M. Zhu, D. Hao, L. Zhang, An empirical study on the scalability of selective mutation testing, in: 25th IEEE International Symposium on Software Reliability Engineering, ISSRE 2014, Naples, Italy, November 3–6, 2014, 2014, pp. 277–287, https://doi.org/10.1109/ISSRE.2014.27.
[119] J. Zhang, Scalability studies on selective mutation testing, in: 37th IEEE/ACM International Conference on Software Engineering, ICSE 2015, Florence, Italy, May 16–24, 2015, vol. 2, 2015, pp. 851–854, https://doi.org/10.1109/ICSE.2015.276.
[120] L. Zhang, M. Gligoric, D. Marinov, S. Khurshid, Operator-based and random mutant selection: better together, in: 2013 28th IEEE/ACM International Conference on Automated Software Engineering, ASE 2013, Silicon Valley, CA, USA, November 11–15, 2013, 2013, pp. 92–102, https://doi.org/10.1109/ASE.2013.6693070.
[121] R. Gopinath, M.A. Alipour, I. Ahmed, C. Jensen, A. Groce, On the limits of mutation reduction strategies, in: Proceedings of the 38th International Conference on Software Engineering, ICSE 2016, Austin, TX, USA, May 14–22, 2016, 2016, pp. 511–522, https://doi.org/10.1145/2884781.2884787.
[122] B. Kurtz, P. Ammann, J. Offutt, M.E. Delamaro, M. Kurtz, N. Gökçe, Analyzing the validity of selective mutation with dominator mutants, in: Proceedings of the 24th ACM SIGSOFT International Symposium on Foundations of Software Engineering, FSE 2016, Seattle, WA, USA, November 13–18, 2016, 2016, pp. 571–582, https://doi.org/10.1145/2950290.2950322.
[123] M. Gligoric, L. Zhang, C. Pereira, G. Pokam, Selective mutation testing for concurrent code, in: International Symposium on Software Testing and Analysis, ISSTA '13, Lugano, Switzerland, July 15–20, 2013, 2013, pp. 224–234, https://doi.org/10.1145/2483760.2483773.
[124] G.K. Kaminski, P. Ammann, Using a fault hierarchy to improve the efficiency of DNF logic mutation testing, in: Second International Conference on Software Testing Verification and Validation, ICST 2009, Denver, CO, USA, April 1–4, 2009, 2009, pp. 386–395, https://doi.org/10.1109/ICST.2009.13.
[125] M. Papadakis, Y.L. Traon, Effective fault localization via mutation analysis: a selective mutation approach, in: Symposium on Applied Computing, SAC 2014, Gyeongju, Republic of Korea, March 24–28, 2014, 2014, pp. 1293–1300, https://doi.org/10.1145/2554850.2554978.
[126] M. Polo, M. Piattini, I.G.R. de Guzmán, Decreasing the cost of mutation testing with second-order mutants, Softw. Test. Verif. Reliab. 19 (2) (2009) 111–131, https://doi.org/10.1002/stvr.392.
[127] M. Papadakis, N. Malevris, M. Kintis, Mutation testing strategies—a collateral approach, in: ICSOFT 2010—Proceedings of the Fifth International Conference on Software and Data Technologies, vol, 2, Athens, Greece, July 22–24, 2010, 2010, pp. 325–328.
[128] P.R. Mateo, M.P. Usaola, J.L.F. Alemán, Validating second-order mutation at system level, IEEE Trans. Softw. Eng. 39 (4) (2013) 570–587, https://doi.org/10.1109/TSE.2012.39.
[129] A. Parsai, A. Murgia, S. Demeyer, A model to estimate first-order mutation coverage from higher-order mutation coverage, in: 2016 IEEE International Conference on Software Quality, Reliability and Security, QRS 2016, Vienna, Austria, August 1–3, 2016, 2016, pp. 365–373, https://doi.org/10.1109/QRS.2016.48.

[130] R. Just, B. Kurtz, P. Ammann, Inferring mutant utility from program context, in: Proceedings of the 26th ACM SIGSOFT International Symposium on Software Testing and Analysis, Santa Barbara, CA, USA, July 10–14, 2017, 2017, pp. 284–294, https://doi.org/10.1145/3092703.3092732.
[131] C. Sun, F. Xue, H. Liu, X. Zhang, A path-aware approach to mutant reduction in mutation testing, Inf. Softw. Technol. 81 (2017) 65–81, https://doi.org/10.1016/j.infsof.2016.02.006.
[132] D. Gong, G. Zhang, X. Yao, F. Meng, Mutant reduction based on dominance relation for weak mutation testing, Inf. Softw. Technol. 81 (2017) 82–96, https://doi.org/10.1016/j.infsof.2016.05.001.
[133] C. Iida, S. Takada, Reducing mutants with mutant killable precondition, in: 2017 IEEE International Conference on Software Testing, Verification and Validation Workshops (ICSTW), 2017, pp. 128–133, https://doi.org/10.1109/ICSTW.2017.29.
[134] M. Patrick, M. Oriol, J.A. Clark, MESSI: mutant evaluation by static semantic interpretation, in: Fifth IEEE International Conference on Software Testing, Verification and Validation, ICST 2012, Montreal, QC, Canada, April 17–21, 2012, 2012, pp. 711–719, https://doi.org/10.1109/ICST.2012.161.
[135] M. Patrick, R. Alexander, M. Oriol, J.A. Clark, Probability-based semantic interpretation of mutants, in: Seventh IEEE International Conference on Software Testing, Verification and Validation, ICST 2014 Workshops Proceedings, March 31–April 4, 2014, Cleveland, OH, USA, 2014, pp. 186–195, https://doi.org/10.1109/ICSTW.2014.18.
[136] M. Sridharan, A.S. Namin, Prioritizing mutation operators based on importance sampling, in: IEEE 21st International Symposium on Software Reliability Engineering, ISSRE 2010, San Jose, CA, USA, November 1–4, 2010, 2010, pp. 378–387, https://doi.org/10.1109/ISSRE.2010.16.
[137] A.S. Namin, X. Xue, O. Rosas, P. Sharma, MuRanker: a mutant ranking tool, Softw. Test. Verif. Reliab. 25 (5–7) (2015) 572–604, https://doi.org/10.1002/stvr.1542.
[138] J. Nam, D. Schuler, A. Zeller, Calibrated mutation testing, in: Fourth IEEE International Conference on Software Testing, Verification and Validation, ICST 2012, Berlin, Germany, March 21–25, 2011, Workshop Proceedings, 2011, pp. 376–381, https://doi.org/10.1109/ICSTW.2011.57.
[139] L. Inozemtseva, H. Hemmati, R. Holmes, Using fault history to improve mutation reduction, in: Joint Meeting of the European Software Engineering Conference and the ACM SIGSOFT Symposium on the Foundations of Software Engineering, ESEC/FSE'13, Saint Petersburg, Russian Federation, August 18–26, 2013, 2013, pp. 639–642, https://doi.org/10.1145/2491411.2494586.
[140] R.H. Untch, A.J. Offutt, M.J. Harrold, Mutation analysis using mutant schemata, in: Proceedings of the 1993 International Symposium on Software Testing and Analysis, ISSTA 1993, Cambridge, MA, USA, June 28–30, 1993, 1993, pp. 139–148, https://doi.org/10.1145/154183.154265.
[141] M. Papadakis, N. Malevris, Automatic mutation test case generation via dynamic symbolic execution, in: IEEE 21st International Symposium on Software Reliability Engineering, ISSRE 2010, San Jose, CA, USA, November 1–4, 2010, 2010, pp. 121–130, https://doi.org/10.1109/ISSRE.2010.38.
[142] B. Wang, Y. Xiong, Y. Shi, L. Zhang, D. Hao, Faster mutation analysis via equivalence modulo states, in: ISSTA 2017 Proceedings of the 26th ACM SIGSOFT International Symposium on Software Testing and Analysis, 2017, pp. 295–306, https://doi.org/10.1145/3092703.3092714.
[143] S. Tokumoto, H. Yoshida, K. Sakamoto, S. Honiden, MuVM: higher order mutation analysis virtual machine for C, in: 2016 IEEE International Conference on Software Testing, Verification and Validation, ICST 2016, Chicago, IL, USA, April 11–15, 2016, 2016, pp. 320–329, https://doi.org/10.1109/ICST.2016.18.

[144] S. Bardin, M. Delahaye, R. David, N. Kosmatov, M. Papadakis, Y.L. Traon, J. Marion, Sound and quasi-complete detection of infeasible test requirements, in: 8th IEEE International Conference on Software Testing, Verification and Validation, ICST 2015, Graz, Austria, April 13–17, 2015, 2015, pp. 1–10, https://doi.org/10.1109/ICST.2015.7102607.
[145] M. Marcozzi, S. Bardin, N. Kosmatov, M. Papadakis, V. Prevosto, L. Correnson, Time to clean your test objectives, in: Proceedings of the 40th International Conference on Software Engineering, ICSE 2018, Gothenburg, Sweden, May 27–3 June, 2018, 2018,
[146] M. Marcozzi, M. Delahaye, S. Bardin, N. Kosmatov, V. Prevosto, Generic and effective specification of structural test objectives, in: 2017 IEEE International Conference on Software Testing, Verification and Validation (ICST), 2017, pp. 436–441, https://doi.org/10.1109/ICST.2017.48.
[147] H. Coles, T. Laurent, C. Henard, M. Papadakis, A. Ventresque, PIT: a practical mutation testing tool for Java (demo), in: Proceedings of the 25th International Symposium on Software Testing and Analysis, ISSTA 2016, Saarbrücken, Germany, July 18–20, 2016, 2016, pp. 449–452, https://doi.org/10.1145/2931037.2948707.
[148] F. Hariri, A. Shi, H. Converse, S. Khurshid, D. Marinov, Evaluating the effects of compiler optimizations on mutation testing at the compiler IR level, in: 27th IEEE International Symposium on Software Reliability Engineering, ISSRE 2016, Ottawa, ON, Canada, October 23–27, 2016, 2016, pp. 105–115, https://doi.org/10.1109/ISSRE.2016.51.
[149] M. Papadakis, N. Malevris, Mutation based test case generation via a path selection strategy, Inf. Softw. Technol. 54 (9) (2012) 915–932, https://doi.org/10.1016/j.infsof.2012.02.004.
[150] V.H.S. Durelli, J. Offutt, M.E. Delamaro, Toward harnessing high-level language virtual machines for further speeding up weak mutation testing, in: Fifth IEEE International Conference on Software Testing, Verification and Validation, ICST 2012, Montreal, QC, Canada, April 17–21, 2012, 2012, pp. 681–690, https://doi.org/10.1109/ICST.2012.158.
[151] X. Devroey, G. Perrouin, M. Cordy, M. Papadakis, A. Legay, P. Schobbens, A variability perspective of mutation analysis, in: Proceedings of the 22nd ACM SIGSOFT International Symposium on Foundations of Software Engineering (FSE-22), Hong Kong, China, November 16–22, 2014, 2014, pp. 841–844, https://doi.org/10.1145/2635868.2666610.
[152] M. Kintis, M. Papadakis, Y. Jia, N. Malevris, Y.L. Traon, M. Harman, Detecting trivial mutant equivalences via compiler optimisations, IEEE Trans. Softw. Eng. PP (99) (2017) 1. ISSN: 0098-5589, https://doi.org/10.1109/TSE.2017.2684805.
[153] M. Kintis, Effective methods to tackle the equivalent mutant problem when testing software with mutation, (Ph.D. thesis), Department of Informatics, Athens University of Economics and Business 2016.
[154] M. Kintis, N. Malevris, Using data flow patterns for equivalent mutant detection, in: Seventh IEEE International Conference on Software Testing, Verification and Validation, ICST 2014 Workshops Proceedings, March 31–April 4, 2014, Cleveland, OH, USA, 2014, pp. 196–205, https://doi.org/10.1109/ICSTW.2014.21.
[155] M. Kintis, N. Malevris, MEDIC: A static analysis framework for equivalent mutant identification, Inf. Softw. Technol. 68 (2015) 1–17, https://doi.org/10.1016/j.infsof.2015.07.009.
[156] D. Holling, S. Banescu, M. Probst, A. Petrovska, A. Pretschner, Nequivack: assessing mutation score confidence, in: Ninth IEEE International Conference on Software Testing, Verification and Validation Workshops, ICST Workshops 2016, Chicago, IL, USA, April 11–15, 2016, 2016, pp. 152–161, https://doi.org/10.1109/ICSTW.2016.29.

[157] M. Kintis, N. Malevris, Identifying more equivalent mutants via code similarity, in: 20th Asia-Pacific Software Engineering Conference, APSEC 2013, Ratchathewi, Bangkok, Thailand, December 2–5, 2013, vol. 1, 2013, pp. 180–188, https://doi.org/10.1109/APSEC.2013.34.

[158] K.A. Foster, Error sensitive test cases analysis (ESTCA), IEEE Trans. Softw. Eng. 6 (3) (1980) 258–264, https://doi.org/10.1109/TSE.1980.234487.

[159] W.E. Howden, Weak mutation testing and completeness of test sets, IEEE Trans. Softw. Eng. 8 (4) (1982) 371–379, https://doi.org/10.1109/TSE.1982.235571.

[160] K. Tai, Predicate-based test generation for computer programs, in: Proceedings of the 15th International Conference on Software Engineering, Baltimore, MD, USA, May 17–21, 1993, 1993, pp, 267–276. http://portal.acm.org/citation.cfm?id=257572.257631.

[161] K. Tai, Theory of fault-based predicate testing for computer programs, IEEE Trans. Softw. Eng. 22 (8) (1996) 552–562, https://doi.org/10.1109/32.536956.

[162] M. Papadakis, Error detection methods in Java programs using the mutation method, (Masters thesis), Athens University of Economics and Business 2005.

[163] G. Kaminski, P. Ammann, J. Offutt, Better predicate testing, in: Proceedings of the 6th International Workshop on Automation of Software Test, AST 2011, Waikiki, Honolulu, HI, USA, May 23–24, 2011, 2011, pp. 57–63, https://doi.org/10.1145/1982595.1982608.

[164] R. Just, G.M. Kapfhammer, F. Schweiggert, Do redundant mutants affect the effectiveness and efficiency of mutation analysis? in: Fifth IEEE International Conference on Software Testing, Verification and Validation, ICST 2012, Montreal, QC, Canada, April 17–21, 2012, 2012, pp, 720–725, https://doi.org/10.1109/ICST.2012.162.

[165] R. Just, F. Schweiggert, Higher accuracy and lower run time: efficient mutation analysis using non-redundant mutation operators, Softw. Test. Verif. Reliab. 25 (5–7) (2015) 490–507, https://doi.org/10.1002/stvr.1561.

[166] L. Fernandes, M. Ribeiro, L. Carvalho, R. Gheyi, M. Mongiovi, A. Santos, A. Cavalcanti, F. Ferrari, J.C. Maldonado, Avoiding useless mutants, in: Proceedings of the 16th ACM SIGPLAN International Conference on Generative Programming: Concepts and Experiences, ACM, New York, NY, USA, 2017, ISBN: 978-1-4503-5524-7, pp. 187–198, https://doi.org/10.1145/3136040.3136053.

[167] T. Laurent, A. Ventresque, M. Papadakis, C. Henard, Y.L. Traon, Assessing and improving the mutation testing practice of PIT, CoRR (2016). http://arxiv.org/abs/1601.02351.

[168] B. Lindström, A. Marki, On strong mutation and subsuming mutants, in: Ninth IEEE International Conference on Software Testing, Verification and Validation Workshops, ICST Workshops 2016, Chicago, IL, USA, April 11–15, 2016, 2016, pp. 112–121, https://doi.org/10.1109/ICSTW.2016.28.

[169] B. Kurtz, P. Ammann, J. Offutt, Static analysis of mutant subsumption, in: Eighth IEEE International Conference on Software Testing, Verification and Validation, ICST 2015 Workshops, Graz, Austria, April 13–17, 2015, 2015, pp. 1–10, https://doi.org/10.1109/ICSTW.2015.7107454.

[170] R.A. DeMillo, A.J. Offutt, Constraint-based automatic test data generation, IEEE Trans. Softw. Eng. 17 (9) (1991) 900–910, https://doi.org/10.1109/32.92910.

[171] G. Fraser, A. Zeller, Mutation-driven generation of unit tests and oracles, IEEE Trans. Softw. Eng. 38 (2) (2012) 278–292, https://doi.org/10.1109/TSE.2011.93.

[172] M. Harman, Y. Jia, W.B. Langdon, Strong higher order mutation-based test data generation, in: SIGSOFT/FSE'11 19th ACM SIGSOFT Symposium on the Foundations of Software Engineering (FSE-19) and ESEC'11: 13th European Software Engineering Conference (ESEC-13), Szeged, Hungary, September 5–9, 2011, 2011, pp. 212–222, https://doi.org/10.1145/2025113.2025144.

[173] S. Anand, E.K. Burke, T.Y. Chen, J.A. Clark, M.B. Cohen, W. Grieskamp, M. Harman, M.J. Harrold, P. McMinn, An orchestrated survey of methodologies for automated software test case generation, J. Syst. Softw. 86 (8) (2013) 1978–2001, https://doi.org/10.1016/j.jss.2013.02.061.
[174] F. Wotawa, M. Nica, B.K. Aichernig, Generating distinguishing tests using the Minion constraint solver, in: Third International Conference on Software Testing, Verification and Validation, ICST 2010, Paris, France, April 7–9, 2010, Workshops Proceedings, 2010, pp. 325–330, https://doi.org/10.1109/ICSTW.2010.11.
[175] S. Nica, On the improvement of the mutation score using distinguishing test cases, in: Fourth IEEE International Conference on Software Testing, Verification and Validation, ICST 2011, Berlin, Germany, March 21–25, 2011, 2011, pp. 423–426, https://doi.org/10.1109/ICST.2011.40.
[176] M. Papadakis, N. Malevris, An effective path selection strategy for mutation testing, in: 16th Asia-Pacific Software Engineering Conference, APSEC 2009, December 1–3, 2009, Batu Ferringhi, Penang, Malaysia, 2009, pp. 422–429, https://doi.org/10.1109/APSEC.2009.68.
[177] M. Papadakis, N. Malevris, Automatically performing weak mutation with the aid of symbolic execution, concolic testing and search-based testing, Softw. Q. J. 19 (4) (2011) 691–723, https://doi.org/10.1007/s11219-011-9142-y.
[178] S. Anand, C.S. Pasareanu, W. Visser, JPF-SE: a symbolic execution extension to Java PathFinder, in: Tools and Algorithms for the Construction and Analysis of Systems, 13th International Conference, TACAS 2007, Held as Part of the Joint European Conferences on Theory and Practice of Software, ETAPS 2007 Braga, Portugal, March 24–April 1, 2007, 2007, pp. 134–138, https://doi.org/10.1007/978-3-540-71209-1_12.
[179] C. Cadar, D. Dunbar, D.R. Engler, KLEE: unassisted and automatic generation of high-coverage tests for complex systems programs, in: 8th USENIX Symposium on Operating Systems Design and Implementation, OSDI 2008, December 8–10, 2008, San Diego, CA, USA, Proceedings, 2008, pp, 209–224. http://www.usenix.org/events/osdi08/tech/full_papers/cadar/cadar.pdf.
[180] H. Riener, R. Bloem, G. Fey, Test case generation from mutants using model checking techniques, in: Fourth IEEE International Conference on Software Testing, Verification and Validation, ICST 2012, Berlin, Germany, March 21–25, 2011, Workshop Proceedings, 2011, pp. 388–397, https://doi.org/10.1109/ICSTW.2011.55.
[181] M. Papadakis, N. Malevris, M. Kallia, Towards automating the generation of mutation tests, in: The 5th Workshop on Automation of Software Test, AST 2010, May 3–4, 2010, Cape Town, South Africa, 2010, pp. 111–118, https://doi.org/10.1145/1808266.1808283.
[182] L. Zhang, T. Xie, L. Zhang, N. Tillmann, J. de Halleux, H. Mei, Test generation via dynamic symbolic execution for mutation testing, in: 26th IEEE International Conference on Software Maintenance (ICSM 2010), September 12–18, 2010, Timisoara, Romania, 2010, pp. 1–10, https://doi.org/10.1109/ICSM.2010.5609672.
[183] S. Bardin, N. Kosmatov, F. Cheynier, Efficient leveraging of symbolic execution to advanced coverage criteria, in: Seventh IEEE International Conference on Software Testing, Verification and Validation, ICST 2014, March 31 2014–April 4, 2014, Cleveland, OH, USA, 2014, pp. 173–182, https://doi.org/10.1109/ICST.2014.30.
[184] K. Jamrozik, G. Fraser, N. Tillmann, J. de Halleux, Augmented dynamic symbolic execution, in: IEEE/ACM International Conference on Automated Software Engineering, ASE'12, Essen, Germany, September 3–7, 2012, 2012, pp. 254–257, https://doi.org/10.1145/2351676.2351716.
[185] F.C.M. Souza, M. Papadakis, Y.L. Traon, M.E. Delamaro, Strong mutation-based test data generation using hill climbing, in: Proceedings of the 9th International Workshop on Search-Based Software Testing, SBST@ICSE 2016, Austin, TX, USA, May 14–22, 2016, 2016, pp. 45–54, https://doi.org/10.1145/2897010.2897012.

[186] K. Ayari, S. Bouktif, G. Antoniol, Automatic mutation test input data generation via ant colony, in: Genetic and Evolutionary Computation Conference, GECCO 2007, Proceedings, London, England, UK, July 7–11, 2007, 2007, pp. 1074–1081, https://doi.org/10.1145/1276958.1277172.
[187] G. Fraser, A. Zeller, Mutation-driven generation of unit tests and oracles, in: Proceedings of the Nineteenth International Symposium on Software Testing and Analysis, ISSTA 2010, Trento, Italy, July 12–16, 2010, 2010, pp. 147–158, https://doi.org/10.1145/1831708.1831728.
[188] M. Papadakis, N. Malevris, Automatic mutation based test data generation, in: 13th Annual Genetic and Evolutionary Computation Conference, GECCO 2011, Companion Material Proceedings, Dublin, Ireland, July 12–16, 2011, 2011, pp. 247–248, https://doi.org/10.1145/2001858.2001997.
[189] M. Papadakis, N. Malevris, Killing mutants effectively a search based approach, in: Knowledge-Based Software Engineering—Proceedings of the Tenth Conference on Knowledge-Based Software Engineering, JCKBSE 2012, Rodos, Greece, August 23–26, 2012, 2012, pp. 217–226, https://doi.org/10.3233/978-1-61499-094-9-217.
[190] M. Papadakis, N. Malevris, Searching and generating test inputs for mutation testing, SpringerPlus 2 (1) (2013) 121. ISSN: 2193-1801, https://doi.org/10.1186/2193-1801-2-121.
[191] M. Patrick, R. Alexander, M. Oriol, J.A. Clark, Using mutation analysis to evolve subdomains for random testing, in: Sixth IEEE International Conference on Software Testing, Verification and Validation, ICST 2013 Workshops Proceedings, Luxembourg, Luxembourg, March 18–22, 2013, 2013, pp. 53–62, https://doi.org/10.1109/ICSTW.2013.14.
[192] G. Fraser, A. Arcuri, Achieving scalable mutation-based generation of whole test suites, Empir. Softw. Eng. 20 (3) (2015) 783–812, https://doi.org/10.1007/s10664-013-9299-z.
[193] A. Shi, A. Gyori, M. Gligoric, A. Zaytsev, D. Marinov, Balancing trade-offs in test-suite reduction, in: Proceedings of the 22nd ACM SIGSOFT International Symposium on Foundations of Software Engineering, (FSE-22), Hong Kong, China, November 16–22, 2014, 2014, pp. 246–256, https://doi.org/10.1145/2635868.2635921.
[194] Y. Lou, D. Hao, L. Zhang, Mutation-based test-case prioritization in software evolution, in: 26th IEEE International Symposium on Software Reliability Engineering, ISSRE 2015, Gaithersbury, MD, USA, November 2–5, 2015, 2015, pp. 46–57, https://doi.org/10.1109/ISSRE.2015.7381798.
[195] L. Zhang, D. Marinov, S. Khurshid, Faster mutation testing inspired by test prioritization and reduction, in: International Symposium on Software Testing and Analysis, ISSTA '13, Lugano, Switzerland, July 15–20, 2013, 2013, pp. 235–245, https://doi.org/10.1145/2483760.2483782.
[196] R. Just, G.M. Kapfhammer, F. Schweiggert, Using non-redundant mutation operators and test suite prioritization to achieve efficient and scalable mutation analysis, in: 23rd IEEE International Symposium on Software Reliability Engineering, ISSRE 2012, Dallas, TX, USA, November 27–30, 2012, 2012, pp. 11–20, https://doi.org/10.1109/ISSRE.2012.31.
[197] Q. Zhu, A. Panichella, A. Zaidman, Speeding-up mutation testing via data compression and state infection, in: 2017 IEEE International Conference on Software Testing, Verification and Validation Workshops (ICSTW), 2017, pp. 103–109, https://doi.org/10.1109/ICSTW.2017.25.
[198] M.E. Delamaro, Proteum - a mutation analysis based testing environmen, (Masters thesis), University of São Paulo, Sao Paulo, Brazil, 1993.
[199] S. Kim, Y. Ma, Y.R. Kwon, Combining weak and strong mutation for a non-interpretive Java mutation system, Softw. Test. Verif. Reliab. 23 (8) (2013) 647–668, https://doi.org/10.1002/stvr.1480.

[200] R. Just, M.D. Ernst, G. Fraser, Efficient mutation analysis by propagating and partitioning infected execution states, in: International Symposium on Software Testing and Analysis, ISSTA '14, San Jose, CA, USA, July 21–26, 2014, 2014, pp. 315–326, https://doi.org/10.1145/2610384.2610388.
[201] P.R. Mateo, M.P. Usaola, Mutant execution cost reduction: through MUSIC (mutant schema improved with extra code), in: Fifth IEEE International Conference on Software Testing, Verification and Validation, ICST 2012, Montreal, QC, Canada, April 17–21, 2012, 2012, pp. 664–672, https://doi.org/10.1109/ICST.2012.156.
[202] P.R. Mateo, M.P. Usaola, Reducing mutation costs through uncovered mutants, Softw. Test. Verif. Reliab. 25 (5–7) (2015) 464–489, https://doi.org/10.1002/stvr.1534.
[203] K.N. King, A.J. Offutt, A Fortran language system for mutation-based software testing, Softw. Pract. Exper. 21 (7) (1991) 685–718, https://doi.org/10.1002/spe.4380210704.
[204] M. Papadakis, M.E. Delamaro, Y.L. Traon, Proteum/FL: a tool for localizing faults using mutation analysis, in: 13th IEEE International Working Conference on Source Code Analysis and Manipulation, SCAM 2013, Eindhoven, Netherlands, September 22–23, 2013, 2013, pp. 94–99, https://doi.org/10.1109/SCAM.2013.6648189.
[205] D. Schuler, A. Zeller, Javalanche: efficient mutation testing for Java, in: Proceedings of the 7th Joint Meeting of the European Software Engineering Conference and the ACM SIGSOFT International Symposium on Foundations of Software Engineering, 2009, Amsterdam, The Netherlands, August 24–28, 2009, 2009, pp. 297–298, https://doi.org/10.1145/1595696.1595750.
[206] P.R. Mateo, M.P. Usaola, Parallel mutation testing, Softw. Test. Verif. Reliab. 23 (4) (2013) 315–350, https://doi.org/10.1002/stvr.1471.
[207] M. Gligoric, V. Jagannath, Q. Luo, D. Marinov, Efficient mutation testing of multithreaded code, Softw. Test. Verif. Reliab. 23 (5) (2013) 375–403, https://doi.org/10.1002/stvr.1469.
[208] M. Gligoric, V. Jagannath, D. Marinov, MuTMuT: efficient exploration for mutation testing of multithreaded code, in: Third International Conference on Software Testing, Verification and Validation, ICST 2010, Paris, France, April 7–9, 2010, 2010, pp. 55–64, https://doi.org/10.1109/ICST.2010.33.
[209] P. Gong, R. Zhao, Z. Li, Faster mutation-based fault localization with a novel mutation execution strategy, in: Eighth IEEE International Conference on Software Testing, Verification and Validation, ICST 2015 Workshops, Graz, Austria, April 13–17, 2015, 2015, pp. 1–10, https://doi.org/10.1109/ICSTW.2015.7107448.
[210] L. Zhang, D. Marinov, L. Zhang, S. Khurshid, Regression mutation testing, in: International Symposium on Software Testing and Analysis, ISSTA 2012, Minneapolis, MN, USA, July 15–20, 2012, 2012, pp. 331–341, https://doi.org/10.1145/2338965.2336793.
[211] C.J. Wright, G.M. Kapfhammer, P. McMinn, Efficient mutation analysis of relational database structure using mutant schemata and parallelisation, in: Sixth IEEE International Conference on Software Testing, Verification and Validation, ICST 2013 Workshops Proceedings, Luxembourg, Luxembourg, March 18–22, 2013, 2013, pp. 63–72, https://doi.org/10.1109/ICSTW.2013.15.
[212] C. Zhou, P.G. Frankl, Inferential checking for mutants modifying satabase states, in: Fourth IEEE International Conference on Software Testing, Verification and Validation, ICST 2011, Berlin, Germany, March 21–25, 2011, 2011, pp. 259–268, https://doi.org/10.1109/ICST.2011.63.
[213] M. Patrick, A.P. Craig, N.J. Cunniffe, M. Parry, C.A. Gilligan, Testing stochastic software using pseudo-oracles, in: Proceedings of the 25th International Symposium on Software Testing and Analysis, ISSTA 2016, Saarbrücken, Germany, July 18–20, 2016, 2016, pp. 235–246, https://doi.org/10.1145/2931037.2931063.

[214] M.J. Rutherford, A. Carzaniga, A.L. Wolf, Evaluating test suites and adequacy criteria using simulation-based models of distributed systems, IEEE Trans. Softw. Eng. 34 (4) (2008) 452–470, https://doi.org/10.1109/TSE.2008.33.

[215] B.J.M. Grün, D. Schuler, A. Zeller, The impact of equivalent mutants, in: Second International Conference on Software Testing Verification and Validation, ICST 2009, Denver, CO, USA, April 1–4, 2009, Workshops Proceedings, 2009, pp. 192–199, https://doi.org/10.1109/ICSTW.2009.37.

[216] D. Schuler, V. Dallmeier, A. Zeller, Efficient mutation testing by checking invariant violations, in: Proceedings of the Eighteenth International Symposium on Software Testing and Analysis, ISSTA 2009, Chicago, IL, USA, July 19–23, 2009, 2009, pp. 69–80, https://doi.org/10.1145/1572272.1572282.

[217] D. Schuler, A. Zeller, (Un-)covering equivalent mutants, in: Third International Conference on Software Testing, Verification and Validation, ICST 2010, Paris, France, April 7–9, 2010, 2010, pp. 45–54, https://doi.org/10.1109/ICST.2010.30.

[218] D. Schuler, A. Zeller, Covering and uncovering equivalent mutants, Softw. Test. Verif. Reliab. 23 (5) (2013) 353–374, https://doi.org/10.1002/stvr.1473.

[219] B. Schwarz, D. Schuler, A. Zeller, Breeding high-impact mutations, in: Fourth IEEE International Conference on Software Testing, Verification and Validation, ICST 2012, Berlin, Germany, March 21–25, 2011, Workshop Proceedings, 2011, pp. 382–387, https://doi.org/10.1109/ICSTW.2011.56.

[220] M. Papadakis, Y.L. Traon, Mutation testing strategies using mutant classification, in: Proceedings of the 28th Annual ACM Symposium on Applied Computing, SAC '13, Coimbra, Portugal, March 18–22, 2013, 2013, pp. 1223–1229, https://doi.org/10.1145/2480362.2480592.

[221] M. Papadakis, M.E. Delamaro, Y.L. Traon, Mitigating the effects of equivalent mutants with mutant classification strategies, Sci. Comput. Program. 95 (2014) 298–319, https://doi.org/10.1016/j.scico.2014.05.012.

[222] M. Kintis, M. Papadakis, N. Malevris, Isolating first order equivalent mutants via second order mutation, in: Fifth IEEE International Conference on Software Testing, Verification and Validation, ICST 2012, Montreal, QC, Canada, April 17–21, 2012, 2012, pp. 701–710, https://doi.org/10.1109/ICST.2012.160.

[223] M. Kintis, M. Papadakis, N. Malevris, Employing second-order mutation for isolating first-order equivalent mutants, Softw. Test. Verif. Reliab. 25 (5–7) (2015) 508–535, https://doi.org/10.1002/stvr.1529.

[224] M.P. Usaola, P.R. Mateo, B.P. Lamancha, Reduction of test suites using mutation, in: Fundamental Approaches to Software Engineering–15th International Conference, FASE 2012, Held as Part of the European Joint Conferences on Theory and Practice of Software, ETAPS 2012, Tallinn, Estonia, March 24–April 1, 2012, 2012, pp. 425–438, https://doi.org/10.1007/978-3-642-28872-2_29.

[225] D. Hao, L. Zhang, X. Wu, H. Mei, G. Rothermel, On-demand test suite reduction, in: 34th International Conference on Software Engineering, ICSE 2012, June 2–9, 2012, Zurich, Switzerland, 2012, pp. 738–748, https://doi.org/10.1109/ICSE.2012.6227144.

[226] M.A. Alipour, A. Shi, R. Gopinath, D. Marinov, A. Groce, Evaluating non-adequate test-case reduction, in: Proceedings of the 31st IEEE/ACM International Conference on Automated Software Engineering, ASE 2016, Singapore, September 3–7, 2016, 2016, pp. 16–26, https://doi.org/10.1145/2970276.2970361.

[227] D.C. Nguyen, A. Marchetto, P. Tonella, Change sensitivity based prioritization for audit testing of webservice compositions, in: Fourth IEEE International Conference on Software Testing, Verification and Validation, ICST 2012, Berlin, Germany, March 21–25, 2011, Workshop Proceedings, 2011, pp. 357–365, https://doi.org/10.1109/ICSTW.2011.50.

[228] D. Tengeri, L. Vidács, Á. Beszédes, J. Jász, G. Balogh, B. Vancsics, T. Gyimóthy, Relating code coverage, mutation score and test suite reducibility to defect density, in: Ninth IEEE International Conference on Software Testing, Verification and Validation Workshops, ICST Workshops 2016, Chicago, IL, USA, April 11–15, 2016, 2016, pp. 174–179, https://doi.org/10.1109/ICSTW.2016.25.
[229] B. Kurtz, P. Ammann, J. Offutt, M. Kurtz, Are we there yet? How redundant and equivalent mutants affect determination of test completeness, in: Ninth IEEE International Conference on Software Testing, Verification and Validation Workshops, ICST Workshops 2016, Chicago, IL, USA, April 11–15, 2016, 2016, pp. 142–151, https://doi.org/10.1109/ICSTW.2016.41.
[230] G. Fraser, A. Zeller, Generating parameterized unit tests, in: Proceedings of the 20th International Symposium on Software Testing and Analysis, ISSTA 2011, Toronto, ON, Canada, July 17–21, 2011, 2011, pp. 364–374, https://doi.org/10.1145/2001420.2001461.
[231] T. Knauth, C. Fetzer, P. Felber, Assertion-driven development: assessing the quality of contracts using meta-mutations, in: Second International Conference on Software Testing Verification and Validation, ICST 2009, Denver, CO, USA, April 1–4, 2009, Workshops Proceedings, 2009, pp. 182–191, https://doi.org/10.1109/ICSTW.2009.40.
[232] G. Fraser, A. Arcuri, EvoSuite: automatic test suite generation for object-oriented software, in: SIGSOFT/FSE'11 19th ACM SIGSOFT Symposium on the Foundations of Software Engineering (FSE-19) and ESEC'11: 13th European Software Engineering Conference (ESEC-13), Szeged, Hungary, September 5–9, 2011, 2011, pp. 416–419, https://doi.org/10.1145/2025113.2025179.
[233] H. Yoshida, S. Tokumoto, M.R. Prasad, I. Ghosh, T. Uehara, FSX: fine-grained incremental unit test generation for C/C++ programs, in: Proceedings of the 25th International Symposium on Software Testing and Analysis, ISSTA 2016, Saarbrücken, Germany, July 18–20, 2016, 2016, pp. 106–117, https://doi.org/10.1145/2931037.2931055.
[234] H. Yoshida, S. Tokumoto, M.R. Prasad, I. Ghosh, T. Uehara, FSX: a tool for fine-grained incremental unit test generation for C/C++ programs, in: Proceedings of the 24th ACM SIGSOFT International Symposium on Foundations of Software Engineering, FSE 2016, Seattle, WA, USA, November 13–18, 2016, 2016, pp. 1052–1056, https://doi.org/10.1145/2950290.2983937.
[235] S. Mirshokraie, A. Mesbah, K. Pattabiraman, PYTHIA: generating test cases with oracles for JavaScript applications, in: 2013 28th IEEE/ACM International Conference on Automated Software Engineering, ASE 2013, Silicon Valley, CA, USA, November 11–15, 2013, 2013, pp. 610–615, https://doi.org/10.1109/ASE.2013.6693121.
[236] S. Mirshokraie, A. Mesbah, K. Pattabiraman, JSEFT: automated javascript unit test generation, in: 8th IEEE International Conference on Software Testing, Verification and Validation, ICST 2015, Graz, Austria, April 13–17, 2015, 2015, pp. 1–10, https://doi.org/10.1109/ICST.2015.7102595.
[237] M. Staats, G. Gay, M.P.E. Heimdahl, Automated oracle creation support, or: how I learned to stop worrying about fault propagation and love mutation testing, in: 34th International Conference on Software Engineering, ICSE 2012, June 2–9, 2012, Zurich, Switzerland, 2012, pp. 870–880, https://doi.org/10.1109/ICSE.2012.6227132.
[238] G. Gay, M. Staats, M.W. Whalen, M.P.E. Heimdahl, Automated oracle data selection support, IEEE Trans. Softw. Eng. 41 (11) (2015) 1119–1137, https://doi.org/10.1109/TSE.2015.2436920.
[239] G. Jahangirova, D. Clark, M. Harman, P. Tonella, Test oracle assessment and improvement, in: Proceedings of the 25th International Symposium on Software Testing and Analysis, ISSTA 2016, Saarbrücken, Germany, July 18–20, 2016, 2016, pp. 247–258, https://doi.org/10.1145/2931037.2931062.

[240] E.T. Barr, M. Harman, P. McMinn, M. Shahbaz, S. Yoo, The oracle problem in software testing: a survey, IEEE Trans. Softw. Eng. 41 (5) (2015) 507–525, https://doi.org/10.1109/TSE.2014.2372785.

[241] M. Papadakis, Y.L. Traon, Using mutants to locate "unknown" faults, in: Fifth IEEE International Conference on Software Testing, Verification and Validation, ICST 2012, Montreal, QC, Canada, April 17–21, 2012, 2012, pp. 691–700, https://doi.org/10.1109/ICST.2012.159.

[242] T.T. Chekam, M. Papadakis, Y.L. Traon, Assessing and comparing mutation-based fault localization techniques, CoRR (2016) http://arxiv,org/abs/1607.05512.

[243] S. Moon, Y. Kim, M. Kim, S. Yoo, Ask the mutants: mutating faulty programs for fault localization, in: Seventh IEEE International Conference on Software Testing, Verification and Validation, ICST 2014, March 31 2014–April 4, 2014, Cleveland, OH, USA, 2014, pp. 153–162, https://doi.org/10.1109/ICST.2014.28.

[244] L. Zhang, L. Zhang, S. Khurshid, Injecting mechanical faults to localize developer faults for evolving software, in: Proceedings of the 2013 ACM SIGPLAN International Conference on Object Oriented Programming Systems Languages & Applications, OOPSLA 2013, Part of SPLASH 2013, Indianapolis, IN, USA, October 26–31, 2013, 2013, pp. 765–784, https://doi.org/10.1145/2509136.2509551.

[245] S. Hong, B. Lee, T. Kwak, Y. Jeon, B. Ko, Y. Kim, M. Kim, Mutation-based fault localization for real-world multilingual programs (T), in: 30th IEEE/ACM International Conference on Automated Software Engineering, ASE 2015, Lincoln, NE, USA, November 9–13, 2015, 2015, pp. 464–475, https://doi.org/10.1109/ASE.2015.14.

[246] S. Hong, T. Kwak, B. Lee, Y. Jeon, B. Ko, Y. Kim, M. Kim, MUSEUM: debugging real-world multilingual programs using mutation analysis, Inf. Softw. Technol. 82 (2017) 80–95, https://doi.org/10.1016/j.infsof.2016.10.002.

[247] S.S. Murtaza, N.H. Madhavji, M. Gittens, Z. Li, Diagnosing new faults using mutants and prior faults, in: Proceedings of the 33rd International Conference on Software Engineering, ICSE 2011, Waikiki, Honolulu, HI, USA, May 21–28, 2011, 2011, pp. 960–963, https://doi.org/10.1145/1985793.1985959.

[248] S.S. Murtaza, A. Hamou-Lhadj, N.H. Madhavji, M. Gittens, An empirical study on the use of mutant traces for diagnosis of faults in deployed systems, J. Syst. Softw. 90 (2014) 29–44, https://doi.org/10.1016/j.jss.2013.11.1094.

[249] V. Musco, M. Monperrus, P. Preux, Mutation-based graph inference for fault localization, in: 16th IEEE International Working Conference on Source Code Analysis and Manipulation, SCAM 2016, Raleigh, NC, USA, October 2–3, 2016, 2016, pp. 97–106, https://doi.org/10.1109/SCAM.2016.24.

[250] V. Debroy, W.E. Wong, Using mutation to automatically suggest fixes for faulty programs, in: Third International Conference on Software Testing, Verification and Validation, ICST 2010, Paris, France, April 7–9, 2010, 2010, pp. 65–74, https://doi.org/10.1109/ICST.2010.66.

[251] V. Debroy, W.E. Wong, Combining mutation and fault localization for automated program debugging, J. Syst. Softw. 90 (2014) 45–60, https://doi.org/10.1016/j.jss.2013.10.042.

[252] W. Weimer, T. Nguyen, C. Le Goues, S. Forrest, Automatically finding patches using genetic programming, in: 31st International Conference on Software Engineering, ICSE 2009, May 16–24, 2009, Vancouver, Canada, Proceedings, 2009, pp. 364–374, https://doi.org/10.1109/ICSE.2009.5070536.

[253] C. Le Goues, M. Dewey-Vogt, S. Forrest, W. Weimer, A systematic study of automated program repair: fixing 55 out of 105 bugs for $8 each, in: 34th International Conference on Software Engineering, ICSE 2012, June 2–9, 2012, Zurich, Switzerland, 2012, pp. 3–13, https://doi.org/10.1109/ICSE.2012.6227211.

[254] W. Weimer, Z.P. Fry, S. Forrest, Leveraging program equivalence for adaptive program repair: models and first results, in: 2013 28th IEEE/ACM International Conference on Automated Software Engineering, ASE 2013, Silicon Valley, CA, USA, November 11–15, 2013, 2013, pp. 356–366, https://doi.org/10.1109/ASE.2013.6693094.
[255] S.H. Tan, A. Roychoudhury, relifix: automated repair of software regressions, in: 37th IEEE/ACM International Conference on Software Engineering, ICSE 2015, Florence, Italy, May 16–24, 2015, vol. 1, 2015, pp. 471–482, https://doi.org/10.1109/ICSE.2015.65.
[256] F. Long, M. Rinard, Staged program repair with condition synthesis, in: Proceedings of the 2015 10th Joint Meeting on Foundations of Software Engineering, ESEC/FSE 2015, Bergamo, Italy, August 30–September 4, 2015, 2015, pp. 166–178, https://doi.org/10.1145/2786805.2786811.
[257] D. Kim, J. Nam, J. Song, S. Kim, Automatic patch generation learned from human-written patches, in: 35th International Conference on Software Engineering, ICSE '13, San Francisco, CA, USA, May 18–26, 2013, 2013, pp. 802–811, https://doi.org/10.1109/ICSE.2013.6606626.
[258] J. Petke, S. Haraldsson, M. Harman, W. Langdon, D. White, J. Woodward, Genetic improvement of software: a comprehensive survey, IEEE Trans. Evol. Comput. PP (99) (2017) 1. ISSN: 1089-778X, https://doi.org/10.1109/TEVC.2017.2693219.
[259] S. Chandra, E. Torlak, S. Barman, R. Bodík, Angelic debugging, in: Proceedings of the 33rd International Conference on Software Engineering, ICSE 2011, Waikiki, Honolulu, HI, USA, May 21–28, 2011, 2011, pp. 121–130, https://doi.org/10.1145/1985793.1985811.
[260] J. Zhang, Z. Wang, L. Zhang, D. Hao, L. Zang, S. Cheng, L. Zhang, Predictive mutation testing, in: Proceedings of the 25th International Symposium on Software Testing and Analysis, ISSTA 2016, Saarbrücken, Germany, July 18–20, 2016, 2016, pp. 342–353, https://doi.org/10.1145/2931037.2931038.
[261] J.M. Rojas, G. Fraser, Code defenders: a mutation testing game, in: Ninth IEEE International Conference on Software Testing, Verification and Validation Workshops, ICST Workshops 2016, Chicago, IL, USA, April 11–15, 2016, 2016, pp. 162–167, https://doi.org/10.1109/ICSTW.2016.43.
[262] J.M. Rojas, T.D. White, B.S. Clegg, G. Fraser, Code defenders: crowdsourcing effective tests and subtle mutants with a mutation testing game, in: Proceedings of the 39th International Conference on Software Engineering, ICSE 2017, Buenos Aires, Argentina, May 20–28, 2017, 2017, pp, 677–688.
[263] M. Harman, Y. Jia, W.B. Langdon, A manifesto for higher order mutation testing, in: Third International Conference on Software Testing, Verification and Validation, ICST 2010, Paris, France, April 7–9, 2010, Workshops Proceedings, 2010, pp. 80–89, https://doi.org/10.1109/ICSTW.2010.13.
[264] Y. Jia, M. Harman, Constructing subtle faults using higher order mutation testing, in: Eighth IEEE International Working Conference on Source Code Analysis and Manipulation (SCAM 2008), September 28–29, 2008, Beijing, China, 2008, pp. 249–258, https://doi.org/10.1109/SCAM.2008.36.
[265] M. Harman, Y. Jia, P.R. Mateo, M. Polo, Angels and monsters: an empirical investigation of potential test effectiveness and efficiency improvement from strongly subsuming higher order mutation, in: ACM/IEEE International Conference on Automated Software Engineering, ASE '14, Vasteras, Sweden, September 15–19, 2014, 2014, pp. 397–408, https://doi.org/10.1145/2642937.2643008.
[266] E. Omar, S. Ghosh, D. Whitley, Constructing subtle higher order mutants for Java and AspectJ programs, in: IEEE 24th International Symposium on Software Reliability Engineering, ISSRE 2013, Pasadena, CA, USA, November 4–7, 2013, 2013, pp. 340–349, https://doi.org/10.1109/ISSRE.2013.6698887.

[267] E. Omar, S. Ghosh, D. Whitley, Subtle higher order mutants, Inf. Softw. Technol. 81 (2017) 3–18, https://doi.org/10.1016/j.infsof.2016.01.016.

[268] E. Omar, S. Ghosh, D. Whitley, Comparing search techniques for finding subtle higher order mutants, in: Genetic and Evolutionary Computation Conference, GECCO '14, Vancouver, BC, Canada, July 12–16, 2014, 2014, pp. 1271–1278, https://doi.org/10.1145/2576768.2598286.

[269] F. Wu, M. Harman, Y. Jia, J. Krinke, HOMI: searching higher order mutants for software improvement, in: Search Based Software Engineering—8th International Symposium, SSBSE 2016, Raleigh, NC, USA, October 8–10, 2016, Proceedings, 2016, pp. 18–33, https://doi.org/10.1007/978-3-319-47106-8_2.

[270] D. Shin, S. Yoo, D. Bae, Diversity-aware mutation adequacy criterion for improving fault detection capability, in: Ninth IEEE International Conference on Software Testing, Verification and Validation Workshops, ICST Workshops 2016, Chicago, IL, USA, April 11–15, 2016, 2016, pp. 122–131, https://doi.org/10.1109/ICSTW.2016.37.

[271] D. Shin, S. Yoo, D.H. Bae, A theoretical and empirical study of diversity-aware mutation adequacy criterion, IEEE Trans. Softw. Eng. PP (99) (2017) 1. ISSN: 0098-5589, https://doi.org/10.1109/TSE.2017.2732347. 1.

[272] C. Henard, M. Papadakis, G. Perrouin, J. Klein, Y.L. Traon, Assessing software product line testing via model-based mutation: an application to similarity testing, in: Sixth IEEE International Conference on Software Testing, Verification and Validation, ICST 2013 Workshops Proceedings, Luxembourg, Luxembourg, March 18–22, 2013, 2013, pp. 188–197, https://doi.org/10.1109/ICSTW.2013.30.

[273] C. Henard, M. Papadakis, Y.L. Traon, MutaLog: a tool for mutating logic formulas, in: Seventh IEEE International Conference on Software Testing, Verification and Validation, ICST 2014 Workshops Proceedings, March 31–April 4, 2014, Cleveland, OH, USA, 2014, pp. 399–404, https://doi.org/10.1109/ICSTW.2014.54.

[274] C. Henard, M. Papadakis, G. Perrouin, J. Klein, Y.L. Traon, Towards automated testing and fixing of re-engineered feature models, in: 35th International Conference on Software Engineering, ICSE '13, San Francisco, CA, USA, May 18–26, 2013, 2013, pp. 1245–1248, https://doi.org/10.1109/ICSE.2013.6606689.

[275] C. Henard, M. Papadakis, Y.L. Traon, Mutation-based generation of software product line test configurations, in: Search-Based Software Engineering—6th International Symposium, SSBSE 2014, Fortaleza, Brazil, August 26–29, 2014, 2014, pp. 92–106, https://doi.org/10.1007/978-3-319-09940-8_7.

[276] R.A.M. Filho, S.R. Vergilio, A mutation and multi-objective test data generation approach for feature testing of software product lines, in: 29th Brazilian Symposium on Software Engineering, SBES 2015, Belo Horizonte, MG, Brazil, September 21–26, 2015, 2015, pp. 21–30, https://doi.org/10.1109/SBES.2015.17.

[277] R.A.M. Filho, S.R. Vergilio, A multi-objective test data generation approach for mutation testing of feature models, J. Softw. Eng. R&D 4 (2016) 4, https://doi.org/10.1186/s40411-016-0030-9.

[278] P. Arcaini, A. Gargantini, P. Vavassori, Generating tests for detecting faults in feature models, in: 8th IEEE International Conference on Software Testing, Verification and Validation, ICST 2015, Graz, Austria, April 13–17, 2015, 2015, pp. 1–10, https://doi.org/10.1109/ICST.2015.7102591.

[279] P. Arcaini, A. Gargantini, P. Vavassori, Automatic detection and removal of conformance faults in feature models, in: 2016 IEEE International Conference on Software Testing, Verification and Validation, ICST 2016, Chicago, IL, USA, April 11–15, 2016, 2016, pp. 102–112, https://doi.org/10.1109/ICST.2016.10.

[280] M.B. Trakhtenbrot, Implementation-oriented mutation testing of statechart models, in: Third International Conference on Software Testing, Verification and Validation, ICST 2010, Paris, France, April 7–9, 2010, Workshops Proceedings, 2010, pp. 120–125, https://doi.org/10.1109/ICSTW.2010.55.

[281] B.K. Aichernig, E. Jöbstl, Towards symbolic model-based mutation testing: pitfalls in expressing semantics as constraints, in: Fifth IEEE International Conference on Software Testing, Verification and Validation, ICST 2012, Montreal, QC, Canada, April 17–21, 2012, 2012, pp. 752–757, https://doi.org/10.1109/ICST.2012.169.
[282] B.K. Aichernig, H. Brandl, E. Jöbstl, W. Krenn, Efficient mutation killers in action, in: Fourth IEEE International Conference on Software Testing, Verification and Validation, ICST 2011, Berlin, Germany, March 21–25, 2011, 2011, pp. 120–129, https://doi.org/10.1109/ICST.2011.57.
[283] B.K. Aichernig, H. Brandl, E. Jöbstl, W. Krenn, R. Schlick, S. Tiran, Killing strategies for model-based mutation testing, Softw. Test. Verif. Reliab. 25 (8) (2015) 716–748, https://doi.org/10.1002/stvr.1522.
[284] F. Belli, M. Beyazit, Exploiting model morphology for event-based testing, IEEE Trans. Softw. Eng. 41 (2) (2015) 113–134, https://doi.org/10.1109/TSE.2014.2360690.
[285] K. El-Fakih, A. Kolomeez, S. Prokopenko, N. Yevtushenko, Extended finite state machine based test derivation driven by user defined faults, in: First International Conference on Software Testing, Verification, and Validation, ICST 2008, Lillehammer, Norway, April 9–11, 2008, 2008, pp. 308–317, https://doi.org/10.1109/ICST.2008.16.
[286] T. Su, G. Meng, Y. Chen, K. Wu, W. Yang, Y. Yao, G. Pu, Y. Liu, Z. Su, Guided, stochastic model-based GUI testing of Android apps, in: Proceedings of the 2017 11th Joint Meeting on Foundations of Software Engineering, ACM, New York, NY, USA, 2017, ISBN: 978-1-4503-5105-8, pp. 245–256, https://doi.org/10.1145/3106237.3106298.
[287] B.K. Aichernig, S. Marcovic, R. Schumi, Property-based testing with external test-case generators, in: 2017 IEEE International Conference on Software Testing, Verification and Validation Workshops (ICSTW), 2017, pp. 337–346, https://doi.org/10.1109/ICSTW.2017.62.
[288] B.K. Aichernig, F. Lorber, Towards generation of adaptive test cases from partial models of determinized timed automata, in: Eighth IEEE International Conference on Software Testing, Verification and Validation, ICST 2015 Workshops, Graz, Austria, April 13–17, 2015, 2015, pp. 1–6, https://doi.org/10.1109/ICSTW.2015.7107409.
[289] B.K. Aichernig, F. Lorber, D. Nickovic, Time for mutants—model-based mutation testing with timed automata, in: Tests and Proofs—7th International Conference, TAP 2013, Budapest, Hungary, June 16–20, 2013, 2013, pp. 20–38, https://doi.org/10.1007/978-3-642-38916-0_2.
[290] K.G. Larsen, F. Lorber, B. Nielsen, U.M. Nyman, Mutation-based test-case generation with ecdar, in: 2017 IEEE International Conference on Software Testing, Verification and Validation Workshops (ICSTW), 2017, pp. 319–328, https://doi.org/10.1109/ICSTW.2017.60.
[291] T. Zhou, H. Sun, J. Liu, X. Chen, D. Du, Improving testing coverage for safety-critical system by mutated specification, in: 21st Asia-Pacific Software Engineering Conference, APSEC 2014, Jeju, South Korea, December 1–4, 2014, vol, 1: Research Papers, 2014, pp. 43–46, https://doi.org/10.1109/APSEC.2014.15.
[292] S.F. Adra, P. McMinn, Mutation operators for agent-based models, in: Third International Conference on Software Testing, Verification and Validation, ICST 2010, Paris, France, April 7–9, 2010, Workshops Proceedings, 2010, pp. 151–156, https://doi.org/10.1109/ICSTW.2010.9.
[293] M. Stephan, M.H. Alalfi, A. Stevenson, J.R. Cordy, Using mutation analysis for a model-clone detector comparison framework, in: 35th International Conference on Software Engineering, ICSE '13, San Francisco, CA, USA, May 18–26, 2013, 2013, pp. 1261–1264, https://doi.org/10.1109/ICSE.2013.6606693.
[294] C.K. Roy, J.R. Cordy, A mutation/injection-based automatic framework for evaluating code clone detection tools, in: Second International Conference on Software

Testing Verification and Validation, ICST 2009, Denver, Colorado, USA, April 1–4, 2009, Workshops Proceedings, 2009, pp. 157–166, https://doi.org/10.1109/ICSTW.2009.18.

[295] M. Stephan, M.H. Alalfi, J.R. Cordy, Towards a taxonomy for simulink model mutations, in: Seventh IEEE International Conference on Software Testing, Verification and Validation, ICST 2014 Workshops Proceedings, March 31–April 4, 2014, Cleveland, OH, USA, 2014, pp. 206–215, https://doi.org/10.1109/ICSTW.2014.17.

[296] I. Pill, I. Rubil, F. Wotawa, M. Nica, SIMULTATE: a toolset for fault injection and mutation testing of simulink models, in: Ninth IEEE International Conference on Software Testing, Verification and Validation Workshops, ICST Workshops 2016, Chicago, IL, USA, April 11–15, 2016, 2016, pp. 168–173, https://doi.org/10.1109/ICSTW.2016.21.

[297] Y. Khan, J. Hassine, Mutation operators for the Atlas Transformation Language, in: Sixth IEEE International Conference on Software Testing, Verification and Validation, ICST 2013 Workshops Proceedings, Luxembourg, Luxembourg, March 18–22, 2013, 2013, pp. 43–52, https://doi.org/10.1109/ICSTW.2013.13.

[298] J. Troya, A. Bergmayr, L. Burgue no, M. Wimmer, Towards systematic mutations for and with ATL model transformations, in: Eighth IEEE International Conference on Software Testing, Verification and Validation, ICST 2015 Workshops, Graz, Austria, April 13–17, 2015, 2015, pp. 1–10, https://doi.org/10.1109/ICSTW.2015.7107455.

[299] M. Tisi, F. Jouault, P. Fraternali, S. Ceri, J. Bézivin, On the use of higher-order model transformations, in: Model Driven Architecture - Foundations and Applications: 5th European Conference, ECMDA-FA 2009, Enschede, The Netherlands, June 23–26, 2009, LNCSvol. 5562, 2009, pp. 18–33, https://doi.org/10.1007/978-3-642-02674-4_3.

[300] V. Aranega, J. Mottu, A. Etien, T. Degueule, B. Baudry, J. Dekeyser, Towards an automation of the mutation analysis dedicated to model transformation, Softw. Test. Verif. Reliab. 25 (5–7) (2015) 653–683, https://doi.org/10.1002/stvr.1532.

[301] A. Bartel, B. Baudry, F. Munoz, J. Klein, T. Mouelhi, Y.L. Traon, Model driven mutation applied to adaptative systems testing, in: Fourth IEEE International Conference on Software Testing, Verification and Validation, ICST 2011 Workshops Proceedings, March 21–March 25, 2011, Berlin, Germany, 2011, pp. 408–413, https://doi.org/10.1109/ICSTW.2011.24.

[302] P. Arcaini, A. Gargantini, E. Riccobene, Using mutation to assess fault detection capability of model review, Softw. Test. Verif. Reliab. 25 (5–7) (2015) 629–652, https://doi.org/10.1002/stvr.1530.

[303] M. Kaplan, T. Klinger, A.M. Paradkar, A. Sinha, C. Williams, C. Yilmaz, Less is more: a minimalistic approach to UML model-based conformance test generation, in: First International Conference on Software Testing, Verification, and Validation, ICST 2008, Lillehammer, Norway, April 9–11, 2008, 2008, pp. 82–91, https://doi.org/10.1109/ICST.2008.48.

[304] G. Fraser, F. Wotawa, Using model-checkers to generate and analyze property relevant test-cases, Softw. Q. J. 16 (2) (2008) 161–183, https://doi.org/10.1007/s11219-007-9031-6.

[305] M. Trakhtenbrot, Mutation patterns for temporal requirements of reactive systems, in: 2017 IEEE International Conference on Software Testing, Verification and Validation Workshops (ICSTW), 2017, pp. 116–121, https://doi.org/10.1109/ICSTW.2017.27.

[306] A. Sullivan, K. Wang, R.N. Zaeem, S. Khurshid, Automated rest generation and mutation testing for alloy, in: 2017 IEEE International Conference on Software Testing, Verification and Validation (ICST), 2017, pp. 264–275, https://doi.org/10.1109/ICST.2017.31.

[307] D. Xu, O. el Ariss, W. Xu, L. Wang, Testing aspect-oriented programs with finite state machines, Softw. Test. Verif. Reliab. 22 (4) (2012) 267–293, https://doi.org/10.1002/stvr.440.
[308] B. Lindström, S.F. Andler, J. Offutt, P. Pettersson, D. Sundmark, Mutating aspect-oriented models to test cross-cutting concerns, in: Eighth IEEE International Conference on Software Testing, Verification and Validation, ICST 2015 Workshops, Graz, Austria, April 13–17, 2015, 2015, pp. 1–10, https://doi.org/10.1109/ICSTW.2015.7107456.
[309] Y. Elrakaiby, T. Mouelhi, Y.L. Traon, Testing obligation policy enforcement using mutation analysis, in: Fifth IEEE International Conference on Software Testing, Verification and Validation, ICST 2012, Montreal, QC, Canada, April 17–21, 2012, 2012, pp. 673–680, https://doi.org/10.1109/ICST.2012.157.
[310] T. Mouelhi, F. Fleurey, B. Baudry, A generic metamodel for security policies mutation, in: First International Conference on Software Testing Verification and Validation, ICST 2008, Lillehammer, Norway, April 9–11, 2008, Workshops Proceedings, 2008, pp. 278–286, https://doi.org/10.1109/ICSTW.2008.2.
[311] A. Bertolino, S. Daoudagh, F. Lonetti, E. Marchetti, F. Martinelli, P. Mori, Testing of PolPA-based usage control systems, Softw. Q. J. 22 (2) (2014) 241–271, https://doi.org/10.1007/s11219-013-9216-0.
[312] T. Mouelhi, Y.L. Traon, B. Baudry, Transforming and selecting functional test cases for security policy testing, in: Second International Conference on Software Testing Verification and Validation, ICST 2009, Denver, CO, USA, April 1–4, 2009, 2009, pp. 171–180, https://doi.org/10.1109/ICST.2009.49.
[313] P.H. Nguyen, M. Papadakis, I. Rubab, Testing delegation policy enforcement via mutation analysis, in: Sixth IEEE International Conference on Software Testing, Verification and Validation, ICST 2013 Workshops Proceedings, Luxembourg, Luxembourg, March 18–22, 2013, 2013, pp. 34–42, https://doi.org/10.1109/ICSTW.2013.12.
[314] J. Hwang, T. Xie, D.E. Kateb, T. Mouelhi, Y.L. Traon, Selection of regression system tests for security policy evolution, in: IEEE/ACM International Conference on Automated Software Engineering, ASE'12, Essen, Germany, September 3–7, 2012, 2012, pp. 266–269, https://doi.org/10.1145/2351676.2351719.
[315] F. Dadeau, P. Héam, R. Kheddam, Mutation-based test generation from security protocols in HLPSL, in: Fourth IEEE International Conference on Software Testing, Verification and Validation, ICST 2011, Berlin, Germany, March 21–25, 2011, 2011, pp. 240–248, https://doi.org/10.1109/ICST.2011.42.
[316] F. Dadeau, P. Héam, R. Kheddam, G. Maatoug, M. Rusinowitch, Model-based mutation testing from security protocols in HLPSL, Softw. Test. Verif. Reliab. 25 (5–7) (2015) 684–711, https://doi.org/10.1002/stvr.1531.
[317] M. Papadakis, C. Henard, Y.L. Traon, Sampling program inputs with mutation analysis: going beyond combinatorial interaction testing, in: Seventh IEEE International Conference on Software Testing, Verification and Validation, ICST 2014, March 31 2014–April 4, 2014, Cleveland, OH, USA, 2014, pp. 1–10, https://doi.org/10.1109/ICST.2014.11.
[318] Z. Zhang, T. Wu, J. Zhang, Boundary value analysis in automatic white-box test generation, in: 26th IEEE International Symposium on Software Reliability Engineering, ISSRE 2015, Gaithersbury, MD, USA, November 2–5, 2015, 2015, pp. 239–249, https://doi.org/10.1109/ISSRE.2015.7381817.
[319] M. Patrick, Y. Jia, Kernel density adaptive random testing, in: Eighth IEEE International Conference on Software Testing, Verification and Validation, ICST 2015 Workshops, Graz, Austria, April 13–17, 2015, 2015, pp. 1–10, https://doi.org/10.1109/ICSTW.2015.7107451.

[320] M. Patrick, Y. Jia, KD-ART: should we intensify or diversify tests to kill mutants? Inf, Softw. Technol. 81 (2017) 36–51, https://doi.org/10.1016/j.infsof.2016.04.009.
[321] J.P. Galeotti, C.A. Furia, E. May, G. Fraser, A. Zeller, Inferring loop invariants by mutation, dynamic analysis, and static checking, IEEE Trans. Softw. Eng. 41 (10) (2015) 1019–1037, https://doi.org/10.1109/TSE.2015.2431688.
[322] C. Andrés, M.G. Merayo, M. Núñez, Passive testing of stochastic timed systems, in: Second International Conference on Software Testing Verification and Validation, ICST 2009, Denver, CO, USA, April 1–4, 2009, 2009, pp. 71–80, https://doi.org/10.1109/ICST.2009.35.
[323] C. Andrés, M.G. Merayo, C. Molinero, Advantages of mutation in passive testing: an empirical study, in: Second International Conference on Software Testing Verification and Validation, ICST 2009, Denver, CO, USA, April 1–4, 2009, Workshops Proceedings, 2009, pp. 230–239, https://doi.org/10.1109/ICSTW.2009.33.
[324] C. Andrés, M.G. Merayo, M. Núñez, Formal passive testing of timed systems: theory and tools, Softw. Test. Verif. Reliab. 22 (6) (2012) 365–405, https://doi.org/10.1002/stvr.1464.
[325] T. Pankumhang, M. Rutherford, Iterative instrumentation for code coverage in time-sensitive systems, in: 8th IEEE International Conference on Software Testing, Verification and Validation, ICST 2015, Graz, Austria, April 13–17, 2015, 2015, pp. 1–10, https://doi.org/10.1109/ICST.2015.7102594.
[326] A. Groce, I. Ahmed, C. Jensen, P.E. McKenney, How verified is my code? Falsification-driven verification (T), in: 30th IEEE/ACM International Conference on Automated Software Engineering, ASE 2015, Lincoln, NE, USA, November 9–13, 2015, 2015, pp. 737–748, https://doi.org/10.1109/ASE.2015.40.
[327] J. Svajlenko, C.K. Roy, S. Duszynski, ForkSim: generating software forks for evaluating cross-project similarity analysis tools, in: 13th IEEE International Working Conference on Source Code Analysis and Manipulation, SCAM 2013, Eindhoven, Netherlands, September 22–23, 2013, 2013, pp. 37–42, https://doi.org/10.1109/SCAM.2013.6648182.
[328] D. Bowes, T. Hall, M. Harman, Y. Jia, F. Sarro, F. Wu, Mutation-aware fault prediction, in: Proceedings of the 25th International Symposium on Software Testing and Analysis, ISSTA 2016, Saarbrücken, Germany, July 18–20, 2016, 2016, pp. 330–341, https://doi.org/10.1145/2931037.2931039.
[329] J. Zhang, Y. Lou, L. Zhang, D. Hao, L. Zhang, H. Mei, Isomorphic regression testing: executing uncovered branches without test augmentation, in: Proceedings of the 24th ACM SIGSOFT International Symposium on Foundations of Software Engineering, FSE 2016, Seattle, WA, USA, November 13–18, 2016, 2016, pp. 883–894, https://doi.org/10.1145/2950290.2950313.
[330] D.D. Nardo, F. Pastore, L.C. Briand, Generating complex and faulty test data through model-based mutation analysis, in: 8th IEEE International Conference on Software Testing, Verification and Validation, ICST 2015, Graz, Austria, April 13–17, 2015, 2015, pp. 1–10, https://doi.org/10.1109/ICST.2015.7102589.
[331] P. Arcaini, A. Gargantini, E. Riccobene, P. Vavassori, Rehabilitating equivalent mutants as static anomaly detectors in software artifacts, in: Eighth IEEE International Conference on Software Testing, Verification and Validation, ICST 2015 Workshops, Graz, Austria, April 13–17, 2015, 2015, pp. 1–6, https://doi.org/10.1109/ICSTW.2015.7107452.
[332] P. Arcaini, A. Gargantini, E. Riccobene, P. Vavassori, A novel use of equivalent mutants for static anomaly detection in software artifacts, Inf. Softw. Technol. 81 (2017) 52–64, https://doi.org/10.1016/j.infsof.2016.01.019.
[333] B. Baudry, S. Allier, M. Monperrus, Tailored source code transformations to synthesize computationally diverse program variants, in: International Symposium on Software Testing and Analysis, ISSTA 2014, San Jose, CA, USA, July 21–26, 2014, 2014, pp. 149–159, https://doi.org/10.1145/2610384.2610415.

[334] B. Lisper, B. Lindström, P. Potena, M. Saadatmand, M. Bohlin, Targeted mutation: efficient mutation analysis for testing non-functional properties, in: 2017 IEEE International Conference on Software Testing, Verification and Validation Workshops (ICSTW), 2017, pp. 65–68, https://doi.org/10.1109/ICSTW.2017.18.
[335] A. Babu, mutatepy: A Mutation Testing Tool for C, http://members.femto-st.fr/pierre-cyrille-heam/mutatepy (accessed May 2017).
[336] I. Moore, Jester and Pester, 2001. (accessed May 2017). http://jester.sourceforge.net/.
[337] M.E. Delamaro, J.C. Maldonado, A.M.R. Vincenzi, Proteum/IM 2.0: an integrated mutation testing environment, in: Springer US, Boston, MA, 2001, ISBN: 978-1-4757-5939-6, pp. 91–101, https://doi.org/10.1007/978-1-4757-5939-6_17.
[338] J.H. Andrews, Y. Zhang, General test result checking with log file analysis, IEEE Trans. Softw. Eng. 29 (7) (2003) 634–648, https://doi.org/10.1109/TSE.2003.1214327.
[339] J.H. Andrews, L.C. Briand, Y. Labiche, Is mutation an appropriate tool for testing experiments? in: G. Roman, W.G. Griswold, B. Nuseibeh (Eds.), 27th International Conference on Software Engineering (ICSE 2005), May 15–21, 2005, St. Louis, MO, USA, ACM, 2005, pp. 402–411, https://doi.org/10.1145/1062455.1062530.
[340] Y. Ma, J. Offutt, Y.R. Kwon, MuJava: a mutation system for Java, in: L.J. Osterweil, H.D. Rombach, M.L. Soffa (Eds.), 28th International Conference on Software Engineering (ICSE 2006), Shanghai, China, May 20–28, 2006, ACM, 2006, pp. 827–830, https://doi.org/10.1145/1134425.
[341] J. Offutt, Y. Ma, Y.R. Kwon, An experimental mutation system for Java, ACM SIGSOFT Softw. Eng. Notes 29 (5) (2004) 1–4, https://doi.org/10.1145/1022494.1022537.
[342] H. Do, G. Rothermel, On the use of mutation faults in empirical assessments of test case prioritization techniques, IEEE Trans. Softw. Eng. 32 (9) (2006) 733–752, https://doi.org/10.1109/TSE.2006.92.
[343] J. Tuya, M.J. Suarez-Cabal, C. de la Riva, SQLMutation: a tool to generate mutants of SQL database queries, in: Second Workshop on Mutation Analysis (Mutation 2006—ISSRE Workshops 2006), 2006, p. 1, https://doi.org/10.1109/MUTATION.2006.13.
[344] Jumble Testing Tool for Java, 2007. (accessed May 2017). http://jumble.sourceforge.net/.
[345] X. Feng, S. Marr, T. O'Callaghan, ESTP: an experimental software testing platform, in: Testing: Academic Industrial Conference—Practice and Research Techniques (taic part 2008), 2008, pp. 59–63, https://doi.org/10.1109/TAIC-PART.2008.8.
[346] R.P. Tan, S. Edwards, Evaluating automated unit testing in Sulu, in: First International Conference on Software Testing, Verification, and Validation, ICST 2008, Lillehammer, Norway, April 9–11, 2008, IEEE Computer Society, 2008, pp. 62–71, https://doi.org/10.1109/ICST.2008.59.
[347] Y. Jia, M. Harman, MILU: a customizable, runtime-optimized higher order mutation testing tool for the full C language, in: Proceedings of the 3rd Testing: Academic and Industrial Conference Practice and Research Techniques (TAIC PART 2008), 2008, pp, 94–98. Windsor, UK.
[348] A. Rajan, M.W. Whalen, M. Staats, M.P.E. Heimdahl, Requirements coverage as an adequacy measure for conformance testing, in: S. Liu, T.S.E. Maibaum, K. Araki (Eds.), Formal Methods and Software Engineering, 10th International Conference on Formal Engineering Methods, ICFEM 2008, Kitakyushu-City, Japan, October 27–31, 2008, Proceedings, Lecture Notes in Computer Science, vol. 5256, Springer, 2008, pp. 86–104, https://doi.org/10.1007/978-3-540-88194-0_8.
[349] M. Weiglhofer, F. Wotawa, "On the fly" input output conformance verification, in: Proceedings of the IASTED International Conference on Software Engineering, ACTA Press, Anaheim, CA, USA, 2008, ISBN: 978-0-88986-716-1, pp. 286–291.
[350] C. Zhou, P.G. Frankl, Mutation testing for Java database applications, in: Second International Conference on Software Testing Verification and Validation, ICST 2009, Denver, CO, USA, April 1–4, 2009, 2009, pp. 396–405, https://doi.org/10.1109/ICST.2009.43.

[351] C. Zhou, P.G. Frankl, JDAMA: Java database application mutation analyser, Softw. Test. Verif. Reliab. 21 (3) (2011) 241–263, https://doi.org/10.1002/stvr.462.

[352] R. Delamare, B. Baudry, S. Ghosh, Y.L. Traon, A test-driven approach to developing pointcut descriptors in aspectJ, in: Second International Conference on Software Testing Verification and Validation, ICST 2009, Denver, CO, USA, April 1–4, 2009, 2009, pp. 376–385, https://doi.org/10.1109/ICST.2009.41.

[353] R. Delamare, B. Baudry, Y.L. Traon, AjMutator: a tool for the mutation analysis of aspectJ pointcut descriptors, in: Second International Conference on Software Testing Verification and Validation, ICST 2009, Denver, CO, USA, April 1–4, 2009, Workshops Proceedings, 2009, pp. 200–204, https://doi.org/10.1109/ICSTW.2009.41.

[354] J.J. Domínguez-Jiménez, A. Estero-Botaro, I. Medina-Bulo, A framework for mutant genetic generation for WS-BPEL, in: M. Nielsen, A. Kucera, P.B. Miltersen, C. Palamidessi, P. Tuma, F.D. Valencia (Eds.), SOFSEM 2009: Theory and Practice of Computer Science, 35th Conference on Current Trends in Theory and Practice of Computer Science, Spindleruv Mlýn, Czech Republic, January 24–30, 2009, Lecture Notes in Computer Science, 5404, Springer, 2009, pp. 229–240, https://doi.org/10.1007/978-3-540-95891-8_23. vol.

[355] E.G. Aydal, R.F. Paige, M. Utting, J. Woodcock, Putting formal specifications under the magnifying glass: model-based testing for validation, in: Second International Conference on Software Testing Verification and Validation, ICST 2009, Denver, CO, USA, April 1–4, 2009, IEEE Computer Society, 2009, pp. 131–140, https://doi.org/10.1109/ICST.2009.20.

[356] K. Dobolyi, W. Weimer, Harnessing web-based application similarities to aid in regression testing, in: ISSRE 2009, 20th International Symposium on Software Reliability Engineering, Mysuru, Karnataka, India, November 16–19, 2009, IEEE Computer Society, 2009, pp. 71–80, https://doi.org/10.1109/ISSRE.2009.18.

[357] M. Ellims, D. Ince, M. Petre, The Csaw C mutation tool: initial results, in: Testing: Academic and Industrial Conference Practice and Research Techniques—MUTATION (TAICPART-MUTATION 2007), 2007, pp. 185–192, https://doi.org/10.1109/TAIC.PART.2007.28.

[358] A. Lakehal, I. Parissis, Structural coverage criteria for LUSTRE/SCADE programs, Softw. Test. Verif. Reliab. 19 (2) (2009) 133–154, https://doi.org/10.1002/stvr.394.

[359] J. Durães, H. Madeira, Definition of software fault emulation operators: a field data study, in: 2003 International Conference on Dependable Systems and Networks (DSN 2003), June 22–25, 2003, San Francisco, CA, USA, IEEE Computer Society, 2003, pp. 105–114, https://doi.org/10.1109/DSN.2003.1209922.

[360] L. Madeyski, N. Radyk, Judy—a mutation testing tool for Java, IET Softw. 4 (1) (2010) 32–42, https://doi.org/10.1049/iet-sen.2008.0038.

[361] R. Just, The major mutation framework: efficient and scalable mutation analysis for Java, in: International Symposium on Software Testing and Analysis, ISSTA '14, San Jose, CA, USA, July 21–26, 2014, 2014, pp. 433–436, https://doi.org/10.1145/2610384.2628053.

[362] P. Madiraju, A.S. Namin, Para(μ) - a partial and higher-order mutation tool with concurrency operators, in: Fourth IEEE International Conference on Software Testing, Verification and Validation, ICST 2012, Berlin, Germany, March 21–25, 2011, Workshop Proceedings, 2011, pp. 351–356, https://doi.org/10.1109/ICSTW.2011.34.

[363] MuBPEL - A Mutation Testing Tool for WS-BPEL, 2011. (accessed May 2017). https://neptuno.uca.es/redmine/projects/sources-fm/wiki/MuBPEL/.

[364] K. Winbladh, A. Ranganathan, Evaluating test selection strategies for end-user specified flow-based applications, in: P. Alexander, C.S. Pasareanu, J.G. Hosking (Eds.), 26th IEEE/ACM International Conference on Automated Software Engineering (ASE 2011), Lawrence, KS, USA, November 6–10, 2011, IEEE Computer Society, 2011, pp. 400–403, https://doi.org/10.1109/ASE.2011.6100083.

[365] A.A. Saifan, J. Dingel, J.S. Bradbury, E. Posse, Implementing and evaluating a runtime conformance checker for mobile agent systems, in: Fourth IEEE International Conference on Software Testing, Verification and Validation, ICST 2011, Berlin, Germany, March 21–25, 2011, IEEE Computer Society, 2011, pp. 269–278, https://doi.org/10.1109/ICST.2011.62.
[366] R. Just, G.M. Kapfhammer, F. Schweiggert, Using conditional mutation to increase the efficiency of mutation analysis, in: Proceedings of the 6th International Workshop on Automation of Software Test, AST 2011, Waikiki, Honolulu, HI, USA, May 23–24, 2011, 2011, pp. 50–56, https://doi.org/10.1145/1982595.1982606.
[367] H. Dan, R.M. Hierons, SMT-C: a semantic mutation testing tools for C, in: Fifth IEEE International Conference on Software Testing, Verification and Validation, ICST 2012, Montreal, QC, Canada, April 17–21, 2012, 2012, pp. 654–663, https://doi.org/10.1109/ICST.2012.155.
[368] M. Schirp, Mutation Testing for Ruby, 2012 (accessed May 2017). https://github.com/mbj/mutant.
[369] M. Kusano, C. Wang, CCmutator: a mutation generator for concurrency constructs in multithreaded C/C++ applications, in: 2013 28th IEEE/ACM International Conference on Automated Software Engineering, ASE 2013, Silicon Valley, CA, USA, November 11–15, 2013, 2013, pp. 722–725, https://doi.org/10.1109/ASE.2013.6693142.
[370] J.S. Bradbury, J.R. Cordy, J. Dingel, Mutation operators for concurrent Java (J2SE 5.0), in: Second Workshop on Mutation Analysis (Mutation 2006–ISSRE Workshops 2006), 2006, p. 11, https://doi.org/10.1109/MUTATION.2006.10.
[371] G.M. Kapfhammer, P. McMinn, C.J. Wright, Search-based testing of relational schema integrity constraints across multiple database management systems, in: Sixth IEEE International Conference on Software Testing, Verification and Validation, ICST 2013, Luxembourg, Luxembourg, March 18–22, 2013, IEEE Computer Society, 2013, pp. 31–40, https://doi.org/10.1109/ICST.2013.47.
[372] A. Bertolino, S. Daoudagh, F. Lonetti, E. Marchetti, XACMUT: XACML 2.0 mutants generator, in: Sixth IEEE International Conference on Software Testing, Verification and Validation, ICST 2013 Workshops Proceedings, Luxembourg, Luxembourg, March 18–22, 2013, IEEE Computer Society, 2013, pp. 28–33, https://doi.org/10.1109/ICSTW.2013.11.
[373] C. Ye, H. Jacobsen, Whitening SOA testing via event exposure, IEEE Trans. Softw. Eng. 39 (10) (2013) 1444–1465, https://doi.org/10.1109/TSE.2013.20.
[374] A. Derezińska, K. Hałas, Analysis of Mutation Operators for the Python Language, in: Springer International Publishing, Cham, 2014, ISBN: 978-3-319-07013-1, pp. 155–164, https://doi.org/10.1007/978-3-319-07013-1_15.
[375] E. Omar, S. Ghosh, D. Whitley, HOMAJ: a tool for higher order mutation testing in aspectJ and Java, in: Seventh IEEE International Conference on Software Testing, Verification and Validation, ICST 2014 Workshops Proceedings, March 31–April 4, 2014, Cleveland, OH, USA, 2014, pp. 165–170, https://doi.org/10.1109/ICSTW.2014.19.
[376] S. Mahajan, W.G.J. Halfond, Finding HTML presentation failures using image comparison techniques, in: I. Crnkovic, M. Chechik, P. Grünbacher (Eds.), ACM/IEEE International Conference on Automated Software Engineering, ASE '14, Vasteras, Sweden, September 15–19, 2014, ACM, 2014, pp. 91–96, https://doi.org/10.1145/2642937.2642966.
[377] K. El-Fakih, A. Simão, N. Jadoon, J.C. Maldonado, On studying the effectiveness of extended finite state machine based test selection criteria, in: Seventh IEEE International Conference on Software Testing, Verification and Validation, ICST 2014 Workshops Proceedings, March 31–April 4, 2014, Cleveland, OH, USA, IEEE Computer Society, 2014, pp. 222–229, https://doi.org/10.1109/ICSTW.2014.25.

[378] K. Maruchi, H. Shin, M. Sakai, MC/DC-like structural coverage criteria for function block diagrams, in: Seventh IEEE International Conference on Software Testing, Verification and Validation, ICST 2014 Workshops Proceedings, March 31–April 4, 2014, Cleveland, OH, USA, IEEE Computer Society, 2014, pp. 253–259, https://doi.org/10.1109/ICSTW.2014.27.
[379] T.A. Walsh, P. McMinn, G.M. Kapfhammer, Automatic detection of potential layout faults following changes to responsive web pages (N), in: M.B. Cohen, L. Grunske, M. Whalen (Eds.), 30th IEEE/ACM International Conference on Automated Software Engineering, ASE 2015, Lincoln, NE, USA, November 9–13, 2015, IEEE Computer Society, 2015, pp. 709–714, https://doi.org/10.1109/ASE.2015.31.
[380] R. Abreu, B. Hofer, A. Perez, F. Wotawa, Using constraints to diagnose faulty spreadsheets, Softw. Q. J. 23 (2) (2015) 297–322, https://doi.org/10.1007/s11219-014-9236-4.
[381] F. Belli, M. Beyazit, A.T. Endo, A.P. Mathur, A. da Silva Simão, Fault domain-based testing in imperfect situations: a heuristic approach and case studies, Softw. Q. J. 23 (3) (2015) 423–452, https://doi.org/10.1007/s11219-014-9242-6.
[382] S.C.P.F. Fabbri, M.E. Delamaro, J.C. Maldonado, P.C. Masiero, Mutation analysis testing for finite state machines, in: 5th International Symposium on Software Reliability Engineering, ISSRE 1994, Monterey, CA, USA, November 6–9, 1994, IEEE, 1994, pp. 220–229, https://doi.org/10.1109/ISSRE.1994.341378.
[383] A. Simao, A. Petrenko, J.C. Maldonado, Comparing finite state machine test coverage criteria, IET Softw, 3 (2) (2009) 91–105.
[384] J. Guan, J. Offutt, A model-based testing technique for component-based real-time embedded systems, in: Eighth IEEE International Conference on Software Testing, Verification and Validation, ICST 2015 Workshops, Graz, Austria, April 13–17, 2015, IEEE Computer Society, 2015, pp. 1–10, https://doi.org/10.1109/ICSTW.2015.7107407.
[385] F.S. Ocariza Jr, K. Pattabiraman, A. Mesbah, Detecting inconsistencies in JavaScript MVC applications, in: A. Bertolino, G. Canfora, S.G. Elbaum (Eds.), 37th IEEE/ACM International Conference on Software Engineering, ICSE 2015, Florence, Italy, May 16–24, 2015, vol. 1, IEEE Computer Society, 2015, pp. 325–335, https://doi.org/10.1109/ICSE.2015.52.
[386] W. Krenn, R. Schlick, S. Tiran, B. Aichernig, E. Jobstl, H. Brandl, MoMut::UML model-based mutation testing for UML, in: 2015 IEEE 8th International Conference on Software Testing, Verification and Validation (ICST), 2015, pp. 1–8, ISSN 2159–4848. https://doi.org/10.1109/ICST.2015.7102627.
[387] E.P. Enoiu, A. Causevic, D. Sundmark, P. Pettersson, A controlled experiment in testing of safety-critical embedded software, in: 2016 IEEE International Conference on Software Testing, Verification and Validation, ICST 2016, Chicago, IL, USA, April 11–15, 2016, IEEE Computer Society, 2016, pp. 1–11, https://doi.org/10.1109/ICST.2016.15.
[388] R. Matinnejad, S. Nejati, L.C. Briand, T. Bruckmann, Automated test suite generation for time-continuous simulink models, in: L.K. Dillon, W. Visser, L. Williams (Eds.), Proceedings of the 38th International Conference on Software Engineering, ICSE 2016, Austin, TX, USA, May 14–22, 2016, ACM, 2016, pp. 595–606, https://doi.org/10.1145/2884781.2884797.
[389] C. Henard, M. Papadakis, M. Harman, Y. Jia, Y.L. Traon, Comparing white-box and black-box test prioritization, in: L.K. Dillon, W. Visser, L. Williams (Eds.), Proceedings of the 38th International Conference on Software Engineering, ICSE 2016, Austin, TX, USA, May 14–22, 2016, ACM, 2016, pp. 523–534, https://doi.org/10.1145/2884781.2884791.

[390] X. Devroey, G. Perrouin, P.Y. Schobbens, P. Heymans, Poster: VIBeS, transition system mutation made easy, in: 2015 IEEE/ACM 37th IEEE International Conference on Software Engineering, ISSN 0270-5257, vol. 2, 2015, pp. 817–818, https://doi.org/10.1109/ICSE.2015.263.
[391] A. Parsai, A. Murgia, S. Demeyer, LittleDarwin: a feature-rich and extensible mutation testing framework for large and complex Java systems, in: Springer International Publishing, Cham, 2017, ISBN: 978-3-319-68972-2, pp. 148–163, https://doi.org/10.1007/978-3-319-68972-2_10.
[392] G. Rothermel, R.H. Untch, C. Chu, M.J. Harrold, Test case prioritization: an empirical study, in: 1999 International Conference on Software Maintenance, ICSM 1999, Oxford, England, UK, August 30–September 3, 1999, 1999, pp. 179–188, https://doi.org/10.1109/ICSM.1999.792604.
[393] A.S. Namin, S. Kakarla, The use of mutation in testing experiments and its sensitivity to external threats, in: Proceedings of the 20th International Symposium on Software Testing and Analysis, ISSTA 2011, Toronto, ON, Canada, July 17–21, 2011, 2011, pp. 342–352, https://doi.org/10.1145/2001420.2001461.
[394] M. Kintis, M. Papadakis, A. Papadopoulos, E. Valvis, N. Malevris, Analysing and comparing the effectiveness of mutation testing tools: a manual study, in: 16th IEEE International Working Conference on Source Code Analysis and Manipulation, SCAM 2016, Raleigh, NC, USA, October 2–3, 2016, 2016, pp. 147–156, https://doi.org/10.1109/SCAM.2016.28.
[395] R.V. Binder, Testing object-oriented systems: models, patterns, and tools, Addison-Wesley Longman Publishing Co,, Inc., Boston, MA, USA, 1999. ISBN: 0-201-80938-9.
[396] W. Visser, What makes killing a mutant hard, in: Proceedings of the 31st IEEE/ACM International Conference on Automated Software Engineering, ASE 2016, Singapore, September 3–7, 2016, 2016, pp. 39–44, https://doi.org/10.1145/2970276.2970345.
[397] P. Ammann, M.E. Delamaro, J. Offutt, Establishing theoretical minimal sets of mutants, in: Seventh IEEE International Conference on Software Testing, Verification and Validation, ICST 2014, March 31 2014–April 4, 2014, Cleveland, OH, USA, 2014, pp. 21–30, https://doi.org/10.1109/ICST.2014.13.
[398] B. Kurtz, P. Ammann, M.E. Delamaro, J. Offutt, L. Deng, Mutant subsumption graphs, in: Seventh IEEE International Conference on Software Testing, Verification and Validation, ICST 2014 Workshops Proceedings, March 31–April 4, 2014, Cleveland, OH, USA, 2014, pp. 176–185, https://doi.org/10.1109/ICSTW.2014.20.
[399] A. Arcuri, L. Briand, A practical guide for using statistical tests to assess randomized algorithms in software engineering, in: ICSE, 2011, ISBN: 978-1-4503-0445-0, pp. 1–10, https://doi.org/10.1145/1985793.1985795.
[400] M. Harman, P. McMinn, J.T. de Souza, S. Yoo, Search based software engineering: techniques, taxonomy, tutorial, in: Springer, Berlin, Heidelberg, 2012, ISBN: 978-3-642-25231-0, pp. 1–59, https://doi.org/10.1007/978-3-642-25231-0_1.
[401] M.E. Delamaro, J. Offutt, Assessing the influence of multiple test case selection on mutation experiments, in: Seventh IEEE International Conference on Software Testing, Verification and Validation, ICST 2014 Workshops Proceedings, March 31–April 4, 2014, Cleveland, OH, USA, 2014, pp. 171–175, https://doi.org/10.1109/ICSTW.2014.22.
[402] M. Kintis, M. Papadakis, A. Papadopoulos, E. Valvis, N. Malevris, Y. Le Traon, How effective mutation testing tools are? An empirical analysis of java mutation testing tools with manual analysis and real faults, Empir. Softw. Eng. (2017). https://doi.org/10.1007/s10664-017-9582-5 (accepted for publication).
[403] R. Gopinath, I. Ahmed, M.A. Alipour, C. Jensen, A. Groce, Does choice of mutation tool matter? Softw, Q. J. (2016) 1–50. ISSN: 1573-1367, https://doi.org/10.1007/s11219-016-9317-7.

[404] A. Márki, B. Lindström, Mutation tools for Java, in: Proceedings of the Symposium on Applied Computing, SAC 2017, Marrakech, Morocco, April 3–7, 2017, 2017, pp. 1364–1415, https://doi.org/10.1145/3019612.3019825.
[405] A. Arcuri, L.C. Briand, Adaptive random testing: an illusion of effectiveness? in: M,B. Dwyer, F. Tip (Eds.), Proceedings of the 20th International Symposium on Software Testing and Analysis, ISSTA 2011, Toronto, ON, Canada, July 17–21, 2011, ACM, 2011, pp. 265–275, https://doi.org/10.1145/2001420.2001452.

ABOUT THE AUTHORS

Mike Papadakis is a research scientist at the Interdisciplinary Centre for Security, Reliability and Trust (SnT) at the University of Luxembourg. He received a PhD diploma in Computer Science from the Athens University of Economics and Business. His research interests include software testing, static analysis, prediction modelling, mutation analysis, and search-based software engineering.

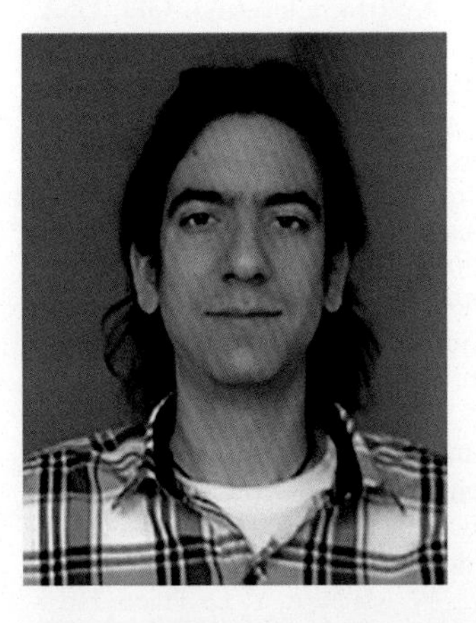

Marinos Kintis is a research associate at the Interdisciplinary Centre for Security, Reliability and Trust (SnT) at the University of Luxembourg. He received the PhD degree from the Department of Informatics of the Athens University of Economics and Business in 2016. The main topic of his dissertation was the introduction of effective techniques to ameliorate the adverse effects of the Equivalent Mutant Problem when testing software with Mutation. His main research interests include software testing, mutation testing, and program analysis. He was awarded a Best Paper Award at the 16th International Working Conference on Source Code Analysis and Manipulation (SCAM 2016) and co-organised the 13th International Workshop on Mutation Analysis (MUTATION 2018).

Jie Zhang is a final-year PhD candidate at the School of Electronics Engineering and Computer Science, Peking University, P.R. China, supervised by Lu Zhang. She is also a research associate in CREST, UCL, supervised by Earl Barr and Mark Harman. She has won the 2016 Fellowship at Microsoft Research Asia, the Top-ten Research Excellence Award of EECS, Peking University, the Lee Wai Wing Scholarship at Peking University, the 2015 National Scholarship, and so on. She served on the program committees of Mutation 2017 and Mutation 2018. Her major research interests are software testing, program analysis, end-user programming, and API mining.

Yue Jia is a lecturer in the Department of Computer Science at University College London. His research interests cover mutation testing, app store analysis, and search-based software engineering. Dr. Jia is director of MaJiCKe, an automated test data generation start up and also co-founder of Appredict, an app store analytics company, spun out from UCL's UCLappA group.

Yves Le Traon is professor at University of Luxembourg where he leads the SERVAL (SEcurity, Reasoning and VALidation) research team. His research interests within the group include (1) innovative testing and debugging techniques, (2) Android apps security and reliability using static code analysis, machine learning techniques and, (3) model-driven engineering with a focus on IoT and CPS. His reputation in the domain of software testing is acknowledged by the community. He has been General Chair of major conferences in the domain, such as the 2013 IEEE International Conference on Software Testing, Verification and Validation (ICST), and Program Chair of the 2016 IEEE International Conference on Software Quality, Reliability and Security (QRS). He serves at the editorial boards of several, internationally known journals (STVR, SoSym, IEEE Transactions on Reliability) and is author of more than 140 publications in international peer-reviewed conferences and journals.

Mark Harman is currently an engineering manager at Facebook and a professor of Software Engineering in the Department of Computer Science at University College London, where he directed the CREST centre for 10 years (2006–2017) and was Head of Software Systems Engineering (2012–2017). He is widely known for work on source code analysis, software testing, app store analysis, and Search Based Software Engineering (SBSE), a field he co-founded and which has grown rapidly to include over 1600 authors spread over more than 40 countries. His SBSE and testing work has been used by many organisations including Daimler, Ericsson, Google, Huawei, Microsoft, and Visa. Prof. Harman is a co-founder (and was co-director) of Appredict, an app store analytics company, spun out from UCL's UCLappA group, and was the chief scientific advisor to Majicke, an automated test data generation start up. In February 2017, he and the other two co-founders of Majicke (Yue Jia and Ke Mao) moved to Facebook, London, in order to develop their research and technology as part of Facebook.

Event-Based Concurrency: Applications, Abstractions, and Analyses

Aditya Kanade
Indian Institute of Science, Bengaluru, India

Contents

Abstract

Due to the increased emphasis on responsiveness, event-based design has become mainstream in software development. Software applications are required to maintain responsiveness even while performing multiple tasks simultaneously. This has resulted in the adoption of a combination of thread and event-based concurrency in modern software such as smartphone applications. In this chapter, we present the fundamental programming and semantic concepts in the combined concurrency model of threads and events. The paradigm of event-based concurrency cuts across programming

Advances in Computers, Volume 112
ISSN 0065-2458
https://doi.org/10.1016/bs.adcom.2017.12.006

languages and application frameworks. We give a flavor of event-driven programming in a few languages and application frameworks. The mix of threads and events complicates reasoning about correctness of applications under all possible interleavings. We discuss advances in the core concurrency analysis techniques for event-driven applications with focus on happens-before analysis, race detection, and model checking. We also survey other analysis techniques and related programming abstractions.

1. INTRODUCTION

The nature of computing has changed dramatically in recent years. Software applications are expected to perform multiple tasks simultaneously while maintaining responsiveness. For instance, a social networking application running on a smartphone must respond to user interactions even while downloading images or videos from the network. Smartphone applications operate in rich environments comprising touch interfaces, network connections, and sensors. These diverse components act as sources of events and applications are required to respond to events triggered by them in a timely manner. Users tend to perceive time lag beyond a few tens of milliseconds of execution time. On the Android platform, the system monitors responsiveness of each application and gives an option to the user to quit an application if it does not respond to an input event within a few seconds.

To meet such stringent latency requirements, software applications rely on a combined concurrency model of threads and events. While foreground threads respond to incoming events, background threads can perform time consuming or blocking computations. Even though both thread-based concurrency and single-threaded event handling are well known in the literature, concurrency models that combine the two have received the due attention only in recent times. The combined model poses unique challenges of its own. We refer to the combined concurrency models of threads and events simply as *event-based concurrency*.

The prevalence of event-based concurrency is not limited to smartphone applications [1]. In general, wherever an application needs to deal with a stream of events in a scalable manner, event-based design becomes necessary. Event-based concurrency is therefore used in embedded systems [2, 3], GUI-driven desktop applications [4], client-side web applications [5], and server-side frameworks [6–8], among others. Some of these applications, such as client-side JavaScript web applications, may use only a single thread; whereas others are multithreaded. The mechanisms for event scheduling may also differ among them. In some, the events are scheduled in the order of their arrival. In others, they are scheduled in a nondeterministic order. In many applications, only the runtime can schedule events; while others permit event handlers to pause themselves and schedule handlers of pending events before resuming again.

In this chapter, we explore the fundamental concepts in event-based concurrency. We introduce the programming abstractions used in development of event-driven applications (Section 2). In particular, we discuss the notions of events, event handlers, and event loops. In addition to the default event loop that a runtime controls, we also illustrate use of programmatic event loops. Programmatic event loops can be entered recursively or can be sequentially composed. These provide more freedom to the developer, but at the cost of increased complexity. We also explain how threads are used in conjunction with events. The paradigm of event-based concurrency cuts across programming languages and application frameworks. We give a flavor of event-driven programming in a few languages and frameworks through examples. Formalizing the scheduling mechanisms in event-based concurrency requires modeling both threads and event queues. An event queue receives the events that a thread is supposed to handle. We present the concurrency semantics of various event scheduling mechanisms popular in practice with emphasis on the concurrency semantics of Android applications (Section 3).

The flexibility of using threads and events brings with it the complexity of dealing with both thread and event interleavings. Similar to purely multithreaded applications (which do not have events), different threads can interleave with each other nondeterministically. In addition, the order in which events arrive at a thread may change across executions. In other words, starting with the same initial state, the behavior of an event-driven application can differ across multiple executions. As software applications are usually large in size, a manual inspection of code is insufficient to ensure that program state and intended semantics are maintained irrespective of scheduling nondeterminism. A number of program analysis techniques have therefore been designed to help developers identify and fix concurrency errors in event-driven applications. In this chapter, we focus on the core techniques of happens-before analysis, race detection, and model checking.

Data races are a common source of errors in concurrent applications. A data race (or race condition) occurs when two conflicting operations do not have a happens-before ordering between them. As a result, they could execute in different orders in different executions, leading to nondeterministic program outcome. The happens-before analysis [9] forms the basis of various concurrency analyses, including data race detection. We define the happens-before relation for Android applications in detail and juxtapose it with the happens-before relations of single-threaded event-driven applications (such as JavaScript programs) and programmatic event loops (Section 4). In contrast to pure multithreaded applications, the happens-before analysis of event-driven applications also involves inferring ordering among different event handlers running on the *same* thread.

An event-driven application receives a large number of events even in a short span of time. In order to perform happens-before analysis over long running executions, it is necessary to efficiently deal with large number of events. Vector clocks [10, 11] is a standard data structure for representing the happens-before relation of multithreaded applications. We survey recent advances that design optimizations to improve space as well as time requirement of happens-before analysis in Section 4.

Using the happens-before analysis, one can reason about a subset of executions of an application. To achieve a formal guarantee of correctness, an application must be analyzed with respect to all possible thread and event interleavings. We discuss model checking approaches designed for systematic state space exploration of event-driven applications. Any reasonable event-driven application has a large state-space which is impractical to store in memory. We discuss the class of stateless model checking algorithms which perform systematic state-space exploration without storing the states. The efficiency of stateless model checking depends on partial order reduction techniques which exploit equivalence of different executions to explore only a few representatives from every equivalence set of executions. We discuss the recent advances in model checking and partial order reduction for event-driven applications (Section 5).

To supplement the detailed discussion of programming abstractions and some of the core analysis techniques for event-driven applications, we also survey-related approaches (Section 6). In event-driven applications, an event is an asynchronous message which is handled at a later point of time. The core abstraction of asynchrony is gaining widespread use under different forms in numerous programming languages and application domains. We briefly discuss these related concepts in Section 6. We also outline future directions of research in the area of event-based concurrency.

2. PROGRAMMING ABSTRACTIONS

In this section, we introduce programming abstractions common in event-based concurrency. Starting with the basic concepts of events and event handlers, we present the scheduling mechanisms of event loops and use of multiple threads. Event-driven programming is supported by several programming languages and application frameworks. We give examples of event-driven programming in a few of them.

2.1 Events and Event Handlers

An *event* is an asynchronous message which consists of an identifier to indicate type of the event and additional data provided by the source of the event. In any modern system, there are numerous sources of events. Table 1 lists some sources of events in the Android environment and example events that they can trigger. In a web browser, events stem from user clicks, actions performed on the parsed representation of the HTML page (called the DOM representation) and responses to web requests. A source of an event posts an event to the recipient component *asynchronously*. That is, the source does not wait for the event to be handled by the recipient of the event. The event gets handled at some future point of time. Compare this to the synchronous method call semantics: when a caller P calls a method Q synchronously, the control transfers immediately to the callee Q until it returns the control back to the caller P. In the case of events, the event generation (analogous to method call) and event handling (analogous to method invocation) are separated in time. Event-driven programming is thus an asynchronous programming paradigm.

For each type of event, an application designates a method, called an *event handler*, which is invoked to handle events of that type. The event handlers are also called *callbacks* or *listeners*. In a sequential or multithreaded application, the control flow is determined by the application logic itself. In comparison, in an event-driven application, the control flow is largely determined by the sequence of events that the application receives. An application responds to the events it receives and thus, the control lies with the environment that triggers the events. This is commonly referred to as

Table 1 Sample Event Sources in the Android Environment

Event Source	**Example Events**
User	Touch, click, swipe, gesture
Network	3G or WiFi network status change
Lifecycle	Creation, pausing and resuming of source code components
System	System boot notification, battery status
Location	Location changes, location provider changes
Code	Intent to start a component, completion of an asynchronous task

inversion of control. The application code must be written in a way that it can work with any legitimate sequence of events.

A simple manifestation of event handling can be seen in HTML pages with JavaScript. Listing 1 shows an HTML element identified as `clktext`. The page uses JavaScript to handle a click event on the `clktext` element. The code of the event handler `myFun` is included inside the `<script>` tag and changes the color of the text contained in the `clktext` element. The handler is registered through the assignment `onclick = ''myFun()''`.

Listing 1 A JavaScript Event Handler That Changes Color of the Text When Clicked

```
<p id=''clktext'' onclick=''myFun()''>Change my color.</p>
<script>
function myFun() {
  document.getElementById(''clktext'').style.
color = ''blue'';
}
</script>
```

In addition to the events triggered by the environment, it is also possible to trigger events programmatically. This helps delegate some computation to the future. As stated in Table 1, starting a component in Android happens asynchronously through a type of event called *intent*. Listing 2 shows use of a timer event in JavaScript to schedule a computation at a later point of time. The event handler `myFun` for the click event of the `clktext` element posts a timer event with a delay of 10ms using the `setTimeout` API. The first argument to `setTimeout`, the `changeColor` method, is registered as the event handler for the timer event.

Listing 2 Timer Event in JavaScript

```
<p id=''clktext'' onclick=''myFun()''>Change my color.</p>
<script>
function myFun() {
  setTimeout(changeColor,10);
}
function changeColor() {
    document.getElementById(''clktext'').style.
color = ''blue'';
}
</script>
```

Events can be posted in different modes. Events can be posted with delays, as shown in Listing 2, or without delays. In Android, an event can be posted in a manner that the event is scheduled before any pending event, called *post to front-of-the-queue*. In some frameworks, events or event sources are assigned different priorities.

2.2 Event Loops

The main thread of an application receives the user interface events. In general, any application thread may be allowed to receive events, either from the environment or from other application threads. To receive events, a thread must be equipped with an *event queue*. The events are enqueued to an event queue in their order of arrival and dequeued in the same order by the thread associated with the queue. We assume that the event queues are unbounded in length. This means that no asynchronous enqueue operation can block. With bounded event queues, it is easy to see that an application can enter a deadlock if two threads block while trying to enqueue events to each others' queues that are already full.

2.2.1 Default Event Loops

An event loop is the basic scheduling mechanism in event-based concurrency. Listing 3 shows what an event loop looks like.

Listing 3 Prototypical Event Loop

```
while(!exit) {
    e = getEvent();
    handleEvent(e);
}
```

The event loop runs until the `exit` flag becomes `true`. Within each iteration, the loop dequeues an event `e` from the associated event queue using the method `getEvent` and invokes its handler through a call to `handleEvent`. An event handler running in the loop, for example, the handler of a shutdown event, can set the `exit` flag. This causes the event loop to terminate and no future events can be scheduled through it.

In most event-driven frameworks such as Android, the event loop is spun by the runtime. We refer to such an event loop as the *default event loop*. An event loop ensures that events are handled one after the other without any interleaving between their handlers.

2.2.2 Programmatic Event Loops

Some frameworks, such as the Qt graphical framework or the web browsers, permit an event handler to spin an event loop programmatically. This improves responsiveness in the cases when an event handler must wait for an input event from the user or the environment to happen. For example, printing a file is a common functionality in many image or text editing applications. The handler that responds to the print event (generated through the GUI or keyboard) lets the user select the printer and its configuration such as simplex or duplex printing, the pages and copies to print, and so on. The print event handler must wait for click of the "OK" or "Cancel" buttons before proceeding.

The print event handler can achieve this by running a programmatic event loop, wherein, it dequeues the pending events in the event queue and invokes their handlers synchronously. Listing 4 shows the use of a programmatic event loop.

Listing 4 Programmatic Event Loop Spun by an Event Handler

```
void PrintHandler(...) {
  // show the print dialog
  ...
  // wait for OK or Cancel button to be pressed
  // spin a programmatic event loop in the meantime
  while(!ok_pressed && !cancel_pressed) {
    e = getEvent();
    handleEvent(e);
  }
  // proceed with printing or cancellation
  ...
}
```

The handlers of "OK" or "Cancel" button clicks contain code to set the loop termination flags `ok_pressed` and `cancel_pressed`, respectively. Thus, the programmatic event loop continues to handle all pending events until one of these events is processed. Programmatic event loops are possible only if the underlying framework exposes the basic methods like `getEvent` and `handleEvent` to the application code. Unlike Qt and web browsers, the Android framework does not expose these primitives to the application code.

2.2.3 Recursive and Cascaded Event Loops

The programmatic event loops can lead to arbitrarily complex event scheduling. In particular, it is possible that an event handler running inside a programmatic event loop spins its own programmatic event loop. This gives rise to *recursion* of programmatic event loops. The event loops are terminated in the reverse order in which they are entered.

Another possibility is that an event handler runs two or more programmatic event loops sequentially. That is, after an event loop finishes, it can start another and so on. These are called *cascaded* event loops. Of course, one may see recursive and cascaded event loops interleaved with each other. Thus, writing code with programmatic event loops is more challenging than doing so with the default event loop. At the same time, it offers more flexibility to the developer.

2.3 Multithreading

In an event-driven application, events are handled on a thread one after the other without preemption. That is, the handler of an event is scheduled only after the previous handler finishes computation. This is true even in the presence of programmatic event loops except that a handler can pause itself to run an event loop, but after the loop exits, the control returns to the paused handler. A limitation of this design is that if a handler blocks a thread then the pending events on the thread are also blocked. Thus, to maintain responsiveness, event handlers should perform only nonblocking and short-running computations. Multithreading can be exploited to offload time-consuming or potentially blocking operations from event-handling threads to other threads.

Listing 5 shows use of multithreading in an Android Activity class `MainActivity`. An Activity class provides the user interface (UI) with which a user can interact. The Android system manages the lifecycle of Activity classes through *lifecycle callbacks*. Lifecycle callbacks are methods declared in the `Activity` class such as `onCreate` and `onStop` methods shown in Listing 5. The Android system calls these methods asynchronously according to a predefined lifecycle state machine. The `onCreate` method is called when the Activity is created, whereas `onStop` is called when the Activity is no longer visible on the screen.

Typically, an Activity creates different UI widgets to populate the screen and registers callbacks for events on the widgets. In this example, we focus only on the multithreading aspect. Suppose the Activity needs to perform a time-consuming computation whose result is to be stored in the field `X`. However, such a time-consuming operation if performed in a lifecycle callback can make the application unresponsive. The Android system can prompt the user to quit the application in that case. The `onCreate` method therefore spawns a thread `t` which performs the time-consuming computation. The same strategy should be employed to delegate blocking operations, such as synchronous download of data from the network, to background threads. Android provides the `AsyncTask` class to simplify such requirements. For brevity, we use a plain thread here.

Listing 5 An Android Activity Class Which Uses Multiple Threads

```
public class MainActivity extends Activity {
  int X;
  protected void onCreate( ){
    Runnable r = new Runnable( ){
      public void run( ){
        X = ..; // perform a time-consuming computation
        runOnUiThread(new Runnable( ){
          public void run( ){
            Log.d(''MainActivity'', ''X: ''+ X);
          }
        });
      }
    };
    Thread t = new Thread(r);
    t.start( );
    X = 1;
  }
  protected void onStop( ){
    X = -1;
  }
}
```

After the thread `t` finishes the computation, it sends an event with a runnable method using `runOnUiThread` API. The `run` method logs value of the field `X`. The `onStop` callback method contains another assignment to `X`. We use this example in Section 4.1.3 to demonstrate different thread and event interleavings which cause data races over the field `X`.

3. CONCURRENCY SEMANTICS

Having introduced the key programming abstractions of event-based concurrency, we now discuss the concurrency semantics. Android is one of the most popular programming platforms today. As seen in Section 2.3, Android applications combine multithreading with event handling. As a concrete case study, we formalize the semantics of event-driven Android applications. Our presentation is based on the semantics formalized in [12]. We contrast the semantics of single-threaded JavaScript applications and applications with programmatic event loops with that of Android applications.

3.1 Android Concurrency

Android is a full-fledged object-oriented language based on Java and provides many concurrency constructs. We simplify the presentation by introducing some low-level primitives that capture the essence of the concurrency behavior of Android applications. We list them as follows:

- *Thread management*: The start and end of a thread `t` are, respectively, denoted by `threadinit(t)` and `threadexit(t)`. The primitive `fork(t,t′)` indicates creation of a thread `t′` by another thread `t`. When a thread `t` consumes a thread `t′` which has finished its execution, we get `join(t,t′)`.
- *Synchronization and memory accesses*: Threads can synchronize with each other using locks. The primitives `acquire(t,l)` and `release(t,l)` show when a thread `t` acquires a lock `l` and when it releases it. We are not interested in the actual values of memory locations. To detect conflicting accesses, we only record whether an access to a memory location `m` by a thread `t` is `read(t,m)` or `write(t,m)` which, respectively, denote read and write accesses.
- *Event processing*: The primitive `loopOnQ(t)` indicates that the thread `t` will start processing events from its event queue. The primitive `post(t,e,t′)` indicates that a thread `t` has posted an event `e` to a thread `t′`. The first and last instructions of the handler of an event `e` executing on a thread `t` are given by `begin(t,e)` and `end(t,e)`.

Fig. 1 shows a partial execution trace of the Android application in Listing 5 in terms of the primitives introduced above. We do not explicitly show the `threadinit`, `threadexit`, and `loopOnQ` operations. The flow of control in a thread is indicated by a dashed vertical line. The operations are executed in the order of their occurrence from top to bottom. The operations executed in an event handler are enclosed in a rectangular box. In this trace, the main thread `mt` first executes the `onCreate` lifecycle callback. The corresponding event is posted by the Android system process which we elide for simplicity. The `onCreate` method forks the thread `t` and then assigns 1 to

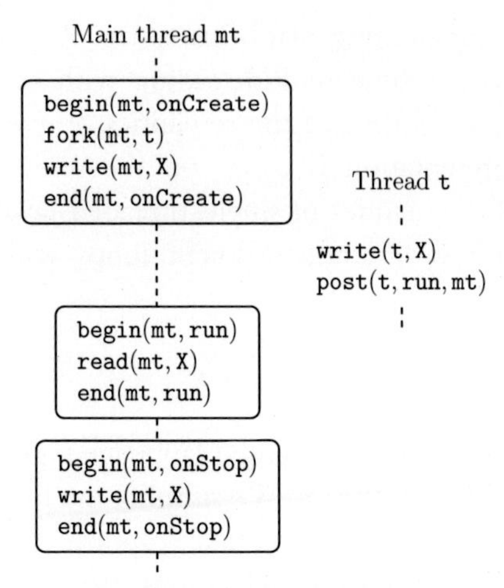

Fig. 1 A partial execution trace of the Android application in Listing 5.

the field `X`. The latter is simply recorded as a `write` operation in the trace. The thread `t` then performs the designated computation and stores the result in `X`. It then posts `run` method to the main thread. The main thread executes the `run` method which logs the value of `X`. This operation is recorded as a `read` operation in the trace. Finally, suppose the user presses the back button due to which the activity goes to background. This results in posting of the `onStop` lifecycle callback to the main thread by the Android system process. The `onStop` callback also writes to `X`. For ease of presentation, we assume unique identifiers for all threads and events. Since we are only interested in the concurrency semantics, we ignore the data associated with an event.

We now give the concurrency semantics of Android applications in terms of the primitives. Let *Meths* be the set of methods defined by an Android application A. Let *Threads* be the set of threads created by the Android runtime for the application by default. The application can dynamically create more threads. In the beginning, we assume that each thread is assigned some method to run. We represent this mapping by *Init*. We now define the *runtime state* of the application A as $\sigma = (\mathcal{C}, \mathcal{R}, \mathcal{F}, \mathcal{E}, \mathcal{Q}, \mathcal{L})$ where

- $\mathcal{C}$, $\mathcal{R}$, and $\mathcal{F}$ are mutually disjoint sets of thread-ids. They, respectively, denote the sets of threads that are created but not yet scheduled, threads that are running and threads that have finished execution.
- The function $\mathcal{E}$ maps each running thread to a method in *Meths* that the thread is currently executing or to an event that it is currently handling.

- The function $\mathcal{Q}$ maps each thread to the event queue associated with it. An event queue Q is an object that supports enqueue and dequeue operations, denoted, respectively, by $\oplus$ and $\ominus$. As noted before, the event queues have unbounded capacity so that no enqueue operation can ever block. For simplicity, we assume that each thread has an associated event queue.
- The function $\mathcal{L}$ records the set of locks held by a thread.

We present the concurrency semantics of Android applications by formalizing how the primitive operations affect the state of the application. Fig. 2 gives inference rules for this. Each rule consists of an antecedent and a

$$(\textsc{Start})\ \frac{\mathcal{C} \text{ is empty}}{\begin{array}{c}\mathcal{C} \leftarrow \{\mathtt{t} \mid \mathtt{t} \in \mathit{Threads}\} \quad \mathcal{R} \leftarrow \emptyset \quad \mathcal{F} \leftarrow \emptyset \quad \mathcal{E} \leftarrow \mathit{Init} \\ \mathcal{Q}(\mathtt{t}) \leftarrow \epsilon, \mathcal{L}(\mathtt{t}) \leftarrow \emptyset \text{ for all } \mathtt{t} \in \mathit{Threads}\end{array}}$$

$$(\textsc{Init})\ \frac{\alpha = \mathtt{threadinit(t)} \quad \mathtt{t} \in \mathcal{C}}{\mathcal{C} \leftarrow \mathcal{C} \setminus \{\mathtt{t}\} \quad \mathcal{R} \leftarrow \mathcal{R} \cup \{\mathtt{t}\}}$$

$$(\textsc{Exit})\ \frac{\alpha = \mathtt{threadexit(t)} \quad \mathtt{t} \in \mathcal{R}}{\mathcal{R} \leftarrow \mathcal{R} \setminus \{\mathtt{t}\} \quad \mathcal{F} \leftarrow \mathcal{F} \cup \{\mathtt{t}\}}$$

$$(\textsc{Fork})\ \frac{\alpha = \mathtt{fork(t, t')} \quad \mathtt{t} \in \mathcal{R} \quad \mathtt{t'} \text{ is a fresh thread-id}}{\mathcal{C} \leftarrow \mathcal{C} \cup \{\mathtt{t'}\} \quad \mathcal{E}(\mathtt{t'}) \leftarrow \mathtt{main} \quad \mathcal{Q}(\mathtt{t'}) \leftarrow \epsilon \quad \mathcal{L}(\mathtt{t'}) \leftarrow \emptyset}$$

$$(\textsc{Join})\ \frac{\alpha = \mathtt{join(t, t')} \quad \mathtt{t} \in \mathcal{R} \quad \mathtt{t'} \in \mathcal{F}}{}$$

$$(\textsc{Acquire})\ \frac{\alpha = \mathtt{acquire(t, l)} \quad \mathtt{t} \in \mathcal{R} \quad \mathtt{l} \notin \mathcal{L}(\mathtt{t'}) \text{ for any } \mathtt{t'} \in \mathcal{R}}{\mathcal{L}(\mathtt{t}) \leftarrow \mathcal{L}(\mathtt{t}) \cup \{\mathtt{l}\}}$$

$$(\textsc{Release})\ \frac{\alpha = \mathtt{release(t, l)} \quad \mathtt{t} \in \mathcal{R} \quad \mathtt{l} \in \mathcal{L}(\mathtt{t})}{\mathcal{L}(\mathtt{t}) \leftarrow \mathcal{L}(\mathtt{t}) \setminus \{\mathtt{l}\}}$$

$$(\textsc{Post})\ \frac{\alpha = \mathtt{post(t, e, t')} \quad \mathtt{t}, \mathtt{t'} \in \mathcal{R}}{\mathcal{Q}(\mathtt{t'}) \leftarrow \mathcal{Q}(\mathtt{t'}) \oplus \mathsf{e}}$$

$$(\textsc{LoopOnQ})\ \frac{\alpha = \mathtt{loopOnQ(t)} \quad \mathtt{t} \in \mathcal{R} \quad \mathtt{t} \text{ has not started handling events}}{\mathcal{E}(\mathtt{t}) \leftarrow \perp}$$

$$(\textsc{Begin})\ \frac{\alpha = \mathsf{begin}(\mathtt{t}, \mathsf{e}) \quad \mathtt{t} \in \mathcal{R} \quad \mathcal{E}(\mathtt{t}) = \perp \quad \mathsf{e} = \mathsf{Front}(\mathcal{Q}(\mathtt{t}))}{\mathcal{Q}(\mathtt{t}) \leftarrow \mathcal{Q}(\mathtt{t}) \ominus \mathsf{e} \quad \mathcal{E}(\mathtt{t}) \leftarrow \mathsf{e}}$$

$$(\textsc{End})\ \frac{\alpha = \mathsf{end}(\mathtt{t}, \mathsf{e}) \quad \mathtt{t} \in \mathcal{R} \quad \mathcal{E}(\mathtt{t}) = \mathsf{e}}{\mathcal{E}(\mathtt{t}) \leftarrow \perp}$$

$$(\textsc{Sequencing})\ \frac{\sigma \xrightarrow{\alpha} \sigma' \quad \sigma' \xrightarrow{\beta} \sigma''}{\sigma \xrightarrow{\alpha;\beta} \sigma''}$$

Fig. 2 Concurrency semantics of Android applications.

consequent, separated by a horizontal line. The meaning of each rule is that if the antecedent holds in a state σ then the successor state σ' is obtained by applying the transformations in the consequent to σ.

The rule START applies in the beginning when the set $\mathcal{C}$ of threads that are created is empty. The rule initializes the set $\mathcal{C}$ to the set *Threads* of threads created by the Android runtime and assigns the *Init* map to $\mathcal{E}$. The sets of running and finished threads are initialized to the empty set. The event queues and locksets of all threads in *Threads* are set to ϵ and the empty set, respectively. The symbol ϵ stands for the empty queue. We refer to the resulting state as the initial state of the application.

For the purposes of semantics, we assume that only one thread executes at any point of time and operations from different threads execute in an interleaved fashion. Let α be the current operation being executed by a thread `t`. The rules INIT and EXIT model the start and end of the thread `t`. They, respectively, move `t` from the set $\mathcal{C}$ to the set $\mathcal{R}$ and from the set $\mathcal{R}$ to the set $\mathcal{F}$. As stated in the rule FORK, when a new thread is forked, it is assigned a fresh thread-id `t'` which is included in the set $\mathcal{C}$. Let `main` be the name of the default method that gets executed on a new thread. The rule FORK initializes $\mathcal{E}(\mathtt{t'})$, $\mathcal{Q}(\mathtt{t'})$, and $\mathcal{L}(\mathtt{t'})$ as shown. The rule JOIN joins a running thread `t` with a finished thread `t'`. The empty consequent means that the state σ of the application (as modeled by us) does not change, except that the thread `t` makes progress. Threads can synchronize with each other through locks. If a lock `l` is not in $\mathcal{L}(\mathtt{t'})$ for any thread $\mathtt{t'} \in \mathcal{R}$, then as per the rule ACQUIRE, the thread `t` can acquire the lock `l`. This is then recorded in $\mathcal{L}(\mathtt{t})$ by adding `l` to it. After a thread `t` has acquired a lock `l`, it can release it through the `release` operation. The rule RELEASE formalizes this by removing `l` from $\mathcal{L}(\mathtt{t})$. Note that Android permits reentrant locks, but we do not model them here.

When a thread `t` posts an event `e` to a thread `t'`, the event `e` is added to the queue $\mathcal{Q}(\mathtt{t'})$ as per the rule POST. In order to start processing events from its event queue, a thread `t` executes the `loopOnQ` operation. According to the rule LOOPONQ, this sets $\mathcal{E}(\mathtt{t})$ to $\perp$. A thread `t` is said to be idle when $\mathcal{E}(\mathtt{t}) = \perp$. Suppose a thread `t` is idle and `e` is the first event in its event queue. Then, the thread can execute the event handler of `e` as shown in the rule BEGIN where `Front` returns the first element of a queue. This results in dequeuing of `e` from $\mathcal{Q}(\mathtt{t})$ and assigning `e` to $\mathcal{E}(\mathtt{t})$. As stated in the rule END, once a thread executes `end` operation of the handler of an event `e`, the thread becomes idle again. Finally, the rule SEQUENCING states that if

σ transitions to σ' on executing α and σ' transitions to σ'' on executing β then the sequential execution of the two operations takes the application from σ to σ''.

This set of rules defines the concurrency semantics of Android applications concisely and precisely, and suffices for our purposes. We have skipped some primitives and details. For a complete treatment, we refer the reader to [12]. These semantics inform design of the happens-before relation we discuss in Section 4. An execution trace ρ of an Android application (in terms of the primitives defined above) is a *valid trace* if starting from the initial state, each of its transitions is feasible according to the rules in Fig. 2. These rules do not model the sequential control flow within each thread explicitly. It is easy to incorporate it by modeling the program counter and call stack.

3.2 Single-Threaded Event Handling

Some event-driven applications are single-threaded. For instance, JavaScript code embedded in HTML pages runs on a single thread. The concurrency semantics of such programs is straightforward. Among the rules in Fig. 2, only POST, BEGIN, END, and SEQUENCING rules apply. If the language or application framework does not guarantee that the events are handled in the same order in which they are received then the rule BEGIN changes. Instead of selecting the event at the front of the queue, it can select any event from the queue. In other words, the event queue, instead of an ordered sequence, is treated as a multiset.

3.3 Programmatic Event Loops

Modeling concurrency semantics of programmatic event loops requires substantial extensions to the rules in Fig. 2. When a programmatic event loop is started, the event handler that starts it pauses itself. As seen in Listing 4, the programmatic event loop is controlled by a Boolean variable (or more generally, by a logical expression over a set of Boolean variables). The loop terminates when some handler running in the loop resets the variable. Once the loop terminates, the control returns to the paused handler. To model the concurrency semantics, it is necessary to extend the set of primitives introduced in Section 3.1 by adding primitives that model pause, resume, and reset operations of programmatic event loops.

As discussed in Section 2.2.3, programmatic event loops can be entered recursively and they are exited in the reverse order to that in which they are entered. This aspect can be modeled by maintaining a stack of frames for each thread with one frame per active programmatic event loop running on the thread. The pause operation pushes a frame object on the stack and the resume operation pops a frame object from it. We refer the reader to [13, 14] for detailed exposition of programmatic event loops.

4. HAPPENS-BEFORE ANALYSIS AND RACE DETECTION

In a multithreaded program, operations of different threads are ordered only partially, which causes nondeterministic scheduling. Different linearizations of the partial order give rise to different schedules. In event-driven applications, the presence of events and event loops further adds to nondeterminism. Analyzing the partial order of operations is therefore a key to identifying nondeterministic concurrency errors. The partial order is formally modeled by the happens-before relation [9]. In this section, we discuss happens-before relations for event-driven applications and describe their use in race detection. We also discuss efficient computation of happens-before relation of event-driven applications.

4.1 Happens-Before Relation for Android Applications

In a pure multithreaded program, even though threads interleave with each other, each thread executes a sequential piece of code. Thus, though operations of different threads are only partially ordered, those belonging to the same thread are totally ordered. The latter observation does not hold for threads which handle events. Operations within each event handler are totally ordered, but two different handlers running on the same thread may not be. If the events are reordered, their handlers too get reordered.

To capture the happens-before relation of Android applications, we define a multithreaded happens-before relation $\preceq_{mt}$ and a single-threaded happens-before relation $\preceq_{st}$. These two together characterize the complete happens-before relation $\preceq$. Our presentation of the happens-before relations is based on [12].

4.1.1 Ordering Among Operations of Different Threads

The relation $\preceq_{mt}$ is the smallest relation closed under the rules in Fig. 3. In these rules, threads `t` and `t′` are two distinct threads. The rule POST-MT states that the `begin` operation of the handler of an event `e` happens after the corresponding `post` operation. Note that the `post` operation α_i executes on thread `t′` and posts the event `e` to another thread `t`. The happens-before

$$(\textsc{Post-mt})\ \frac{\alpha_i = \texttt{post(t', e, t)} \qquad \alpha_j = \texttt{begin(t, e)}}{\alpha_i \preceq_{mt} \alpha_j}$$

$$(\textsc{Fork})\ \frac{\alpha_i = \texttt{fork(t, t')} \qquad \alpha_j = \texttt{threadinit(t')}}{\alpha_i \preceq_{mt} \alpha_j}$$

$$(\textsc{Join})\ \frac{\alpha_i = \texttt{threadexit(t')} \qquad \alpha_j = \texttt{join(t, t')}}{\alpha_i \preceq_{mt} \alpha_j}$$

$$(\textsc{Lock})\ \frac{\alpha_i = \texttt{release(t, l)} \qquad \alpha_j = \texttt{acquire(t', l)}}{\alpha_i \preceq_{mt} \alpha_j}$$

$$(\textsc{Trans-mt})\ \frac{\alpha_i \preceq \alpha_k \qquad \alpha_k \preceq \alpha_j \qquad \alpha_i \text{ and } \alpha_j \text{ belong to different threads}}{\alpha_i \preceq_{mt} \alpha_j}$$

Fig. 3 Multithreaded happens-before rules.

rules are well-grounded in the concurrency semantics discussed in Section 3.1. It can be observed that the `begin` operation of the handler of an event `e` is feasible on a thread `t` only if the thread is idle and `e` is at the front of the event queue of thread `t`. The event `e` can only be added into the event queue by a prior `post` operation. Thus, it is sound to infer a happens-before order between a `post` operation and the matching `begin` operation. The rest of the rules in Fig. 3 are same as those for purely multithreaded programs. The rules Fork and Join, respectively, state that a `fork` operation happens before the matching `threadinit` operation and a `threadexit` operation happens before the matching `join` operation.

The rule Lock states that a lock acquire can happen only after the lock is released. Unlike other rules, this rule induces a may-happens-before relation. Strictly speaking, locks ensure mutual exclusion and not ordering. It is not necessary that if a thread `t` acquires a lock `l` before another thread `t'` in one interleaving, then it will do so in all interleavings. Some race detection techniques choose not to use this rule as it can give false negatives and use alternate techniques like locksets [15]. Finally, the relation $\preceq_{mt}$ is transitively closed as per the rule Trans-mt.

4.1.2 Ordering Among Operations of the Same Thread

We now discuss the rules in Fig. 4 which order operations on the same thread. An event handler running on a thread `t` can post another event to the same thread. The rule Post-st captures the ordering in such a case. It is analogous to the rule Post-mt in Fig. 3 but for operations running on the same thread. The rule Fifo captures the FIFO semantics of the event

$$(\text{POST-ST}) \frac{\alpha_i = \texttt{post(t,e,t)} \qquad \alpha_j = \texttt{begin(t,e)}}{\alpha_i \preceq_{st} \alpha_j}$$

$$(\text{FIFO}) \frac{\alpha_i = \texttt{end(t,e}_1\texttt{)} \qquad \alpha_j = \texttt{begin(t,e}_2\texttt{)} \qquad \texttt{post(\_,e}_1\texttt{,t)} \preceq \texttt{post(\_,e}_2\texttt{,t)}}{\alpha_i \preceq_{st} \alpha_j}$$

$$(\text{NOPRE}) \frac{\alpha_i = \texttt{end(t,e}_1\texttt{)} \qquad \alpha_j = \texttt{begin(t,e}_2\texttt{)} \qquad \texttt{begin(t,e}_1\texttt{)} \preceq_{st} \texttt{end(t,e}_2\texttt{)}}{\alpha_i \preceq_{st} \alpha_j}$$

$$(\text{PO}) \frac{\begin{array}{c}\alpha_i \text{ and } \alpha_j \text{ occur before the thread starts to handle events or belong to the same handler} \\ \alpha_i \text{ occurs before } \alpha_j\end{array}}{\alpha_i \preceq_{st} \alpha_j}$$

$$(\text{TRANS-ST}) \frac{\alpha_i \preceq_{st} \alpha_k \qquad \alpha_k \preceq_{st} \alpha_j}{\alpha_i \preceq_{st} \alpha_j}$$

Fig. 4 Thread-local happens-before rules.

queue. Consider two events e_1 and e_2 posted to the same thread t. If their post operations are ordered by the happens-before relation $\preceq$ then e_2 is dequeued only after the handler of e_1 finishes execution. This is because the event queue dequeues events in their order of arrival. We denote the do not care values by symbol '_' in the rules. As noted before, events can be posted in different modes such as with delays and posting to front of the event queue. We have considered the most common and simple case when events are posted with zero delay. The FIFO rule gives the event ordering in that case. For other extensions, we refer the reader to [12, 16–18].

According to the rule BEGIN in Fig. 2, an event handler runs to completion on a thread before another event handler can be scheduled on the same thread. The happens-before order arising from this restriction of non-preemption of event handlers is captured in the rule NOPRE in Fig. 4. Let α_i and α_j, respectively, be the end operation of an event e_1 and the begin operation of an event e_2. Both e_1 and e_2 are handled on the same thread t. If e_1 is dequeued before any operation of e_2, including the last operation of its event handler end(t,e_2), then $\alpha_i \preceq_{st} \alpha_j$ [16]. The code executed by a thread before it starts handling events from its event queue is run sequentially. Any two operations executed in this sequential code (that occurs before loopOnQ) are ordered by their order of occurrence, as stated in the rule PO. Further, by the rule PO, all operations of an event handler are ordered by their order of occurrence. The rule PO is referred to as the program order rule. Finally, the relation $\preceq_{st}$ is transitively closed as per the rule TRANS-ST.

4.1.3 Happens-Before-Based Race Detection

If two operations α_i and α_j access the same memory location such that at least one of them is a write operation then they are said to be *conflicting*. If there is no happens-before order between a pair of operations then the operations can occur in different orders in different interleavings. If this is the case for a pair of conflicting operations (α_i, α_j) then we say that there is a *data race* between α_i and α_j. Data races are usually indicative of concurrency errors and result in nondeterministic outcomes even for the same program input. Data race detection is therefore a very active research topic. In recent times, several techniques for detection [12, 16, 17] and verification of harmfulness [19] of data races have been proposed for Android applications.

To detect data races dynamically, we analyze an execution trace ρ of an application. For each pair of conflicting operations in ρ, we check whether they are ordered by the happens-before relation $\preceq$ presented above. Let us consider the execution trace from Fig. 1. By the multithreaded happens-before rules of Fig. 3, the `fork` operation in the `onCreate` callback happens before the `write` operation of thread `t`. In the trace, the `write` operation to `X` in the `onCreate` callback occurs before the `write` operation of thread `t`. However, as they do not have a happens-before ordering between them, in another interleaving, the two can be reordered. Fig. 5A shows such a trace. This is an example of a multithreaded race in an Android application.

Android applications can also exhibit single-threaded data races. In Fig. 1, the `run` method is posted programmatically by the thread `t`, whereas the `onStop` callback is posted by the environment (not shown explicitly in the trace). These two `post` operations do not have a happens-before ordering and consequently, their handlers also do not have any fixed order of

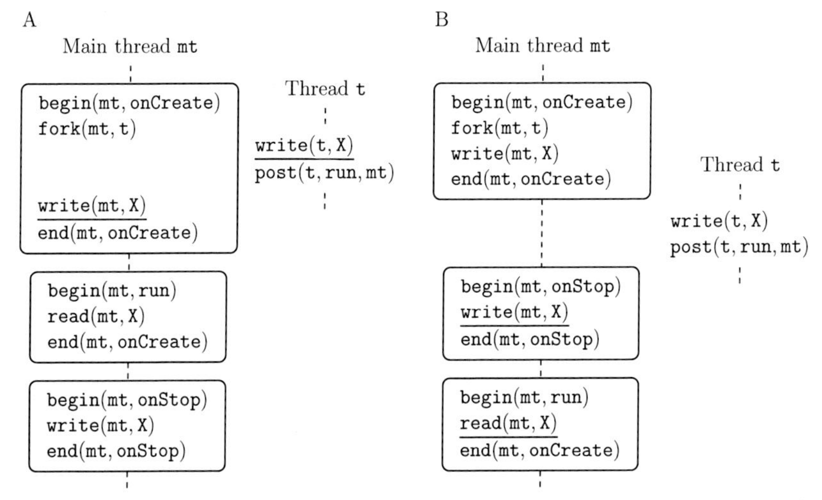

Fig. 5 Examples of multithreaded and single-threaded data races. (A) A multithreaded data race. (B) A single-threaded data race.

execution. The two handlers both access the field `X` with one of them being a `write` operation. The trace in Fig. 5B shows an alternate interleaving of the trace in Fig. 1 in which the execution order of `onStop` and `run` methods is changed. This exhibits a data race between two conflicting accesses made by the same thread. In a purely multithreaded program, the operations of a single thread are totally ordered. Therefore, they do not exhibit single-threaded races. In event-driven applications, as different event handlers running on the same thread may not be totally ordered, we may encounter single-threaded races.

4.2 Single-Threaded Event Handling

The happens-before analysis of single-threaded event-driven applications is characterized by the rules in Fig. 4. JavaScript web applications form a prominent class of single-threaded event-driven applications. JavaScript does not guarantee the FIFO semantics. That is, events may be dequeued in a nondeterministic order from the event queue. In such cases, the FIFO rule of Fig. 4 is not applicable.

In a webpage with JavaScript, the asynchronous execution of JavaScript code may interleave nondeterministically with the parsing and loading of HTML elements by the web browser, and user actions. The script element inside the `<script>` tag in Listing 1 is an inline script which is executed synchronously when the browser parses it. To improve responsiveness, the script elements could be deferred until the whole webpage is loaded or can be scheduled to execute asynchronously. JavaScript provides means to configure this. HTML itself is a rich markup language and supports a variety of elements. The web browser triggers various events to indicate the progress in the loading of the contents of a webpage and JavaScript code can be attached to handle them. Most modern web applications use asynchronous JavaScript with XML (AJAX) to communicate asynchronously with the network. Callbacks are registered so that the page responds when the network data arrives. An additional source of nondeterminism is the actions by the user. Therefore, the happens-before analysis of JavaScript web applications requires analyzing the order in which HTML elements are processed by the web browser, the asynchronous execution of JavaScript code and order of user (and other environment) events.

Apart from data races over JavaScript variables, races in web applications manifest in several different ways. For example, an event may be triggered before the event handler registered for it is loaded. The event is permanently lost in such a case. Another common case is that a webpage tries to initialize contents of a text-input field programmatically. However, as soon as the field is visible, the user can enter text. If the code to initialize the field runs after it, it can overwrite the text entered by the user, causing inconvenience to the user. The happens-before analysis and race detection of JavaScript web applications is studied in [20]. The single-threaded nature of JavaScript makes it difficult to synchronize across different code fragments that may execute asynchronously. Developers resort to use of ad hoc synchronization through shared variables. EventRacer [21] defines a notion of covering races to distinguish between races over synchronization variables from those over variables protected through synchronization. A recent work [22] addresses the problem of repair of races in JavaScript web applications. It requires a repair policy which states when certain events should be postponed or discarded. Given a repair policy, the JavaScript code is instrumented with an event controller which enforces the repair policy at runtime.

4.3 Programmatic Event Loops

As discussed so far, event-driven applications can exhibit both multithreaded data races and data races between two handlers executing on the same thread. The cause for the former is nondeterministic thread interleavings and for the latter, it is nondeterministic order in which events are handled on a thread. With only the runtime spinning the default event loop, the event handlers running on the same thread can only be reordered in their entirety. In the presence of programmatic event loops, event handlers on the same thread can actually interleave. In particular, an event handler running inside a programmatic event loop can interfere with the paused event handler that is spinning the programmatic event loop, leading to a data race. For instance, suppose a handler allocates an object and spins a programmatic event loop. If another handler runs inside the loop and frees the object, the application crashes when the paused handler resumes and tries to access the object. If the handler that frees the object does not run in the programmatic event loop, the application runs fine.

The problem of happens-before analysis and race detection in the presence of programmatic event loops is studied in [13, 14]. In Section 3.3, we discussed the pause, reset and resume primitives. The happens-before analysis of programmatic event loops gives rules that establish happens-before ordering between the pause and resume operations, as well as pause-reset and reset-resume pairs. The FIFO rule in Fig. 4 deals only with event scheduling through the default event loop. This rule needs to be instantiated for programmatic event loops also. Consider an event handler H which resets the control variable of a programmatic event loop. If another event handler G happens-before H then G may run before H within the programmatic event loop. The happens-before rules in [13, 14] also analyze cascaded and recursive event loops, and thread interleavings.

4.4 Efficient Computation of Happens-Before Relation

We have seen the rules that govern the happens-before relation in event-based concurrent applications. The efficiency of computation of the happens-before relation depends on the choice of data structures and algorithms used for implementing the happens-before analysis. We discuss two main representations, graphs and vector clocks, that have been explored in the literature for happens-before analysis. The evaluation of event-ordering rules has been identified as a bottleneck in happens-before analysis of event-driven applications. We discuss the recent advances that aim at speeding it up.

4.4.1 Graph Representations

Detecting data races requires computing the happens-before relation. Several approaches [12, 13, 16, 20] used graph-based representations of the happens-before relation. In a graph-based representation, nodes of the graph are individual operations and the directed edges indicate happens-before ordering between them. To compute the happens-before relation, the happens-before rules are applied until no more edges can be added to the graph. To reduce the number of nodes, DroidRacer [12] represents a contiguous sequence of memory operations without intervening synchronization operations by a single node. SparseRacer [13, 14] exploits the structure and semantics of traces of applications with programmatic event loops to define a sparse representation of the happens-before relation. The sparse representation identifies blocks of operations that execute atomically with respect to other blocks on the same thread. Instead of representing the ordering between every pair of operations, it suffices to represent the ordering between every pair of blocks.

4.4.2 Vector Clocks

As the happens-before relation is transitively closed, graph-based representations of the relation need to saturate the edges transitively. Transitive closure takes subcubic time in the number of nodes. This limits scalability of the happens-before analysis over long traces. Vector clocks [10, 11] is a popular data structure which helps overcome the cost of transitive closure. A *vector clock* consists of a vector of logical clocks, one per thread or event. Let D be the set of threads and events. A vector clock VC is a map from $D \rightarrow \mathbb{N}$. Given vector clocks VC_1 and VC_2, (1) $VC_1 \sqsubseteq VC_2$ iff $VC_1(x) \leq VC_2(x)$ for all $x \in D$ and (2) $VC_1 \sqcup VC_2 = \lambda x \cdot max(VC_1(x), VC_2(x))$. The former defines the partial order over vector clocks and the latter defines the join operation. Consider an operation α in a trace ρ. Let α execute as part of a thread or event d. The vector clock of the operation α is computed by taking join of all operations that happen before it and then incrementing value of the clock corresponding to d. Let VC_1 and VC_2 be vector clocks of operations α_1 and α_2. The happens-before order $\alpha_1 \preceq \alpha_2$ can be checked by comparing VC_1 and VC_2. The transitive closure of the happens-before relation is modeled implicitly through the comparison over logical clocks in the $\sqsubseteq$ operation.

An event-driven application receives thousands of events even within a short-time span. Assigning one logical clock for each event becomes prohibitively expensive. EVENTRACER [21] presents a scheme of clock assignment to event handlers based on the chain decomposition [23] mechanism. Let C be a finite set of discrete values called chains. The chain decomposition assigns each operation in a trace ρ to a chain such that two operations are assigned the same chain only if they are totally ordered. EVENTRACER exploits this to reuse logical clocks across events that are totally ordered with each other.

4.4.3 Event Orderings

The FIFO and NOPRE rules in Fig. 4 add an ordering between a pair of events executed on the same thread. To compute all orderings derivable by these rules, one may naïvely iterate over every pair of events (executed on the same thread). However, as the number of events grows, this becomes impractical. For every event e, EVENTRACER [17] makes reverse depth-first traversal (DFS) over an HB graph to compute only those events that are possible candidates for evaluation of the event-ordering rules such as FIFO and NOPRE. The HB graph models the happens-before ordering among operations of a trace but without transitively saturating the graph. The DFS traversal performs transitive closure dynamically. It uses some rules for pruning the search so that the DFS traversal does not have to explore the entire HB graph, which is linear in the size of the trace.

A limitation of the HB graph-based algorithm is that it may redundantly compute some events as candidates due to inadequate pruning and non-deterministic order of DFS traversal. Further, separate DFS traversals may be needed for different event-ordering rules. Not only does this result in additional vector clock operations, it also wastes time in unnecessary DFS traversals. Recently, event graph [18] and AsyncClock [24] data structures have been introduced as alternatives to the HB graph. Event graph [18] is a data structure that tracks a subset of HB edges between events and tries to overcome these limitations. By maintaining these edges, it can identify candidate events of an event e by simply looking up the incoming edges to e in the event graph (instead of performing DFS traversals over the HB graph). Event graph transitively closes edges but only if necessary. While transitively closing the HB edges between events, it checks whether the HB information captured by the edge is already present in the vector clocks of the two events. It does not propagate the edge transitively in that case. It also defines an order over the incoming edges to efficiently perform vector clock joins. The AsyncClock data structure [24] maintains a vector for each event queue where each component tracks the most recent event in a chain of events. It evaluates event-ordering rules by consulting the AsyncClock values. It introduces optimizations to reclaim the data structures associated with events that may no longer be needed to reduce the memory requirement of the algorithm.

5. MODEL CHECKING

Given a model M of a system and a specification φ, model checking asks whether the model satisfies the specification, written more formally as $M \models \varphi$. Model checking has been applied successfully to models of hardware and software systems with specifications written in various temporal logics [25]. We discuss a flavor of model checking called explicit-state model checking. The objective is to explore the state-space of an event-driven program by explicitly enumerating the states reachable under different interleavings of threads and reorderings of events through repeated executions. The specifications to be checked are in the form of state assertions. If during the exploration we reach a state s which violates an assertion (or causes a deadlock) then a sequence of state transitions from the initial state leading to the state s is obtained for further debugging. If we traverse the entire state-space without reaching any assertion violation then the program is verified to be correct.

Let us consider an Android application A. A transition by a thread t of the application is a sequence of contiguous operations executed by t such that only the first operation in the sequence accesses the shared memory and all the succeeding operations in the sequence access only local memory. We have discussed the semantics of concurrency-relevant operations in Section 3.1. The semantics of other statements such as assignments and conditionals are standard. Recall that not only can different threads access shared memory, memory can be shared even by different event handlers executing on the same thread. Let $T = (S, s_0, \Delta)$ be the *transition system* modeling the state-space of A. The set S comprises of the states of the program with s_0 as the initial state. A state $s \in S$ is essentially a valuation of all variables of the application including the program counter. Even though there can be an exponential number of interleavings of transitions by various threads and event handlers, in practice, many of them result in the same states being visited repeatedly. To avoid redundant explorations, the states of the application can be stored explicitly. If all the outgoing transitions of a state s are already explored then the search need not proceed further from s next time it is visited. This kind of exploration is called a stateful search as the states are explicitly stored during exploration. However, concurrent programs experience state-space explosion due to large number of interleavings possible. An individual state of a real-world application can itself be very large. Due to this, a stateful search quickly runs out of memory. A number of optimizations such as state compression, abstraction, and symmetry reduction have been devised for multithreaded programs to overcome this problem [26].

An alternative to stateful search is called stateless search [27]. Instead of exploring states of the transition system, stateless search works by iterating over the space of interleavings without storing the visited states. Since the states visited by an interleaving are not stored explicitly, pruning of redundant interleavings depends on partial order reductions (POR) [28–31]. The key idea in partial order reduction is to visit only a representative interleaving from each class of equivalent interleavings called a Mazurkiewicz trace [32]. The exploration of such a reduced state-space still provides some formal guarantees such as preservation of all deadlocks and local assertion violations.

POR-based model checking reorders all pairs of dependent transitions systematically. Two transitions are said to be *dependent* with each other if (1) they may enable or disable each other, or (2) distinct states may be reached if they are explored in different orders starting from the same state.

These partial order reduction techniques can be easily extended to work with event-driven multithreaded programs like Android applications by suitably defining the notion of dependent transitions. In addition to the usual dependence between transitions of multithreaded programs (such as synchronization operations), operations that post events to the same thread should also be treated as mutually dependent. While two `post` operations to the same thread cannot enable or disable each other (as we assume the event queues to be unbounded), with ordered event queues, we reach distinct queue configurations if the `posts` occur in different orders. A POR technique can then explore both orders of execution for every pair of `posts` to the same thread. These in turn induce reordering of corresponding event handlers on the same thread.

Maiya et al. [33] observed that the simple extension to POR for event-driven programs outlined earlier may result in exploring many interleavings unnecessarily. If the handlers of two events e_1 and e_2 do not contain dependent transitions which access shared memory then reordering them is not essential. As we have discussed previously, event-driven programs receive a large number of events and treating every pair of handlers as dependent by default is undesirable. They propose a new partial order reduction technique in which two `posts` are considered as dependent only if reordering them can help reorder a pair of truly dependent transitions. R^4 [34] is a stateless model checker which adapts dynamic partial order reduction [35] to single-threaded JavaScript applications and combines it with conflict-bounded reversal. It also features approximate replay of event sequences which allows exploration to continue even when some events in the sequence are disabled or new events appear.

6. RELATED AND FUTURE DIRECTIONS

In the preceding sections, we have studied common event-driven concurrency idioms and recent advances in analysis techniques for them. In this section, we discuss related work and possible future directions of research. This section is not meant to be an exhaustive bibliography of all relevant programming abstractions and analysis techniques. The goal is to provide some pointers to the readers interested in further study.

Automated analysis of Android applications has received significant attention in the past few years. AsyncDroid [36] adapts delay bounding [37] to Android applications and supports replay of UI events for systematic bug exploration. Bouajjani et al. [38] define robustness against concurrency as a correctness criterion for Android applications. An application is robust if its multithreaded semantics is a refinement of an idealized single-threaded semantics. They show that the problem of verifying robustness can be reduced in linear time to reachability in sequential programs. In addition to several race detection and model checking approaches we discussed, there has also been work on deadlock detection and prevention. Dimmunix [39] provides platform-wide deadlock immunity to Android applications by storing deadlock signatures and preventing them from recurring in future.

The approaches we have seen in this chapter are dynamic in nature as they analyze or explore runtime behaviors of applications. There are several static analyses designed specifically for event-driven or asynchronous programs. Unlike nonevent-driven programs, event-driven programs have multiple entry points. Julia [40] is a static analysis framework which identifies entry points of Android applications so that the entire code of the application is reachable from them. It also handles reflection used by applications for inflating UI but does not handle multithreading soundly. For precision of an analysis, it is essential to model ordering among events of an Android application but this makes the analysis intractable. Blackshear et al. [41] present a selective control flow abstraction of event orderings for goal-directed static analyses. They apply it to prove safety of pointer dereferences in Android applications. Mishra et al. [42] give an intermediate representation of Android applications which models the asynchronous call semantics and lifecycle state machine semantics accurately and use it in a typestate analysis. Kahlon et al. [43] consider static race detection for concurrent C programs with asynchronous calls. The main challenge in their setting is to build a concurrent control flow graph in the presence of function pointers.

Abstractions of asynchronous programs where the asynchronous procedures are assumed to run atomically and in arbitrary order have been considered in the literature and subjected to verification [44–46] and data flow analysis [47]. Sen and Viswanathan [44] cast the problem as reachability in multiset pushdown systems. Jhala and Majumdar [47] capture the potentially unbounded pending events through a counter-based abstraction and extend the framework of interprocedural finite distributive subset (IFDS) analyses to asynchronous programs. Ganty and Majumdar [45] show decidability of safety and liveness verification for asynchronous programs operating over finite data. The decidability of safety verification for a richer class of asynchronous programs is considered in [48]. Gavran et al. [46] present a rely-guarantee framework for modular verification of asynchronous programs. Bouajjani and Emmi [49] propose a phase bounding approach to reduce verification of asynchronous programs to that of sequential programs and combine it with delay bounding for asynchronous programs executing on multiple threads.

Actors [50] is a paradigm for asynchronous message passing concurrency where no two actors share memory. Actors have been implemented in various languages such as Erlang, Java, and Scala. λActor [51] models actor-style concurrency and performs infinite-state model checking. Sen and Agha [52] use DPOR for exploring different orders of message sends along with mixed concrete and symbolic execution for test generation. Basset [53] extends a stateful model checker, Java PathFinder [26], for actor programs compiled to Java bytecode. TransDPOR [54] exploits transitivity of dependence relation pertaining to actor programs for selective state-space exploration. Concuerror [55] adapts iterative context bounding [56] to actor-based Erlang programs and avoids exploration of redundant process blocks. P [57] is a domain-specific language of interacting state machines with deferred events. P programs can be model checked to ensure that all events are handled in a timely manner. Desai et al. [58] propose an algorithm for verification of P programs by searching for almost-synchronous invariants. These invariants are derived by selectively exploring a set of interleavings that keep the size of event queues to the minimal size needed while covering all local states. P# [59] allows C# constructs to be used in P code, and features static race detection and systematic concurrency testing.

A drawback of event-driven programming is the need for explicit management of callbacks or handlers. The application logic must be split carefully across multiple callbacks by considering the order of their execution. This problem is popularly referred to as "callback hell." To overcome this problem, C# has started supporting async/await primitives which permit coding in the familiar sequential style. The task of coordinating callbacks is delegated to the compiler. The async/await primitives are coming to many languages like C++, Dart, ECMAScript, F#, PHP, Python, and Scala. Bierman et al. [60] present the core semantics of async/await primitives used by C#. Santhiar and Kanade [61] elaborate on their scheduling aspects. Okur et al. [62] present a refactoring tool to convert callback-based programs to use async/await. It also employs pattern matching to detect some likely misuses of async/await and to potentially correct them. Santhiar and Kanade [63] propose continuation scheduling graphs (CSGs) as an intermediate representation to capture both synchronous and asynchronous control flow in async/await programs. CSGs also capture the scheduling constraints and have been used for static deadlock detection.

Event-based programming is widely used in graphical user interface (GUI) design. Many approaches have been developed to explore input events to explore GUI of Android applications. AndroidRipper [64] does GUI ripping for automated testing of Android applications. Dynodroid [65] generates system events in addition to GUI events. SwiftHand [66] learns an approximate model of the GUI on-the-fly and generates user inputs based on the learned model for testing. EventBreak [67] is a test generation engine for testing responsiveness of user interfaces of JavaScript applications. It generates tests to identify event handlers whose execution time may increase with running time of the application.

Event-based concurrency is a very flexible and scalable mechanism to deal with potentially blocking operations while retaining responsiveness. As we have seen in this chapter, there are several ways in which events are handled and threads are used in event-driven concurrent programs. We expect more innovations in the design space of event-based programming as richer applications with varied architectures and performance requirements emerge in the future. Reactive extensions [68–70] is one such paradigm which is gaining widespread use for real-time processing of streams of events. Though there is much work on analysis of asynchronous event-driven programs, their applicability in practice is still a work in progress. Many programs operate in safety or performance critical settings. Finding concurrency bugs in them and verifying their correctness will continue to be topics of interest in the foreseeable future. General-purpose hardware architectures have been designed with optimizations that take advantage of instruction and data locality of programs based on synchronous style of execution. As observed in [71], asynchronous event-driven programs have different execution characteristics. Development of specialized hardware accelerators for executing event-driven programs can help improve system performance.

7. SUMMARY

In the concurrency world, traditionally, multithreaded applications and their analysis has received significant attention. A new generation of languages and application frameworks combines multithreading with event handling. Applications developed using this concurrency model offer better responsiveness and scalability. In this chapter, we presented programming abstractions common in event-driven multithreaded programming. We studied the notion of event loops which provides the basic scheduling mechanism of event handling. We also discussed use of single-threading and programmatic event loops.

Events may arrive at an application in any nondeterministic order. This compounds the nondeterminism inherent in scheduling of multiple threads. We gave formal semantics of Android applications and discussed other variants in terms of use of threads and event loops. Designing scalable and precise analysis techniques for concurrency bug detection and verification of event-driven applications has become a very active research area. We discussed advances in the core analysis techniques such as happens-before analysis, race detection, and model checking. Event-based concurrency and related asynchronous programming paradigms will continue to dominate software design in future. We hope that this chapter will serve as a gentle yet substantial introduction to this area.

ACKNOWLEDGMENTS

I would like to thank Anirudh Santhiar and Pallavi Maiya for their helpful comments and suggestions on the material covered in this chapter.

REFERENCES

[1] Z. Mednieks, L. Dornin, G.B. Meike, M. Nakamura, Programming Android, O'Reilly Media, Inc., 2012.
[2] E. Cheong, J. Liebman, J. Liu, F. Zhao, TinyGALS: a programming model for event-driven embedded systems, in: SAC, ACM, 2003, pp. 698–704, https://doi.org/10.1145/952532.952668.
[3] K. Klues, C.-J.M. Liang, J. Paek, R. Musăloiu-E, P. Levis, A. Terzis, R. Govindan, TOSThreads: thread-safe and non-invasive preemption in TinyOS, in: SenSys, ACM, 2009, pp. 127–140.
[4] J. Blanchette, M. Summerfield, C++ GUI Programming With Qt 4, Open Source Software Development Series, Prentice Hall, 2008.
[5] S. Souders, Even Saster Web Sites: Performance Best Practices for Web Developers, O'Reilly Media, 2009.
[6] V.S. Pai, P. Druschel, W. Zwaenepoel, Flash: an efficient and portable web server, in: -USENIX, USENIX Association, 1999, pp. 199–212.
[7] M. Welsh, D. Culler, E. Brewer, SEDA: an architecture for well-conditioned, scalable internet services, in: ACM SIGOPS Operating Systems Review, vol. 35, ACM, 2001, pp. 230–243.
[8] M. Cantelon, M. Harter, T. Holowaychuk, N. Rajlich, Node.js in Action, Manning Publications, 2013.

[9] L. Lamport, Time, clocks, and the ordering of events in a distributed system, Commun. ACM 21 (7) (1978) 558–565, https://doi.org/10.1145/359545.359563.
[10] C.J. Fidge, Timestamps in message-passing systems that preserve the partial ordering, in: The 11th Australian Computer Science Conference, University of Queensland, 1988.
[11] F. Mattern, Virtual time and global states of distributed systems, Parallel Distr. Algorithms 1 (23) (1989) 215–226.
[12] P. Maiya, A. Kanade, R. Majumdar, Race detection for Android applications, in: PLDI, ACM, 2014, pp. 316–325.
[13] A. Santhiar, S. Kaleeswaran, A. Kanade, Efficient race detection in the presence of programmatic event loops, in: ISSTA, ACM, ISBN: 978-1-4503-4390-9, 2016, pp. 366–376, https://doi.org/10.1145/2931037.2931068.
[14] A. Santhiar, S. Kaleeswaran, A. Kanade, Efficient race detection for multi-threaded programs with programmatic event loops, Technical report, 2017.
[15] S. Savage, M. Burrows, G. Nelson, P. Sobalvarro, T. Anderson, Eraser: a dynamic data race detector for multithreaded programs, ACM Trans. Comput. Syst. 15 (4) (1997) 391–411, https://doi.org/10.1145/265924.265927.
[16] C.-H. Hsiao, J. Yu, S. Narayanasamy, Z. Kong, C.L. Pereira, G.A. Pokam, P.M. Chen, J. Flinn, Race detection for event-driven mobile applications, in: PLDI, ACM, ISBN: 978-1-4503-2784-8, 2014, pp. 326–336, https://doi.org/10.1145/2594291.2594330.
[17] P. Bielik, V. Raychev, M. Vechev, Scalable race detection for Android applications, in: OOPSLA, ACM, 2015, pp. 332–348.
[18] P. Maiya, A. Kanade, Efficient computation of happens-before relation for event-driven programs, in: ISSTA, ACM, 2017, pp. 102–112.
[19] Y. Hu, I. Neamtiu, A. Alavi, Automatically verifying and reproducing event-based races in Android apps, in: ISSTA, ACM, ISBN: 978-1-4503-4390-9, 2016, pp. 377–388, https://doi.org/10.1145/2931037.2931069.
[20] B. Petrov, M. Vechev, M. Sridharan, J. Dolby, Race detection for web applications, in: PLDI, ACM, 2012, pp. 251–262, https://doi.org/10.1145/2254064.2254095.
[21] V. Raychev, M. Vechev, M. Sridharan, Effective race detection for event-driven programs, in: OOPSLA, ACM, 2013, pp. 151–166.
[22] C.Q. Adamsen, A. Møller, R. Karim, M. Sridharan, F. Tip, K. Sen, Repairing event race errors by controlling nondeterminism, in: ICSE, IEEE Press, 2017, pp. 289–299.
[23] H.V. Jagadish, A compression technique to materialize transitive closure, ACM Trans. Database Syst. 15 (4) (1990) 558–598, ISSN: 0362-5915, https://doi.org/10.1145/99935.99944.
[24] C.-H. Hsiao, S. Narayanasamy, E.M.I. Khan, C.L. Pereira, G.A. Pokam, AsyncClock: scalable inference of asynchronous event causality, in: ASPLOS, ACM, 2017, pp. 193–205.
[25] E.M. Clarke, O. Grumberg, D. Peled, Model Checking, MIT Press, 1999.
[26] W. Visser, K. Havelund, G. Brat, S. Park, Model checking programs, in: ASE, IEEE, 2000, pp. 3–11.
[27] P. Godefroid, Model checking for programming languages using VeriSoft, in: POPL, ACM, 1997, pp. 174–186.
[28] A. Valmari, Stubborn sets for reduced state space generation, in: Advances in Petri Nets, Springer, 1989, pp. 491–515, https://doi.org/10.1007/3-540-53863-1_36.
[29] D. Peled, All from one, one for all: on model checking using representatives, in: CAV, Springer, 1993, pp. 409–423, https://doi.org/10.1007/3-540-56922-7_34.
[30] P. Godefroid, Partial-Order Methods for the Verification of Concurrent Systems - An Approach to the State-Explosion Problem, LNCS, 1032, Springer, 1996.
[31] E.M. Clarke, O. Grumberg, M. Minea, D. Peled, State space reduction using partial order techniques, Int. J. Softw. Tools Technol. Transfer 2 (3) (1999) 279–287.

[32] A.W. Mazurkiewicz, Trace theory, in: W. Brauer, W. Reisig, G. Rozenberg (Eds.), Advances in Petri Nets, Springer, 1986, pp. 279–324, https://doi.org/10.1007/3-540-17906-2_30.

[33] P. Maiya, R. Gupta, A. Kanade, R. Majumdar, Partial order reduction for event-driven multi-threaded programs, in: TACAS, Springer, 2016, pp. 680–697, https://doi.org/10.1007/978-3-662-49674-9_44.

[34] C.S. Jensen, A. Møller, V. Raychev, D. Dimitrov, M. Vechev, Stateless model checking of event-driven applications, in: OOPSLA, ACM, 2015, pp. 57–73, https://doi.org/10.1145/2814270.2814282.

[35] C. Flanagan, P. Godefroid, Dynamic partial-order reduction for model checking Software, in: POPL, ACM, 2005, pp. 110–121.

[36] B.K. Ozkan, M. Emmi, S. Tasiran, Systematic asynchrony bug exploration for Android apps, in: CAV, Springer, 2015, pp. 455–461, https://doi.org/10.1007/978-3-319-21690-4_28.

[37] M. Emmi, S. Qadeer, Z. Rakamarić, Delay-bounded scheduling, in: POPL, ACM, 2011, pp. 411–422.

[38] A. Bouajjani, M. Emmi, C. Enea, B.K. Ozkan, S. Tasiran, Verifying robustness of event-driven asynchronous programs against concurrency, in: ESOP, Springer, 2017, pp. 170–200, https://doi.org/10.1007/978-3-662-54434-1_7.

[39] H. Jula, T. Rensch, G. Candea, Platform-wide deadlock immunity for mobile phones, in: DSN-W, IEEE, 2011, pp. 205–210.

[40] E. Payet, F. Spoto, Static analysis of Android programs, in: CADE, Springer-Verlag, ISBN: 978-3-642-22437-9, 2011, pp. 439–445. http://dl.acm.org/citation.cfm?id=2032266.2032299.

[41] S. Blackshear, B.-Y.E. Chang, M. Sridharan, Selective control-flow abstraction via jumping, in: OOPSLA, ACM, ISSN: 978-1-4503-3689-5, 2015, pp. 163–182, https://doi.org/10.1145/2814270.2814293.

[42] A. Mishra, A. Kanade, Y.N. Srikant, Asynchrony-aware static analysis of Android applications, in: MEMOCODE, IEEE, 2016, pp. 163–172, https://doi.org/10.1109/MEMCOD.2016.7797761.

[43] V. Kahlon, N. Sinha, E. Kruus, Y. Zhang, Static data race detection for concurrent programs with asynchronous calls, in: ESEC/FSE, ACM, 2009, pp. 13–22.

[44] K. Sen, M. Viswanathan, Model checking multithreaded programs with asynchronous atomic methods, in: CAV, Springer, 2006, pp. 300–314.

[45] P. Ganty, R. Majumdar, Algorithmic verification of asynchronous programs, ACM Trans. Program. Lang. Syst. 34 (1) (2012) 6:1–6:48, https://doi.org/10.1145/2160910.2160915.

[46] I. Gavran, F. Niksic, A. Kanade, R. Majumdar, V. Vafeiadis, Rely/guarantee reasoning for asynchronous programs, in: CONCUR, Schloss Dagstuhl - Leibniz-Zentrum fuer Informatik, 2015, pp. 483–496, https://doi.org/10.4230/LIPIcs.CONCUR.2015.483.

[47] R. Jhala, R. Majumdar, Interprocedural analysis of asynchronous programs, in: POPL, ACM, 2007, pp. 339–350.

[48] M. Emmi, P. Ganty, R. Majumdar, F. Rosa-Velardo, Analysis of asynchronous programs with event-based synchronization, in: ESOP, Springer, 2015, pp. 535–559, https://doi.org/10.1007/978-3-662-46669-8_22.

[49] A. Bouajjani, M. Emmi, Bounded phase analysis of message-passing programs, Int. J. Softw. Tools Technol. Transfer 16 (2) (2014) 127–146, https://doi.org/10.1007/s10009-013-0276-z.

[50] G. Agha, Actors: a model of concurrent computation in distributed systems, Tech. rep., MIT Artificial Intelligence Lab, 1985.

[51] E. D'Osualdo, J. Kochems, C.-H.L. Ong, Automatic verification of Erlang-style concurrency, in: SAS, Springer-Verlag, 2013, pp. 454–476. http://mjolnir.cs.ox.ac.uk/soter/papers/sas13.pdf.

[52] K. Sen, G. Agha, Automated systematic testing of open distributed programs, in: FASE, Springer, 2006, pp. 339–356, https://doi.org/10.1007/11693017_25.
[53] S. Lauterburg, M. Dotta, D. Marinov, G. Agha, A framework for state-space exploration of java-based actor programs, in: ASE, ACM, 2009, pp. 468–479, https://doi.org/10.1109/ASE.2009.88.
[54] S. Tasharofi, R.K. Karmani, S. Lauterburg, A. Legay, D. Marinov, G. Agha, TransDPOR: a novel dynamic partial-order reduction technique for testing actor programs, in: FMOODS/FORTE, Springer, 2012, pp. 219–234, https://doi.org/10.1007/978-3-642-30793-5_14.
[55] M. Christakis, A. Gotovos, K.F. Sagonas, Systematic testing for detecting concurrency errors in Erlang programs, in: ICST, IEEE Computer Society, 2013, pp. 154–163.
[56] M. Musuvathi, S. Qadeer, Iterative context bounding for systematic testing of multithreaded programs, in: PLDI, ACM, 2007, pp. 446–455.
[57] A. Desai, V. Gupta, E. Jackson, S. Qadeer, S. Rajamani, D. Zufferey, P: Safe asynchronous event driven programming, in: PLDI, ACM, 2013, pp. 321–332.
[58] A. Desai, P. Garg, P. Madhusudan, Natural proofs for asynchronous programs using almost-synchronous reductions, in: OOPSLA, ACM, 2014, pp. 709–725.
[59] P. Deligiannis, A.F. Donaldson, J. Ketema, A. Lal, P. Thomson, Asynchronous programming, analysis and testing with state machines, in: PLDI, ACM, ISSN 0362-1340, 2015, pp. 154–164, https://doi.org/10.1145/2813885.2737996.
[60] G. Bierman, C. Russo, G. Mainland, E. Meijer, M. Torgersen, Pause 'N' play: formalizing asynchronous C#, in: ECOOP, Springer-Verlag, Berlin, Heidelberg, 2012, pp. 233–257.
[61] A. Santhiar, A. Kanade, Semantics of asynchronous C#, 2017, http://www.iisc-seal.net/publications/asyncsemantics.pdf.
[62] S. Okur, D.L. Hartveld, D. Dig, A. van Deursen, A study and toolkit for asynchronous programming in C#, in: ICSE, ACM, 2014, pp. 1117–1127, https://doi.org/10.1145/2568225.2568309.
[63] A. Santhiar, A. Kanade, Static deadlock detection for asynchronous C# programs, in: PLDI, ACM, 2017, pp. 292–305, https://doi.org/10.1145/3062341.3062361.
[64] D. Amalfitano, A.R. Fasolino, P. Tramontana, S.D. Carmine, A.M. Memon, Using GUI ripping for automated testing of Android applications, in: ASE, ACM, 2012, pp. 258–261.
[65] A. Machiry, R. Tahiliani, M. Naik, Dynodroid: an input generation system for Android apps, in: FSE, ACM, 2013, pp. 224–234.
[66] W. Choi, G. Necula, K. Sen, Guided GUI testing of Android apps with minimal restart and approximate learning, in: OOPSLA, ACM, ISBN: 978-1-4503-2374-1, 2013, pp. 623–640, https://doi.org/10.1145/2509136.2509552.
[67] M. Pradel, P. Schuh, G.C. Necula, K. Sen, EventBreak: analyzing the responsiveness of user interfaces through performance-guided test generation, in: OOPSLA, ACM, 2014, pp. 33–47, https://doi.org/10.1145/2660193.2660233.
[68] L. Campbell, Introduction to Rx: A Step by Step Guide to the Reactive Extensions to .NET, Amazon Asia-Pacific Holdings Private Limited, 2012.
[69] T. Nurkiewicz, B. Christensen, Reactive Programming With RxJava: Creating Asynchronous, Event-Based Applications, O'Reilly Media, 2016.
[70] T. Subonis, Reactive Android Programming, Packt Publishing, 2017.
[71] G. Chadha, S. Mahlke, S. Narayanasamy, Accelerating asynchronous programs through event sneak peek, in: ISCA, ACM, ISBN: 978-1-4503-3402-0, 2015, pp. 642–654, https://doi.org/10.1145/2749469.2750373.

ABOUT THE AUTHOR

Aditya Kanade is an Associate Professor of Computer Science and Automation at the Indian Institute of Science. His research spans programming languages, software engineering, and artificial intelligence. He has received an ACM best paper award (EMSOFT 2008), a teaching excellence award, and research grants/gifts from NVidia, Microsoft Research India, IBM, and Mozilla. He has served as a PC cochair of ICSE 2017 Demo Track and on the Review Board of IEEE Transactions on Software Engineering.

A Taxonomy of Software Integrity Protection Techniques

Mohsen Ahmadvand*, Alexander Pretschner*, Florian Kelbert†
*Technical University of Munich, Munich, Germany
†Imperial College London, London, United Kingdom

Contents

Advances in Computers, Volume 112
ISSN 0065-2458
https://doi.org/10.1016/bs.adcom.2017.12.007

Abstract

Tampering with software by man-at-the-end (MATE) attackers is an attack that can lead to security circumvention, privacy violation, reputation damage, and revenue loss. In this model, adversaries are end users who have full control over software as well as its execution environment. This full control enables them to tamper with programs to their benefit and to the detriment of software vendors or other end users. Software integrity protection research seeks for means to mitigate those attacks. Since the seminal work of Aucsmith, a great deal of research effort has been devoted to fight MATE attacks, and many protection schemes were designed by both academia and industry. Advances in trusted hardware, such as TPM and Intel SGX, have also enabled researchers to utilize such technologies for additional protection. Despite the introduction of various protection schemes, there is no comprehensive comparison study that points out advantages and disadvantages of different schemes. Constraints of different schemes and their applicability in various industrial settings have not been studied. More importantly, except for some partial classifications, to the best of our knowledge, there is no taxonomy of integrity protection techniques. These limitations have left practitioners in doubt about effectiveness and applicability of such schemes to their infrastructure. In this work, we propose a taxonomy that captures protection processes by encompassing system, defense and attack perspectives. Later, we carry out a survey and map reviewed papers on our taxonomy. Finally, we correlate different dimensions of the taxonomy and discuss observations along with research gaps in the field.

1. INTRODUCTION

A well-known attack in computer security is man-in-the-middle in which adversaries intercept network messages and potentially tamper with their content. In this model attackers only have control over the communication channel and hence utilizing secure communication protocols can (to a great extend) mitigate the risk.

Beside network attacks, there is another type of attacks that could potentially lead to violation of safety and security of software systems. Given that end users have full control over the host on which programs are being executed, they can manifest serious attacks on software systems by inspecting a program's execution or tampering with the host's hardware/software [1]. These attacks are known as *man-at-the-end* (MATE) attackers. As the name suggests, in this model the attacker resides at the end of line in which communication is secured and firewalls are bypassed. This is where the protection offered by security protocols for communication ends.

MATE attackers have direct access on the host on which the program is being executed. In most cases, there is no limitation whatsoever on what they can control or manipulate on the host. In effect, a user who installs an application on his or her machine has all controls over the computation unit. In this setting, perpetrators can inspect programs' execution flow and tamper with anything in program binaries or during runtime, which, in turn, enable them to extract confidential data, subvert critical operations and tamper with the input and/or output data. Cheating in games, defeating license checks, stealing proprietary data (musics and movies), and displaying extra ads in browsers are typical examples of MATE attacks.

1.1 Attacker Goals and Motives

While there are different goals that MATE attackers may pursue, we particularly focus on attacks violating software integrity; other attacker goals are out of our scope.

Reputation, financial gains, sabotage, and terrorism are nonexhaustive motivations for attackers to target the integrity of software systems. Disabling a license check is a classical example of MATE attacks in which perpetrator can potentially affect the revenue of the software vendor, for instance, by publicly releasing a patch for a popular software. Adversaries may harm the reputation of a company by manipulating program behavior, for instance, to show inappropriate ads. Perpetrators may further target the safety and/or security of critical systems. Corresponding consequences of attacks on a nuclear power plant system could be dire. Akhundzada et al. [2] elaborate on MATE attacker motives in more details.

1.2 Local Attacks

Local attackers normally have the full privilege on the software system as well as physical access to the hardware. Such attackers can readily tamper with softwares at any stage (at rest, in-memory and in-execution). Moreover, they could potentially load a malicious kernel driver to bypass security mechanisms employed by software. For instance, [3, 4] present two attacks to defeat protection schemes on a modified Linux kernel. Physical access to the host has the same effect. It enables attackers to tamper with systems' hardware and/or configurations. Garfinkel et al. [5] show that perpetrators can install a malicious hardware to disclose confidential data.

1.3 Remote Attacks

MATE attackers do not necessarily require physical access to the system of interest. Many harmful attacks can be carried out remotely. This type of attacks is known as *remote-man-at-the-end* (RMATE) [6]. In this model, attackers need to have either remote access to the target system or deploy their attacks by tricking end users. For example, Banescu et al. [7, 8] discuss two RMATE attacks on Google Chromium which are actually carried out by deployed malicious payloads into the Google Chromium browser. These payloads (in form of plugins) alter the original behavior of the browser to execute malicious activities, e.g., showing extra ads or collecting user information.

Intruders are another type of attackers that are very similar to RMATE attackers. Intruders penetrate software systems normally through their public interfaces (e.g., websites) in order to find vulnerabilities. A successful exploit may enable them to tamper with the integrity of the system, e.g., by materializing a buffer overflow or SQL injections. While RMATE attacks have much in common with intruders, the main difference lies in the access privilege that RMATE attackers possess before and in order to manifest their attacks. Simply put, RMATE attacks exclude exploiting vulnerabilities and hence starts by a granted access to the host or a program that it contains. Nevertheless, an intruder (after successfully compromising the security) can carry out RMATE attacks. Therefore, the borderline between the two is blurry.

1.4 Further Attack Types

Another realization of MATE attacks is through *repackaged software* [9]. In this attack, perpetrators obtain software bundles (normally popular ones) and modify them with malicious codes to create counterfeit versions. Later, these repackaged softwares are shipped to software hubs to be installed by victims. Since malicious operations are normally dormant, counterfeit softwares appear to be the same as the original ones to end users. That is, they can remain on user devices for a period of time and eventually harm their assets, for instance, by deleting user files after 3 h of program usage. Nevertheless, attackers first need to get users to install repackaged softwares. This, however, does not seem to be an obstacle for attackers: a recent study has shown that 77% of the popular applications available on Google play store, which is a trusted software repository for Android application, have repackaged versions [10].

Targeted malware is another form of MATE attacks in which a sophisticated malware is designed to violate the integrity of a particular system. Stuxnet [11] is a malware that manipulated programmable logic controllers' code causing severe damage to the Iranian nuclear program. ProjectSauron [12] is another example of targeted malware where governmental sensitive information was covertly collected and subsequently transmitted to the attacker's server.

In the light of these threats, researching, developing, and deploying protection mechanisms against MATE attackers is of paramount importance. In particular, the integrity of software demands protection.

1.5 Approaches to Protect Software Integrity

Generally, integrity protection refers to mechanisms that protect the logic and/or data of particular software. Integrity protection is a part of the *Software Protection* field, which is also known as *tamperproofing*. Collberg defines tamperproofing as "[a way] to ensure that [a program] executes as intended, even in the presence of an adversary who tries to disrupt, monitor, or change the execution" [13].

At a high level, integrity protection techniques are comprised of two main mechanisms: `monitor` and `response`. While the former *detects* inconsistencies by monitoring a system's representation, the latter acts upon the detection of such inconsistencies by punishing the attacker [14]. In case of repackaged software and RMATE attacks, these reactions will be limited to terminating the process, informing users about violation of integrity or notifying a home server (phoning home).

Unlike cryptographic protocols, protection schemes against MATE fail to provide hard security guarantees, e.g., by difficulty of solving a computationally complex problem such as the discrete logarithm in a secure multiplicative group [15]. In fact, Dedić et al. [16] argue that there is no tamper resistant method that resists against polynomial time adversaries. This means all the protection schemes, given enough time and resources, are eventually defeated by attackers. However, protection schemes can raise the bar against perpetrators by increasing the cost and effort needed to violate system integrity. The introduced cost and effort is often sufficient to mitigate RMATE, repackaged software or targeted malwares. However, we are not aware of any studies that measures the resilience thoroughly.

Tamper-proofing is commonly used in combination with *obfuscation* [1] software protection technique. Obfuscation aims to reduce the understandability of adversaries by complicating programs. In contrast to tamper-proofing, obfuscation remains unaware of the program modifications. Moreover, tamper-proofing can detect the occurrence of tampering attacks and respond to such attacks in some way punishing to actors [14].

Advances in hardware security has positively impacted integrity protection research. *Trusted platform module* (TPM) (https://trustedcomputinggroup.org/) is an approach in which software protection meets hardware security [17, 18]. TPM is a tamper-resilient microcontroller designed to securely carry out sensitive operations, such as secure boot, encryption, attestation, and random number generation.

Recently, Intel has also been working on secure hardware modules. Their software guard extensions (SGX), introduced in 2015, are finding their way in academic research [19].

1.6 Gap

Despite the existence of a multitude protection schemes that mitigate certain attacks on different system assets, we are not aware of a comprehensive study that compares advantages and disadvantages of these schemes. No holistic study was done to measure completeness and effectiveness of such schemes. How these schemes operate and what components they are comprised of not identified nor plotted in context.

Current classifications lack the level of detail required to evaluate and select schemes by practitioners. We are not aware of any classification beyond what was proposed by Collberg et al. [14, chapter 7]. In effect, classifications were done only at an abstract level, i.e., monitor and response mechanisms. While integrity protection schemes comprise far more components and impact various parts of the system, they introduce unique constraints which may turn them completely inapplicable in certain application contexts. These limitations have left practitioners in doubt about the effectiveness and applicability of such schemes to their infrastructure.

1.7 Contribution

We propose a taxonomy encompassing system, defense and attack views and extract relevant criteria in each view. These three views aim at capturing a holistic view of protection mechanisms. The system view captures the components of interest of the system to protect. The defense view elaborates on characteristics of defense mechanism. The attack view draws the attacker model and resilience of the schemes to certain attacks.

We evaluate our taxonomy by mapping over 49 reviewed research papers and discuss their advantages and disadvantages. We further correlate different dimensions of our taxonomy and discuss insightful observations with regard to security, resilience, performance, applicability, and usage constraints in practice.

1.8 Structure

This document is structured as follows. Section 2 presents a motivating data right management example to introduce further integrity threat samples, which supports the understandability of the taxonomy. In Section 3 we propose our taxonomy and define all its elements. In Section 4 we evaluate the taxonomy by applying it to the reviewed literature. To evaluate advantages and disadvantages of different schemes in Section 5 we correlate interesting dimensions of the taxonomy and discuss our findings. In Section 6 we review the related work. Section 7 discusses the future of integrity protection research. Finally, in Section 8 we conclude our study.

2. MOTIVATING EXAMPLE

In this section, we present a simplified digital rights management (DRM) system as a motivating example for our taxonomy. DRM enables authors to protect their digital proprietary data while sharing them with other people. To do so, DRM enforces a set of *usage policies* at the user end (client side) that limits what users can actually do with the protected content.

The goal of our sample DRM system is to enable secure lending of proprietary content to end users. For the sake of simplicity, we assume that two usage policies are employed in this system: users can lease a protected media (digital content) (a) for a certain period of time or (b) for a particular number of views.

The system as depicted in Fig. 1 is comprised of two compartments: a server (which runs on trusted commodity) and a client (which runs on untrusted commodity and hence is exposed to MATE attacks). On the server side, three components, viz., `license server`, `data provider`

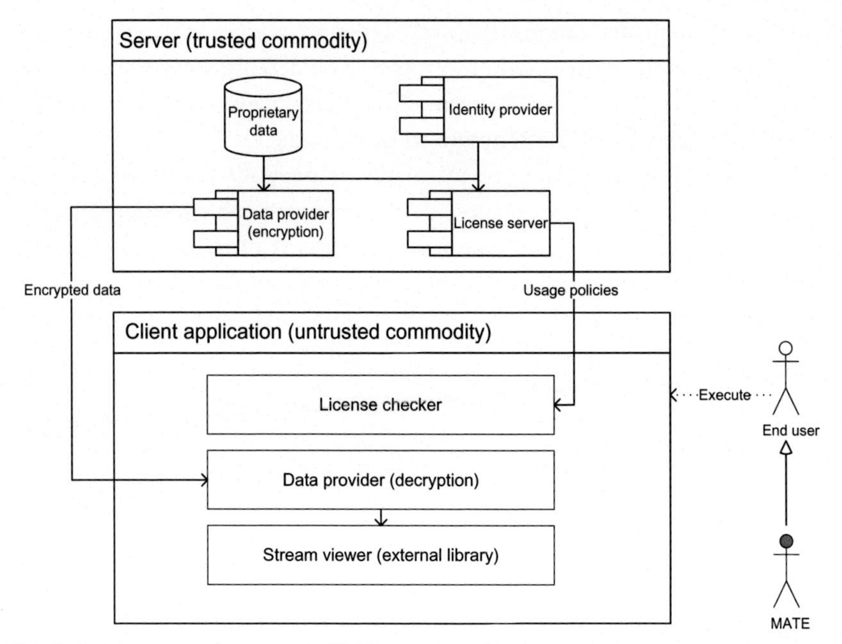

Fig. 1 Architecture of a example DRM system to lend proprietary content.

`(encryption)`, and `proprietary data`, are collaborating to deliver the desired functionality. On the client side, three components, viz., `license checker`, `data provider (decryption)`, and `stream viewer`, are handling user requests.

The `license server` is in charge of generating usage polices according to the purchased license by the users. It sends usage polices to the client applications. The `data provider` and `proprietary database` work closely together. The `data provider` retrieves the requested protected content from the database and encrypts it with client keys.

On the client side, the `license checker` enforces the policies that are specified by the `license server`. That is, the date of expiry and number of views are tracked by the license checker. As soon as the limit is met, the restriction is applied, for instance, by removing the protected content. To add more resilience, the data provider avoids to decrypt the file completely, instead it decrypts it in fragments and gradually feeds them to the stream viewer. In this way, the entire content is never exposed as a whole on the client side.

The entire system is developed by the *Sample DRM* company, except for the stream viewer. As stated in the figure, stream viewer is an external library, for which Sample DRM only has access to the compiled binaries.

2.1 Threats

In this system the protected content is transmitted to the client machines on which end users have full control over program execution. This immediately gives rise to MATE attacks. In Fig. 2 we present a set of potential threats from MATE attackers in form of an attack tree [20].

Starting from the root node, a perpetrator's goal is to use the protected content without the restriction which was specified in the usage policy. For this purpose, she has three options: extract data from the stream viewer, exfiltrate the content, or circumvent the license checker.

Extracting protected content from the viewer requires the attacker to dump memory fragments and later reassemble them. She can (among other possible attacks) tamper with the viewer to perform this malicious activity automatically. Consequently, an attack can manipulate the viewer to dump and merge decrypted content behind the scene until the entire file is extracted.

Exfiltrating the content by decrypting requires the attacker to obtain the $client_{private}$ key as well as encrypted contents. Since the data provider (decryption) has access to this key, it can be targeted by reverse engineering attacks to first locate the key and subsequently to extract it.

Circumventing the license checker is yet another way to illegitimately use the protected content. The license checker is effectively in charge of verifying the usage policies on access requests. Attackers can manipulate the sensitive data, i.e., the usage variables for number of views and expiry date. If this is done, then the DRM protection is defeated. Furthermore, they can

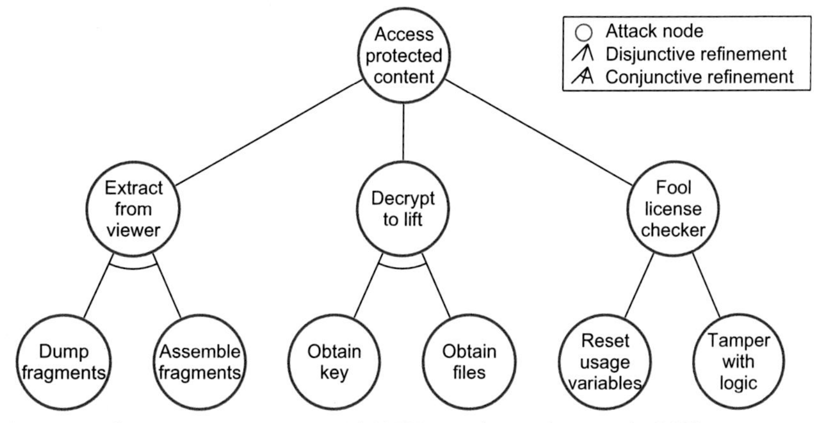

Fig. 2 Attack tree capturing potential MATE attacks on the sample DRM system.

tamper with the license checking logic, for instance, to accept any license as valid regardless of the actual validity of the license.

The aforementioned threats are a representative set of MATE threats that requires integrity protection in applications. Later in Section 4 we will map the elements of our taxonomy to the aforementioned example and show how our taxonomy serves as a blueprint for integrity protection practitioners. To avoid wordiness we refer to the aforementioned case study as the sample DRM in the remainder of this chapter.

3. PROPOSED TAXONOMY

Current classifications of integrity protection techniques lack a level of detail that is required by practitioners to evaluate and select schemes. To the best of our knowledge, the existing classifications remain at an abstract level, i.e., describing monitor and response mechanisms. However, integrity protection schemes comprise far more components and impact various parts of the system. For example, they may introduce unique constraints which make them inapplicable in certain application contexts. Such limitations have left practitioners in doubt about the effectiveness and applicability of integrity protection schemes to their infrastructure.

In this section we propose a taxonomy for software integrity protection techniques that provides a holistic classification and subsequently facilitates the usage of protection schemes in different contexts. We build the taxonomy on the basis of the protection process that a user follows when aiming to protect the integrity of a program. This process starts with identifying system assets and continues with the identification of possible attacks and defense mechanisms.

As a consequence, our taxonomy is comprised of three dimensions: (i) a *system view*, (ii) an *attack view*, and (iii) a *defense view*. The *system view* describes the system as a whole, as well as the encapsulated assets, their granularity, and representations in different program life cycles. It is the integrity of the assets that ought to be protected. Examples include license checkers and, in our DRM example scenario, data providers. The *attack view* captures actions that perpetrators may carry out to undermine the integrity of aforementioned assets. For example, Fig. 2 shows that tampering with the logic of license checker could harm system assets. Finally, the *defense view* encompasses mechanisms to prevent or detect attacks on different representations of the assets. For instance, we can utilize logic protection measures to mitigate the risk of tampering attacks on the license checker. Unlike attacks, defense mechanisms may have implications on the system life cycle, since they may alter a system in various ways, e.g., in terms of performance, integration, and incident handling.

In the course of building the taxonomy we noticed strong dependencies between the elements within and across the three dimensions. Because such relations are widely captured using class diagrams [21, 22], we make use of UML class diagrams [23] to model and depict such relations. Concretely, our model uses *association*, *inheritance*, *aggregation*, and *composition* relations with the same meaning as defined in the UML specification. Association indicates two elements are related, simply put, one can access an attribute or a method of the other. Aggregation depicts the part-of relationship between two elements. Inheritance expresses specialization of a particular element, while composition expresses the set of elements that compose a particular element. In the following, we first introduce the taxonomy along with a few examples. Then, we discuss the dependencies between different views. Section 4 will later map related literature onto the classified attack and defense mechanisms.

3.1 System

The system dimension captures the characteristics of the system to be protected. This dimension is comprised of `asset`, `life cycle activity`, `representation`, and `granularity` main classes along with some more specific subclasses for each class. In the following, we first describe these main classes, their relation and then elaborate on each of them and their subclasses. Fig. 3 presents the system view of the taxonomy.

`Asset` is the core class of this view as it captures the elements whose integrity need to be protected against attacks. In our taxonomy, assets include `behavior` and `data` the tampering of which may harm system users or software producers. Tampering with the license checking behavior and manipulating usage count variables data (number of views and expiry date) in the sample DRM are examples of behavior and data assets, respectively.

`Life cycle activities` capture the different stages that a program undergoes in its life cycle. Depending on the stage of a program, assets have different `representations`. Simply put, assets are exposed and presented in different representations from program source code all the way to the program code pages in the main memory. Each of these stages has access to different representation of the assets, however. That is, particular asset representations are only available in particular life cycle activities.

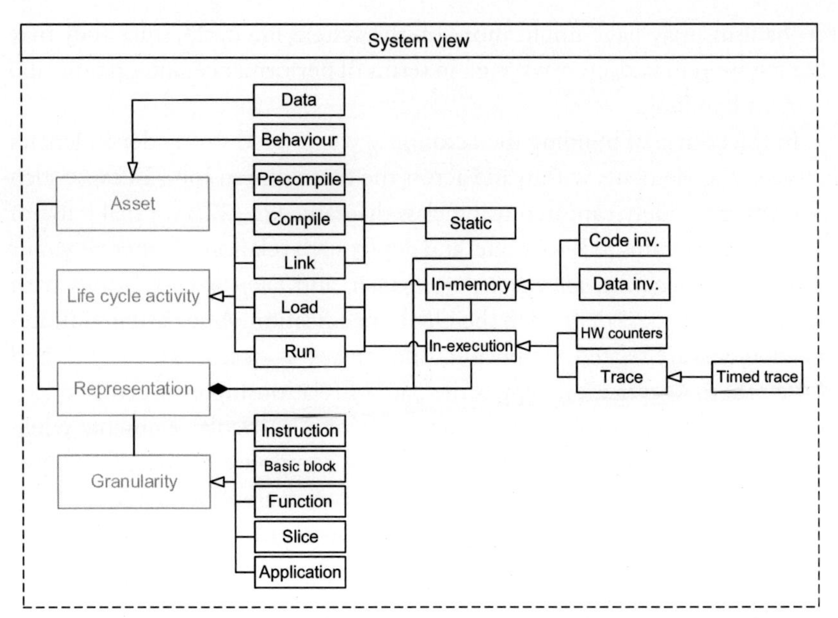

Fig. 3 System view dimension of the taxonomy depicting the system main classes as well as subclasses and their relations.

An asset can have various `granularities`. Granularities correspond to different abstraction levels, e.g., a license check may correspond to a source code C function along with its set of statements, or, more fine-grained, a basic block within one function.

1. *Integrity assets* include valuable *data* or sensitive *behavior(s)* of the system, tampering with which renders the system's security defeated.
   (a) *Data* refers to any sensitive information that is produced or processed by the application. Application input, configuration, and database files are examples of such data assets. For instance, in the sample DRM, usage count variables are data assets.
   (b) *Behavior* is the effect of program execution. Similar to tampering with data, subverting an application's behavior (logic) can have obnoxious consequences. For instance, in the sample DRM, stream viewing decrypted fragments as well as license checking are behavioral assets.
2. *Representation.* Assets can have multiple representations, viz., *static*, *in-memory*, and *in-execution*, depending on the program state from start to finish.

(a) *Static* captures assets when they are at rest, i.e., stored on disk. These assets representations are accessible via the file system.
(b) *In-memory* represents assets in memory, i.e., RAM or virtual memory. Process memory captures this representation which includes *code* and *data* invariants of processes.
(c) *In-execution* captures tangible effects of assets during execution. *Trace* and *hardware counters* are subclasses of the in-execution representation. Meanwhile, trace has a subclass that comes with timing data, i.e., *timed trace*.

3. *Granularity*. Assets have different scales and levels of detail that vary from an instruction to the entire application. For example, a license check is usually represented as a function, whereas control flow integrity depends on a large set of branching instructions in the application.
(a) *Instruction*. A single instruction is evaluated as security critical.
(b) *Basic block (BB)*. A set of consecutive instructions with exactly one entry point and one exit point.
(c) *Slice*. A set of consequential instructions scattered in a program, e.g., the branching instructions that dictate the control flow of a program.
(d) *Function*. A complete function, e.g., *licenceCheck()*, is the goal of protection.
(e) *Application*. The entire application or library is supposed to be protected, i.e., the entire program is security critical. For instance, a power plants controller application is in its entirety sensitive.

4. *Life cycle*. The life cycle indicates a series of different states that each program undergoes from development all the way to execution, viz., *precompile*, *compile*, *postcompile*, *link*, *load*, and *run*. Every program has to go through these activities strictly in the mentioned order in order to eventually get executed. However, if a program is not developed in-house, i.e., source codes are not at hand, the first two activities (precompile and compile) are out of reach on the protector side, as these activities require access to source codes.

   As depicted in the system view (Fig. 3) by association links, the first four states (precompile, compile, postcompile, and link) deal with the static representation of the assets. The last two states (load and run) are concerned with both the in-memory and in-execution representations. In the following we describe these states:
(a) *Precompile*. This is the state in which a program's artifact, which are represented statically, e.g., in the form of application source code

and other bindings such as testing, database and setup scripts, etc., are delivered. Source-level transformations could be triggered at this stage to add protection routines.

(b) *Compile.* In this stage, a compiler transforms the program's (static) source code into the targeted machine code. It runs several passes starting from lexical analysis and finishing with binary generation. Additional protection passes can be integrated into the compiler pipeline.

(c) *Postcompile.* After compilation, static program artifacts are transformed into executable binaries or libraries. Since the source code is no longer available in this phase, protection schemes operating at this level have to utilize disassembler tools to recover a representation (normally in assembly language) of the application on which they could carry out protection transformations.

(d) *Link.* This refers to the state in which a *linker* obtains a set of static compiler-generated artifacts, combines them into a single binary, and relocates address layout accordingly. Protections in this phase not only could carry out transformation on the program and all its (static) dependencies,[a] but can also mediate (and potentially secure) the process of binary combination and address space management.

(e) *Load.* This is the stage in which the *loader* (i.e., provided by OS) loads a previously combined executable into the memory, resolves dynamic dependencies, and finally carries out the required address relocations. Unlike the previous states, this state deals with both the in-memory and in-execution representations of assets. More importantly, it is visited every time a program is subject to execution. Therefore, load time protection transformations can turn protection into a running target to render attacker's previous knowledge about the protection irrelevant. Loader transformations, apart from being executed on every program start, can also verify (and potentially protect) dynamically linked dependencies.

(f) *Run.* This stage begins as soon as the loader triggers the program's *entry point* instructions and lasts until the program is terminated. Run also operates at the in-memory and in-execution representations of the assets. Moreover, the run state can constantly mutate a program to alter off-board or on-board protections. Although at

[a] Except for dynamic dependencies.

runtime dynamic dependencies are already resolved, still a protection mechanism can authenticate the loaded dependencies and decide upon unloading or reloading them.

The system view captures critical elements along with their various representation in the system that attacker has interest in undermining their integrity. Both the attack and defense views are applied on a system and hence the system view is the base of the taxonomy. In the following, we first elaborate on the attack view aiming for identifying potential threats and techniques to violate system integrity assets. Later on we discuss the defense view that provides means to mitigate or raise the bar against identified threats (from the attack view) and common attack tools.

3.2 Attack

The attack dimension expands over the *attacker* view, encompassing high-level attacks, tools and their relation to other dimensions of the taxonomy, viz., system view and defense view. As depicted in Fig. 4, the attack view is comprised of the `Attacker`, `Reverse engineering`, `Tools`, `Discovery`, and `Attack` main classes. Attacker represent the perpetrator whose goal is to harm system assets by violating system integrity. To do so, attackers use reverse engineering techniques. Reverse engineering is a process in which an attacker utilizes (offensive) tools to discover assets and possibly protection mechanisms. The process normally ends with manifestation of a concrete attack that harms system integrity after defeating protection measures (if any).

In the following, we discuss each main class of the attack view in more details along with their subclasses.

1. *Reverse engineering* (*RE*). RE is located at the heart of the attack view. It encompasses attack, attacker, discovery, and tools. RE aims at harming system assets.
2. *Attack*. Tampering attacks themselves can have multiple forms (inheritance) based on which representation of the assets they are applied to. Attacks are carried out on a representation of the assets; thus there is a dependency between attacks and system representations. In the following, we elaborate on different attacks on different asset representations.
   - **(a)** *Binary patching*. A successful exploit at the file system level can tamper with the static representation of the assets.
   - **(b)** *Process memory patching*. An exploit at the program level or OS level may enable attackers to directly tamper with the process memory.

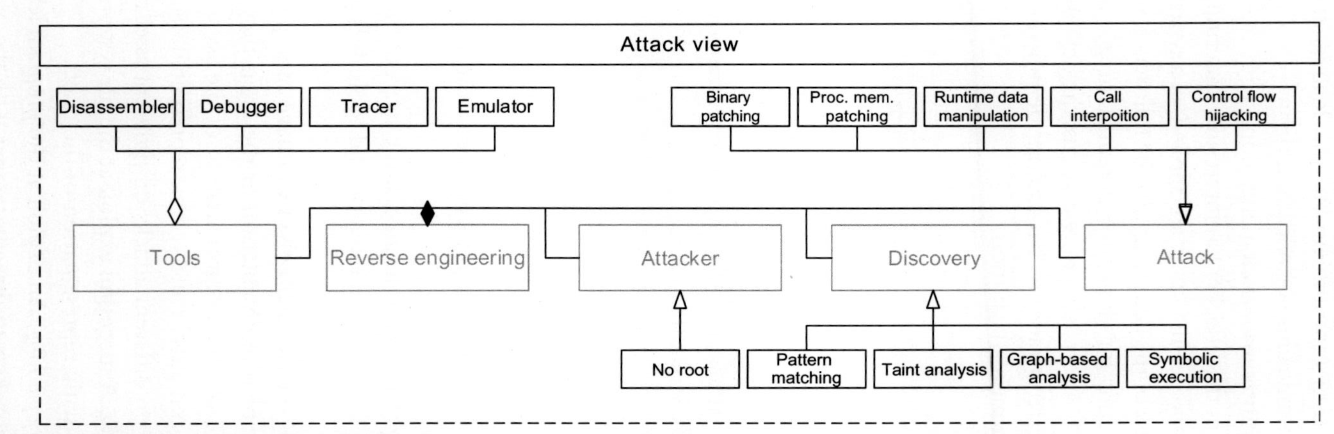

Fig. 4 Attack view dimension of the taxonomy showing the attack main classes, subclasses and relations.

(c) *Runtime data manipulation.* Similar to memory patching, program or OS level compromises can enable an attacker to tamper with the program's dynamic data that includes the input to a program, the produced output by a program. The runtime data are allocated in system stack and heap.
(d) *Call interposition.* Once a system is compromised, attackers may, for instance, intercept system calls to inject malicious behavior. Call interpositions are out of the scope of software protection, they rather fall under infrastructure security. For this reason, we did not include them in our paper survey. But we still believe it is a potential threat to software integrity and hence should be listed in the taxonomy.
(e) *Control flow hijacking.* Calls and branches define the execution flow of the program. Attackers target the program control flow in a wide range of attacks (e.g., return-oriented programming [24] and buffer overflows).

3. *Discovery.* In order for attackers to violate the integrity of a protected system they need to identify assets or protection routines in a given representation of a program. We name this phase the *discovery* phase. This implies a relation between the system representation and discovery.

   Banescu et al. [25] formulated attacks as search problems (e.g., identifying protection guards in the application can be formulated as a search problem). To the best of our knowledge, no study has addressed the difficulty of discovering protection techniques. In our survey, however, we found four techniques that are commonly referred in the literature, viz., `pattern matching`, `taint analysis`, `graph-based analysis`, and finally `symbolic execution`. Therefore, we resort to these four common approaches in discovering protection measures.

   (a) *Pattern matching.* Manually analyzing a large and complex program is labor and resource intensive. Therefore, normally attackers try to identify and defeat protection mechanism by employing automated attacks based on pattern matching, using, for example, *grep*. Pattern matching is not limited to search for strings, it also can search for properties of an artifact (e.g., entropy). Such attacks are plausible if and when an application commits to the usage of a recognizable pattern in their protection. Note that pattern matching can be applied on all representations.
   (b) *Taint analysis.* Tainting analysis is a technique in which the influence of a particular input on program instructions can be examined.

This tool can facilitate the detection of protection routines for perpetrators. For instance, taint analysis can help attackers to find the connection between check and response functions by following the influence of check variables on response instructions [26].

(c) *Graph-based analysis.* A protected program can be represented as a graph in which basic blocks are the graph nodes and jumps express edges. Dedić et al. in [16] argue that in most cases protection nodes are weakly connected as opposed to the other nodes and hence easier to detect.

(d) *Symbolic execution.* This enables adversaries to see which inputs to the program triggers the execution of which part of the program. Since the program is actually being executed to discover execution paths, nothing can remain unseen for symbolic execution engines as long as the constrain solver is successful. This unique feature of symbolic execution can enable attackers to visit all hidden (obfuscated/encrypted/dynamically loaded or generated) instructions of the protection scheme.

4. *Tools.* An attacker can utilize a set of tools to support reverse engineering activity, e.g., to carry out attacks or to identify assets or defense measures in place. To the best of our knowledge, the resilience of different schemes against reverse engineering tools has not been studied. Therefore, in our taxonomy we made a set of assumptions to decide whether to mark a scheme resilient against a particular tool or not. We will discuss these assumptions with substantial details in Section 5.7. In the following we discuss some generic tools that reverse engineers normally use.

(a) *Disassembler.* An attacker may utilize dissembler to disassemble a binary potentially to analyze the protection logic.

(b) *Debugger.* Another tool that enables the attacker to monitor the execution of a program in a slow paced sequential manner. In this attack, debugger can access or alter any runtime data.

(c) *Tracer.* Analyzing program's execution traces could potentially reveal protection mechanism. This becomes more useful when a program employs obfuscation/encryption to hide its logic. However, the traces can reveal the executed instruction (after decryption and deobfuscation). In this event, attackers may employ more intelligent analysis to defeat a stealthy protection.

(d) *Emulator.* Attackers can employ emulators to study program execution in a lab manner. All system calls and executed instruction can be closely monitored and thus deepen the knowledge of the program internals. With the help of snapshots, steps could be reversed to recover from faulty states, which in turn facilitates the attack.

5. *Attacker.* The actor who carries out disrupting actions to violate the integrity of the system is the attacker. When classifying attackers, distinguishing RMATE and MATE attackers, although appealing, is out of reach. The reason being that the two have very much in common; the main difference is the physical access that MATE attackers possess. Therefore, we classify our attackers to two groups: *attackers with root privileges* and *attackers without root privileges*. This is indicated with `no root` title (corresponding to attackers without root accesses) in our taxonomy in Table 2.

The attack view captures potential threats along with the tools and techniques that attackers can use to violate system integrity. The same attacks could be used by adversaries to circumvent or even defeat protection mechanisms as well. Therefore, in order for defense mechanisms to be effective, it is crucial that they resist against attacks and tools.

3.3 Defense

This dimension can be seen as the bridge between the system and the attack dimensions. It serves as the core of the taxonomy by capturing the activities in which system assets are protected using a set of measures against potential threats. As can be seen in Fig. 5, the defense dimension is comprised of four main classes: `measure`, `protection level`, `trust anchor`, and `overhead`.

The measure refers to a method that is employed to mitigate integrity attacks on programs. Protection level indicates the level of abstraction at which protection is employed.

Trust anchor specifies whether the measure relies on any root of trust, e.g., trusted hardware, or not. Finally, the overhead reports on the performance impact of a measure on the system.

1. *Measure.* A protection measure is the foundation of the protection activity. It can be done either completely locally (i.e., a program verifies its own integrity) or remotely (i.e., a trusted party remotely attests integrity). The inheritance relation between the measure, *local* and *remote verification*, stands for this matter. Each measure itself is composed of five distinctive actions: *transform*, *monitor*, *check*, *response*, and *harden*. These classes, although pursuing different objectives, contribute in protection a program. In the following, we describe the role of each action in the protection process.

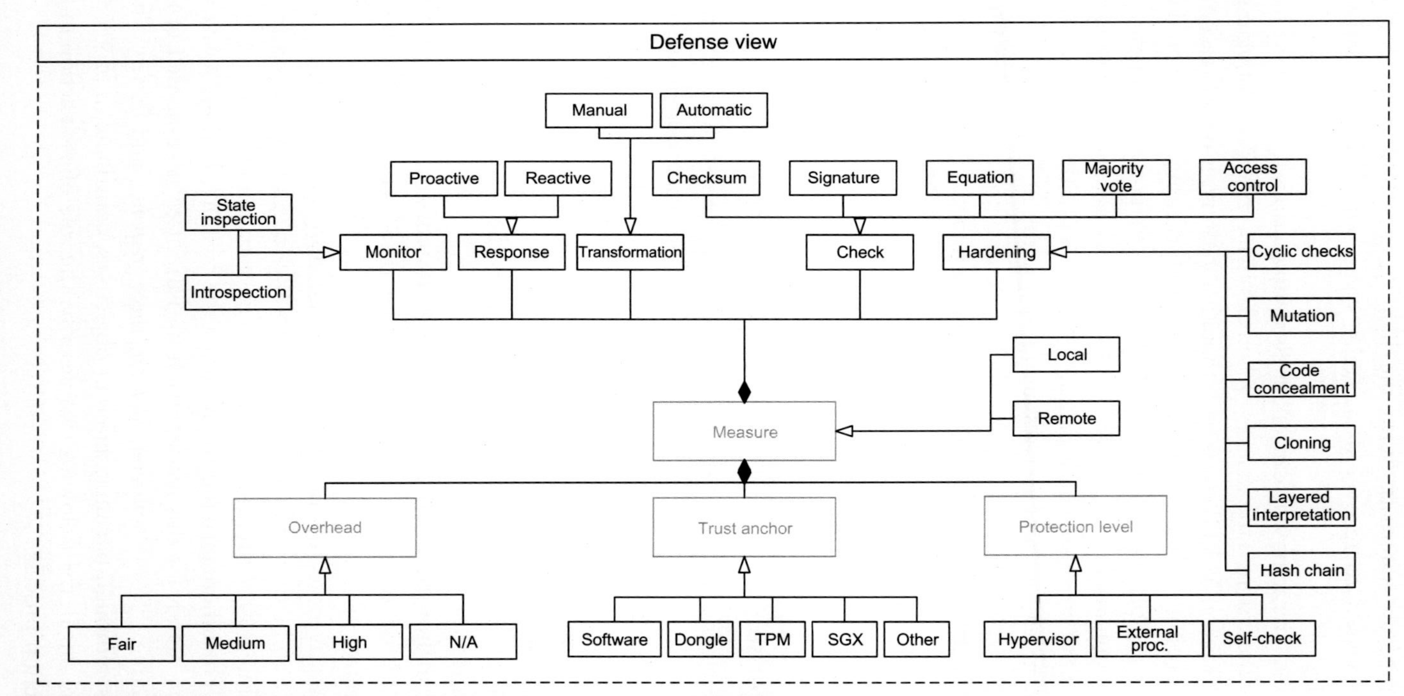

Fig. 5 Defense view capturing the main classes, subclasses, and relations of the defense dimension.

(a) *Transform*. This action applies transformations on representations of assets. As aforementioned, these transformations can be carried out at different program life cycles depending on the targeted representation and the artifact at hand. This indicates a link between the transform and the life cycle activity.

(b) *Monitor*. This component can actively inspect different representations of the system assets. Obviously, this component has to have an access to the system, otherwise it cannot audit anything. Hence, there is a tight binding between monitoring and representation. For monitoring system representations, there are two approaches that are widely used in the literature, viz., introspection and state inspection.

  i. *Introspection* monitors a set of static invariant properties of a program, e.g., code blocks, to detect tampering attacks.

  ii. *State inspection* monitors the result/effect of the execution of a program, e.g., function return values, to reason about its integrity.

(c) *Check*. This component provides a mean to reason about the collected data by the monitor element and decides whether assets are compromised or not. The output of this component, based on the employed technique, could be a binary (Yes/No) or accompanied with a confidence number. In the following we discuss the checking mechanisms that are commonly used in protection schemes.

  i. *Checksum*. We classify any mathematical operation that converts large block of data into compact values, which is ideal for comparisons, as checksum-based techniques. CRC32 and hash functions fall under this category of checking mechanism.

  ii. *Signature*. This refers to a cryptographic protocol for integrity verification by means of verifying the digital signature of artifacts.

  iii. *Equation*. The result of a mathematical equation evaluation (with program runtime features) defines whether program integrity was violated or not.

  iv. *Majority vote*. A set of functionally equivalent components disjointly carry out a computation and then collectively decide upon the integrity of the outcome.

  v. *Access control*. A policy enforcement point beyond the control of the program and (possibly) attacker mediates the access to the critical resources.

(d) *Response*. Based on the check component's decision, the response component reacts in a punishing way to attackers. This reaction could vary from process termination, performance degradation, or even attempting to recover compromised assets. Technically speaking, there are two classes of response mechanism: *proactive* and *reactive*.

i. *Proactive*. Schemes utilizing this class of response act intrusively. That is, upon the report of integrity violating, immediate actions are taken to prevent attacks.

ii. *Reactive*. In some cases less intrusive and stealthier responses are desirable. In this model of schemes, the detection of integrity violation does not result in obvious reactions such as program terminations. A reactive response may, for instance, silently report on the violation without making the attacker or user realize.

(e) *Harden*. Since the measure itself can be subject to attacks, a consolidation technique is employed by the measure. For example a simple routine (say `monitor()`) that is solely responsible for auditing program state can easily get manipulated by an attacker. The same goes for response mechanisms. An open termination of the program as a response to tampers only adds a weak security.

Thus, it is crucial to add strength to protection measure using hardening techniques. Hardening aims to impede discovery process and thus raises the bar against attacks. This directly relates to the discovery and attack activities in the attacker view.

The hardening can be seen as adding a hard problem for the attacker to solve. This does not necessarily have to be a *np-complete* problem, in some cases even a quadratic one will cause enough trouble for attackers and exhausts their resources, specially when the attacker is forced to manually solve the problem. For instance when an automated pattern matching fails to defeat the mechanism, attackers are doomed to manually analyze a large portion of the code, which, in turn, presumably deteriorates their success rates. In our literature review, we have identified six different hardening techniques that are commonly used by integrity protection measures. We discuss these techniques as follows.

i. *Cyclic checks*. In this model, protection is strengthened by using a network of overlapping protection nodes. Therefore, a particular asset (representation) may get inspected by more than one checker. In some variation of the cyclic checks, the checkers themselves are also protected by the very same mean.

ii. *Mutation.* Defeating a protection technique normally is the result of a process in which an attacker first has to acquire necessary knowledge about the scheme and its hardening problem, and then, utilizing the knowledge to break the protection. To this end, mutation techniques try to render attacker's prior knowledge irrelevant by frequently evolving the protection scheme.

iii. *Code concealment.* In the course of scheme analysis, attackers often refer to the program code and base their studies on it. Code concealment tries to impede this process by concealing a particular representation of the code. Nonetheless, program instructions have to get translated into legitimate machine code right before the execution. The later this translation occurs, the more an attacker is troubled. Some approaches tend to flush translated instructions after execution so at no point in time the attacker gets a chance to glance over to the entire application code at once. Program obfuscation and encryption are examples of this technique.

iv. *Cloning.* Multiple copies of sensitive parts are shipped in the application and at runtime a random predicate defines which clone shall be executed. This enables the protection scheme to remain partially functional even after successful attacks. In Section 5.8 we will discuss this measure in the context of concrete attacks.

v. *Layered interpretation.* To utilize layered protection a program has to be run within an emulator or a virtual machine. Some schemes entirely virtualized the target application, while others virtualize some parts of the application. This technique enables employment of protection measures at a higher level of abstraction. Since programs has to be run by the host (hypervisor/emulator), they can be verified before execution and executed only if they pass the verifications.

vi. *Hash chain.* Evolving keys and hash chain assures past events cannot be forged or forgotten. This technique is widely utilized to maintain an unforgeable evidence of the system state.

2. *Trust anchor.* In an abstract sense, this criterion specifies whether the scheme is entirely implemented in *software* or it is assisted by a trusted hardware module. Hardware-based approaches are generally assumed to be harder and more costly to circumvent, due to the required equipments and knowledge at the attacker end. On the other hand, hardware-based schemes add more cost, to acquire the required modules, for each client. In the following we discuss different trust anchors.

  (a) *Software.* This indicates that the scheme security is purely based on software hardening mechanisms without any trusted hardware whatsoever.
  (b) *Dongle.* This refers to hardware peripherals that could be used to store small chunks of data (e.g., keys) or program with more resilience against tampering attacks. None of the reviewed schemes in our survey utilizes dongles. Thus, we exclude it from our further classifications. However, according to [27] dongles are popular in practice.
  (c) *TPM.* Trusted platform module is a quite mature hardware chip that is used by plenty of hardware-assisted integrity protection schemes.
  (d) *SGX.* Intel SGX offers dynamic process isolation to mitigate runtime integrity attacks.
  (e) *Other.* An indicator for reliance on hardware modules other than the explicitly stated ones.

3. *Protection level.* This indicates the enforcement locality of a protection scheme. It can happen (stated with inheritance relation) in three different abstraction levels:

  (a) *Self-check.* This level is also known as internal protection in which a protection scheme is integrated into the program to be protected. Simply put, this process is in charge of carrying out its own integrity verifications and further decisions about how to react to compromises. Applying this technique to a distributed architecture adds more resilience. Mainly because the protection schemes react disjointly, a single point of failure is prevented. On the negative side, however, all the decisions are solely based on the state of the program-to-protect, thereby these decisions may lack contextual information.
  (b) *External process.* A dedicated process, integrity protection process (IPP), monitors protected programs, and responds accordingly. Unlike self-check approaches, IPP can combine multiple sources

of contextual information to improve reasoning and thus responses. Also, it enables a mean for high-level policy enforcement. These features, in turn, enhance both security and usability of a protection measure. However, there are two drawbacks in this model: (a) it is challenging to capture the actual state of the programs via an external process, e.g., a malicious program can forge good states for IPP, and (b) granted the high permission level of IPP, it can become a critical component to attack; single points of failure.

(c) *Hypervisor*. Protection is a part of a hypervisor logic. This entails that all processes are being virtualized and executed on top of the security hypervisor. The current state of the industry highly advocates this model. The hypervisor, in contrast to external process, does not have the problem of state forgery, as it (in theory) can capture all stealthy actions. Nevertheless, the problem of single point of failure remains as a concern.

4. *Overhead*. Performance overhead is an important aspect of integrity protection schemes. Practitioners indeed need an estimation of the overhead on their services. However, a great deal of the protection schemes has not been thoroughly evaluated, so performance bounds are unknown. We classified performance bounds into four classes as follows.

(a) Fair ($0 < \textit{overhead} < 100\%$).
(b) Medium ($101\% < \textit{overhead} < 200\%$).
(c) High ($\textit{overhead} > 201\%$).
(d) N/A. This class represents schemes with lack of experiment results or unjustifiable numbers due to limited experiments.

The defense view elaborates on techniques to protect integrity against attacks along with implications of such protections on system. Hence defense acts as a bridge between system and attack views. Moreover, defense view can be seen as a classification for integrity protection techniques, which can facilitate scheme selection by users.

3.4 View Dependencies

Previously, we discussed all the three views of the taxonomy, their classes and dependencies in each view. In this section we provide an overview of the dependencies between the classes from different views. The high-level overview of these dependencies is depicted in Fig. 6.

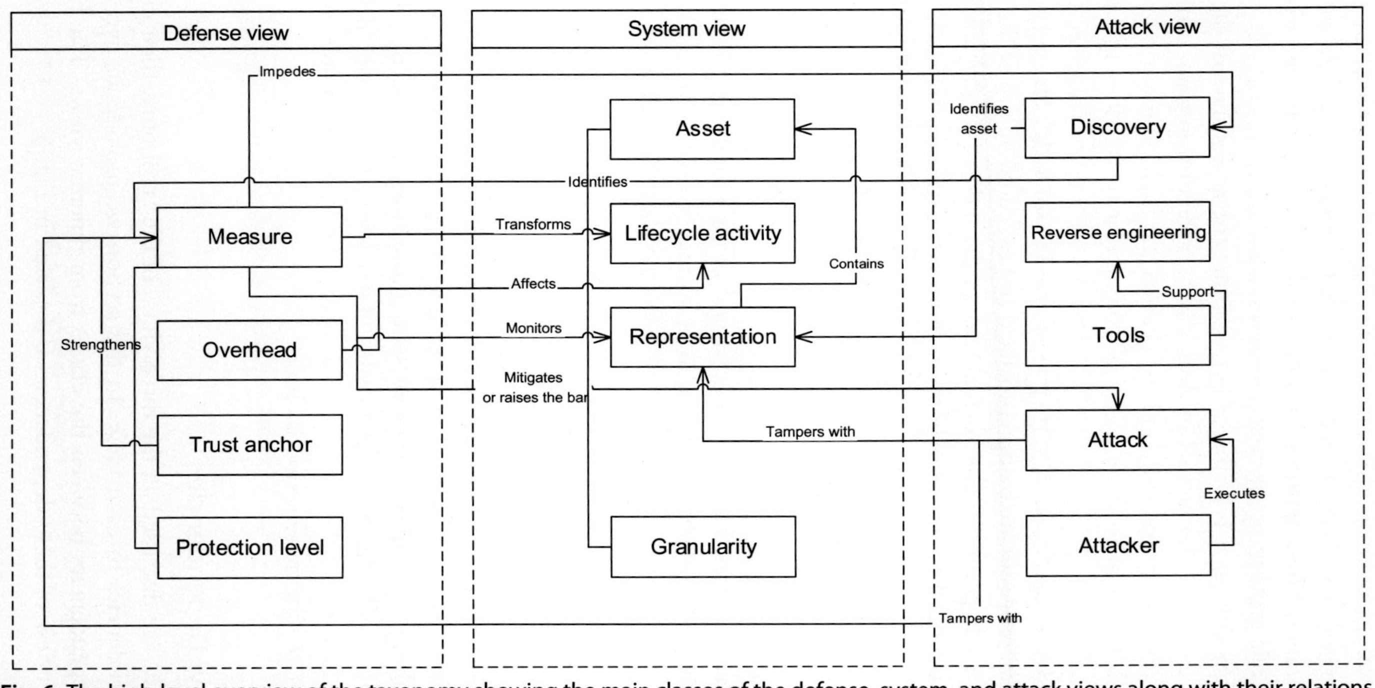

Fig. 6 The high-level overview of the taxonomy showing the main classes of the defense, system, and attack views along with their relations.

In the system view (in the middle), an association between the representation and asset (Contains) expresses the fact that system assets are exposed in different representations.

In the attacker view (on the right), the support of the tools in the whole process of reverse engineering is expressed by the link between the tools and reverse engineering (the "Supports" link). Attackers are the actor to execute an attack, this is also highlighted by an association between attacker and attack. The asset identification, which is the first objective of attackers, is depicted in form of the association between the discovery and representation classes (the "Identifies assets" association). The second objective of attackers, after identifying assets, is to tamper with the representation of interest. This is captured by the link between the attack and the representation (captioned with "Tampers with").

The defense view (on the left) abstractly captures the defense process. A protection measure may rely on a trust anchor for additional strength, which is shown with the link captioned with "Strengthens." In order to enable protection, a measure needs to transform a representation of the system in one (or more) of the life cycle activities. This is illustrated by the association between the measure and the life cycle activity (Transforms). Once a protection measure is in place, one (or multiple) representation of a system is actively or passively monitored, the "Monitors" association between the measure and the representation shows this relation. Protection measures could potentially affect system performance and introduce overheads, which is shown with the "affects" link between the overhead and life cycle activity. Protection measures based on their hardening measure and response mechanism impede asset discovery and mitigate or raise the bar against attacks. The "Impedes" link from the measure to discovery (in the attack view) and "Mitigates or raises the bar" between the measure and the attack depict these two interrelations.

However, protection measure is not the end of the game for attackers. They can target the protection mechanism itself, discover its logic and disable its protection. The "Identifies" link between the discovery and the measure as well as "Tampers with" between the attack and the measure captures this matter.

At this point we have covered all the three views of the taxonomy and their interrelations. For a complete view, Table 1 depicts the entire taxonomy. To support readability, for each element of the taxonomy we give an example in the context of sample DRM system. All the examples relate to the DRM system, and thus we avoid repeating the context in each example.

Table 1 Software Integrity Protection Taxonomy Along With Examples in the Context of the Sample DRM

System View	Assets	Behavior		There are some sensitive logics that need to be integrity protected, e.g., license checker routine and stream viewer
		Data		Usage variables need to be protected against manipulation attacks. Otherwise, attackers can readily circumvent the usage policies
		D&B		The usage variables and any routine that has deals with these variables, in this case content provider, need to be protected
	Representation	Static		Malware or attacker may tamper with the DRM binaries to carry out their attacks
		Mem.	Code Invar.	The static shape of the license checker code blocks in the memory is a target to attack for attackers
			Data Invar.	Usage count variables in memory are the targets of tampering attacks
		Exe.	Trace	Tampering attacks causes identifiable changes in the program traces. For instance, dumping decrypted fragments in the stream viewer introduces additional calls in the trace
			Timed Trace	Tampering with program routine will alter the program response time. For instance disabling license check routine by simply returning true could introduce new timing bounds
			HW Counters	Hardware performance counters capture some technical details about the executing routine, such as number of branches, and indirect calls. Tampering with the license checker inevitably affects these numbers.
	Granularity	Instructions		All the individual instructions that access $client_{private}$ key need to be protected
		BB		The basic block in which license checking takes place need to be protected
		Function		The licenseChecker() and getExpiryDate() methods need to be protected
		Slice		The chain of instructions that read from or write to the usage count variables need to be protected
		Application		The entire stream viewer component (application) need to be protected

	Life Cycle a.		Precompile	License checker and data provider can be protected at the source code level
			Compile	License checker and data provider can be protected at the compiler level
			Postcompile	Stream viewer can only be protected in a postcompile process, because it is an external library
			Load	Every time the license checker is loaded in the memory a (new) protection is applied
			Run	The license checker routine is updated on the fly (during execution)
Attack View	Reverse Engineering	Tools	Disassembly	Attackers may disassemble the DRM client to understand the logic
			Debugger	Attackers could inspect the license checking control flow using a debugger
			Tracer	Attackers could deepen their knowledge about the fragmented decryption in the stream viewer by dumping a trace of the program
			Emulator	An emulator could be utilized to facilitate program analysis
		Discovery	Taint Analysis	Attacker can use tainting analysis to detect the checks that enforce usage policies
			Sym. Exec.	Symbolic execution could be utilized to find instances of usage variables for which the program delivers the protected contents
			Memory Split	Attackers could execute memory split attack to defeat self-checksumming-based protections
			Patt. Match.	Attackers could use pattern matching to detect license checks, protection routines, and usage policy enforcement points
		Attack	Binary	Attackers can tamper with the program executables at rest
			Proc. Mem.	Attackers could leverage in-memory patching to carry out their attacks
			Run. Data	Program input and output data is under control of the attackers.
			Ctrl Flow	Attackers can subvert program control flow to for instance decrypt content without license check

Continued

Table 1 Software Integrity Protection Taxonomy Along With Examples in the Context of the Sample DRM—cont'd

Defense View	Meas.	Local	The DRM client is fully in charge of the integrity protection
		Remote	Security state of the DRM client is to be reported to the main server
	Mon.	State Inspection	Monitor and check the results of the program execution
		Introspection	Read code pages of the license checker compare it to a known value
	Resp.	Proactive	Any detected attack raises the alarm and a delayed or an immediate punishing response is triggered
		Reactive	All accesses to the protected content (genuine or forged) is permitted; however, the server is notified about the violation of the usage policies in a postmortem verification. Server can add such users into a blacklist or file a lawsuit against them
	Trans.	Manual	Protectors have to execute manual tasks in the course of protecting a program, for instance, by developing nontrivial clones for the license checker
		Automatic	The protection transformation requires a limited user intervention, and hence it is to a great extent automatic
	Check	Checksum	A checksum is computed over the code blocks of the license check
		Signature	Signature of the components are verified at runtime
		Equation Eval.	Access data form a set of verifiable equation, e.g., total number of accesses is less than total number of views specified in the usage policy
		Majority Vote	License checker is cloned and executed simultaneously by different threads/processes. Afterwards, a majority vote decides whether the license is valid or not
		Access Control	All access requests need to pass through a master node which securely enforces usage policies

Hardening	Cyclic Checks	A network of integrity checkers collectively prevents software manipulation
	Mutation	The license checker code blocks are constantly modified at runtime to mitigate passive attacks
	Code Conceal.	The License checker code is encrypted and only decrypted at runtime
	Cloning	The license checker is duplicated and every time at runtime a randomly chosen clone is executed
	Layered Interp.	The client application utilizes a random instruction set that can only be executed on a custom secure virtual machine
	Hash Chain	Usage event logs are securely chained to maintain an unforgeable evidence of the system usage. Such information could be used in postmortem verifications
Overhead	Fair	Target clients have limited computation resources, and thus a low overhead is desired
	Medium	Target clients have fairly good computation resources
	High	Sensitive parts of the program shall be protected strictly, high overheads are tolerated
	N/A	No information about overhead constraints
T. Anchor	TPM	Client systems do have a TPM chip
	SGX	Client systems support Intel SGX
	Other	Other trusted modules are decided to be used, e.g., a custom-built hardware
	Software	The protection is completely software based
Prot. Lvl.	Internal	The client application is in charge of verifying its own integrity
	External	A dedicated integrity protecting process frequently verifies the integrity of client applications
	Hypervisor	A secure hypervisor is shipped with the client installation bundle. The client application can only be executed using the provided hypervisor

Table 2 Applied Taxonomy on the Surveyed Literature

PaperID	System View																			Attack View					Defense View																													
	Assets				Representation					Granularity					Life Cycle Activity					No Root	Attack				Meas.		Mon.		Resp.		Trans.		Check				Hardening							Overhead				Trust Anchor				Prot. Lvl.		
					In Mem.		In Exec.																																																
	Behavior	Data	Data and Behav.	Static	Code Shape	Data Shape	Trace	Timed Trace	HW Counters	Instructions	BB	Function	Slice	Application	Precompile	Compile	Postcompile	Load	Run		Binary	Process Memory	Runtime Data	Control Flow	Local	Remote	State Inspection	Introspection	Proactive	Reactive	Manual	Automatic	Checksum	Signature	Equation Eval.	Majority Vote	Access Control	Cyclic Checks	Mutation	Conceal.	Cloning	Layered Inter.	Hash Chain	Fair	Medium	High	N/A	TPM	SGX	Other	Software	Internal	External	Hypervisor
[28]	x						x			x							x			x				x	x		x		x			x												x							x	x		
[29]	x				x						x						x					x			x			x	x		x		x	x						x							x				x	x		
[8]		x					x			x			x		x					x			x	x	x		x		x		x						x			x					x						x	x		
[7]	x				x						x						x			x		x			x			x	x			x	x					x		x					x						x	x		
[19]	x	x	x		x									x			x					x	X			x		x	x			x					x					x		x					x				x	
[30]	x						x			x														x	x		x		x			x															x				x		x	
[31]	x	x	x		x							x					x			x		x			x		x		x			x					x					x	x			x		x		x				x
[32]		x				x							x		x					x			x		x		x		x			x															x				x	x		
[33]		x				x				x					x					x			x		x		x		x		x																x				x	x		
[34]	x			x										x			x			x	x				x			x	x		x			x													x				x		x	
[35]	x				x						x						x					x			x			x	x		x		x					x								x					x	x		
[36]	x						x						x			x							x	x	x		x		x		x																x				x	x		
[37]	x													x			x			x					x			x	x		x						x					x		x							x			x
[38]	x			x								x							x		x					x			x			x							x					x							x		x	
[16]	x			x							x										x				x			x	x		x					x											x				x	x		

[39]	x			x									x			x			x	x					x		x	x		x		x													x		x		x	
[40]	x	x	x		x								x	x							x			x			x	x		x		x								x	x	x				x				x
[41]	x										x					x								x			x	x		x										x			x				x			x
[42]		x				x					x					x			x			x			x	x		x		x					x										x		x		x	
[43]	x				x					x						x					x			x			x	x		x						x				x		x					x	x		
[44]	x				x					x						x	x				x			x			x	x		x		x				x	x			x		x					x	x		
[45]	x				x							x				x					x			x			x	x		x		x													x		x	x		
[46]	x					x					x			x								x		x		x		x		x					x	x									x		x	x		
[47]	x						x			x		x				x							x	x		x		x		x									x					x			x	x		
[48]	x							x					x			x					x				x	x		x			x	x												x			x		x	
[49]	x							x					x			x					x				x	x		x			x	x											x				x		x	
[50]	x						x						x	x									x		x	x			x		x	x	x		x						x		x				x	x		
[51]	x				x							x			x						x			x			x	x			x	x								x					x		x	x		
[52]	x				x								x			x					x			x			x	x		x		x									x					x			x	
[53]		x				x					x			x								x			x	x		x			x													x			x		x	
[54]		x				x						x		x					x			x			x	x		x			x			x							x	x				x			x	
[55]	x			x									x			x			x	x				x			x	x		x			x												x		x		x	
[56]	x				x								x			x			x		x			x			x	x			x	x	x							x					x		x		x	
[57]	x						x				x			x									x	x				x		x									x						x		x	x		
[58]	x				x							x						x			x			x			x	x			x						x	x					x				x	x		
[59]	x								x							x			x				x	x		x		x		x				x							x	x				x		x		

Continued

Table 2 Applied Taxonomy on the Surveyed Literature—cont'd

	System View																			Attack View					Defense View																													
	Assets				Representation					Granularity					Life Cycle Activity					No Root	Attack				Meas.		Mon.		Resp.		Trans.		Check					Hardening						Overhead				Trust Anchor				Prot. Lvl.		
					In Mem.		In Exec.																																															
PaperID	Behavior	Data	Data and Behav.	Static	Code Shape	Data Shape	Trace	Timed Trace	HW Counters	Instructions	BB	Function	Slice	Application	Precompile	Compile	Postcompile	Load	Run		Binary	Process Memory	Runtime Data	Control Flow	Local	Remote	State Inspection	Introspection	Proactive	Reactive	Manual	Automatic	Checksum	Signature	Equation Eval.	Majority Vote	Access Control	Cyclic Checks	Mutation	Conceal.	Cloning	Layered Inter.	Hash Chain	Fair	Medium	High	N/A	TPM	SGX	Other	Software	Internal	External	Hypervisor
[60]	x							x				x					x									x	x		x		x		x														x				x		x	
[18]	x				x									x			x					x			x		x		x		x		x									x	x	x				x		x				x
[17]		x		x										x							x					x		x		x	x		x										x				x	x					x	
[61]							x				x						x			x				x	x		x		x			x												x							x	x		
[62]							x							x			x			x				x	x		x		x			x					x							x							x		x	
[63]	x			x								x					x			x		x				x		x	x		x												x	x				x						x
[64]	x				x							x					x					x			x			x	x		x		x							x						x					x	x		
[65]	x							x				x					x									x	x		x		x		x											x							x		x	
[66]	x				x			x	x			x			x							x			x		x		x		x																x				x		x	
[67]		x				x								x						x			x			x	x		x		x					x											x				x		x	
[68]	x			x									x			x				x	x				x		x		x		x															x					x	x		
[69]	x				x						x						x					x			x				x		x		x							x					x						x	x		
[70]	x				x									x			x					x			x			x	x		x		x									x		x							x		x	x

4. APPLYING THE PROPOSED TAXONOMY

We conducted a survey on integrity protection and tamper resistant techniques for software systems. This by no means is exhaustive, but we believe it serves as a representative of different techniques in the literature. Table 2 demonstrates the mapped literature on the proposed taxonomy.

Since there is no single research work that analyzes the resilience of protection schemes against MATE tools and discovery methods, we were not able to directly map the reviewed works to these criteria (in the attack dimension). To address this limitation, in Sections 5.7 and 5.8 we reason about schemes' resilience based on their hardening methods. This enables us to first understand the role of hardening measures in the resilience of schemes and subsequently to complete the mapping. The space limitation hinders detailed discussion of the reviewed schemes. Instead, in the next section we correlate various aspects of the reviewed literature and discuss our findings.

5. OBSERVATION AND ANALYSIS

As part of our analysis we correlate interesting elements in the taxonomy from different views and discuss our findings. These correlations are those that we particularly find interesting for practitioners.

However, there are far more possible correlations that one can carry out. To address this concern, we published a web-based tool that enables end users to correlate any arbitrary pair of elements from different dimensions. The tool is accessible at http://www22.in.tum.de/tools/integrity-taxonomy/. In the following we report on the dependencies of the taxonomy dimensions.

5.1 Asset Protection in Different Representations

We defined integrity as a property of software that applies to both data and behavior. From the user perspective, it is a daunting task to find protection schemes that protect data and/or logic at a particular representation.

Signature verification is one of the techniques that users commonly utilize to protect integrity. The static signature verification is natively supported by almost all operating systems. These schemes are rather a nonpreventive security measure to protect users, but not softwares. Clearly, this does not match the MATE attacker model, where the user is also the attacker.

Besides, these techniques have two shortcomings: (i) their security relies on operating system settings, which could be easily manipulated, and

(ii) they suffer from the *time of check–time of use* (*ToCToU*) limitation in which signatures are verified on the static representation of programs upon execution, while many attacks (including in-memory patching) could potentially occur after execution is started. Kanuparthi et al. [71] reported that over 60% of attacks published by CERT were either buffer overflows or of ToCToU attacks.

Due to the aforementioned limitations, the static signature verification is not an optimal technique as it fails to protect other asset representations during execution.

A natural question that may come to ones mind is why protection for a certain representation instead of protecting assets in all the representations. The answer is threefold. First, attackers might not have the right privileges on some representations, e.g., accessing in-execution representation of processes requires *root* privilege, except when attackers find a way to inject their attacks into the program of interest. Reasonably, to avoid unnecessary overhead we aim to protect the representations that are at risk. Second, some risks might be acceptable for a certain use case, e.g., in IoT programs naively protecting static representation of a program might be sufficient, given that runtime protection imposes unacceptable overhead. Finally, 100% protection of assets in all representations is technically infeasible. Therefore, to shed light on this concern, we correlate software integrity assets to different representations in order to identify corresponding protection schemes.

The desired correlation is depicted in Table 3. In the following we discuss the relevant schemes according to their asset-to-protect.

5.1.1 Behavior Protection Schemes

As the correlation suggests, a majority of behavior protection schemes with a total number of 18 [7, 18, 19, 29, 31, 35, 40, 43–45, 51, 52, 56, 58, 64, 66, 69, 70] target the *code invariants* representation in their protection mechanism.

Meanwhile, [46] referred to the shape (invariants) of computed data to reason about the behavior integrity. Refs. [59, 66] took hardware performance counters (e.g., the number of indirect calls and elapsed times) into account to evaluate programs' integrity.

Some efforts were found in the literature to raise the bar against program manipulation when programs are at rest (binary). Refs. [34, 39, 68] are examples of hardening binaries by statically analyzing them and subsequently adding resilience against tampering attacks. Dedić et al. [16] rather presented a

Table 3 The Correlation Between the Integrity Asset Protection and Representation

Representation	Integrity Assets		
	Behavior	**Data**	**Data and Behavior**
Static	[16, 34, 38, 39, 55, 63, 68]	[17]	
Code inv.	[7, 18, 19, 29, 31, 35, 40, 43–45, 51, 52, 56, 58, 64, 66, 69, 70]	[19, 31, 40]	[19, 31, 40]
Data inv.	[46]	[32, 33, 42, 53, 54, 67]	—
HWC	[59, 66]	—	—
Timed trc.	[48, 49, 60, 65, 66]	—	—
Trace	[28, 30, 36, 47, 50, 57]	[8]	—

theoretical concept to design more resilient integrity protection schemes. Refs. [38, 55, 63] proposed techniques to protect program binaries after distribution.

In order to manifest some tampering attacks, attackers inevitably need to change, inject, or reorder instructions. In this situation, the order in which program instructions are being executed along with the data that they read from or write to, i.e., a program trace, is a source of knowledge to detect such attacks. This information could be extracted form a program's execution trace. Refs. [36, 47] incorporated memory accesses and branching conditions in the program trace into a hash variable to later match it against the expected trace hash. Jin and Lotspiech [50] passively verified the execution traces that are reflected into log records. Kulkarni and Metta [57] duplicated sensitive regions in the program to add more resilience to attacks by forcing perpetrators to tamper with all clones of sensitive codes.

The correlation also indicates that timing analysis (timed trace) is utilized in [28, 30, 48, 49, 59, 60, 65, 66] to verify integrity of the program behavior. The assumption in the mentioned schemes is that a certain untampered routine expresses identifiable execution time characteristics.

5.1.2 Data Protection Schemes

Tampering with a program's data can enable attackers to manifest integrity violating attacks. The simplest way to do so is by attaching a debugger and subverting the program flow, for instance, by flipping a register's value. As the correlation suggests, several schemes protect the program's data integrity. Karapanos et al. [53] introduced a hash chain alike concept for data integrity with the help of a semitrusted entity. Sun et al. [67] equipped system data flow analyzer nodes and routed the traffic to them for integrity analysis. Gao et al. [42] proposed to duplicate sensitive processes and compared their computed values based on a majority vote scheme. Castro et al. [33] used data flow graph to verify whether the application conforms to the genuine data flow model at runtime. Kil [54] collected data invariants in a program and verified them at runtime. Chen et al. [36] computed a cumulative hash of memory accesses at the instruction level to form a hash trace of the program execution.

5.1.3 Data and Behavior Protection Schemes

Protecting data and behavior together might be a more appealing option when the program to protect possesses both sensitive data and logic. In our survey, we found three schemes that aim for data and logic protection. Dewan et al. [40] proposed a technique in which a program's logic as well as data are isolated from external processes by a secure hypervisor. They also designed a process to encrypt data before persistence, so that the data is never exposed to adversaries in plain text. Brasser et al. [31] in addition to logic and data isolation, introduced a secure interprocess data propagation in a (nearly) real-time manner. Baumann et al. [19] designed a fully isolated execution environment for the process as a whole on Intel SGX hardware commodity. All the three schemes rely on trusted hardware.

The presented correlation singles out relevant schemes for protecting different integrity assets (behavior, data, and data and behavior) in different representations. This facilitates the scheme selection based on the assets and representations at the user end.

5.2 MATE Attack Mitigation on Different Asset Representations

A program possesses different representations depending on its state, viz., static, in-memory and in-execution. Each of theses representations is subject to tampering attacks. Protections on representations closer to the time of use (i.e., execution) could prevent or detect a wider range of tampering attacks. That is, the static representation of a program is implicitly protected, if the very same program utilizes in-memory protections.

In this chapter, we limit our adversary model, regardless of how attacks are executed, to four generic attack goals: binary patching, process memory patching, runtime data modification, and control flow hijacking. Constrained by their permissions, attackers can choose different program representations as their attack targets.

Table 4 Defense Mechanisms Against Tampering Attacks for Different (Asset) Representations

Tampering Attacks	**Representation**					
	Code Inv.	**Data Inv.**	**HWC**	**Static**	**Timed Trace**	**Trace**
Binary	—	—	—	[16, 17, 34, 38, 39, 55, 68]	—	—
Control flow	—	—	[59]	—	—	[8, 28, 47] [30, 36, 50, 57, 61, 62]
Process memory	[7, 18, 19, 29, 31, 35, 40, 43–45, 51, 52, 56, 58, 64, 66, 69, 70]	—	[66]	[63]	[48, 49, 66]	—
Runtime data	—	[32, 33, 42, 46, 53, 54, 67]	—	—	—	[8, 36]

In this correlation we aim to identify protection techniques that mitigate the aforementioned generic attacks on different representations. For this purpose, we relate representations in the system view to attacks in the attack view. This is depicted in Table 4. We discuss our findings classified by attacks, viz., binary, control flow, and process memory as follows.

5.2.1 Binary

Static manipulation (aka static patching) targets a program while it is at rest (prior to execution). To prevent these sort of attacks the surveyed literature suggests static representation verification as well as a set of hardening measures to raise the bar against tampering attacks in the host. Note that we do not include signature verifications supported by operating systems, as they do not target MATE attackers. In the following we review the schemes that protect the static representation of programs.

5.2.1.1 Static Representation Verification

Kim and Spafford [55] and Deswarte et al. [39] proposed remote file integrity protection using file system hashing. The main difference lies in the fact that the latter initiates the hash with a remote challenge. However, this technique can easily be defeated if the attacker keeps a copy of the genuine files just for the sake of hash computations.

5.2.1.2 Hardening Measures

Neisse et al. [17] suggested to plug (hardware-assisted) configuration trackers in the system to monitor modifications on the file system. All modifications are then signed by a TPM within the host and then reported to a third party verifier to judge about the maliciousness of the actions.

Catuogno and Visconti [34] designed a protection scheme to defend against dependency injecting and tampers with the update scripts in a system by utilizing dynamic signature verification and system call interposition in the kernel. They argued these two can mitigate the persistence of a compromise in the system. Meanwhile, [16] proposed to strongly connect integrity protection code with the normal program control flow and discussed theoretical complexity using a graph-based game.

To add resilience against dynamic attacks such as buffer overflows in the static representation, [68] proposed an efficient means to detect potential integrity pitfalls in a distributed system by first merging all network wired nodes into a unified control flow graph, so-called *Distributed Control Flow Graph* (*DCFG*) and then running a static analyzer to find all user-input reachable buffer overflows. In their way, intuitively, pitfalls can be identified with less false positives as opposed to analyzing each node individually.

Collberg et al. in [38] proposed to protect client programs' integrity by forcing constant and frequent updates. In their approach, a trusted server constantly uses software diversification techniques to change a server's method signatures, which in turn coerces the client to update all the program artifacts right after each mutation.

5.2.2 Control Flow

As the correlation suggests, there are two representations that were utilized to prevent a program's control flow manipulation, viz., HW counters and traces.

Hardware performance counters were used in [59] to reason about the program's control flow integrity based on performance factors such as the number of calls and the elapsed time on each branch.

The trace representation was used by nine protection schemes. These schemes are listed as follows. Banescu et al. in [8] proposed a technique to ensure only genuine internal calls can access assets. This is done via a runtime integrity monitoring that traces stack's return addresses and compare them to a white list of call traces. Stack inspection works well for synchronous calls, however, due to the incomplete trace information, trigger, and forget calls cannot be handled in this way. To cope with asynchronous calls, caller threads are verified using a message authentication code (MAC mechanism).

Abadi et al. in [28] proposed a scheme to protect the integrity of programs' control flow by injecting a set of caller reachability assertions in the beginning of program branches.

Chen et al. in [36] proposed to compute a checksum over a program's execution trace by hashing its assignment and branching instructions. These hashes are later verified at desired locations in the protected program. Similarly, [47] proposed to compute a hash over a program's assignments and executed branches. These hashes later are used to dynamically jump to the right memory locations. Mismatches in hashes crash a program as they potentially cause jumps to illegitimate addresses. Dynamic jumps are created by the mean of interleaving multiple basic blocks in super blocks with multiple entry and exit points. In effect, the hash values determine which entry/exit points the program shall take at runtime.

To capture an unforgeable evidence of the traversed program's control flow, Jin et al. in [50] proposed a protection mechanism based on secure logging (forward integrity). Their scheme is detective. That is, tampering attacks are detected in a postmortem analysis after they took place. Kulkarni and Metta [57] mitigated call graph manipulation by utilizing opaque predicates to randomly switch between a set of nontrivial clones of sensitive program logic (input by human experts). Pappas et al. [62] designed a mitigation for ROP attacks by monitoring branching behavior and size of fragments executed before each branch. Blietz and Tyagi [30] proposed a mechanism in which a program-to-protect subscribes to an external process which dictates the genuine control flow of the application. The external process can be protected by other expensive integrity protections without introducing extensive overhead on the main application.

5.2.3 Process Memory

Protecting process memory based on *code invariants* is the most common approach in the literature. In the following, we briefly introduce different schemes based on this technique.

Self-encryption is a defense mechanism utilized by a number of protection schemes. Aucsmith [29] proposed a scheme that utilizes code block encryption and signature matching to mitigate tampering attacks on process memory. The author refers to the integrity checking codes as *integrity verification kernel* (*IV K*). These IVKs, for better stealth, are interleaved with the program functionality. Using this technique, the entire application code except for the starting block is encrypted. From the memory layout of each executed block a key is derived which can decrypt the next block. Ref. [69] is another encryption-based protection in which the key to decrypt the next basic block is derived from a chain of hashes of previous blocks. Thus, tampering with blocks breaks the chain and results in an unrecoverable program. Protsenko et al. [64] applied the self-encryption with reencryption after each invocation to protect Android intermediate code. Protected programs using this scheme, however, cannot use reflection nor host service contracts (otherwise unreachable) due to the code encryption.

Self-hashing is another mechanism that is employed by a number of process memory protection schemes. Spinellis [66] used reflection to obtain program codes at runtime. Program codes are subsequently verified using a hash function. In addition to hash verification, HW counters are also utilized to increase the resilience. Chang and Atallah in [35] proposed a software protection technique that builds up a network of compact protection blocks (aka Guards) that along with performing integrity checking of the application blocks (basic blocks) can also verify other protection guards. These protection guards, provided that an untampered version of the code is available, can potentially heal the tampered regions of the program. Banescu et al. [7] proposed a technique that utilizes a network of cyclic checkers with multiple hash functions to prevent memory tampering attacks. This scheme can cope with the relocation problem in binaries and hence it protects instructions that contain absolute addresses. To the best of our knowledge, this scheme is the only self-checksumming protection measure whose source code is publicly available.[b] Horne et al. [45] proposed a check and fix protection scheme that utilizes a set of linear testers that compute a hash over certain regions in the process memory. Upon mismatch detection correctors attempt to heal the tampered program.

Ghosh et al. [43] proposed a technique combining cyclic checks and self-encryption. In this scheme, first cyclic guards are injected and then the entire program is encrypted. An emulator is shipped with the program which can decrypt and eventually execute it. The decryption key is shipped into the binary using white box cryptography, so attackers can not easily extract it

[b] https://github.com/google/syzygy/tree/integrity.

from the program. Ghosh et al. [44] did a similar protection with the difference of using virtualization obfuscation instead of self-encryption. In addition to that, a set of self-checking guards are generated on the fly to protect the dynamically translated program in the cache. Junod et al. [51] applied control flow flattening obfuscation transformation in combination with self-hashing protection guards. In their scheme, random intervals of a program are checked using CRC32 checksums.

Madou et al. [58] used constant mutation at runtime so that the same memory address is used for multiple code and data blocks in one execution.

Secure hypervisor is used by some protection mechanisms to protect the process memory. Dewan et al. [40] used a secure hypervisor to verify code pages of the protected programs. Kimball and Baldwin [56] proposed a measure that utilizes a secure emulator, code singing, and encryption. First programs code blocks are signed and encrypted. Later, the secure emulator can decrypt and execute the genuine code blocks. Yao et al. [70] proposed a hypervisor monitoring tool called VMI to introspect the memory of running virtual machines. VMI has to, however, be executed with root permission on the host. Brasser et al. [31] aimed for integrity protection in IoT devices by proposing a concept of secure (tamper resistant) tasks enabled by a trusted hardware. Furthermore, a secure interprocess communication protocol is also designed. In order to pass a secure message, a source process loads the message m and destination id, d_{id}, into CPU registers and then triggers a secure IPC interrupt. The interrupt eventually makes m along with the s_{id} accessible for destination process/task. Morgan et al. [18] proposed a hardware-assisted hypervisor that enables users to define their custom integrity routines. Similarly, [52] proposed to equip hosts with custom attestation and measurement probes to detect violation of integrity.

Baumann et al. [19] proposed a technique to use SGX secure enclaves to place the entire application in isolated containers. This prevents attackers form runtime process memory manipulations.

Static protection to isolate process memory was done in [63]. Their goal was to ensure genuine geolocation information by starting the service in a lightweight hypervisor which is secured by a TPM.

Timed traces are another representation which was used to build protection schemes. In effect, [48, 49] based their schemes completely on timing analysis to verify process memory. In this approach, once the attestation is launched, all active memory contents are copied to the flash memory. Then, the verifier sends a random seed that the attestation client uses for proof of computation. At different check-points the verifier sends further seeds and defines the memory addresses that need to be incorporated in the proof computation. Meanwhile, timing analysis could be used as an additional source of information. For instance, [66] used reflection as its core of integrity protection, but also utilized timing analysis for additional hardening.

5.2.4 Runtime Data

To protect runtime data, data invariants and trace representations were utilized by the reviewed protection schemes. The correlation indicates that seven schemes operate based on the data shape representation. Karapanos et al. [53] relied on an authenticated data structure (Merkle tree) to enable users to verify correctness and freshness of the access control policies and user data. Carbone et al. [32] used a combination of static analyzer and memory analyzer to detect data tampering that leads to kernel exploits. Sun et al. [67] used SDN to route data to an extra node to verify network data integrity. Ibrahim and Banescu [46] proposed to utilize a set of checkers are added to the program to check return values of the sensitive functions. Inputs for these checkers are generated using symbolic execution. Gao et al. [42], similar to [46], proposed to feed the return values of a given service to a set of clones and have a majority vote-based scheme to decide upon integrity of the service. Kil [54] proposed to first run a program using the Daikon tool [72] to capture dynamic data invariants. From these invariants later a set of guards are generated and injected into the program of interest. At runtime these invariant should hold, otherwise response function is triggered. Castro et al. [33] proposed a technique which uses data flow graph to build a data model. At runtime the conformance of the application data flow to the model is frequently verified.

Trace representation was used in two protection schemes to authenticate data integrity. Banescu et al. [8] utilized white box cryptography to protect the integrity of a configuration file in the Chromium browser. A white box proxy enables Chromium to sign and verify genuine configuration files. While attackers have difficulties extracting the key from white box crypto, they might be able to subvert the execution flow and misuse the proxy to sign malicious configuration files. Therefore, this scheme authenticates any call to the proxy.

Chen et al. [36] incorporated a selected set of memory references into hash variables. The constant program data can be protected using this measure. However, incorporating input data will make it impossible to precompute expected hashes, due to nondeterminism in programs.

In this correlation we identified and discussed the different schemes to mitigate particular MATE attacks, viz., binary, control flow, process memory and runtime data manipulations, at different representations. Among the reviewed schemes only [68] targeted distributed systems. This work, however, focused on adding resilience to the static representation against buffer overflows. All the other reviewed schemes focused on protecting assets in a centralized application, which might be in contact with a remote server. However, these schemes clearly fall short in protecting a software system that is comprised of a set of distributed nodes, communicating over network. Protecting program representations in a distributed architecture appears to be a major gap.

5.3 Defense Integration to Program Life Cycle

Each defense comes with a mean in which monitoring probes are hooked, connections to response algorithm are established and hardening is applied. In a defense mechanism *Transformation* is the component that employs protection in a program by applying necessary modifications. Transformations may target different stages of a program varying from source codes to executable binaries. This will impose some constraints on the applicability of defenses to different contexts. For instance, one may have no access to the source code for a third party component. That is, for such cases, all the defenses based on source code transformations become irrelevant. The stream viewer in the DRM sample is an example of such components.

In addition to the constrains imposed by life cycle activities, program representations on which protections are applied impose further constrains. For instance, monitoring memory can impose intolerable overheads in IoT devices. Therefore, it is of major importance to identify integration constraints of protection schemes in different application contexts. To capture this concern, we analyze the correlation between transformation target and system representation. Table 5 illustrates the outcome of this correlation. We discuss the correlation by iterating over life cycle activities as follows.

5.3.1 Precompile

In the following we have a closer look at the schemes that operate on the source code level and discuss their correlation to the asset representations.

5.3.1.1 Code Invariants

Spinellis [66] proposed a scheme for remote verification of integrity protection. The idea is to retrieve a program's code shape using reflection and subsequently computing a hash over it. This hash is then reported to the remote party. These reflection calls are injected into the program source code.

Table 5 Defense Mechanisms for Different (Asset) Representations and Their Correlation to Program Life Cycles

Life Cycle	Representation					
	Static	Code Inv.	Data Inv.	HWC	Trace	Timed Trace
Precompile	—	[40, 66]	[32, 33, 46, 53, 54]	[66]	[8, 50, 57]	[66]
Compile	[68]	[51]	—	—	[36]	—
Postcompile	[34, 39, 55, 63]	[7, 18, 19, 29, 31, 35, 43–45, 52, 56, 64, 69, 70]	[42]	[59]	[28, 47, 61, 62]	[48, 49, 60, 65]
Load	—	[44]	—	—	—	—
Run	[38]	[58]	—	—	—	—

Dewan et al. [40] proposed a scheme that requires developers to generate integrity manifests for all programs-to-protect. Integrity manifest contain program measurements and a unique program identity. These manifests need to be shipped to a secure hypervisor which follows the measurements to evaluate the integrity of programs at runtime.

5.3.1.2 Data Invariants

According to the correlation outcome there are four schemes that operate on data shape to verify software integrity. We will discuss these schemes, focusing on their integrability to systems. Data invariants verification is the technique that is used in [54]. In order to capture these invariants, the *Daikon* tool is utilized, which requires programs to be instrumented and subsequently executed. Afterwards, in the protection phase, these invariants are verified by a set of protection guards.

Karapanos et al. [53] proposed a mechanism to protect program inputs using a set of data integrity polices specified by users. These policies are enforced by the mean of an authenticated data structure [73].

In order to protect program functionality at runtime, [46] proposed a scheme for C# programs in which return values of program functions are tested in a network of cyclic checks. To construct checkers, the PEX symbolic execution tool [74] is used. PEX generates test cases (input–output pairs) for the functions of interest. Checkers are generated based on the test cases. They invoke a protected function with the test's input arguments and subsequently match the output of the function with the expected result of the test case.

Conversely, [32] designed a scheme that inspects kernel's dynamic data to detect tampering attacks. They claim that their scheme can protect up to 99% of such attacks on kernel.

5.3.1.3 Trace

The following schemes rely on trace data to evaluate program integrity. Given that log records intrinsically capture enough information about the executed trace, [50] based their scheme on recovering a program trace from its logs. In this scheme program code needs to be augmented with a comprehensive logging mechanism. A hash chain mechanism is utilized to preserve the forward security properties. The logs are supposed to be verified by a trusted external entity Since attackers cannot forge logs, the verifier can detect malicious activities during verifications. [57] proposed a scheme which aims at diversifying program's execution trace by randomly executing nontrivial clones of the sensitive functions. The downside is that all the nontrivial clones have to manually be implemented. Banescu et al. [8] aimed at defeating control flow hijackers by a set of stack trace introspection guards. These guards are injected in the prologue of sensitive functions in the source code.

5.3.2 Compile

Compiler-level protection could potentially generate far more optimized protection routines thanks to built-in optimizations. Despite the benefits, compiler-based protections require access to the source code. Also, compiler transformations will target a specific compiler and hence restrict the choice of compilers at the user end.

Furthermore, since compiler optimization passes are oblivious to protections implemented, they could potentially break them and thus cause false alarms in protected programs. For instance, some compilers utilize enhanced memory caching by block reordering. This relocates program blocks after protection is laid out and hence breaks the hash checks in self-checksumming-based protections, resulting in false alarms [7].

In our survey we only found three compiler-based protections. Junod et al. [51] implemented a scheme that aims at protecting *code shape* by a combination of control flow flattening and tampering protection in a compiler pass (implemented in the LLVM infrastructure). On the negative side, no evaluation results on the tamper resistance overhead were published in this work. In an effort for securing distributed programs against buffer overflow exploits, [68] proposed a LLVM-based analysis tool that combines the CFG of all individual programs to construct a distributed CFG. The distributed CFG is to improve the accuracy of vulnerability detection. Oblivious hashing [36] is another scheme that is implemented in abstract syntax trees. It operates on in-execution representation to authenticate program *trace*. In this transformation, program assignments and branching conditions are reflected in a hash variable that serves as a trace hash. Later, this hash could be verified at any desired point in the program.

5.3.3 Load

In our survey, we only found one research work that transforms program instructions at load time. Ghosh et al. in [44] proposed a tamper-proofing technique on top of instruction virtualization obfuscation.[c] In this scheme, a mechanism was designed to protect translated instructions (code invariants) within the emulator in a program by safeguarding them using a network of checkers (similar to [35]) along with frequent cache flushes.

5.3.4 Run

Transforming applications at runtime is another option. These transformations render adversaries' knowledge obsolete by turning the protection into a moving target. In this setting, attackers have a limited time before the next mutation takes place, and hence their success rates deteriorate. Collberg et al. [38] proposed a scheme for client–server applications in which the server API is constantly mutated using diversification obfuscation techniques, forcing clients to update frequently. Assuming that the server is secure, attackers have a limited time to carry out reverse engineering attacks on client applications (*static*). Soon after the server API is mutated, client applications have to be updated to reflect API usage changes, which renders attackers knowledge obsolete.

The downside of this scheme, however, is the obligation of having the server and the clients constantly connected to receive highly frequent updates.

Similarly, [58] proposed a tamper resistant scheme based on *dynamic code mutation* in which program's data and code in the process memory are regularly relocated such that code and data memory cells are indistinguishable. That is, a particular memory cell (which is allocated for the process memory) serves both data and code during the course of an execution. Dynamic code

[c] It is a technique in which program instructions are randomized such that only the shipped emulator can execute them [75].

mutation aims for keeping the code unreadable until execution time. Therefore, we classified this scheme as *code invariants* protector.

The basic assumptions in this technique are that reverse engineering (and hence tampering with programs) is harder (a) when program code and data are indistinguishable, and (b) when there is no fixed mapping between instruction and memory cells so that the same memory region is used by multiple code blocks at a course of execution.

The general observation is that runtime transformations are technically much more challenging to implement, which justifies the limited number of schemes in this category.

5.3.5 Postcompile

Postcompilation transformation operates at the binary level. That is, compiled programs and libraries, without presence of their source code, can be protected. Because of their applicability on majority of product line, plenty of schemes utilize postcompile transformation. In fact, we found protection schemes for all the system representations which carry out their transformations in a postcompile process. In the following, we elaborate on these schemes classified by their target representations.

5.3.6 Code Invariants

Schemes that protect the code invariants (introspection) in a postcompilation transformation process are listed in the following. To protect integrity of Android applications [64] proposed a self-encrypting scheme that operates on the intermediate representation of Android programs, i.e., Dalvik executable files. Although the key is shipped with the program, this scheme adds resilience against tampering attacks. Banescu et al. [7] applied self-checksumming transformation on windows portable executable (PE) binaries. Chang and Atallah [35] used cyclic network of checkers with repairers to recover tampered blocks to protect WIN32 (DLL and EXE) binaries. Similarly, [45] protected binaries based on self-checksumming technique. Refs. [18, 31, 52, 56, 70] proposed schemes that operate at the hypervisor level to protect program binaries. Baumann et al. [19] used an isolation supported by Intel SGX to place the entire applications inside protected enclaves and protect them from attackers. Refs. [29, 43, 56, 69] proposed to encrypt program binaries in order to protect it. Ghosh et al. [44] designed a method that transforms program binary into a random instruction set to carry out protection.

5.3.7 Data Invariants

We identified two schemes that protect data integrity in after compilation. In the scheme proposed by [42] multiple clones of a program are simultaneously executed to detect instances that diverge from the majority of the clones. Malone et al. in [59] proposed a scheme that has two phases: *calibration* and *protection*. In the first phase a program is executed while hardware performance counter (HPC) values are continuously collected. A mathematical tool is then utilized to identify hidden relations among the gathered data in form of equations, which essentially captures the effect of genuine execution of the program on HPC. Finally, a set of guards are injected in the program to evaluate those equations at different intervals. However, both Windows and Linux operating systems prohibit any access to HPC instructions in user-mode. This essentially requires applications to invoke a kernel call to read HPC values at runtime, which opens the door for call interceptions.

5.3.8 Static (File)

To protect the integrity of static files checksum-based techniques were proposed in the literature. Kim and Spafford [55] introduced a tool to verify signatures of a large number of binaries on a system. Deswarte et al. [39] enabled a remote verifier to compute and verify checksums of static files residing on a server. In this model, the verifier sends a challenge with which the server has to initialize checksum variables. To protect operating system static services, [34] proposed a technique to guarantee the integrity of standard operating system application by safeguarding core software package providers.

As a mean to protect specific services in a system, [63] developed a genuine geolocation provider service that is enclosed in a statically secured lightweight hypervisor.

5.3.9 Timed Trace

Refs. [60, 65] are two consecutive schemes that authenticated the integrity of legacy systems (no multi core CPU) by measuring the time differences between genuine and forged execution of programs (timed traces). Similarly, [48, 49] also relied on timing analysis to detect malicious behaviors in remote programs. In these schemes, the verifier requests a checksum over certain regions of the process memory (memory printing) and at the same time measures the elapsed time. The checksum value and timing analysis enable the verifier to reason about integrity of the system. Nevertheless, during the memory printing process the system must stop functioning, otherwise the timing data will be inconsistent.

5.3.10 Trace

Trace representation is utilized by a set of protection schemes to protect control flow integrity. Abadi et al. in [28] proposed a scheme in which the integrity of control flow is verified by a simple token matching scheme. For this purpose, before every jump instruction a unique token is pushed into the stack. Subsequently, at all destinations (jump targets) a predicate is added to verify the token which should have been previously added to the stack. Jacob et al. [17] leveraged invariant instruction length on X86 architecture to design a scheme with which tampering leads to an inevitable crash at runtime. Pappas et al. [62] verified branching patterns to detect ROP attacks at the Windows 7 kernel, which can transparently protect all executing applications without any modification in them.

In this section, we correlated representation-to-protect and program life cycle dimensions. This correlation suggests that there are gaps in compiler-based, load-time and run-time protections.

5.4 Correlation of Transformation Life Cycle and Protection Overhead

Another interesting direction to look into is the influence of the transformation life cycle on the overhead of schemes. Compiler-level transformations leverage optimization passes in the compilation pipe line. Naturally, applying transformation at compiler-level should impose less overhead in comparison to postcompile transformation schemes. Mainly because binary level modifications occur at the end of the build pipe line, where no further optimizations are carried out. To capture this view we correlate overhead classes with the transformation life cycles that is presented in Table 6. We map our surveyed papers on four classes of overhead indicators, viz., N/A, fair, medium, and high.

5.4.1 Fair

As can be seen in Table 6, a majority of schemes with fair overhead are directly applied on the binaries, i.e., postcompile transformation. Precisely speaking, [44] with 10% overhead (in addition to the virtualization overhead which is reported to be about 30%) from the load category, [40, 54] with 8%–10% overhead from the precompile category, and [38] with 4%–23% overhead from the run category.

Table 6 The Correlation Between the Transformation Life Cycle and Overhead

Overhead	Life Cycle				
	Precompile	**Compile**	**Postcompile**	**Load**	**Run**
Fair	[40, 54]	—	[18, 19, 28, 37, 43, 44, 59, 61–63, 65, 70]	[44]	[38]
Medium	[8, 50]	—	[7, 41, 49, 69]	—	[58]
High	[53]	[68]	[31, 35, 47, 48, 64]	—	—
N/A	[32, 33, 46, 57, 66]	[36, 51]	[29, 34, 39, 42, 45, 55, 56, 60]	—	—

5.4.2 Medium

In our survey we identified six schemes with a medium overhead. In postcompile category [41] with 128%, [7] with 134% for non-CPU intensive applications, and [69] with 107% are classified as schemes with a medium overhead. In the precompile category, [8] with 130% and [50] are in medium class. In the run category, [58] with an overhead of 107% (for a program with 70 protected basic blocks) falls under this overhead class.

5.4.3 High

In the high overhead class, we found seven schemes for precompile and postcompile transformation activities. Karapanos et al. [53] impose 5× slowdowns on reads and 8× slowdowns on writes. Similarly, [64] introduces an overhead of 500%, [47] and [31] both are reported to impose 3× slowdown in protected applications. Teixeira et al. [68] utilize an algorithm with complexity of $O(N^2 + 2^N)$.

5.4.4 N/A

We mapped all the schemes without performance evaluations or insufficient results into this class. A major problem with the performance evaluation of the reviewed protection schemes is lack of comprehensiveness. Thus, we are incapable of estimating the overhead of over 16 protection schemes. Surprisingly, the two compiler-level protection schemes [36, 51] also fall under this category.

This correlation once again shows that there is a gap in benchmarking protection schemes. To the extent that from our results, we cannot approve nor reject the hypothesis that compiler-level transformations perform better.

We would like to emphasize here that this classification is just an estimation based on the authors' claim on the efficiency of their proposed schemes. Therefore, depending on what dataset was used in the evaluation process by the authors, the overhead might differ. A fair evaluation would be to use a constant dataset to measure overhead of different schemes.

Unfortunately, this is not possible due to two reasons. First, to the best of our knowledge, very few of the reviewed schemes were made open source. Secondly, even if the source codes were available we would not be able to compare their performance without further classifications. Because these schemes are designed for different operating systems and different representations. Thus, they need to be clustered technologically and then evaluated.

5.5 Check and Monitor Representation

Protection schemes, as stated earlier, operate on a particular representation of the assets. This implies a sort of monitoring mechanism on the representation of interest. The collected data (in the monitoring process) need to be analyzed and ultimately verified by the scheme's *Check* compartment. Studying the different check methods for verifying different asset representations is particularly interesting. Table 7 depicts the correlation of check and representation items. Based on this correlation we report on different check methods that were used in the literature as follows.

Table 7 The Correlation of the Representation and Check Mechanism

Check Mechanism	**Asset Representation**					
	Static	**Code Inv.**	**Data Inv.**	**HWC**	**Timed Trace**	**Trace**
Access control	—	[19, 31]	—	—	—	[8, 62]
Checksum	[17, 39]	[7, 18, 29, 35, 40, 44, 45, 51, 52, 56, 64, 69, 70]	—	—	[48, 49, 60, 65]	[50]
Equation eval.	—	—	[54]	[59]	—	—
Majority vote	[16]	—	[42, 46, 67]	—	—	[50]
Signature	[34, 55]	[29, 56]	—	—	—	[50]

HWC, hardware counters.

5.5.1 Code Invariants

Expectedly, the majority of code shape verifiers with a total number of 13 use checksum-based techniques, which also includes hash functions. In addition to checksums, [29, 56] employed signature matching as a secondary measure for tamper detection. Refs. [19, 31] maintained code invariants security by enforcing access control at a trusted hardware level.

5.5.2 Data Invariants

In order to check the data shape, three schemes [42, 46, 67] used majority vote principle. On the other hand, [54, 59] used equation system to verify the integrity.

5.5.3 Hardware Counters

Refs. [59, 76] used hardware performance counters in their checking mechanism.

5.5.4 Timed Trace

Refs. [48, 49, 60, 65] computed a checksum alike routine and measure the elapsed time.

The correlation between checking mechanism in different protection schemes suggests that checksumming is the most used technique for integrity verification. Other checking measures such as signature verification, access control, equation evaluation, and majority votes are equally unpopular.

5.6 Monitoring Representation Correlation With Protection Level (Enforcement Level)

In this section, we are aiming to find out which representations could be monitored at which protection level. For this matter, we correlate *protection level* and *representation* in Table 8. Since we are interested in identifying the possibility of monitoring representations at different levels, we only indicate the number of citations instead of citing the references.

Table 8 The Correlation of the Representation and Protection Level

	Static	Code Invariants	Data Invariants	HW Counters	Timed Trace	Trace
Internal	2	10	3	1	—	7
External	5	5	4	1	5	2
Hypervisor	1	4	—	—	—	—

The correlation indicates that both the internal and the external protection levels have no limitations on monitoring system representations, with the exception of traces which were not targeted by any of external protectors in the reviewed literature.

Furthermore, the correlation shows a gap in the hypervisor-based protection techniques in which none of the data shape, hardware counters, and traces were used in the reviewed schemes. It is not clear whether this is due to a technical limitation of hypervisors or simply a research gap. Further studies should be conducted to address this gap.

5.7 Hardening vs Reverse Engineering Attacks and Tools

Attacks perhaps are the least studied (and published) part of the integrity protection schemes. This is due to two main factors: (a) plenty of security measures in industrial protection schemes such as Themida and VMProtect are not publicly disclosed, which, in turn, has left researcher without enough knowledge about their security. (b) Breaking schemes is unethical as it opens the gate for the attackers to compromise system's security. For example, the CIA Vault7 windows file sharing exploit has led to WannaCry ransomware which infected 213,000 windows machines in 112 countries [77].

Due to the lack of comprehensive research on attacks on integrity protection schemes, evaluating their security against various attacks is impossible. Therefore, as an initial step toward analyzing the security of these schemes, we made assumptions regarding the resilience of protection schemes against different attacks. In the following we state our assumptions:

1. *Disassembler*: code concealment and layered interpretation hardening measures hinder disassembly. Code concealment techniques commonly affect the correctness of static disassemblers [78].
2. *Debugger*: in a strict sense only schemes that directly utilize antidebugger measures impede debug-based attacks. These technique are, however, ad hoc [79]. In this work, we consider hardening measures that impose some difficulties on the ability of debugging programs as counter debug measures. Cyclic checks, layered protection, and mutation are the three hardening measures that we believe impede debug-based attacks. Cyclic checks confuse attackers and exhaust their resources. Layered protection complicates the program execution flow and thus has an impact on the effectiveness of debugging. Mutation-based techniques renders debug knowledge useless after each mutation. Furthermore, SGX protected application cannot be debugged in production.

3. *Tracer*: this enables attackers to monitor what instructions are actually being executed by the program. With the help of this tool, attackers could effectively bypass code concealment and virtualization as program instructions have to be translated and eventually executed. This is the point that a tracer can dump the plain instructions. Nevertheless, frequent program mutation renders captured data useless. Furthermore, code clones are normally harder to capture using a tracer, and thus it hinders tracer-based attacks. *Intel SGX isolation*, by design, resist against tracers.
4. *Emulator*: this enables adversaries to analyze protected programs in a revertible and side-effect free environment. In effect, this defeats response mechanism and gives unlimited attempts to attackers. Beside antiemulation measures, which are again ad hoc, only hardware-based security appears to impede emulators. It is noteworthy that emulation is a technique that facilitates attacks, and on its own it does not break a scheme. That is, perpetrators still need to utilize other tools to detect and ultimately defeat the protection. Therefore, we exclude emulator resilience from our analysis.

With the given assumption, we have classified the reviewed literature based on their hardening measures. Table 9 represents the mapping between different schemes and their resilience to MATE tools.

5.8 Resilience Against Known Attacks

As mentioned earlier, the security of integrity protection schemes has not been thoroughly evaluated, due to the obscurity and lack of access to the resources. This has led to very few attempts on breaking such schemes.

To the best of our knowledge, three generic attacks on integrity protection schemes were published, two technical attacks and one conceptual work for manifesting attacks. In the following, we discuss these attacks and map them to the hardening measures that could potentially address them.

1. *Pattern matching*: there are no paper on pattern matching attacks on different protections schemes. Additionally, most of the papers mention obfuscation as the golden hammer against pattern matching and entropy measurements. Unfortunately, we cannot evaluate resilience of the reviewed schemes against pattern matching as their source code

Table 9 MATE Tools Resilience in the Reviewed Schemes

Disassembler	Debugger	Tracer
[7, 8, 29, 58, 64, 69]	[7, 18, 19, 31, 35, 37, 40, 41, 43, 44, 46, 51, 56, 58, 70]	[19, 38, 44, 47, 57, 58, 70]

is not available. Therefore, we do not consider this attack in the mapping process.

2. *Memory split*: Wurster et al. in [3] proposed an attack to defeat self-hashing protection schemes by redirecting self-reads to an untampered version of the program loaded in a different memory address. For this purpose, they modified the kernel to establish two distinct memory pipelines for self-read and instruction fetch (execution). In this setting, the execution pipeline can readily be tampered with by attackers and self-reads remain unaware of such modifications.

   To address this generic attack, Giffin et al. in [80] proposed self-modifying code defense mechanism. The idea is to add a token to the program code at runtime and verify it accordingly. Dynamically generated tokens can detect redirection of self-memory reads in the protected processes.

   Self-modifying code, despite its effectiveness in thwarting memory split attacks, opens the door for other integrity violating attacks. The reason being that enabling self-modification requires flagging writable memory pages as executable pages in the program. This raises the concerns about code injection attacks [81], in which maliciously prepared memory blocks could potentially get executed.

   This attack only applies to introspection-based techniques, i.e., techniques in which a program's code blocks have to be read at runtime. Thus, state inspection-based techniques are resilience against memory split attack. In addition to that, self-mutating protection schemes express the same effect as the defense that was proposed by [81]. All in all, our survey shows state inspection-based and mutation-based techniques mitigate the memory split attack.

3. *Taint analysis to detect self-checksumming*: Qiu et al. in [26] proposed a technique to detect self-checksumming guards in programs. Their attack relies on the fact that these class of protections forces a protected program to read its own memory at runtime. This obligation enabled them to utilize dynamic information flows in detection of self-checksumming guards and check conditions (which trigger the response mechanism). Technically speaking, their approach captures a trace of the program (using Intel Pin[d]) and then searches for a value *X* which is tainted by self-memory values (backward taint analysis) and then detects those *X*s that are used in a condition (forward taint analysis).

[d] Intel pin is a dynamic program instrumentation tool.

This attack appears to be the ultimate attack on self-checking protection schemes as it can in theory detect all checks in a given program. However, tracing medium to large sized programs will generate massive trace logs that have to be analyzed. Banescu et al. [7] reported analyzing a small trace (captured in 10 min) of Google Chromium using the tainting technique requires a long time in which only 1% of the traces were analyzed in a day.

Since this attack relies on a Tracer tool (Intel Pin), we can mark schemes that resist against tracer as resilient to the attack presented in [26]. Needless to say that larger programs, due to the complexity of the backward and forward taint analysis, in general cannot be targeted by this attack in a reasonable time.

4. *Graph-based analysis*: Dedic et al. in [16] indicate that flow graph analysis such as pattern matching and connectivity analysis (given that checker nodes are weakly connected to other graph nodes in a program) can help defeat the existing protection schemes. They developed a graph-based game to formally present these attacks.

   Against these attacks, they proposed three design principles, viz., *strongly connected checks*, *distributed checks*, and *threshold detection schemes*, that can significantly harden the identification and disabling of a tamper protection technique. Strongly connected checks requires checkers to be strongly connected to other nodes of the program. Distributed checks refers to a network of checkers which an attacker has to disable them all in order to successfully tamper with an application. Threshold detection schemes suggest to call a response function only when a *k* number of checks failed. This prevents attackers from sequentially detecting checkers in the program.

   This work can be seen as a conceptual work with no actual implementation of the defenses or attacks. It would be interesting to employ the suggested measures and carry out the attack to verify their claim. None of the reviewed schemes appear to utilize all the defenses together. Therefore, it might be the case that in theory all of the schemes are defeated by the graph-based attack. Since this attack has not been implemented in practice, we rather leave it for further evaluations.

In Table 10 we map the schemes that resist against the two practical attacks.

Table 10 Resilient Schemes Against the Two Known Technical Attacks

Taint-Based Attack	Memory Split Attack
[19, 38, 44, 47, 57, 58, 70]	[8, 18, 28, 30–33, 36, 38, 42, 44, 46–50, 53, 54, 58–62, 65–68]

6. RELATED WORK

The closest and yet distant research to our work is ASPIRE (www.aspire-fp7.eu) project. It is a project funded by European Union for software protection. ASPIRE is designed to facilitate software protection by introducing a reference architecture for protection schemes. Conformance to this architecture enables practitioners to compose a chain of protections simultaneously, which offers more resilience [82]. Data hiding, code hiding, tamper-proofing, remote attestation, and renewability are the core protection principles that are addressed in ASPIRE.

ASPIRE introduces a security policy language expressed by annotations. To protect a program using this framework, end users have to annotate programs with their desired security properties. A compiler tool chain is designed which analyzes these annotations and carries out necessary protection steps. To do so, the tool chain comes with three components: a source code transformation engine powered by TXL [83], a set of routines to link external protection schemes compatible with standard compilers (LLVM and gcc), and a link-time binary rewriting infrastructure for postcompilation protections powered by Diablo (http://diablo.elis.ugent.be/). All these enable rapid development of protection tools which could be used by different programs in a customizable and yet composable manner.

The ASPIRE framework focuses on a technical architecture for design and employment of a wide range of protection mechanisms (not only integrity protection). However, both the programs to protect and the protection schemes have to conform to the requirements of the tool chain. Specifically, the source code (in C/C++ language) of programs is mandatory. Furthermore, constraints of each scheme need to be specified in its manifest. The framework uses these manifests to report potential conflicts to users based on which they can identify and subsequently single out conflicting measures.

The ASPIRE project has not provided a taxonomy of integrity protection schemes and does not include comparison of different schemes. In fact, the outcome of this work can contribute in extending the ASPIRE architecture to accommodate a wider range of integrity protection schemes, namely, by means of introducing new requirements for the annotation language (such as desired resilience against certain attacks), and extensions for the support of further protection schemes (such as hypervisor-based and hardware-assisted ones).

To the best of our knowledge, our work is the first taxonomy of integrity protection schemes. Therefore, reviewing existing taxonomies was not an option for us. Instead, we reviewed scheme classifications and surveys on integrity protection. In the following we report on the reviewed publications.

6.1 Existing Classifications

Mavrogiannopoulos et al. in [84] propose a taxonomy for self-modifying obfuscation techniques, which essentially is based on the concept of mutation.

Collberg et al. in [13] classify integrity protection techniques into four main categories: *self-checking*, *self-modifying*, *layered interpretation*, and *remote tamper-proofing*. In a similar effort, Bryant et al. [85] classify tamper resistant systems into six categories, viz., hardware assisted, encryption, obfuscation, watermarking, fingerprinting, and guarding. Considering the fact that watermarking on its own does not contribute to integrity protection, we exclude it from our analysis. Guarding also fits the self-checking category, thus it can be omitted. Likewise encryption is a substance of self-modifying primitive. After we resolved the conflicts, we study the structure of the five remaining categories of integrity protection techniques as follows.

6.1.1 Self-Checking

This refers to a technique in which self-unaware programs are transformed into somewhat self-conscious equivalent versions. The transformation is done via equipping programs with a set of monitoring probes (checkers/testers), which monitor *some authentic features* of the target program, along with a group of assertions, who compare probe results to the corresponding *known expected values* and take *appropriate actions* against tampering attacks.

Depending on which features of a program are being monitored, self-checking itself has two subcategories: *introspection* and *state inspection*. The former exercises the shape of a program, for instance, code blocks, whereas the latter is after monitoring the program's execution effects, for example, the sum of a program's constant variables. Among the two, state inspection is more appealing because, unlike the introspection, it reflects the actual execution of a program, and thus it is harder to counterfeit. However, monitoring dynamic properties of a program is more difficult compared to the static code shape verification.

6.1.2 Self-Modifying

Turning a program into a running target by mutating it at runtime is the idea behind self-modifying techniques. The majority of schemes that are based on this technique use encryption as the mean to mutate and ultimately to protect a program. However, there are some schemes that use instruction reordering, instead of encryption, to mutate the program.

Mavrogiannopoulos et al. in [84] define four main criteria in their proposed taxonomy for self-modifying obfuscation techniques: *concealment* (the size of program slices that are being modified at each mutation interval), *encoding* (the technique that is used to mutate the code, e.g., instruction reordering or encryption), *visibility* (whether the code is entirely or partially protected) and *exposure* (whether the actual code is permanently or temporarily obtainable at runtime). Nevertheless, in a more recent survey, virtualization was also proposed as a relevant technique in [86], but we rather consider virtualization as a subcategory of layered interpretation techniques.

Sasirekha et al. in [87] review 11 software protection techniques. Their research confirms the trade off between and security and performance for all the reviewed schemes based on which they suggested compiler-level protection as possible direction for further research. However, their work lacks any classification of techniques.

6.1.3 Layered Interpretation

In this approach a program's instructions are replaced with a set of seemingly random byte codes. These codes can only be executed with a program specific emulator that is normally shipped with the program itself. Tampering with the program byte code may result in unrecognized byte code in the emulator and may eventually causes failures. However, some techniques (for instance [56]) perform integrity verifications at the emulator prior to executions. In general, layered interpretation techniques add resilience against static program analysis attacks. To the best of our knowledge, there is also no published taxonomy on layered interpretation protection methods.

6.1.4 Remote Tamper-Proofing

Protection schemes that enable a software systems to be verified externally fall under this category. In an abstract sense, remote tamper-proofing has very much in common with *remote attestation* concept. Since remote attestation is a vast field of research that deserves a taxonomy on its own, its classification falls out of the scope of this chapter.

6.2 Comparison of Schemes and Surveys

Dedic et al. in [16] define *distributed check* and *threshold-based detection* as significant resilience improver factors in integrity protection schemes. They use a graph-based game and theoretically prove the enhanced complexity of the scheme with the aforementioned factors. From this work two interesting features are extracted: *detection* and *response algorithms*.

In an effort for security classification and unification of protection notation, Collberg et al. after defining a notation use natural science and human history to derive a set of 11 unique defense primitives in [88]. These primitives state the defense concepts rather than security guarantees.

In an effort for MATE attack classification, Akhunzada et al. in [2] study different types and motives behind these attacks. As another outcome of their work, they stressed that performance factor is one of the main pain points of protection techniques. However, it appears that no study tackled this matter and there is a lack of a benchmark for protection schemes [2]. The focus of their work is not the techniques-to-protect themselves, but rather the attackers. Consequently, the security guarantees are not studied.

The main gap here is lack of a holistic view software protection process. The process itself is not well defined, business requirements are not elicited, constraints are not studied, and more importantly, security guarantees that different schemes offer are not studied.

7. FUTURE OF INTEGRITY PROTECTION

Integrity protection is a crucial subject in cyber security. We believe this subject will attract more attention as the trend of digitalization expands into more domains of our daily life. As a matter of fact, the Internet of Things (IoT) is one of the major domains which requires immediate action w.r.t. integrity protection. The reason is the exposure of IoT devices to MATE attackers.

Integrity protection will be impacted by three dimensions: (a) trusted hardware, (b) computer networks, and (c) complexity of software systems. As a consequence of advances in trusted hardware, software-based integrity protection will perhaps be slowly replaced by more hardware-based or hybrid methods. Simply put, advancements in secure and yet cheap hardware equipments will be a major influence on the transition from software based to hardware aided integrity protection. However, this transition will require a new adaptation of the current trust model utilized by trusted hardware.

Network advances can enable a far more flexible software systems with respect to connectivity requirements. Currently, enforcing constant connections between a remote trusted party and a client system is undesirable in many contexts, e.g., the automotive industry. Lower latency and higher availability in communication protocols can foster a new model of integrity protection that far more relies on remote parties.

Beside hardware and network improvements, software systems are getting more complex as a natural response to complex problems. This negatively impacts integrity protection, as it becomes more challenging for users to apply protections. Therefore, integrity protection requires a solution to reduce the complexity of protection.

In the following we have a closer look at the aforementioned influencing dimensions.

7.1 Trusted Hardware and Collective Trust Model

Advances in hardware aided integrity could offer better program isolation and hence better integrity. Moving toward future, secure hardware should become more and more common in new computers. Dedicated hardware can improve integrity protection. However, trusted hardware is and will not be silver bullet. Side channels remain a problem in such systems. Recent attack proposed in [89] on SGX protected program enabled attackers to extract RSA keys from protected (fully isolated) enclaves. This proves that integrity protection against MATE attacks remains a cat–mouse game.

Hardware security can potentially improve integrity protection in software systems. Nevertheless, this is a double-edged sword, as malware can also benefit from such technologies to hide their malicious behavior from antiviruses or malware analysts. To cope with this problem, Intel SGX requires protected programs (enclaves) to be signed by Intel in production machines. This implies trusting Intel with holding signing keys. Intel argues this is necessary for two reasons. First, they can sell licenses to use their security technology, without purchasing a license they will not sign anything. Second, they can control who can use their security system, in an effort to ensure malware producers are unable to misuse Intel's ecosystem.

The current model, however, clearly resembles a single point of failure. If the current trend of block chain and collective trust model persists, it is likely that Intel will not remain the only player. Most probably a block chain alike system will replace the current model. One naive realization would be to have a group of semitrusted entities (with conflicting interests) to sign protected applications. Mapping this to SGX means that n entities will sign enclaves in addition to Intel. At the user end, the protected program will be executed, if and only if k signatures match, where $0 < k \leq n$.

A more interesting realization would be to enable users to report any misbehavior of the protected program by broadcasting a message to a set of disjoint logging entities. Then a set of third party verifiers (similar to miners in block chain) can verify these broadcast list and subsequently warn new users about the malicious protected packages. The main benefit of this approach is that there is no single point of failure.

7.2 Goal-Based Protection

Currently, most protection schemes require the user to specify sensitive regions/data in the program to be protected. The protection scheme then runs a transformation in order to safeguard the specified items. For instance, Intel SGX requires users to implement their application in two regions: trusted and untrusted. The trusted region will access sensitive data and hence shall be protected by secure enclaves. Software advances along integrity protecting hardwares can enable a smooth transition from manual software transformation to automatic or at least guided transformations. One possibility is to have users specify high-level integrity goals, instead of marking code regions, then some static analyzers scan the program and automatically mark regions of the program to be protected. This not only reduces the effort on the user end but also improves the precision of the protection.

Once these sensitive regions are identified, then hardware integrity could be utilized to protect them. For instance, sensitive regions could be automatically moved into SGX trusted zone and necessary signature changes could be applied to minimize the user intervene.

7.3 Tactile Internet

Real-time constraints are one of the main reasons to deliver applications to untrusted clients. Recent advancement in cloud computing has greatly extended this model by migrating application to client–server paradigm. Servers implement functionality of the system and deliver results to clients as per their requests. Today, we have fully functional word processing applications hosted on the cloud and enabled users to interact with them via their browser. In this setting, software vendors need not to worry about license check circumvention as end users do not have access to the program binaries nor execution environment. A secure authentication mechanism in effect handles license verifications.

However, given the current latency and availability of Internet networks, shifting to enforcing clients to stay in direct and constant connections with servers imposes substantial overheads. In a system with real-time constraint, e.g., the automotive industry, such delays are unacceptable. Thus, a thick layer of software systems has to reside on the client side to deliver the functionality within an acceptable time frame.

Advances in tactile Internet, which guarantees insignificant latency and high availability [90], may render the need for distributing softwares obsolete, by enabling the execution of softwares on trusted servers and communicating results to end users. The tactile Internet can change the current model, by moving the control system away into a trusted secure environment. Moving the sensitive code into trusted servers does not fully mitigate integrity attacks, however. For instance, attackers may manage to intercept and forge control commands send to a connected car. Another attack could be to manipulate internal equipments to behave faulty. Nevertheless, this shift reduces the attack surface and effectively protects proprietary data.

All in all, integrity protection will move toward a more hardware protected system with a major dependency on network reachability in real time. Moreover, user's role in the protection process will be as much as defining high-level integrity goals after. The rest of the protection will be handled by the system automatically.

8. CONCLUSIONS

MATE attackers have successfully executed serious integrity attacks far beyond disabling license checks in software system, to the extent that the safety and security of users is at stake.

Software integrity protection offers a wide range of techniques to mitigate a variety of MATE attacks. Despite the existence of different protection schemes to mitigate certain attacks on different system assets, there is no comprehensive study that compares advantages and disadvantages of these schemes. No holistic study was done to measure completeness and more importantly effectiveness of such schemes in different industrial contexts.

In this work, we therefore presented a taxonomy for software integrity protection which is based on the protection process. This process starts by identifying integrity assets in the system along with their exposure in different asset representations throughout the program life cycle. These steps are captured in the system view, which is the first dimension of our taxonomy. In the next dimension, the attack view, potential threats to assets' integrity are analyzed and desirable protections against particular attacks and tools are singled out. Then, the defense dimension sheds light on different techniques to satisfy the desired protection level for assets at different representations. This dimension also serves as a classification for integrity protection techniques by introducing eight unique criteria for comparison, viz., protection level, trust anchor, overhead, monitor, check, response, transformation, and hardening. To support the understandability of the taxonomy we used a fictional DRM case study and mapped it to the taxonomy elements.

We also evaluated our taxonomy by mapping over 49 research articles in the field of integrity protection. From the mapped articles we correlated different elements in the taxonomy to address practical concerns with protection schemes, and to identify research gaps.

In the correlation of assets and representations on which protections are applied we identified relevant schemes for protecting different integrity assets (behavior, data, and data and behavior). This facilitates the scheme selection based on the assets and representations at the user end. A major gap was identified in protecting integrity of distributed software systems.

In MATE vs representation correlation we have shown different schemes that mitigate different MATE attacks, viz., binary, control flow, process memory, and runtime data, at different asset representations.

In the correlation of transformation life cycle and protected representations, we classified schemes based on their representation-to-protect and their applicability on different program life cycles. Our results indicate a gap in compiler-based, load, and runtime protection schemes.

To study the impact of the transformation life cycle on the protection overhead, we correlated transformation life cycles and scheme overheads. Despite the general belief that compiler-based protection should perform better, our results suggest that postcompilation transformations perform quite optimal. Nevertheless, we were unable to verify the performance of compiler-based protections due to lack of benchmark results. We believe the lack of reliable benchmark data as well as a benchmarking infrastructure is a significant gap in software integrity protection. This is due to the fact that only implementation of a few protection schemes were made public, for instance, in self-checking schemes only [7] has published their source code. Thus, evaluating their performance, without reimplementing them, is infeasible.

The correlation between checking mechanism in different protection schemes suggests that checksumming is the most used technique for integrity verification. Other checking measures are equally unpopular. The correlation of representation and protection level shows a gap in hypervisor-based protections.

Another major gap in the literature is to analyze resilience of protection schemes against attacks. Based on our classification data we evaluated the resilience of reviewed schemes against MATE tools and known attacks on integrity protections. In the resilience against MATE tools analysis, we proposed a set of assumptions for declaring whether a scheme resists again a certain MATE tool or not. The assumptions are based on the hardening techniques that are used by protection schemes. These assumptions were then used to classify schemes based on resilience against MATE tools.

In the last correlation we analyzed the resilience of reviewed schemes against known attacks on protection schemes. Our survey indicated that there is a gap in studying the resilience of protection schemes against pattern matching attacks. This is again due to a lack of any benchmark of different techniques whatsoever. Regarding memory split and taint-based attacks, we first identified hardening measures that impedes such attacks. Subsequently, we marked schemes as resilient to memory split and taint-based attacks according to their hardening measures.

All in all, a big gap in integrity protection research is the lack of a benchmark of performance and security guarantees, both of which require a dataset of integrity protection mechanisms.

8.1 Limitation

The proposed taxonomy only serves as a help for practitioners. We do not claim that it is exhaustive nor complete w.r.t. industry standard protection processes. A good direction for future work is to map the taxonomy on standard processes such as Microsoft Secure Development life cycle [91].

Our taxonomy puts more emphasis on the defense mechanisms, as the number of published attacks on integrity protections is incomparably fewer than published defenses. Mapping concrete attacks will be a good extension to our taxonomy.

Although we included some research papers on infrastructural security, the main focus of our analysis was application level integrity protection. In reality, there are plenty of ways in the infrastructure that adversaries can potentially exploit. Such exploits could lead to circumvention of defense mechanisms and potentially violation of the application integrity. For instance, one major attack that we did not address in this work is call interception (also known as call hooking), which enable attackers to run malicious code by swapping the address of external calls. Another interesting direction is to map infrastructural integrity protection measures along with known attacks on them to the taxonomy.

REFERENCES

[1] C.S. Collberg, C. Thomborson, Watermarking, tamper-proofing, and obfuscation-tools for software protection, IEEE Trans. Softw. Eng. 28 (8) (2002) 735–746.
[2] A. Akhunzada, M. Sookhak, N.B. Anuar, A. Gani, E. Ahmed, M. Shiraz, S. Furnell, A. Hayat, M.K. Khan, Man-at-the end attacks: analysis, taxonomy, human aspects, motivation and future directions, J. Netw. Comput. Appl. 48 (2015) 44–57.

[3] G. Wurster, P.C. Van Oorschot, A. Somayaji, A generic attack on checksumming-based software tamper resistance, in: Proceedings - IEEE Symposium on Security and Privacy, ISBN 0769523390, ISSN 10816011, 2005, pp. 127–135, https://doi.org/10.1109/SP.2005.2.
[4] P.C. Van Oorschot, A. Somayaji, G. Wurster, Hardware-assisted circumvention of self-hashing software tamper resistance, IEEE Trans. Dependable Secure Comput. 2 (2) (2005) 82–92.
[5] T. Garfinkel, B. Pfaff, J. Chow, M. Rosenblum, D. Boneh, Terra: a virtual machine-based platform for trusted computing, in: ACM SIGOPS Operating Systems Review, vol. 37 (5), ACM, 2003, pp. 193–206.
[6] C. Collberg, Defending Against Remote Man-At-The-End Attacks, 2017 https://www2.cs.arizona.edu/projects/focal/security/project1.html.
[7] S. Banescu, M. Ahmadvand, A. Pretschner, R. Shield, C. Hamilton, Detecting patching of executables without system calls, in: Proceedings of the Conference on Data and Application Security and Privacy, ISBN 978-1-4503-4523-1/17/03, 2017, https://doi.org/10.1145/3029806.3029835.
[8] S. Banescu, A. Pretschner, D. Battré, S. Cazzulani, R. Shield, G. Thompson, Software-based protection against changeware, in: Proceedings of the 5th ACM Conference on Data and Application Security and Privacy, ACM, 2015, pp. 231–242.
[9] L. Luo, Y. Fu, D. Wu, S. Zhu, P. Liu, Repackage-proofing Android Apps, in: 2016 46th Annual IEEE/IFIP International Conference on Dependable Systems and Networks (DSN), IEEE, 2016, pp. 550–561.
[10] S. Luo, P. Yan, Fake Apps: Feigning Legitimacy, Trend Micro, 2014. http://www.trendmicro.de/cloud-content/us/pdfs/security-intelligence/white-papers/wp-fake-apps.pdfTechnical report.
[11] S. Karnouskos, Stuxnet worm impact on industrial cyber-physical system security, in: IECON 2011-37th Annual Conference on IEEE Industrial Electronics Society, IEEE, 2011, pp. 4490–4494.
[12] Kaspersky Lab, ProjectSauron: Top Level Cyber-Espionage Platform Covertly Extracts Encrypted Government Comms, 2017, https://securelist.com/75533/faq-the-projectsauron-apt/.
[13] C.S. Collberg, C. Thomborson, Watermarking, tamper-proofing, and obfuscation - tools for software protection, IEEE Trans. Softw. Eng. 28 (8) (2002) 735–746.
[14] C. Collberg, J. Nagra, Surreptitious Software: Obfuscation, Watermarking, and Tamperproofing for Software Protection, Addison-Wesley Professional, 2009.
[15] S. Pohlig, M. Hellman, An improved algorithm for computing logarithms over GF (p) and its cryptographic significance (Corresp.), IEEE Trans. Inf. Theory 24 (1) (1978) 106–110.
[16] N. Dedić, M. Jakubowski, R. Venkatesan, A graph game model for software tamper protection, in: International Workshop on Information Hiding, Springer, 2007, pp. 80–95.
[17] R. Neisse, D. Holling, A. Pretschner, Implementing trust in cloud infrastructures, in: Proceedings of the 2011 11th IEEE/ACM International Symposium on Cluster, Cloud and Grid Computing, IEEE Computer Society, 2011, pp. 524–533.
[18] B. Morgan, E. Alata, V. Nicomette, M. Kaâniche, G. Averlant, Design and implementation of a hardware assisted security architecture for software integrity monitoring, in: 2015 IEEE 21st Pacific Rim International Symposium on Dependable Computing (PRDC), IEEE, 2015, pp. 189–198.
[19] A. Baumann, M. Peinado, G. Hunt, Shielding applications from an untrusted cloud with haven, ACM Trans. Comput. Syst. 33 (3) (2015) 8.
[20] B. Kordy, P. Kordy, S. Mauw, P. Schweitzer, ADTool: security analysis with attack-defense trees, in: International Conference on Quantitative Evaluation of Systems, Springer, 2013, pp. 173–176.

[21] P. Johnson, A. Vernotte, M. Ekstedt, R. Lagerström, pwnPr3d: an attack-graph-driven probabilistic threat-modeling approach, in: 2016 11th International Conference on Availability, Reliability and Security (ARES), IEEE, 2016, pp. 278–283.

[22] S. Brockmans, R. Volz, A. Eberhart, P. Löffler, Visual modeling of OWL DL ontologies using UML, in: International Semantic Web Conference, vol. 3298, Springer, 2004, pp. 198–213.

[23] J. Rumbaugh, I. Jacobson, G. Booch, The Unified Modeling Language Reference Manual, Pearson Higher Education, 2004.

[24] H. Shacham, E. Buchanan, R. Roemer, S. Savage, Return-Oriented Programming: Exploits Without Code Injection, Black Hat USA Briefings, 2008.

[25] S. Banescu, C. Collberg, V. Ganesh, Z. Newsham, A. Pretschner, Code obfuscation against symbolic execution attacks, in: Proceedings of the 32nd Annual Conference on Computer Security Applications, ACM, 2016, pp. 189–200.

[26] J. Qiu, B. Yadegari, B. Johannesmeyer, S. Debray, X. Su, Identifying and understanding self-checksumming defenses in software, in: Proceedings of the 5th ACM Conference on Data and Application Security and Privacy, ISBN 9781450331913, 2015, pp. 207–218, https://doi.org/10.1145/2699026.2699109. http://www.cs.arizona.edu/debray/Publications/checksumming-attack.pdf.

[27] U. Piazzalunga, P. Salvaneschi, F. Balducci, P. Jacomuzzi, C. Moroncelli, Security strength measurement for dongle-protected software, IEEE Secur. Priv. 5 (6) (2007) 32–40, ISSN: 1540-7993, https://doi.org/10.1109/MSP.2007.176.

[28] M. Abadi, M. Budiu, U. Erlingsson, J. Ligatti, Control-flow integrity, in: Proceedings of the 12th ACM conference on Computer and communications security, ACM, 2005, pp. 340–353.

[29] D. Aucsmith, Tamper resistant software: an implementation, in: Proceedings of the First International Workshop on Information Hiding, ISBN 3-540-61996-8, 1996, pp. 317–333. http://link.springer.com/chapter/10.1007/3-540-61996-8_49.

[30] B. Blietz, A. Tyagi, Software tamper resistance through dynamic program monitoring. in: LNCS, Lecture Notes in Computer Science (including subseries Lecture Notes in Artificial Intelligence and Lecture Notes in Bioinformatics), LNCS, vol. 3919, ISBN 3540359982, ISSN 16113349, 2006, pp. 146–163, https://doi.org/10.1007/11787952_12.

[31] F. Brasser, B. El Mahjoub, A.-R. Sadeghi, C. Wachsmann, P. Koeberl, TyTAN: tiny trust anchor for tiny devices, in: 2015 52nd ACM/EDAC/IEEE Design Automation Conference (DAC), IEEE, 2015, pp. 1–6.

[32] M. Carbone, W. Cui, L. Lu, W. Lee, M. Peinado, X. Jiang, Mapping kernel objects to enable systematic integrity checking. in: *Proceedings of the 16th ACM Conference on Computer and Communications Security*, CCS'09, Chicago, IL, USA, 2009, ISBN: 978-1-60558-894-0, pp. 555–565, https://doi.org/10.1145/1653662.1653729, ACM, New York, NY, USA.

[33] M. Castro, M. Costa, T. Harris, Securing software by enforcing data-flow integrity, in: Proceedings of the 7th symposium on Operating systems design and implementation, ISBN 1-931971-47-1, 2006, pp. 147–160. http://dl.acm.org/citation.cfm?id=1298455.1298470 http://www.usenix.org/event/osdi06/tech/full_papers/castro/castro_html/.

[34] L. Catuogno, I. Visconti, A format-independent architecture for run-time integrity checking of executable code, in: International Conference on Security in Communication Networks, Springer, 2002, pp. 219–233.

[35] H. Chang, M.J. Atallah, Protecting software code by guards, in: ACM Workshop on Digital Rights Management, Springer, 2001, pp. 160–175.

[36] Y. Chen, R. Venkatesan, M. Cary, R. Pang, S. Sinha, M.H. Jakubowski, Oblivious hashing: a stealthy software integrity verification primitive, in: International Workshop on Information Hiding, Springer, 2002, pp. 400–414.

[37] M. Christodorescu, R. Sailer, D.L. Schales, D. Sgandurra, D. Zamboni, Cloud security is not (just) virtualization security: a short paper, in: Proceedings of the 2009 ACM workshop on Cloud computing security, ACM, 2009, pp. 97–102.
[38] C. Collberg, S. Martin, J. Myers, J. Nagra, Distributed application tamper detection via continuous software updates, in: Proceedings of the 28th Annual Computer Security Applications Conference, ACM, 2012, pp. 319–328.
[39] Y. Deswarte, J.-J. Quisquater, A. Saïdane, Remote integrity checking, in: Integrity and internal control in information systems VI, Springer, 2004, pp. 1–11.
[40] P. Dewan, D. Durham, H. Khosravi, M. Long, G. Nagabhushan, A hypervisor-based system for protecting software runtime memory and persistent storage, in: Proceedings of the 2008 Spring Simulation Multiconference, Society for Computer Simulation International, 2008, pp. 828–835.
[41] J. Gan, R. Kok, P. Kohli, Y. Ding, B. Mah, Using virtual machine protections to enhance whitebox cryptography, in: Software Protection (SPRO), 2015 IEEE/ACM 1st International Workshop on, 2015, pp. 17–23, https://doi.org/10.1109/SPRO.2015.12.
[42] Z. Gao, N. Desalvo, P.D. Khoa, S.H. Kim, L. Xu, W.W. Ro, R.M. Verma, W. Shi, Integrity protection for big data processing with dynamic redundancy computation, in: 2015 IEEE International Conference on Autonomic Computing (ICAC), 2015, pp. 159–160, https://doi.org/10.1109/ICAC.2015.34.
[43] S. Ghosh, J.D. Hiser, J.W. Davidson, A secure and robust approach to software tamper resistance, in: Lecture Notes in Computer Science (including subseries Lecture Notes in Artificial Intelligence and Lecture Notes in Bioinformatics), vol. 6387 LNCS, ISBN 364216434X, 2010, pp. 33–47.
[44] S. Ghosh, J. Hiser, J.W. Davidson, Software protection for dynamically-generated code, in: Proceedings of the 2nd ACM SIGPLAN Program Protection and Reverse Engineering Workshop, ACM, 2013, p. 1.
[45] B. Horne, L. Matheson, C. Sheehan, R. Tarjan, Dynamic Self-Checking Techniques for Improved Tamper Resistance, in: Security and Privacy in Digital Rights Management, ISBN 978-3-540-43677-5, 2002, pp. 141–159, https://doi.org/10.1007/3-540-47870-1_9. http://citeseerx.ist.psu.edu/viewdoc/summary?doi=10.1.1.13.3308.
[46] A. Ibrahim, S. Banescu, StIns4CS: a state inspection tool for C, in: Proceedings of the 2016 ACM Workshop on Software Protection, ACM, 2016, pp. 61–71.
[47] M. Jacob, M.H. Jakubowski, R. Venkatesan, Towards integral binary execution: implementing oblivious hashing using overlapped instruction encodings, in: Proceedings of the 9th Workshop on Multimedia & Security, ACM, 2007, pp. 129–140.
[48] M. Jakobsson, K.-A. Johansson, Retroactive detection of malware with applications to mobile platforms, in: Proceedings of the 5th USENIX Conference on Hot Topics in Security, Washinton, DC, HotSec'10, USENIX Association, Berkeley, CA, USA, 2010, pp. 1–13.
[49] M. Jakobsson, K.-A. Johansson, Practical and secure software-based attestation, in: 2011 Workshop on Lightweight Security & Privacy: Devices, Protocols and Applications (LightSec), IEEE, 2011, pp. 1–9.
[50] H. Jin, J. Lotspiech, Forensic analysis for tamper resistant software, in: 14th International Symposium on Software Reliability Engineering, 2003. ISSRE 2003, IEEE, 2003, pp. 133–142.
[51] P. Junod, J. Rinaldini, J. Wehrli, J. Michielin, Obfuscator-LLVM: software protection for the masses, in: Proceedings of the 1st International Workshop on Software Protection, IEEE Press, 2015, pp. 3–9.
[52] T. Kanstrén, S. Lehtonen, R. Savola, H. Kukkohovi, K. Hätönen, Architecture for high confidence cloud security monitoring, in: 2015 IEEE International Conference on Cloud Engineering (IC2E), IEEE, 2015, pp. 195–200.

[53] N. Karapanos, A. Filios, R.A. Popa, S. Capkun, Verena: end-to-end integrity protection for web applications, in: Proceedings of the 37th IEEE Symposium on Security and Privacy (IEEE S&P), 2016.
[54] C. Kil, Remote attestation to dynamic system properties: towards providing complete system integrity evidence, in: IEEE/IFIP International Conference on Dependable Systems & Networks, ISBN 9781424444212, 2009, pp. 115–124.
[55] G.H. Kim, E.H. Spafford, Experiences With Tripwire: Using Integrity Checkers for Intrusion Detection, 1994.
[56] W.B. Kimball, R.O. Baldwin, Emulation-Based Software Protection, 2012,US Patent 8,285,987.
[57] A. Kulkarni, R. Metta, A new code obfuscation scheme for software protection, in: 2014 IEEE 8th International Symposium on Service Oriented System Engineering (SOSE), IEEE, 2014, pp. 409–414.
[58] M. Madou, B. Anckaert, P. Moseley, S. Debray, B. De Sutter, K. De Bosschere, Software protection through dynamic code mutation, in: International Workshop on Information Security Applications, Springer, 2005, pp. 194–206.
[59] C. Malone, M. Zahran, R. Karri, Are hardware performance counters a cost effective way for integrity checking of programs, in: Proceedings of the sixth ACM workshop on Scalable trusted computing - STC '11 (2011) 71, ISSN 15437221, https://doi.org/10.1145/2046582.2046596. http://www.scopus.com/inward/record.url?eid=2-s2.0-80755143408&partnerID=40&md5=ad5db1f8e5c0131a2a17f457ba1b0497 http://dl.acm.org/citation.cfm?doid=2046582.2046596.
[60] L. Martignoni, R. Paleari, D. Bruschi, Conqueror: tamper-proof code execution on legacy systems, in: Lecture Notes in Computer Science (Including Subseries Lecture Notes in Artificial Intelligence and Lecture Notes in Bioinformatics), vol. 6201 LNCS, 2010, pp. 21–40, https://doi.org/10.1007/978-3-642-14215-4_2.
[61] V. Pappas, M. Polychronakis, A.D. Keromytis, Smashing the gadgets: hindering return-oriented programming using in-place code randomization, in: 2012 IEEE Symposium on Security and Privacy, IEEE, 2012, pp. 601–615.
[62] V. Pappas, M. Polychronakis, A.D. Keromytis, Transparent ROP exploit mitigation using indirect branch tracing, in: Presented as Part of the 22nd USENIX Security Symposium (USENIX Security 13), 2013, pp. 447–462.
[63] S. Park, J.N. Yoon, C. Kang, K.H. Kim, T. Han, TGVisor: a tiny hypervisor-based trusted geolocation framework for mobile cloud clients, in: 2015 3rd IEEE International Conference on Mobile Cloud Computing, Services, and Engineering (MobileCloud), IEEE, 2015, pp. 99–108.
[64] M. Protsenko, S. Kreuter, T. Müller, Dynamic self-protection and tamperproofing for android apps using native code, in: 2015 10th International Conference on Availability, Reliability and Security (ARES), 2015, pp. 129–138, https://doi.org/10.1109/ARES.2015.98.
[65] A. Seshadri, M. Luk, E. Shi, A. Perrig, L. van Doorn, P. Khosla, Pioneer: Verifying Code Integrity and Enforcing Untampered Code Execution on Legacy Systems, ACM SIGOPS Oper. Syst. Rev. (2005), ISSN 01635980, https://doi.org/10.1145/1095809.1095812. http://dl.acm.org/citation.cfm?id=1095809.1095812.
[66] D. Spinellis, Reflection as a mechanism for software integrity verification, ACM Trans. Inf. Syst. Secur. 3 (1) (2000) 51–62. ISSN: 10949224, https://doi.org/10.1145/353323.353383.
[67] Y. Sun, S. Nanda, T. Jaeger, Security-as-a-service for microservices-based cloud applications, in: 2015 IEEE 7th International Conference on Cloud Computing Technology and Science (CloudCom), IEEE, 2015, pp. 50–57.
[68] F.A. Teixeira, G.V. Machado, F.M. Pereira, H.C. Wong, J. Nogueira, L.B. Oliveira, SIoT: securing the internet of things through distributed system analysis, in: Proceedings

of the 14th International Conference on Information Processing in Sensor Networks, ACM, 2015, pp. 310–321.
[69] P. Wang, S.-K. Kang, K. Kim, Tamper resistant software through dynamic integrity checking, in: Proc. Symp. on Cyptography and Information Security (SCIS 05), 2005. ISBN: 8242866627.
[70] F. Yao, R. Sprabery, R.H. Campbell, CryptVMI: a flexible and encrypted virtual machine introspection system in the cloud, in: Proceedings of the 2nd International Workshop on Security in Cloud Computing, ACM, 2014, pp. 11–18.
[71] A.K. Kanuparthi, M. Zahran, R. Karri, Architecture support for dynamic integrity checking, IEEE Trans. Inf. Forensics Secur. 7 (1) (2012) 321–332.
[72] M.D. Ernst, J.H. Perkins, P.J. Guo, S. McCamant, C. Pacheco, M.S. Tschantz, C. Xiao, The Daikon system for dynamic detection of likely invariants. Sci. Comput. Program. 69 (1-3) (2007) 35–45, ISSN: 01676423, https://doi.org/10.1016/j.scico.2007.01.015.
[73] R. Merkle, A digital signature based on a conventional encryption function, in: Advances in Cryptology CRYPTO 87, Springer, 2006, pp. 369–378.
[74] N. Tillmann, P. de Halleux, Pex–white box test generation for .NET, in: Proc. of Tests and Proofs (TAP'08), 4966, Springer Verlag, Prato, Italy, 2008, pp. 134–153. https://www.microsoft.com/en-us/research/publication/pex-white-box-test-generation-for-net/.
[75] B. Anckaert, M. Jakubowski, R. Venkatesan, Proteus: virtualization for diversified tamper-resistance, in: Proceedings of the ACM Workshop on Digital Rights Management, ACM, 2006, pp. 47–58.
[76] D. Spinellis, Reflection as a mechanism for software integrity verification, ACM Trans. Inf. Syst. Secur. 3 (1) (2000) 51–62.
[77] Avast Lab, WannaCry Ransomware, 2017) https://blog.avast.com/ransomware-that-infected-telefonica-and-nhs-hospitals-is-spreading-aggressively-with-over-50000-attacks-so-far-today.
[78] C. Linn, S. Debray, Obfuscation of executable code to improve resistance to static disassembly, in: Proceedings of the 10th ACM Conference on Computer and Communications Security, ACM, 2003, pp. 290–299.
[79] B. Abrath, B. Coppens, S. Volckaert, J. Wijnant, B. De Sutter, Tightly-coupled self-debugging software protection. in: Proceedings of the 6th Workshop on Software Security, Protection, and Reverse Engineering, SSPREW '16, ACM, New York, NY, USA, ISBN: 978-1-4503-4841-6, 2016, pp. 7:1–7:10, https://doi.org/10.1145/3015135.3015142.
[80] J.T. Giffin, M. Christodorescu, L. Kruger, Strengthening software self-checksumming via self-modifying code, in: 21st Annual Computer Security Applications Conference (ACSAC'05), IEEE, 2005, p. 10.
[81] P. Akritidis, C. Cadar, C. Raiciu, M. Costa, M. Castro, Preventing memory error exploits with WIT, in: IEEE Symposium on Security and Privacy, 2008. SP 2008, IEEE, 2008, pp. 263–277.
[82] B. De Sutter, P. Falcarin, B. Wyseur, C. Basile, M. Ceccato, J. DAnnoville, M. Zunke, A reference architecture for software protection. in: 2016 13th Working IEEE/IFIP Conference on Software Architecture (WICSA), 2016, pp. 291–294, https://doi.org/10.1109/WICSA.2016.43.
[83] J.R. Cordy, TXL—a language for programming language tools and applications, Electron. Notes Theor. Comput. Sci. 110 (2004) 3–31.
[84] N. Mavrogiannopoulos, N. Kisserli, B. Preneel, A taxonomy of self-modifying code for obfuscation, Comput. Secur. 30 (8) (2011) 679–691.
[85] E.D. Bryant, M.J. Atallah, M.R. Stytz, A survey of anti-tamper technologies, CrossTalk 17 (2004), 12–16.
[86] M. Xianya, Z. Yi, W. Baosheng, T. Yong, A survey of software protection methods based on self-modifying code, in: 2015 International Conference on Computational Intelligence and Communication Networks (CICN), IEEE, 2015, pp. 589–593.

[87] N. Sasirekha, M.O. Hemalatha, A survey on software protection techniques, Glob. J. Comput. Sci. Technol. 12 (1) (2012) 53–58.
[88] C. Collberg, J. Nagra, F.-Y. Wang, Surreptitious software: models from biology and history, in: International Conference on Mathematical Methods, Models, and Architectures for Computer Network Security, Springer, 2007, pp. 1–21.
[89] M. Schwarz, S. Weiser, D. Gruss, C. Maurice, S. Mangard, Malware guard extension: using SGX to conceal cache attacks, in: M. Polychronakis, M. Meier (Eds.), *Detection of Intrusions and Malware, and Vulnerability Assessment*, Springer International Publishing, Cham, 2017. ISBN: 978-3-319-60876-1. pp. 3–24.
[90] M. Maier, M. Chowdhury, B.P. Rimal, D.P. Van, The tactile internet: vision, recent progress, and open challenges, IEEE Commun. Mag. 54 (5) (2016) 138–145.
[91] Microsoft, Simplified Implementation of the Microsoft SDL. 2017) https://www.microsoft.com/en-us/download/details.aspx?id=12379.

ABOUT THE AUTHORS

Mohsen Ahmadvand is a researcher and PhD student at the Chair of Software Engineering, headed by Prof. Pretschner, at TU Munich. He holds a master's degree (with distinction) in computer science majoring in software security also from TU Munich. His research interest lies primary in the area of software integrity protection against Man-At-The-End (MATE) attackers. Other areas of his interest are microservice security, insider threat mitigation, trusted computing, and cryptography.

Alexander Pretschner is a full professor of software and systems engineering at the Technical University of Munich, scientific director at the fortiss research and technology transfer institute in Munich, and speaker of the board of the Munich Center for Internet Research. His research focuses on all aspects of systems quality, specifically testing and security. He received the Diploma and MS degrees in computer science from RWTH Aachen University and the University of Kansas, and PhD from Technical University of Munich. His prior positions include those of a senior researcher at ETH Zurich; of a group manager at the Fraunhofer Institute for Experimental

Software Engineering; of an adjunct associate professor at the Technical University of Kaiserslautern; of a full professor at Karlsruhe Institute of Technology; program or general (co)chair of ICST, MODELS, ESSOS, and CODASPY; associate editor of the *Journal of Software Testing, Verification and Reliability* (Wiley), the *Journal of Computer and System Sciences* (Elsevier), and the *Journal of Software Systems Modeling* (Springer); and former associate editor of the *IEEE Transactions on Dependable and Secure Computing.*

Florian Kelbert is a postdoctoral researcher at Imperial College London. His current research focuses on trusted execution environments and their application in distributed systems. Previous research includes distributed data usage control and interdisciplinary projects on internet privacy and accountability. Florian obtained his PhD from Technical University of Munich and his Diploma in computer science from Ulm University